GLENCOE
Digital Communication Tools

Includes Keyboarding Review!
digicom.glencoe.com

Kathryn J. Gust, M.Ed.
Burke County Public Schools
Technology Department
Morganton, North Carolina

McGraw Hill Glencoe

New York, New York   Columbus, Ohio   Chicago, Illinois   Peoria, Illinois   Woodland Hills, California

Photography Credits

Banana Stock/Image State 35; Paul Barton/Corbis vii, 261, 435; Brand X Pictures (Royalty-Free)/ Getty Images 231, 419; Harry Bradley/Corbis 182; Ron Chapple/Getty Images 300; Comstock Images (Royalty-Free)/Getty Images 49(tr); Gary Conner/PhotoEdit 54; Corbis 124(b), 124(m), 124(t); Francisco Cruz/Superstock 72; Dex Images/Corbis 288; Kevin Dodge/Masterfile 430; Jon Feingersh/Corbis 389; John Foxx/Imagestate vi, 331; Jules Frazier/PhotoDisc 58; Tim Fuller Photography 1, 2, 9, 83, 130, 153(bl), 192, 227(bl), 313, Tim Fuller Photography 379, 384, 393, 428, 429, 468, 488; Getty Images 122, 123, 132(3), 146, 149, 203, 153(tl), 484; Getty Images News/Getty Images 112, 451; Spencer Grant/PhotoEdit 19; Jeff Greenberg/Index Stock 463; Charles Gupton/Corbis 276; Hewlett Packard 190(tl); Courtesy of ImagiWorks, Inc. 246; Image Source/ Superstock 442; Images Source/Getty Images 167; David Katzenstein/Corbis 79; Rob Lewine/Corbis 183; Liu Liquin/Corbis 53(tr); LWA-JDC/Corbis 262; LWA-Sharie Kennedy/Corbis 77; Masterfile (Royalty-Free) 323; Masterfile/Masterfile 463; Joe McBride/Corbis 379; Courtesy of Microsoft 238; MTPA Stock/Masterfile 439; Michael Newman/PhotoEdit v, xi, 133, 108, 160, 354, 447; Patrick Olear/PhotoEdit 190(bl), 229; Courtesy of Palm One xii, 126, 136, 304; Tim Pannell/Corbis 454; Jose Luis Pelaez/Corbis iv, 49(bl); Brian Pieters/Masterfile 3; Pixtal/Superstock 78; Photodics Blue/ Getty Images 490; David Pollack/Corbis 260; Royalty Free/Corbis 12; Royalty Free/Imagestate 150; Chuck Savage/Corbis 359; Walter Schmid/Stone 182; Strauss/Curtis/Masterfile 219; Taxi/Getty Images 477; ThinkStock/Superstock 370; Sylvia Torres/Corbis 155; Courtesy of Toshiba 227(tl); Lito C. Uyan/Corbis 478

Screen Capture Credits

Abbreviation Key: MS = Screen shots used by permission from Microsoft Corporation.

All rights reserved, ©2001 MS Windows XP, 6, 11, 12, 13, 14, 16, 17, 25, 26, 27, 38, 39, 174, 199, 260, 265, 310, 311, 312, 313, 356, 357, 358, 359, 388, 390, 391, 392, 393, 406, 423, 424, 425; ©1983-2003 MS Outlook 2003, 41, 42, 43, 44; ©1983-2003 MS Explorer 2003, 61, 62, 64, 65, 88, 89, 93, 95, 97, 102, 106, 110, 111, 422; ©2005 by Yahoo! Inc. YAHOO! and the YAHOO! logo are trademarks of Yahoo! Inc. All rights reserved, 66; ©2005 Google Inc., 67, 88, 93, 95, 97, 102, 105, 106, 110, 111, 388; ©1981-2001 MS Notepad, 93, 94, 95, 114; ©2004 palmOne, Inc. All rights reserved, 126, 127, 129, 131, 134, 135, 137, 138, 140, 141, 144, 145, 147, 228; ©1983-2003 MS Word 2003, 170, 199, 248, 251, 265, 266, 267, 268, 271, 272, 273, 274, 278, 279, 280, 281, 284, 285, 286, 287, 290, 291, 298, 299, 302, 308; ©1983-2003 MS Journal 2003, 192, 193, 194, 195, 198, 205, 206, 207, 208, 209, 210, 213, 214, 215, 216, 217, 221, 240, 241, 242, 244, 256; ©1985-2002 MS Windows XP Tablet PC Edition, 200, 201, 202, 221, 222; ©1983-2003 MS OneNote 2003, 251, 252, 253; ©1983-2004 MS Excel 2003, 316, 317, 318, 319, 320, 321, 322, 325, 326, 327, 329, 333, 334, 336, 337, 338, 340, 341, 345, 346, 347, 348, 352, 485, 501; ©1983-2003 MS Access, 362, 363, 364, 367, 368, 371, 372, 373, 374, 377, 378, 382, 385, 386, 387; ©1983-2003 MS PowerPoint, 396, 397, 398, 402, 403, 404, 405, 409, 413, 414, 418, 421; ©2004 DataViz, Inc. All rights reserved, 235, 236.

McGraw Hill Glencoe

The *McGraw-Hill* Companies

Copyright © 2006 by the McGraw-Hill Companies, Inc. All rights reserved. Except as permitted under the United States Copyright Act, no part of this publication may be reproduced or distributed in any form or by any means, or stored in a database or retrieval system, without prior written permission of the publisher.

Send all inquiries to:
Glencoe/McGraw-Hill
21600 Oxnard Street, Suite 500
Woodland Hills, CA 91367

ISBN 0-07-865692-3 (Student Edition)
ISBN 0-07-867636-3 (Teacher Resource Manual)

Microsoft, Microsoft Office, Microsoft Word, Microsoft Excel, Microsoft Access, Microsoft PowerPoint, Internet Explorer, and Windows and all other Microsoft names and logos are trademarks or registered trademarks of Microsoft Corporation in the United States and/or other countries.

All other brand names are trademarks or registered trademarks of their respective companies.

Between the time that Web site information is gathered and published, it is not unusual for some sites to have changed URLs or closed. URLs will be updated in reprints or on the Online Resource Center when possible.

Printed in the United States of America

2 3 4 5 6 7 8 9 0   027   09 08 07 06 05

Views
- Design View, 366
- Outline View, 398–399
- Page Break View, 347
- Report View, 373
- Slide View, 398–399

Virus, 70, 71
Voice chat, 40
Voice Command mode, 161–165
Volunteering, 440, 441

W

Web editors, 85
Web page, 83–116
- alignment, 104–107
- color, 91
- Excel documents saved as, 94
- from databases, 385–388
- font, 91
- images added to, 100–102
- from presentations, 420–423
- PowerPoint documents saved as, 94
- previewing, 87–88
- printing, 65
- reasons for creating, 83
- reopening, 90
- saving, 87–88
- Word documents saved as, 94

Web site designers, 288
Web site
- dictating, 185
- links to pages within, 115
- personal, 28
- planning, 113

Webcam, 40, 45
Weblogs, 40, 44–45
Web-safe color palette, R15
Weekly View, in PDAs, 145
Wi-Fi, 182
Windows, working in multiple, 88
WinZip, 351, 377, 378
Wireless connection, 203
Wireless devices, 182
Wizards, 374

Word, 262–312
- Access data integrated with, 381–382
- alignment, 262
- bold, 266–267
- chart added to, 342
- copy, 267–268
- cut, 267–268
- delete, 267
- drawing in, 215–217
- font format, 268–269
- Format Painter, 268
- form/template, 307–308
- footers, 277–280
- grammar, 272–273
- graphics in, 302–303
- headers, 277–280
- hide/show formatting, 272
- italics, 266–267
- importing into Journal, 206
- line spacing, 271
- lists, 280–282
- margins, 274
- memo, 381
- orientation of page, 274
- paste, 267–268
- pasting spreadsheet into, 347–348
- print preview/print, 268
- save/save as, 267
- show/hide formatting, 272
- spelling, 272–273
- synchronizing documents on, 236
- tables in, 284–287, 290–293
- tabs, 266
- template/form, 307–308
- text boxes in, 298–299
- thesaurus, 272–273
- underline, 266–267
- views in, 265

Word processing, 262–312. *See also* Word
- charts integrated into, 339–343
- editing features of, 264–275
- footers, 277–280
- formatting features of, 264–275
- forms, 289–293
- graphics, 301–303
- headers, 277–280
- lists, 280–282
- online forms/templates, 306–308
- PDA for, 304–305
- reasons for using, 264
- speech recognition dictation with, 294–296
- spreadsheets integrated into, 344–349
- tables, 283–287
- text boxes, 297–299

WordArt, 403–404
WordPad, 6
Workbook, 315
Workplace
- accountability at, 354
- good judgment in the, 439
- integrity in the, 369
- negotiation in the, 463

Works cited, R3
Worksheets, 315
- data entry into, 317
- deleting, 322
- inserting, 322
- renaming, 337
- selecting, 317–318

Wrapping text, 340
Write Anywhere, 200, 247–248
Writing pad, 193
Writing Tools tab, 199–200

Y

Yahoo! search, 66–67

Z

Zip disk, 23
Zipped files, 351, 377, 378
Zoom, 347

Index I11

About the Author

Kathryn (Kathy) Gust is an Instructional Technology Specialist at Freedom High School, Morganton, North Carolina. In her current role, she helps teachers integrate computer technology with their curriculum. Kathy holds a B.A. from the University of South Florida and her Business Education Teacher Certification from Appalachian State University. She completed her M.A. in Educational Media with Instructional Technology at Appalachian State University where she is currently working on her Ed.S. in Higher Education. Additional certifications include IC3 and Distance Education.

Kathy has taught middle school computer skills classes and computer applications courses at the community college level. She also wrote the curriculum for online business education courses for the Web Academy. Kathy has presented at NCAECT, NCETC, and ACBMITEC Conferences and participated in the development of two new North Carolina curriculums for the Department of Public Instruction, Business and Information Technology Department.

Kathy lives in Lenoir, North Carolina, with her husband and four sons. When not teaching and writing or working on professional development, Kathy enjoys spending time with her family and reading.

Academic Reviewers

Cara Cavin
Verona Area High School
Verona, Wisconsin

Marcia Ellis
Cobb County Schools
Powder Springs, Georgia

Pat Eloranta
Medford High School
Medford, Wisconsin

Jeff Fuller
Southwest High School
Raleigh, North Carolina

Matt Gehrett
Ridgeview High School
Bakersfield, California

Linda Jacobi
Twin Rivers High School
Genoa, Nebraska

Julie Jaehne
University of Houston
Houston, Texas

Peter Lorenz
Badger High School
Lake Geneva, Wisconsin

Carole Payne
Woodland High School
Cartersville, Georgia

Karen Pflugh
Wesley Chapel High School
Dade City, Florida

Rochelle Reed
East St. Louis High School
Bellville, Illinois

Linda Rounsavall
Edison High School
Alexandria, Virginia

LuAnn Seely
Sturgis High School
Sturgis, Michigan

David C. Soltesz
John Marshall High School
Peach Bottom, West Virginia

Emily Thompson
Burke Alternative
Morganton, North Carolina

Michelle Whalen
JEB Stuart High School
Falls Church, Virginia

Contributors

Jack E. Johnson, Ph.D.
Department of Management and Business Systems
State University of West Georgia
Carrollton, Georgia

Judith Chiri-Mulkey
Department of Computer Information Systems
Pikes Peak Community College
Colorado Springs, Colorado

Dolores Sykes Cotton
Business Education
Detroit Public Schools
Detroit, Michigan

Carole G. Stanley, M.Ed.
Rains Junior High School
Emory, Texas

Educational Consultant

Terrie Gray, Ed.D.
Paradise, CA

Spreadsheets, 313–353. *See also* Excel
    charts, 335–343
    formatting, 315–323, 328–330
    formulas, 324–327
    online templates, 350–353
    PDA synchronization of, 346
    planning, 315–316
    printing, 320–321
    reasons for using, 315
    templates, 332–334, 350–353
    in Word documents, 339–349
Square. *See* Drawing toolbar
Statement of confidentiality, 482
Stationery, 243, 244
Sticky Notes, 247
Storage, 15–17
    capacity, 23
    devices, 16–17, 23–27
Stylus, 126
SUM function, 324, 326, 327
Symbols, 184–185
Synchronization, 130, 304, 305, 396
Synchronizing, 235–237

T

Tab
    command, 177, 178
    gesture, 221
    key, 266
Tables
    format for, R6
    and queries, 371
Tables, Word, 284–287, 290–293
    borders, 286
    columns, 285, 292
    deleting, 284
    drawing, 284–285
    inserting, 284
    resizing, 292
    rows, 285, 292
    shading, 287
    split, 292
Tablet PCs, 5, 8, 190–222
    control panel, 196–197
    creating images with, 239–242
    data entry with, 193–195
    drawing, 212–217
    gestures, 219–221
    Input Panel text entry, 198–202
    Journal. *See* Journal, Tablet PC
    reasons for using, 192
    Write Anywhere feature of, 247–248
Tabs, 251
Tags, HTML
    capital letters used in, 86
    closing, 85, 86, 93–94, 104
    definition of, 85
    hypertext reference, 113, 114
    lowercase letters used in, 86
    opening, 85, 86
    paragraph, 95–96
Templates
    Access, 377–378
    customizing, 351–352
    downloading, 351
    Excel, 332–334
    online spreadsheet, 350–353
    reasons for using, 350
    Tablet PC Journal, 243, 244
    Word, 306–308
Text
    with gestures, 219–221
    on handheld computers, 126–127, 140
    Journal, 204–210
    OneNote, 251
    PowerPoint, 398
    on Tablet PCs, 198–202
Text boxes
    PowerPoint, 403, 424
    Word, 297–299
Text Correction window, 207
Thesaurus, 273
Thumb drive. *See* USB flash drive
Tilde (~) command, 185
Time management, 133
Title page, R4
Titles
    PowerPoint, 397, 398, 422
    spreadsheet, 345–346
To Do Lists, 146–147
Toolbars
    drawing, 403–404
    formatting, 205
    in PowerPoint, 396
Training
    careers in, 30
    job, 31, 54
    planning your, 452–454
    required, 456
Transportation, Distribution, and Logistics career cluster, 30, 54

U

Underline button, 266–268
Underline command, 162
Underlining, 264
Underscore (_) command, 185
Undo button, 209, 265, 319
Undo command, 170
Undocking, 196
Uniform resource locators (URLs), 60, 113
U.S. Bureau of Labor Statistics, 219
U.S. Department of Defense, 78
U.S. Department of Education, 30
U.S. National Recycling Coalition, 261
Unordered lists, 109–110
URL (Uniform Resource Locator), 60, 113
USB flash drive, 23, 26, 27, 204
User profile, 155
    creating your, 156
    training your, 157

V

Value (number), 316
Values, personal, 432, 434
Verbal communication skills, 300
Vertical Bar (|) command, 185
Video, digital, 46
Video files, 250
Viewing
    Access Web page, 388
    Journal, 210
    source code, 110

Table of Contents

Why Develop Skills with Digital
Communication Tools? xiii
Be an Active Reader xiv
Technology Standards xvi

UNIT 1 — Impact of Digital Communication Tools — 1

Digital Dimension → Language Arts — 2

CHAPTER 1 — Digital Communication Tools in Our Lives — 3

Lesson 1.1	New Ways to Enter Text with Digital Tools	5
Lesson 1.2	Operating Systems, Hardware, and Software	10
Lesson 1.3	Storage and File Management	15
Lesson 1.4	Input and Output Devices	20
Lesson 1.5	External Storage Devices	23
Lesson 1.6	Digital Skills in the 21st Century Workplace	28

Chapter Application and Assessment
- Digital Dimension—Language Arts — 32
- Review — 33
- Self-Assessment — 34

CHAPTER 2 — Communicating with Digital Technology — 35

Lesson 2.1	Digital Image Technology	37
Lesson 2.2	Communicating Online	40
Lesson 2.3	Internet Security and Acceptable Use	47
Lesson 2.4	Ethics and Netiquette	50

Chapter Application and Assessment
- Digital Dimension—Language Arts — 55
- Review — 56
- Self-Assessment — 57

CHAPTER 3 — Using the Internet — 58

Lesson 3.1	Browse the Internet	60
Lesson 3.2	Customize Your Browser and Search Online	63
Lesson 3.3	Security and Viruses	70
Lesson 3.4	Copyright, Fair Use, and Plagiarism	73
Lesson 3.5	Computer Crimes	75

Chapter Application and Assessment
- Digital Dimension—Language Arts — 80
- Review — 81
- Self-Assessment — 82

CHAPTER 4 — Create a Web Page with HTML — 83

Lesson 4.1	Get Started with HTML	85
Lesson 4.2	Format Text	91
Lesson 4.3	Insert Images and Backgrounds	100
Lesson 4.4	Text and Graphic Alignment	104
Lesson 4.5	Bulleted and Numbered Lists	109
Lesson 4.6	Links to Other Pages and Web Sites	113

Chapter Application and Assessment
- Digital Dimension—Language Arts — 117
- Review — 118
- Self-Assessment — 119

Unit Application and Assessment

Digital Dimension → Language Arts Lab — 120

R

Read-only memory (ROM), 23
Records, 361
    adding, 368
    deleting, 366
Redo button, 265, 319
Reference, 482
Renaming worksheets, 337
Reply to e-mail, 43–44
Report View, 373
Report, 370, 372–374, 380
    bibliography for, R4
    contents for, R6
    citations in, R3
    multipage, R3, R4
    one-page, R3
    outline for, R2
    title page for, R4
Resizing
    image, 302
    table, 292
    WordArt, 404
Resolution, 37, 46
Restore button, 11
Results, search, 66
Résumés
    tips for, 478
    writing, 470–473, 475
Retail/Wholesale Sales and Service career cluster, 30, 69
Return voice command, 157
Right alignment, 164
Right Brace (}) voice command, 185
Right Bracket (]) voice command, 185
Right-click, 195
Right-hand setting, 199
Right Parenthesis command, 185
ROM (read-only memory), 23
Row height, 292
Rows, 283, 285, 341

S

Safety, workplace, 503
Salary, questions about, 489
Save, 6
    to CD, 25–26
    to floppy disk, 25
    Journal, 206
    to USB flash drive, 27
    Web page, 87
    in Word, 267
Save As, 6
    in PowerPoint, 415
    template, 333
    in Word, 267
Save command, 165
Scaling, 349
Scientific Research and Engineering career cluster, 30, 31
Scratch Out gesture, 219–220
Scratch That command, 177, 178
Screen brightness, 197
Screen orientation, 196
Search engines, 63
Searching
    advanced, 66–67
    Boolean, 67
    Google, 67
    online, 63, 66–67
    with Yahoo!, 66–67
Security, Internet, 47–48, 70–71
Select All command, 157, 184
Select command, 163
Select Paragraph command, 157
Select Word command, 157
Selecting
    cell, 329
    cell range, 329
    text, 201, 266
    words, 178
Semicolon (;), 172
Shading, 287, 328, 329
Shapes, 401
Shareware, 229
Show/hide, 272
Single spacing, 294
Sizing handles, 302
Skills, 433
    definition of, 432
    required, 449
Skills résumé, 470, 472
Slash (/), 172, 173
Slash command, 180, 181
Slide View, 398–399
Slides
    adding, 398
    animation in, 409
    definition of, 395
    graphics in, 402–403
    printing, 417
    rearranging, 397
Smartphones, 203
Soft keyboard, 126, 127
Soft skills, 432
Software, 7, 10
    anti-virus, 70
    copyright license for, 75
    identifying, 12–13
    installation of, 12, 17
    speech recognition, 161
Sorting, database, 370, 371
Source code, viewing, 91, 110
Space gesture, 220
Spaces between HTML tags and lines, 83
Spacing, 114, 294
Speak command, 158
Special characters, 184
Speech recognition, 5, 7, 153–186
    correction, making, 172–175
    emoticons, 184–185
    letter dictation with, 294–296
    navigating and editing text, 168–171
    number entry, 179–182
    preparing to use, 155–158
    punctuation, 172–175
    reasons for using, 161
    Scratch That commands, 177–178
    symbols, 184–185
    on Tablet PCs, 200
    Tab command, 177–178
    tips for improving, 165
    Voice Command mode of, 161–165
Spell checking, 19, 272–273, 320
Split cells, 291
Split table, 292

UNIT 2: Digital Communication Tools and Skills 122

Digital Dimension → Social Studies 123

CHAPTER 5: Handheld Computers—PDAs 124

Lesson 5.1	Enter Text in Memo Pad	126
Lesson 5.2	Explore Memo Pad	128
Lesson 5.3	Beam a Memo	130
Lesson 5.4	Use, Create, and Beam Categories	134
Lesson 5.5	Create and Beam a Note with Note Pad	137
Lesson 5.6	Create an Address Book and Business Card	139
Lesson 5.7	Use Date Book	143
Lesson 5.8	Create a To Do List	146

Chapter Application and Assessment
- Digital Dimension—Social Studies 150
- Review 151
- Self-Assessment 152

CHAPTER 6: Speech Recognition Tools 153

Lesson 6.1	Prepare to Use Speech Recognition	155
Lesson 6.2	Use Voice Command Mode	161
Lesson 6.3	Navigate and Edit Text	168
Lesson 6.4	Add Punctuation and Make Corrections	172
Lesson 6.5	Use the Scratch That and Tab Commands	177
Lesson 6.6	Dictate Numbers	179
Lesson 6.7	Dictate Special Symbols and Emoticons	184

Chapter Application and Assessment
- Digital Dimension—Social Studies 187
- Review 188
- Self-Assessment 189

CHAPTER 7: On-Screen Writing—Tablet PCs 190

Lesson 7.1	Get Started with the Tablet PC	192
Lesson 7.2	Enter Text with Input Panel	198
Lesson 7.3	Write and Edit On-Screen in Journal	204
Lesson 7.4	Draw On-Screen in Journal and in Word	212
Lesson 7.5	Edit Text with Gestures	220

Chapter Application and Assessment
- Digital Dimension—Social Studies 224
- Review 225
- Self-Assessment 226

CHAPTER 8: Additional Features of Digital Tools 227

Lesson 8.1	Add Applications to a PDA	229
Lesson 8.2	Solve Mathematical Problems Using a PDA	232
Lesson 8.3	Synchronize Documents with a PDA	235
Lesson 8.4	Create Document Images with a Tablet PC	239
Lesson 8.5	Create Templates and Stationery in Journal	243
Lesson 8.6	Use Write Anywhere on the Tablet PC	247
Lesson 8.7	Explore OneNote	250

Chapter Application and Assessment
- Digital Dimension—Social Studies 255
- Review 256
- Self-Assessment 257

Unit Application and Assessment

Digital Dimension → Social Studies Lab 258

Oral presentation skills, 167
Order
   alphabetical, 17
   chronological, 456
Ordered lists, 109, 111
Organizing
   with To Do Lists, 146–147
   Favorites, 62
   files, 16
   handheld information, 134–135
Orientation
   page, 274, 321
   screen, 196
   title, 346
OS. *See* Operating system
Outline, for report, R2
Outline button, 329
Outline View, 398–399, 417
Output devices, 20–22

P

Page Break View, 347
Page layout, 104
Page orientation, 274, 321
Paragraph tag in HTML, 95–96
Paragraph command, 157, 173, 184
Passwords, 47
Paste command, 161
Pasting
   events, 144
   spreadsheet into Word, 347–348
   in Word, 267–268
   into Word, 215
Pauses, 165, 179–181
PDAs. *See* Personal digital assistants
Pen
   -eye coordination, 194
   settings, 205
   tool, 213
Percent Sign (%) command, 185
Period (.) command, 157–158, 172, 173
Personal business letter, R2

Personal digital assistants (PDAs), 5, 7, 229–237. *See also* Handheld computers
   adding applications to, 229–230
   deleting applications from, 232
   mathematical problem solving with, 232–234
   and probes, 246
   on smartphones, 203
   spreadsheet synchronization with, 346
   synchronizing documents with, 235–237
   for word processing, 304–305
Personal journal, 235, 236
Personal letter, R2
Personal Web sites, 28
Phone numbers, in speech recognition, 179, 181
Photographs, digital, 38, 46
Pixel, 37
Plagiarism, 73
Plain text, 85
Planning a career, 53
Plus Sign command, 181
Point command, 180
Points per inch, 269
Portrait orientation, 274
PowerPoint, 393–425
   animation in, 408–410
   AutoShapes, 404–405
   charts in, 413–415
   clip art in, 402
   Drawing toolbar/WordArt in, 403–404
   formatting graphics in, 402–403
   master slide view, 398–399
   new presentation, 397
   new slide added to, 398
   outline view, 398–399
   on PDAs, 235
   printing options for, 418–419
   slide layout view, 398–399
   synchronizing PDAs with, 236
   text entry, 398

   title, 397
   toolbar display in, 396
   Web pages from, 421–423
   WordArt in, 403–404
Preferences, To Do List, 147
Presentation skills, 167
Presentations, 393–423. *See also* PowerPoint
   animation in, 408–410
   charts in, 412–415
   graphics in, 401–406
   printing options for, 417–419
   starting, 395–399
   Web pages from, 420–423
Previewing
   Print. *See* Print Preview
   Web page, 88
Primary key, 364, 381
Print command, 165
Print Layout view, 265
Print Preview,
   Excel, 320–321, 338
   PowerPoint, 418
   Word, 268
Printers, 20
Printing
   picture, 39
   presentation options for, 417–419
   Web page, 65
   from Word, 268
Prioritize, 459
Priority, e-mail, 42
Probes, 246
Program folders, 17
Project management, 79
Proofreaders' marks, R6
Proofreading, 19, 270, R6
Punctuation speech commands, 172–173
Punctuation and usage, R7

Q

Query, 370–372
Question mark (?), 172, 173
Quick Keys, 198, 201–202
Quotation marks (" "), 172, 173

UNIT 3: Digital Communication Tools in the World of Work — 260

Digital Dimension → Science — 261

CHAPTER 9: Word Processing — 262

- Lesson 9.1 Use Basic Text Editing and Formatting — 264
- Lesson 9.2 Use More Advanced Editing and Formatting — 270
- Lesson 9.3 Add Headers, Footers, Bullets, and Numbering — 277
- Lesson 9.4 Create and Edit Tables — 283
- Lesson 9.5 Create Business Forms — 289
- Lesson 9.6 Dictate Letters with Speech Recognition — 294
- Lesson 9.7 Create and Edit Text Boxes — 297
- Lesson 9.8 Insert and Format Graphics — 301
- Lesson 9.9 Use a PDA for Word Processing — 304
- Lesson 9.10 Use Online Forms and Templates — 306

Chapter Application and Assessment
- Digital Dimension—Science — 310
- Review — 311
- Self-Assessment — 312

CHAPTER 10: Spreadsheets — 313

- Lesson 10.1 Format Basic Spreadsheets — 315
- Lesson 10.2 Create Formulas in Spreadsheets — 324
- Lesson 10.3 Apply Advanced Spreadsheet Formatting — 328
- Lesson 10.4 Use Business Form Templates in Spreadsheets — 332
- Lesson 10.5 Use Spreadsheet Data to Create Charts — 335
- Lesson 10.6 Integrate Charts with Word Processing — 339
- Lesson 10.7 Integrate Spreadsheets with Word Processing — 344
- Lesson 10.8 Use Online Spreadsheet Templates — 350

Chapter Application and Assessment
- Digital Dimension—Science — 356
- Review — 357
- Self-Assessment — 358

CHAPTER 11: Databases — 359

- Lesson 11.1 Get Started with a Database — 361
- Lesson 11.2 Sort Data and Create Queries and Reports — 370
- Lesson 11.3 Use Online Database Forms — 376
- Lesson 11.4 Integrate Databases with Applications — 380
- Lesson 11.5 Create a Web Page with a Database — 384

Chapter Application and Assessment
- Digital Dimension—Science — 390
- Review — 391
- Self-Assessment — 392

CHAPTER 12: Presentations — 393

- Lesson 12.1 Begin a Presentation — 395
- Lesson 12.2 Add Graphics to Presentations — 401
- Lesson 12.3 Add Animation to Presentations — 408
- Lesson 12.4 Add Charts to Presentations — 412
- Lesson 12.5 Investigate Print Options for Presentations — 417
- Lesson 12.6 Convert a Presentation into a Web Page — 420

Chapter Application and Assessment
- Digital Dimension—Science — 423
- Review — 424
- Self-Assessment — 425

Unit Application and Assessment

Digital Dimension → Science — Lab 426

L

Label (text), 316
Landscape orientation, 274
Lasso selection tool, 206–208, 215
Law and Public Safety career cluster, 30, 355
LCD (Liquid Crystal Diode) projector, 20
Left alignment, 165, 295
Left Brace ({) command, 185
Left Bracket ([) command, 185
Left-hand setting, 199
Left Parenthesis command, 185
Legend, 412
Less Than (<) command, 185
Lifelong learning, 464
Line length, 96–97
Line space gesture, 221
Line spacing, 208–209, 271
Links, 113–115
Liquid Crystal Diode (LCD) projector, 20
Lists
    To Do, 146–147
    HTML, 109–111
    speech recognition, 173–174
    in tables, 283
    Word, 277, 280–281
Listservs, 40
Loan calculator, 230

M

Manufacturing career cluster, 30, 116
Margins
    Excel, 321
    Word, 265, 274
Marking To Do items, in PDAs, 147
Mathematical operators, 179, 181
Mathematical problem solving, 232–234
Maximize button, 11
.mdb extension, 362
Mediation, 504

Memo Pad, 126–128
Memos
    Access databases in, 381–382
    beaming, 130–131
    creating, 129
    editing, 129
Menu access in speech recognition, 162
Merge, 290–291
Merge cells, 340
Microphone
    external noise minimized with, 165
    and keyboards, 177
    position of, 156
    turning off/on, 157
Microsoft Access. *See* Access
Microsoft Excel. *See* Excel
Microsoft PowerPoint. *See* PowerPoint
Microsoft Word. *See* Word
Minimize button, 11
Mode keys, 164
Modified-block style, 294
Monthly View in PDAs, 145
Mouse, 13–14, 168, 170
Move to commands (speech recognition)
    Beginning of Document, 168
    Beginning of Line, 168
    End of Document, 168
    End of Line, 168
    Next Line, 168
    Previous Line, 168
Moving
    cell contents, 319–320
    in speech recognition software, 168
    windows, 88
Multipage report, R3
    with side headings, R4
Mute button, 157
My Documents, 6, 16, 267

N

Naming files, 129
National Computer Crime Squad (NCCS), 75
Navigation speech commands, 168–169
NCCS (National Computer Crime Squad), 75
Netiquette, 50, 51
Networking, 440
New Line command, 157–158, 174
New Paragraph command, 173
Normal view, 265
NOT keyword, 63
Note Pad, 137–138
Notes, 147
Notes pages, 417
Numbered lists, 173, 277, 280–281
Number expression, R13
Numbers
    dictating, 179–182
    formatting spreadsheet, 316, 318

O

Occupational Outlook Handbook, 369, 379, 450
On-the-job training, 452
OneNote, 238, 250–253
One-page report, R3
Online communication, 40–45
Online forms/templates, 306–308
    database, 376–378
    spreadsheet, 350–353
Online Help, 67
Onscreen writing. *See* Tablet PCs
Open Angle (<) command, 185
Open Brace ({) command, 185
Open Bracket ([) command, 185
Open command, 169
Open Parenthesis command, 185
Open quote (") command, speech recognition, 172, 173
Opening tags, 85, 86
Operating system (OS), 10, 11
Operators, mathematical, 179, 181
Optical mouse, 14
Optical storage device, 23
OR keyword, 63

UNIT 4 Developing 21st Century Employability Skills — 428

Digital Dimension → Math — 429

CHAPTER 13 Explore 21st Century Careers — 430

Lesson 13.1 Match Careers to You — 432
Lesson 13.2 Explore Career Clusters—Focus on a Career Path — 436
Lesson 13.3 Find Out What People Do on the Job — 440

Chapter Application and Assessment
Digital Dimension—Math — 444
Review — 445
Self-Assessment — 446

CHAPTER 14 Plan Your Career — 447

Lesson 14.1 Evaluate Job Descriptions — 449
Lesson 14.2 Plan Your Training and Education — 452
Lesson 14.3 Develop Your Career Plan — 458
Lesson 14.4 The Career Decision-Making Process — 459

Chapter Application and Assessment
Digital Dimension—Math — 465
Review — 466
Self-Assessment — 467

CHAPTER 15 Get the Job You Want — 494

Lesson 15.1 Write Your Résumé — 470
Lesson 15.2 Find Job Openings — 479
Lesson 15.3 Fill Out Job Application Forms — 482
Lesson 15.4 Manage Your Job Search — 485
Lesson 15.5 Succeed in Interviews — 487

Chapter Application and Assessment
Digital Dimension—Math — 493
Review — 494
Self-Assessment — 495

Unit Application and Assessment

Digital Dimension → Math Lab — 496

Appendix A Keyboarding Skills — A1

Appendix R Reference — R1

Glossary — GI

Index — II

"Hits," in Internet searches, 66
Home page, 63, 64
Home row keys, 264
Horizontal rule, 91
Hospitality and Tourism career cluster, 30, 223
Hovering, 241
HTML. See Hypertext Markup Language
.html extension, 15
Human Services career cluster, 30, 160
Hyperlinks, 60, 61, 113–115, 421
Hypertext Markup Language (HTML), 85
   background color, 88–89
   body section of, 86
   bold, 92
   break tag, 95–97
   closing tags, 85, 86, 93–94, 104
   creating new document in, 86–87
   font color, 92
   font size, 92
   fonts, 91–92
   head section of, 86
   italics, 92
   lists, 109–111
   opening tags, 85, 86
   paragraph tag, 95–97
   previewing, 87–88
   reopening, 90
   saving, 87–88
   tags. See Tags, HTML
Hypertext reference tag, 113, 114
Hypertext transfer protocol (http), 60
Hyphen (-) command, 181, 185

I

Ibsen, Henrik, 354
Images
   editing, 302–303
   layers of, 303
   printing, 39
   resizing, 302

Tablet PC used for creating, 239–242
   on Web pages, 100–102
Import, a document, 206, 239, 240
Increase Decimal button, 318
Indentation, 266
Information Technology career cluster, 30, 419
Ink, Tablet PC, 240
Input devices, 20–22
Input panel, Tablet PC, 192–196
   docking, 196
   gesture used to open, 221
   sending text, 194–195
   text entry, 198–202
   undocking, 196
   viewing options, 199
   writing text, 194–195
Inserting
   AutoShapes, 404–405
   columns, 285, 341
   database records, 368
   rows, 285, 341
   tables, 284
   WordArt, 403–404
   worksheets, 322
Internet, 58–76
   advertising on the, 48
   applications for, 58
   browsing the, 60–67
   copyright and, 73–74
   cybercrime and, 75–77
   fair use and, 73–74
   job postings on the, 479, 480
   plagiarism and, 73–74
Internet security, 47–48, 70–71
Internships, 440, 452
Interviews, job, 470, 487–489
Invoices, 290–292
Italics, 92–93, 264, 268
Italics button, 266–267
Italics command, 163

J

Job applications, 482–484
Job banks, 479
Job description, 449–451

Job fairs, 479
Job hunting, 28
Job interviews, 470, 487–489
Job objective, 449
Job openings, finding, 479–481
Job search, 468–489
     interviews, 470, 487–489
     job applications, 482–484
     job listings, 479–481
     managing your, 485–486
Job shadow, 440
Job-specific skills, 432
Job tasks, 449
Job title, 449
Jobs online, 479, 480
Journal, personal, 235, 236
Journal, Tablet PC, 193, 195, 200, 204–210
   drawing, 212–215
   erase, 209
   flag, 208
   formatting toolbar, 205
   handwriting, 206–208
   highlighter settings, 205–206
   importing Word documents into, 206
   line spaces, 208–209
   opening, 205
   pensettings, 205–206
   saving in, 206
   templates/stationery in, 243–244
   text entry, 204–210
   undo, 209
   viewing, 210
   writing in, 206
Journal Viewer, 204
Joystick, 20
.jpg extension, 15, 39, 301
Jump drive. See USB flash drive
Junk e-mail, 44

K

Keyboard(s), 20
   and microphones, 177
   soft, 126, 127
Keying, accurate, 101

Activity Contents

CHAPTER 1

1	Enter Text and Save Files	6
2	Investigate Handheld Computers	7
3	Investigate Speech Recognition	7
4	Investigate Tablet PCs	8
5	Start the Computer and Open the Control Panel	11
6	Explore Drives and Folders	11
7	Identify Software	12
8	Identify Hardware	13
9	Explore Folders	16
10	Create a New Folder on the C: Drive	16
11	Create a New Folder on a Floppy Disk	17
12	Input and Output	21
13	Save to a Floppy Disk	25
14	Save to a Compact Disk (CD)	25
15	Save to a USB Flash Drive	26
16	Plan an E-Portfolio	29

CHAPTER 2

1	Take and Upload Digital Photographs	38
2	View and Rename Photos	38
3	Create a Scrapbook Page	39
4	Print Pictures	39
5	Create an E-mail	41
6	Set Priority	42
7	Insert an Attachment	42
8	Delete an E-Mail Message	43
9	Read and Reply to an E-Mail Message	43
10	Forward an E-mail Message	44
11	Explore Weblogs	44
12	Explore Webcams	45
13	Investigate Security and Firewalls	48
14	Explore Acceptable Use Policies	48
15	Explore Ethics	51
16	Explore Netiquette	51

CHAPTER 3

1	Use Hyperlinks	61
2	Add Favorites to Your Browser	61
3	Organize Favorites	62
4	Set a Home Page	64
5	Print Web Pages	65
6	Do a Yahoo! Search	66
7	Do an Advanced Search	66
8	Do a Google search	67
9	Do a Boolean Search	67
10	Security and Viruses	71
11	Learn About Viruses	71
12	Learn About Fair Use and Copyright	74
13	Explore the Anti-Cybercrime Effort	76

CHAPTER 4

1	Create a New HTML Document	86
2	Save and Preview a Web Page	87
3	Add Background Color	88
4	Use Color Codes to Change Background Color	89
5	Open the FBLA Document as a Web Page and as a Notepad Document	90
6	Apply Font Color Codes	92
7	Apply Header Codes	92
8	Apply Bold and Italics	92
9	Add Closing Tags	93
10	Use Paragraph and Break Tags	94
11	Control Line Length with Break Tags	96
12	Add a Horizontal Rule	96
13	Add Clip Art to Your Web Page	101
14	Add a Digital Photo to Your Web Page	101
15	Use Background Images	102
16	Align Text	105
17	Align Graphics	106
18	Create an Unordered List	110
19	Create an Ordered List	111
20	Create a Hyperlink	114
21	Link to a Page Within Your Site	115
22	Add a Link to an E-Mail Address on Your Web Page	115

Favorites, 61–62
FBI (Federal Bureau of Investigation), 75
Federal Bureau of Investigation (FBI), 75
Fields, 361–364
File extensions, 15
File management, 15–18
File names, 129, 213, 278
File sharing, 8
File transfer, 235, 236
Fill Series, 341
Finance career cluster, 30, 309
Firewalls, 47, 70
Flags
    document, 239–241
    Journal, 208
    OneNote, 252–253
Flaming, 50
Flash drive. *See* USB flash drive
Floppy disk, 23, 25, 267
Floppy disk drive, 16, 17
Folders, 6, 15–17
    Favorites, 62
    OS, 12
    program, 17
Font color, 92, 269
Font size, 92, 269
Fonts, 91–92, 268–269
Footers
    Excel chart, 348–349
    PowerPoint, 408, 409, 418
    Word, 277–280
Footnotes, 278
Force Num command, 180, 181
Format for envelopes, R5
Format Painter button, 268
Formatting
    Access Web page, 387
    graphics in presentation slide, 402–403
    report in Access, 380
Formatting toolbar in Journal, 205
Formatting, Excel, 315–323, 328–330
    alignment, 330, 345–346
    borders, 329, 346
    chart, 348

    font, 321–322
    numbers, 318
    shading, 329, 346
    title, 345–346
Formatting, Word
    alignment, 272
    font, 268–269
    footers, 277–280
    headers, 277–280
    line spacing, 271
    margins, 274
    page orientation, 274
    with PDAs, 304
    show/hide, 271–272
Forms, business. *See* Business forms
Formula bar, 319, 326
Formulas, 315, 324–327
Forward slash (/), 128
Forwarding e-mail, 44
Fractions, 179, 180
Freeware, 229, 230
FrontPage, 85
Function, 324

G

Gesture, 219–221
.gif extension, 15, 301
Global Positioning System (GPS), 78
Goal setting, 456–458
Goals, 379
Goethe, Johann Wolfgang von, 379
Google search, 67
Government and Public Administration career cluster, 30, 254
GPS (Global Positioning System), 78
Graffiti, 127
Grammar, R10
Grammar checking, 19, 272–273
Graphics, 301–303
    aligning, 106–107
    PowerPoint, 401–406
    Word, 297, 301–303
Grayscale, 417, 418

Greater Than (>) command, 185
Gridlines, 320–321

H

Hackers, in Internet security, 47
Handheld computers, 5, 7, 124–149. *See also* Personal digital assistants
    address books on, 139, 140
    applications for, 136
    beaming/synchronization with, 130–131
    business cards on, 139, 141
    date book on, 143–145
    To Do Lists on, 146–147
    Memo Pad on, 128–129
    Note Pad on, 137–138
    organizing information on, 134–135
    PowerPoint synchronized with, 396
    reasons for using, 126
    synchronization with, 130–131
    text input on, 126–127
Handouts, slide, 417
Hard copy, 417
Hard drive, 15
    saving to, 267
Hardware, 10
    identifying, 13–14
Head section (HTML), 86
Headers
    Excel chart, 348–349
    PowerPoint, 418
    Word, 277–280
Headset, 165
Headset controls, 157
Health Science career cluster, 30, 149
Help
    online, 67
    speech recognition, 170
Hiding
    of formatting marks, 272
    of Quick Keys pad, 198
Highlighter button, 214
Highlighter settings, 206

CHAPTER 5

#	Title	Page
1	Enter Text	127
2	Create and Edit a Memo	129
3	Beam and Receive a Memo	131
4	Use and Create a Category	135
5	Move a Memo to a Different Category	135
6	Beam a Category	135
7	Create and Beam a Note	138
8	Add and Edit Addresses	140
9	Create and Beam a Business Card	141
10	Use Date Book in Daily View	144
11	Create a Future Event	144
12	Copy and Paste an Event	144
13	Delete an Event	145
14	Choose Weekly and Monthly View Format	145
15	Add Items to the To Do List Application	147
16	Use the Details Feature	147
17	Select Preferences	147
18	Add a Note to a To Do Item	147
19	Mark and Delete an Item	147
19	Cut and Paste Text	170
20	Clear Text	170
21	Copy Text	170
22	Use the Help Menu and Add Words to the Dictionary	170
23	Entering Punctuation	173
24	Add Bullets	173
25	Use the Correction Feature	174
26	Correct Capitalization	175
27	Change Case	175
28	Use the Scratch That Command	178
29	Use the Tab Command	178
30	Enter Numerals	180
31	Enter Decimals	180
32	Enter Fractions	180
33	Enter Dates	181
34	Enter Phone Numbers	181
35	Enter Currency and Time	181
36	Enter Mathematical Operators	181
37	Dictate Special Symbols	185
38	Dictate Emoticons	185

CHAPTER 6

#	Title	Page
1	Position the Microphone	156
2	Create a User Profile	156
3	Turn the Microphone Off and On	157
4	Continue Training Your Profile	157
5	Use the Period and New Line Commands	157
6	Use the Backspace Command	158
7	Access Menus	162
8	Use the Underline Command	162
9	Use the Bold Command	163
10	Use the Italics Command	163
11	Use the Center Command	164
12	Use the Align Right Command	164
13	Use the Align Left Command	165
14	Save a Document	165
15	Print a Document	165
16	Dictate and Proofread	169
17	Navigate a Document	169
18	Open a File	169

CHAPTER 7

#	Title	Page
1	Write and Add text	194
2	Open Journal	195
3	Calibrate the Digital Plan	196
4	Dock and Undock Input Panel	196
5	Change Screen Orientation	196
6	Change Screen Brightness	197
7	Choose Left- or Right-Hand Settings	199
8	View Input Panel Options	199
9	Choose the Writing Pad Tab	199
10	Change Options in Writing Tools	200
11	View the Write Anywhere Feature	200
12	View the Speech Recognition Feature	200
13	Explore Advanced Tab Options	201
14	Practice Data Entry	201
15	Select and Edit Text	201
16	Use Quick Keys	201

Denial of service (DoS), 47
Descending sort, 370
Design View, 366
Details feature, in To Do List, 147
Dictate/dictation
    definition of, 155
    improving, 168
    of numbers, 179–182
Dictation mode, 161
Dictionary
    speech recognition, 170, 171, 184
    Word, 273
Digital camcorders, 37
Digital cameras, 37, 38, 46
Digital devices, input/output, 20–21
Digital image technology, 37–39
Digital pen, 5, 191, 196
Digital photographs, 38, 100–102
Digital technology applications, 35
Digital tools, 3–8
Digital video disc (DVD), 23, 24
Disclaimers, 482
Divider tabs, OneNote, 251
.doc extension, 15
Docking, 196
Docking station, 193
Dollar Sign ($) command, 185
Domain name, 60
DoS (denial of service), 47
Double-click speed, 14
Double Dash (--) command, 185
Double spacing, 270, 294
Draw Table, 284–285
Drawing, 212–217
    in Journal, 212–215
    in Word, 215–217
Drawing toolbar, 403–404
Dreamweaver, 85
Drives, 12
DVD. *See* Digital video disc
DVD drive, 23, 24
DVD recorders, 37

E

E-cycling, 261
E-learning, 452
E-mail, 40–44
    attachments to, 42–43
    creating, 41–42
    deleting, 43
    forwarding, 44
    junk, 44
    links to, 115
    priority of, 42
    read, 43–44
    reply to, 43–44
E-mail addresses, 185
E-mail cover letters, 473
E-portfolio, 28–29
Edit command, 170
Editing, 140, 235, R6–R14
    formulas, 327
    images, 302–303
    spelling and grammar, 272–273
    spreadsheet cells, 318–319
    text, 201
    thesaurus, 273
Education and Training career cluster, 30, 407
EEOC (Equal Employment Opportunity Commission), 411
Electronic résumés, 473
Ellipsis (...) command, 185
Emoticons, 184, 185
Employee rights and obligations, 411
Employment-at-will, 482
End Curly Brace (}) command, 185
End of Line command, 157
Endnotes, 278
Enter command, 157
Enter key, 184
Entrepreneur, 99
Enunciate, 155
Envelope format, R5
Equal Employment Opportunity Commission (EEOC), 411
Equal sign (=), 325

Equal Sign (=) command, 185
Equals command, 181
Erase, 157, 158, 170, 177–178, 209
Ergonomics, 503
ESCAPE command, 184
Ethics, 50–51, 73–74
Events, Date Book, 143–145
Excel, 313–358
    Access data integrated with, 383
    charts in, 336–338
    columns, 341
    data entry in, 317
    editing cells, 318–319
    fill series, 341
    formatting numbers, 318
    formatting text, 317–318, 321–322
    formulas, 324–327
    merge cells, 340
    moving cell contents, 319–320
    Page Break View, 347
    pasting spreadsheet into Word, 347–348
    printing, 320–321
    rows, 341
    synchronizing PDAs with, 236
    templates in, 333–334
    title, 345–346
    wrapping text, 340
    zoom, 347
Exclamation mark (!), 172, 173, 185
Expense log, 232
Extensions
    domain, 60
    file, 15
External hard drive, 23
External storage devices, 23–27

F

Fact checking, 19
Fair use, 73
Family benefits, 400

Index 14

17	Open Journal	205
18	Change Pen and Highlighter Settings	205
19	Write and Save a Journal Note	206
20	Select Handwriting and Convert It to Typed Text	206
21	Add a Flag	208
22	Insert Line Spaces in a Journal Note	208
23	Erase and Undo	209
24	View Journal Notes	210
25	Create and Edit a Drawing in Journal	213
26	Copy the Chart and Paste It into Word	215
27	Draw in Word	215
28	Use the Scratch Out Gesture	221
29	Format a Document with Gestures	221
30	Insert a Letter Space with the Space Gesture	221
31	Remove a Character with the Backspace Gesture	221
32	Insert a Line Space with the Enter Gesture	222
33	Insert a Tab with the Tab Gesture	222

CHAPTER 8

1	Identify the PDA Operating System	230
2	Download Freeware from the Internet	230
3	Synchronize to a PDA	230
4	Use a Loan Calculator	230
5	Identify Math Applications for a PDA	233
6	Add Expenses	233
7	Transfer Office Documents	236
8	Personal Journal on a PDA	236
9	Import and Mark Up a Microsoft Word Document	240
10	Add Flags	240
11	Find Flags	241
12	Use Templates	244
13	Personalize Stationery	244
14	Use Write Anywhere	248
15	Enter and Move Text	251
16	Rename Divider Tabs	251
17	Flag Notes	252

CHAPTER 9

1	Explore Print Layout and Normal View	265
2	Indent Text Using the Tab Key	266
3	Select Text and Make Text Bold	266
4	Make Text Italic and Underlined	266
5	Delete Text	267
6	Save and Save As	267
7	Copy, Cut, and Paste Text	267
8	Use the Format Painter	268
9	Print Preview and Print	268
10	Format Text Font	268
11	Change Line Spacing	271
12	Show and Hide Formatting Marks	272
13	Align Text	272
14	Use the Spelling and Grammar Feature	272
15	Use the Thesaurus Language Feature	273
16	Set Margins	274
17	Set Page Orientation	274
18	Create Headers and Footers	278
19	Create Bulleted or Numbered Lists	280
20	Insert and Delete a Table	284
21	Draw a Table and Enter Text	284
22	Add and Delete Rows	285
23	Change the Border	286
24	Add Shading to Cells	287
25	Create a Table for an Invoice	290
26	Merge Cells	290
27	Split Cells	291
28	Complete an Invoice	291
29	Change Row Height	292
30	Change Column Width	292
31	Automatically Resize to Fit Contents	292
32	Split a Table	292
33	Dictate a Letter with Speech Recognition	295
34	Create a Text Box	298
35	Change a Border and Fill with Color	299

Cells
- alignment of, 330
- color of, 329
- definition of, 283
- editing, 318–319
- merge, 340
- moving contents of, 319–320

Center voice command, 164
Center for Safe and Responsible Internet Use, 72
Central processing unit (CPU), 10
Certifications, job, 54
Change case, 175
Chart Wizard, 336–338, 348
Charts, 335–343
- creating, 341–342
- data selection for, 336
- drawing in Journal, 213–215
- footer, 348–349
- formatting, 348
- header, 348–349
- options, 337–338
- presentation, 412–415
- scaling, 349
- in Word, 342

Children's Internet Protection Act (CIPA), 52
Chronological order, 456
Chronological résumé, 470, 471
CIPA (Children's Internet Protection Act), 52
Clear Text voice command, 170
Clip art, 100–102, 301, 302, 401, 402
Close Angle (>) voice command, 185
Close Brace (}) voice command, 185
Close Bracket (]) voice command, 185
Close button, 11
Close Parenthesis voice command, 185
Close quote (") voice command, 172, 173
Closing tags, 85, 86, 93–94, 104
College Is Possible program, 452
Colon (:) voice command, 172, 173

Color
- background, 88–89
- contrasting, 91
- spreadsheet, 329

Color codes, 89
Column width, 292, 322
Columns, 283, 285, 341
Comma (,) voice command, 172, 173
Comments, in document, 239
Communication
- careers in, 30
- integrity in, 369
- methods of, 4
- online, 40–45
- successful, 300

Compact disc (CD), 23, 25–26
Compact disc recordable (CD-R), 24
Compact disc recordable read-only memory (CD-ROM), 24
Compact disc rewritable (CD-RW), 24
Compressed files, 351, 377, 378
Computers, recycling of, 261
Confidentiality, statement of, 482
Contact list, for job leads, 486
Contents, table of, R6
Contrasting colors, 91
Control Panel, 11–14
- digital pen, 196
- left-/right-hand settings, 199
- screen brightness, 197
- screen orientation, 196

Cookies, 65
Copy command, 161, 170
Copying
- events, 144
- from Journal, 215
- in Word, 267–268

Copyright, 73–75
Correction feature, 172, 174–175
Cover letters, 473, 474, 476
Crackers, and Internet security, 47
CTRL + A (select all), 184
CTRL + HOME, 327
Curly Brace ({) command, 185

Currency, 179, 181
Currency Style button, 327
Cut command, 161, 170
Cutting, in Word, 267–268
Cyberbullies, 72, 218
Cybercrime, 75–77

D

Daily View, in PDAs, 144
Dash (–), 172
Data type, 363
Database(s), 359–388
- AutoForm, 367–368
- creating, 362–366
- definition of, 361
- integrating, 380–384
- online forms, 376–378
- queries, 370–375
- reports, 370–375
- sort, 370–375
- structure of, 361
- uses of, 361
- Web page creation from, 385–388

Date Book, 143–145
Decimal places, 318, 348
Decimals, 179, 180
Decision making, career-planning, 459–462
Decrease Decimal button, 318
Default layout, 398
Default setting, 270
Delete command, 157
Delete Next Word command, 157
Delete Previous Word command, 157
Deleting
- columns, 285
- database records, 366
- dictionary words, 171
- To Do items, 147
- e-mails, 43
- events, 145
- objects, 403
- rows, 285
- tables, 284
- text, 267
- worksheets, 322

#	Title	Page
36	Enter and Edit Text in a Text Box	299
37	Add Clip Art to a Document	302
38	Resize an Image	302
39	Edit an Image	302
40	Move an Image in Front of Text	303
41	Synchronize a Document	305
42	Edit a Word Document on a PDA	305
43	Download Business Templates	307
44	Fill in a Business Template	307
45	Create a New Folder from the Save As Window	308

CHAPTER 10

#	Title	Page
1	Enter Data in a New Worksheet	317
2	Select a Worksheet and Format Text	317
3	Format Numbers	318
4	Edit Cell Contents	318
5	Move Cell Contents	319
6	Print a Spreadsheet	320
7	Format Text in a Spreadsheet	321
8	Change Column Width	322
9	Insert and Delete Worksheets	322
10	Enter a Formula into a Spreadsheet	325
11	Use the AutoSum Feature	325
12	Enter Additional Formulas into a Spreadsheet	326
13	Apply Borders and Shading to the Cells in a Spreadsheet	329
14	Align Cell Contents	330
15	Apply an AutoFormat	330
16	Open and Save a Template	333
17	Enter Text into a Template	333
18	Create a Spreadsheet	336
19	Select Data to Put into a Chart and Create a Chart Using the Chart Wizard	336
20	Rename Your Worksheets	337
21	Change Chart Options	337
22	Wrap Text in Cells	340
23	Merge Cells in a Spreadsheet	340
24	Insert Rows and Columns	341
25	Use the Fill Series Feature	341
26	Create a Chart	341
27	Add a Chart to a Word Document	342
28	Format a Spreadsheet Title	345
29	Align Title Text at an Angle	346
30	Apply Formatting to the Spreadsheet Data	346
31	Use Page Break View	347
32	Magnify Your Spreadsheet	347
33	Paste an Excel Spreadsheet into a Word Document	347
34	Apply Formatting to a Chart	348
35	Add a Header and a Footer to a Chart	348
36	Download an Excel Template	351
37	Customize a Template	351

CHAPTER 11

#	Title	Page
1	Create a Database File	362
2	Create a Database Table and Define the Fields	362
3	Choose the Data Type for a Field	363
4	Enter Data	364
5	Switch to Design View and Change a Field Title	366
6	Delete a Record	366
7	Create an AutoForm	367
8	Add a Record with AutoForm	368
9	Sort Data	371
10	Create a Query	371
11	Create a Report	372
12	Modify Report Design	373
13	Change Text Formatting	374
14	Locate and Download an Online Access Database Form	377
15	Customize the Access Template	377
16	Create a Memo	381
17	Insert Access Data into the Memo	381
18	Insert Access Data into Excel	382
19	Format the Spreadsheet and Create a Chart	382
20	Create a Web Page from an Access Database	385
21	Format the Data Access Page	387
22	View the Data Access Page in a Web Browser	388

AutoShapes, 404–405
AutoSum, 326, 327, 340

B

Background color, 88–89
Background images, 102
Backgrounds
    slide, 397, 418
    Web page, 100
Backslash (\), 172
Backspace voice command, 157, 158
Backspace gesture, 220
Backup plan, 459, 462
Beaming
    categories, 135
    memos, 130–131
    notes, 138
    Word documents, 304
Beginning of Line voice command, 157
Benefits, company, 400
Bibliography, R4
Block style, 294, 295
.bmp extension, 15
Body language, 490
Body section, HTML, 86
Bold, 92–93, 264, 266, 268
Bold voice command, 163
Bookmarks, 61
Boolean operators, 63
"Boot" instructions, 11
Borders,
    Excel, 328, 329
    PowerPoint chart, 415
    Word, 286, 299
    Boxed table, R6
Breaks, 95–97
Brightness, Tablet PC screen, 197
Browsers, 60–67
    customizing, 63, 64
    Favorites added to, 61
    and hyperlinks, 61
    organizing Favorites, 62
    viewing Access page in, 388
Bullets, 173–174, 277, 280–281
Business and Administration career cluster, 30, 79

Business cards, 139, 141
Business forms
    database, 376–378
    Word, 306
    word processing, 289–293
Business letters, 294, 295, R2
Button(s)
    close, 11
    currency style, 327
    decrease decimal, 318
    Format Painter, 268
    highlighter, 214
    increase decimal, 318
    italics, 266–267
    maximize, 11
    minimize, 11
    mute, 157
    outline, 329
    redo, 265, 319
    restore, 11
    underline, 266–268
    undo, 209, 265, 319

C

C: drive, 16, 17
Calendars, 143
Calibrate digital pen, 196
Cameras, digital, 37, 38, 46
Capital letters, use of
    in e-mail, 50
    in HTML tags, 86, 115
Capitalization, R12
Capitalize voice command, 175
Career clusters, 30
    Agriculture and Natural Resources, 30, 339
    Architecture and Construction, 30, 176
    Arts, Audio/Video Technology, and Communication, 30, 288
    Business and Administration, 30, 79
    Education and Training, 30, 407
    Finance, 30, 309
    Government and Public Administration, 30, 254

    Health Science, 30, 149
    Hospitality and Tourism, 30, 223
    Human Services, 30, 160
    Information Technology, 30, 419
    Law and Public Safety, 30, 355
    Manufacturing, 30, 116
    Retail/Wholesale Sales and Service, 30, 69
    Scientific Research and Engineering, 30, 31
    Transportation, Distribution, and Logistics, 30, 54
Career planning, 447–464
    decision-making process, 459–462
    education, 452–454
    goal setting, 456–458
    job description evaluation, 449–451
    mentors, 464
    negotiation, 463
    overcoming obstacles, 455
    training, 452–454
Careers
    accountability and success of, 354
    and interests/values/skills/aptitudes, 432–434
    and leadership, 435
    learning more about, 440–442
Caret (^) voice command, 185
Case, change, 175
Categories, in a PDA, 134–135
Category box, 135
CD. *See* Compact disc
CD-R (compact disc recordable), 24
CD-ROM (compact disc recordable read-only memory), 24
CD-RW (compact disc rewritable), 24
Cell address, 316, 317
Cell name, 316, 317

CHAPTER 12

1	Display and Hide Toolbars	396
2	Create a New Presentation	397
3	Title Your Presentation	397
4	Add a New Slide to Your Presentation	397
5	Enter Text Into the New Slide	398
6	Explore PowerPoint Views	398
7	Use Two Methods to Delete a Slide	399
8	Insert Clip Art into a Slide	402
9	Format a Graphic in a Slide	402
10	Use the Drawing Toolbar and Insert WordArt	403
11	Insert and Format AutoShapes	404
12	Add Animation to Slides	409
13	Add a Footer to a Presentation	409
14	Add a Chart to a Presentation	413
15	Edit a Chart in a Presentation	414
16	Add a Border to a Chart	415
17	Save as a New Presentation	415
18	Print a Presentation	418
19	Add a Hyperlink to a Presentation	421
20	Save a Presentation As a Web Page	421

CHAPTER 13

1	What Types of Jobs Are Out There?	433
2	What Are Your Skills?	433
3	What Are Your Interests?	434
4	What Are Your Aptitudes?	434
5	What Are Your Values?	434
6	Use Your Interests to Focus on Jobs in Different Career Clusters	437
7	Identify People Who Work in a Career That Interests You	441
8	Find Internship Positions	441
9	Find Volunteer Positions	441

CHAPTER 14

1	Evaluate Job Descriptions	450
2	Plan Your Education and Training	453
3	Categorize Career Goals	457
4	Make Decisions	461

CHAPTER 15

1	Develop Two Résumés	475
2	Write a Cover Letter	476
3	Search a Local Newspaper	480
4	Find a Job Online	480
5	Fill Out a Job Application Form	483
6	Evaluate Disclaimers	483
7	Job Lead Contact List	486
8	Research Companies to Prepare for an Interview	486
9	Practice Interview Responses	489
10	Follow-Up after an Interview	489

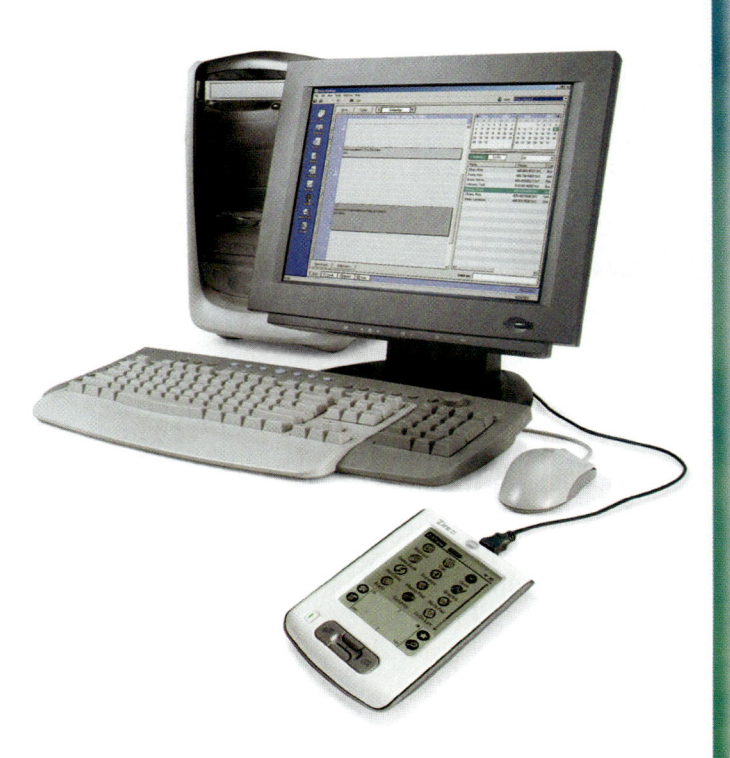

Index

Individual Keys

A, A2
B, A17
C, A9
D, A2
E, A5
F, A2
G, A14
H, A5
I, A7
J, A2
K, A2
L, A2
M, A7
N, A9
O, A5
P, A23
Q, A19
R, A7
S, A2
T, A9
U, A17
V, A11
W, A14
X, A23
Y, A25
Z, A27

1, A42
2, A39
3, A36
4, A33
5, A45
6, A45
7, A33
8, A36
9, A39
0, A42

Symbols

: (colon), 172, 173
, (comma), 172, 173
… (ellipsis command), 185
. (period), 172, 173
; (semicolon), 172
-- (double dash command), 185
– (dash), 172
/ (forward slash), 128
- (hyphen command), 185
/ (slash), 172, 173
& (ampersand command), 185
* (asterisk), 327
* (asterisk command), 185
@ (at sign command), 185
\ (backslash), 172
^ (caret command), 185
} (close brace command), 185
] (close bracket command), 185
) (close parenthesis command), 185
$ (dollar sign command), 185
= (equal sign), 325
= (equal sign command), 185
(error in Excel), 383
! (exclamation mark), 172, 173
! (exclamation mark command), 185
> (greater than command), 185
< (less than command), 185
{ (open brace command), 185
[(open bracket command), 185
((open parenthesis command), 185
% (percent sign command), 185
? (question mark), 172, 173
" " (quotation marks), 172, 173
~ (tilde command), 185
_ (underscore command), 185
| (vertical bar command), 185

A

A: drive, 16, 17
Abbreviations, R14
Acceptable use policy (AUP), 47
Access, 359–392
    AutoForm, 367–368
    creating database in, 362
    data entry, 364–366
    deleting records, 366
    fields, 362–364
    forms, 377–378
    integrating, 381–384
    query, 370–372
    reports, 370, 372–374
    sort, 370, 371
    tables, 362
    templates, 377–378
    Web page creation from, 385–388
Accessing menus (speech recognition), 162
Accountability, 354
Address Book, in PDAs, 139, 140
Address line, in browser, 88
Agriculture and Natural Resources career cluster, 30, 339
Alarm, setting in PDA, 144
Align Left voice command, 165
Align Right voice command, 164
Alignment, 270, 272
    in block style letter, 295
    footer, 277
    header, 277
    speech recognition, 164–165
    spreadsheet, 330
    title, 345–346
    Web page, 104–107
All Aspects of Industry, 53
American Council on Education, 452
Ampersand (&) voice command, 185
Anchor, 113, 114
AND keyword, 63
Animated clip art, 100, 101
Animation, 408–410
Anti-virus software, 70
Apprenticeships, 452
Aptitudes, 432, 434
Architecture and Construction career cluster, 30, 176
Arts, Audio/Video Technology, and Communication career cluster, 30, 288
Ascending sort, 370
Asterisk (*), 327
Asterisk (*) voice command, 181, 185
At Sign (@) voice command, 185
Attachments, e-mail, 42–43
Attribute, 88
Audio files, 250
AUP (acceptable use policy), 47
Author, Web site, 108
AutoFit Selection, 322
AutoForm, 367–368
AutoFormat, 328, 330
Automatically Resize to Fit Contents, 292

To The Student

Why Develop Skills with Digital Communication Tools?

For many students, digital tools are becoming a basic part of daily life. By understanding how to use digital tools and applications, you will learn skills that will help you in school and in your career.

Digital Communication Tools is intended to help you develop skills needed to succeed in school and throughout your life. This textbook was written and designed to help you achieve each of the following goals:

Become a 21st Century Digital Citizen

- Understand how to use technology wisely and safely.
- Understand how computers, digital tools, and the Internet work.
- Evaluate the accuracy and usefulness of information on the Web.
- Find and share information quickly, safely, and ethically.
- Understand the norms of behavior as a citizen in the digital age.

Become an Effective User of Technology Tools

- Become a skilled user of PDAs, speech recognition, and Tablet PCs.
- Become a skilled and creative user of Microsoft Word, Excel, PowerPoint, and Access.
- Create interesting projects using both individual applications and integrated applications.
- Become an effective researcher using the resources of the World Wide Web.
- Become an expert at navigating the Web and evaluating Web sites.

Develop Learning and Study Skills for All Subjects

- Improve reading comprehension with both guided and independent reading strategies.
- Develop critical thinking skills.
- Build teamwork skills.
- Integrate technology skills across the curriculum.
- Offer constructive feedback to improve your own and others' projects.

plagiarism Using other people's ideas and words without acknowledging that the information came from them.

primary key A field that contains a unique identification for each record in a database.

prioritize To put in order from most important to least important.

Q

query A database object that enables you to locate records that meet a certain criteria.

Quick Keys A group of keys that provide an easy way to move the insertion point through a document.

R

read-only memory (ROM) Memory that holds data that can be read by a digital device but cannot be changed or deleted.

record A collection of fields for one item in a database.

reference The name of a person an employer can contact to find out about a job applicant.

report A database object used to present the information in a database in a way that is easy to read.

resolution The amount of detail in a digital image, measured in pixels.

résumé A summary of your skills, abilities, education, work history, and achievements.

row A part of a table that extends across the table.

S

search engine A Web site that finds Web pages that contain words you tell it to search for.

Selection Tool A tool that enables you to select text, handwriting, or graphics in Journal.

shareware Software that is distributed free on a trial basis and may later be purchased.

skills Things you can do as a result of your talents, training, and experience.

slide An image created in PowerPoint that is displayed on a screen as part of a presentation and can contain text, graphics, sound, video, and animation.

software A set of instructions, also called a program or application, that tells a computer how to perform tasks.

spreadsheet A software application used to list, analyze, and perform calculations on data.

stylus A penlike device used to select icons on a PDA screen or to write on a PDA screen.

synchronization The process of transferring data between a PDA and a computer.

T

tag A piece of HTML code that tells a Web browser how to display with a particular section of text, a graphic, or other Web page element.

template A document that is set up with formatting that can be used over and over again.

To Do List An application on a PDA that allows you to create, keep track of, and prioritize important tasks and events.

U

uniform resource locator (URL) A unique Web address that directs a browser to a Web page.

unordered list A list that has bullet points before each item and is used when the items can be in any order.

user profile A record kept on a computer of the way a user pronounces words and the tone and speed with which the user speaks.

V

value Any number entered in a cell.

values Beliefs and ideas you live by and think are important.

virus A small computer program that can copy itself repeatedly in a computer system and cause a variety of problems.

Voice Command mode The mode in which the computer uses spoken words as instructions rather than as literal text.

W

webcam A small camera that connects to a computer and sends an image over the Web.

weblog An online journal that is updated almost every day by its author.

Writing Pad An area of Input Panel that recognizes handwritten input.

To The Student

BE AN ACTIVE READER!

When you read this textbook, you will be learning about technology and how it is used in the world around you. This textbook is a good example of non-fiction writing—it describes real-world ideas and facts. It is also an example of technical writing because it tells you how to use technology.

Here are some reading strategies that will help you become an active textbook reader. Choose the strategies that work best for you. If you have trouble as you read your textbook, look back at these strategies for help.

Before You Read

SET A PURPOSE
- Why am I reading the textbook?
- How might I be able to use what I learn in my own life?

PREVIEW
- Read the chapter title to find out what the topic will be.
- Read the subtitles to see what I will learn about the topic.
- Skim the photos, charts, graphs, or maps.
- Look for vocabulary words that are boldfaced. How are they defined?

DRAW FROM YOUR OWN BACKGROUND
- What do I already know about the topic?
- How is the new information different from what I already know?

firewall A software program that acts as a barrier between a computer and the Internet to protect the computer from unauthorized access.

font A combination of specific visual characteristics of text, including size, typeface style, bold, and italics.

footer An area for text or other objects that will appear at the bottom of every page in a document.

formula bar Displays the data within a cell and allows you to edit the contents of the cell.

formula A mathematical expression, such as adding or averaging, that performs calculations on data in a spreadsheet.

freeware A software application that you can download and share at no charge.

function A predefined formula in a spreadsheet.

G

gesture A simple shape made with the digital pen that sends a command to the Tablet PC to edit or format text.

H

hard copy A paper printout of what is shown on a computer screen.

hard drive The built-in storage device inside a computer.

hardware All of the physical parts of a computer.

header An area for text or other objects that will appear at the top of every page in a document.

horizontal rule A line that runs across the width of a page.

hyperlink Any text or graphic on a Web page that will take you to a new location when clicked.

Hypertext Markup Language (HTML) A set of codes used to create documents for the Web.

I

import To open a document in an application other than the one in which it was originally created.

Input Panel A window where you can enter data using an on-screen keyboard or handwriting and then send the data to a document.

input Information that is entered into a computer.

interests Your favorite activities, the subjects you like best in school, and your hobbies.

internship A temporary position, paid or unpaid, where you can gain practical work experience in a career field.

J

job description An explanation of the major areas of a job necessary for an employee to perform the job successfully.

job interview A formal meeting between an employer and a job applicant.

job shadow To follow someone for a day or two on the job.

Journal A standard Tablet PC application that saves data as handwriting, drawings, or typed text.

L

label Text that is entered in a cell.

legend A table that lists and explains the symbols used in a chart.

M

margin The blank edge that borders the area in which text and objects can be placed on a page.

Memo Pad An application that allows you to create simple documents.

merge To combine two or more cells to create one cell.

N

netiquette The manners people use in electronic communications.

networking Things you can do as a result of your talents, training, and experience.

Notepad The application on a PDA that allows you to create short notes or simple diagrams.

O

operating system Software that controls all of the other software programs and allows the computer to perform basic tasks.

ordered list A list that has numbers before each item, and the items are usually in priority order.

output Information that a computer produces or processes.

P

pixel A single point in a digital image.

To The Student

As You Read

QUESTION
- What is the main idea?
- How well do the details support the main idea?
- How do the photos, charts, graphs, and maps support the main idea?

CONNECT
- Think about people, places, and events in my own life. Are there any similarities with those in your textbook?

PREDICT
- Can I predict events or outcomes by using clues and information that I already know.

VISUALIZE
- Can I imagine the settings, actions, and people that are described.
- Can I use/create graphic organizers to help me see relationships found in the information.

IF YOU DON'T KNOW WHAT A WORD MEANS...
- think about the setting, or context, in which the word is used.
- check if prefixes such as *un-*, *non-*, or *pre-* can help you break down the word.
- look up the word's definition in a dictionary or glossary.

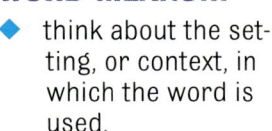

READING DOs

Do...
- ✓ establish a purpose for reading.
- ✓ think about how your own experiences relate to the topic.
- ✓ try different reading strategies.

READING DON'Ts

Don't...
- ⊘ ignore how the textbook is organized.
- ⊘ allow yourself to be easily distracted.
- ⊘ hurry to finish the material.

After You Read

SUMMARIZE
- What have I learned from this text?
- How can I apply what I have learned?

ASSESS
- What was the main idea?
- Did the text clearly support the main idea?
- Can I use this new information in other school subjects or at home?

Glossary

A

alignment The way that text or objects are lined up across a page.

animation A sound or visual effect that can be added to text and graphics in a presentation.

aptitude Your potential for learning a skill.

ascending sort A sort of data arranged in alphabetical (A–Z) order or numerical (0–9) order.

attribute A specific instruction that tells the browser how to display the text or graphics enclosed by the HTML tags.

AutoFormat Predefined shading and border formats that can be applied to cells in a spreadsheet.

B

backup plan An alternative course of action.

beam The process of sending data from one device to another via an infrared beam.

Boolean operator A word that can make an Internet search more specific and effective.

browser An application that reads the contents of Web pages and displays them on your computer screen.

C

career clusters Groups of similar occupations and industries developed by the U.S. Department of Education.

categories Categories are used to organize data, and are similar to folders on a desktop computer or to files in a file drawer.

cell The intersection of a column and a row in a table.

chronological order The order in which events happen.

clip art Electronic illustrations that can be in-serted into a document.

column A part of a table that extends up and down the table.

copyright A type of legal protection for works that are created or owned by a person or company.

Correction feature A feature that allows a user to correct mistakes in dictated text while also recording the corrections as part of the user profile settings.

cover letter A short letter you send along with a résumé, specifying the job you are interested in and highlighting specific skills or accomplishments you feel make you a good candidate for the job.

D

database A software application that stores a collection of related information and allows easy retrieval of the information.

default A setting that was determined at the time the program was created.

descending sort A sort of data arranged in high to low alphabetical (Z–A) order or numerical (9–0) order.

dictate To speak words for a computer to enter as text in a document.

Dictation mode The mode in which the computer enters words and characters exactly as they are spoken.

digital Information that is represented as individual pieces of data using the numbers 1 and 0, rather than as a continuous stream.

domain name The part of a URL that usually identifies the owner or the subject of a Web site.

E

employment-at-will Employment that can be terminated with or without cause by the employer.

enunciate To pronounce words clearly, including the beginning and ending sounds of the words.

e-portfolio A collection of the best documents and projects you have created and saved in an electronic format.

ethics Society's rules for what is right and what is wrong.

expense log A software application in which you can enter and track your daily expenses.

external storage device A storage device that can be easily removed from a computer and is often small and portable.

F

fair use Describes when you may use a copyrighted work without permission.

field A category that holds one piece of information in a database.

Technology Standards

Most educators today believe that in order to live, learn, and work successfully in an increasingly complex society, students must be able to use technology effectively.

What Are Standards?

Standards give educators a benchmark that lets them and their students know what skills every student should master in a particular subject area. Students who know what standards they are expected to meet before they begin a course are better prepared to succeed and meet expectations. The technology standards described below and on the next pages are designed to help students understand and apply the skills needed to be technologically literate in today's society.

ISTE and NETS

The International Society for Technology in Education (ISTE) has developed National Educational Technology Standards for students (NETS-S). The ISTE standards identify skills that students can practice and master in school, but the skills are also used outside of school, at home and at work. The activities in this book are designed to meet ISTE standards. For more information about ISTE and the NETS, please visit www.iste.com

The Web-Safe Color Palette

Colors in the chart are notated in both their hexademical values for use in Web design programs and their RGB (red, green, blue) values for use in graphics programs.

The first number/letter combination shown represents the color's hexadecimal value, which is commonly used for programming and Web design programs. The second set of numbers shown represents the color's RGB value; each of these numbers correlates to an intensity value of red, green, and blue as displayed on a monitor. For example, an orange color would have the RGB color designation of red 255, green 102, and blue 51.

These colors may appear differently on your monitor than they do here in print.

Technology Standards

Technology Foundation Standards for Students

The NETS are divided into six broad categories that are listed below. Activities in the book are specifically designed to meet the standards within each category.

1. **Basic operations and concepts**
   - Students demonstrate a sound understanding of the nature and operation of technology systems.
   - Students are proficient in the use of technology.

2. **Social, ethical, and human issues**
   - Students understand the ethical, cultural, and societal issues related to technology.
   - Students practice responsible use of technology systems, information, and software.
   - Students develop positive attitudes toward technology uses that support lifelong learning, collarboration, personal pursuits, and productivity.

3. **Technology productivity tools**
   - Students use technology tools to enhance learning, increase productivity, and promote creativity.
   - Students use productivity tools to collaborate in constructing technology-enhanced models, prepare publications, and produce other creative works.

4. **Technology communications tools**
   - Students use telecommunications to collaborate, publish, and interact with peers, experts, and other audiences.
   - Students use a variety of media and formats to communicate information and ideas effectively to multiple audiences.

5. **Technology research tools**
   - Students use technology to locate, evaluate, and collect information from a variety of sources.
   - Students use technology tools to process data and report results.
   - Students evaluate and select new information resources and technological innovations based on the appropriateness of specific tasks.

6. **Technology problem-solving and decision-making tools**
   - Students use technology resources for solving problems and making informed decisions.
   - Students employ technology in the development of strategies for solving problems in the real world.

Mechanics (continued)

3. Spell out:

    - Numbers used as the first word in a sentence.
      Seventy people attended the conference in San Diego last week.

    - The smaller of two adjacent numbers.
      We have ordered two 5-pound packages for the meeting.

    - The words millions and billions in even amounts (do not use decimals with even amounts).
      The lottery is worth 28 million this month.

    - Fractions.
      About one-half of the audience responded to the questionnaire.

ABBREVIATIONS:

1. In nontechnical writing, do not abbreviate common nouns (such as *dept.* or *pkg.*), compass points, units of measure, or the names of months, days of the week, cities, or states (except in addresses).
   The Sales Department will meet on Tuesday, March 7, in Tempe, Arizona.

2. In lowercase abbreviations made up of single initials, use a period after each initial but no internal spaces.
   We will be including several states (e.g., Maine, New Hampshire, Vermont, Massachusetts, and Connecticut).

3. In all-capital abbreviations made up of single initials, do not use periods or internal spaces. (Exception: Keep the periods in most academic degrees and in abbreviations of geographic names other than two-letter state abbreviations.)
   You need to call the EEO office for clarification on that issue.

Technology Standards

Educational Technology Performance Indicators for Students

In this text, all students should have opportunities to demonstrate the following performance indicators for technological literacy. Each performance indicator refers to the NETS Foundation Standards category or categories (listed on previous page) to which the performance is linked.

1. Identify capabilities and limitations of contemporary and emerging technology resources and assess the potential of these systems and services to address personal, lifelong learning, and workplace needs. (2)
2. Make informed choices among technology systems, resources and services. (1, 2)
3. Analyze advantages and disadvantages of widespread use and reliance on technology in the workplace and in society as a whole. (2)
4. Demonstrate and advocate for legal and ethical behaviors among peers, family, and community regarding the use of technology and information. (2)
5. Use technology tools and resources for managing and communicating personal/professional information (e.g. finances, schedules, addresses, purchases, correspondence). (3, 4)
6. Evaluate technology-based options, including distance and distributed education, for lifelong learning. (5)
7. Routinely and efficiently use online information resources to meet needs for collaboration, research, publications, communications, and productivity. (4, 5, 6)
8. Select and apply technology tools for research, information analysis, problem-solving, and decision-making in content learning. (4, 5)
9. Investigate and apply expert systems, intelligent agents, and simulations in real-world situations. (3, 5, 6)
10. Collaborate with peers, experts, and others to contribute to content-related knowledge base by using technology to compile, synthesize, produce, and disseminate information, models, and other creative works. (4, 5, 6)

Mechanics (continued)

3. Capitalize the names of the days of the week, months, holidays, and religious days (but do not capitalize the names of the seasons).
   On Thursday, November 25, we will celebrate Thanksgiving, the most popular fall holiday.

4. Capitalize nouns followed by a number or letter (except for the nouns *line, note, page, paragraph,* and *size*).
   Please read Chapter 5, but not page 94.

5. Capitalize compass points (such as *north, south,* or *northeast*) only when they designate definite regions.
   The Crenshaws will vacation in the Northeast this summer.
   We will have to drive north to reach the closest Canadian border.

6. Capitalize common organizational terms (such as *advertising department* and *finance committee*) when they are the actual names of the units in the writer's own organization and when they are preceded by the word *the*.
   The quarterly report from the Advertising Department will be presented today.

7. Capitalize the names of specific course titles but not the names of subjects or areas of study.
   I have enrolled in Accounting 201 and will also take a marketing course.

NUMBER EXPRESSION:

1. In general, spell out numbers 1 through 10, and use figures for numbers above 10.
   We have rented two movies for tonight.
   The decision was reached after 27 precincts had sent in their results.

2. Use figures for:
   - Dates (use *st, d,* or *th* only if the day precedes the month).
     We will drive to the camp on the 23d of May.
     The tax report is due on April 15.

   - All numbers if two or more related numbers both above and below ten are used in the same sentence.
     Mr. Carter sent in 7 receipts; Ms. Cantrell sent in 22 receipts.

   - Measurements (time, money, distance, weight, and percentage).
     At 10 a.m. we delivered the $500 coin bank in a 17-pound container.

   - Mixed numbers.
     Our sales are up $9\frac{1}{2}$ percent over last year.

UNIT 1
Impact of Digital Communication Tools

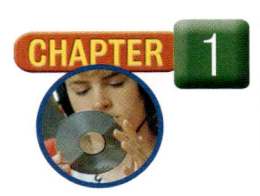

CHAPTER 1 Digital Communication Tools in Our Lives

CHAPTER 2 Communicating with Digital Technology

CHAPTER 3 Using the Internet

CHAPTER 4 Create a Web Page with HTML

Curriculum Connections
- Language Arts—Reading
- Language Arts—Writing
- Technology

Grammar (continued)

- *Principal* means "primary"; *principle* means "rule."
  The **principal** means of research were interviewing and surveying.
  They must not violate the **principles** under which our country was founded.

- *Passed* means "went by"; *past* means "before now."
  We **passed** another car from our home state.
  In the **past**, we always took the same route.

- *Advice* means "to provide guidance"; *advise* means "help."
  The **advice** I gave her was simple.
  I **advise** you to finish your project.

- *Council* is a group; *counsel* is a person who provides advice.
  The student **council** met to discuss graduation.
  The court asked that **counsel** be present at the hearing.

- *Then* means "at that time"; *than* is used for comparisons.
  He read for a while; **then** he turned out the light.
  She reads more books **than** I do.

- *Its* is the possessive form of it; *it's* is a contraction for it is.
  We researched the country and **its** people.
  It's not too late to finish the story.

- *Two* means "one more than one"; *too* means "also"; *to* means "in a direction."
  There were **two** people in the boat.
  We wished we were on board, **too**.
  The boat headed out **to** sea.

- *Stationery* means "paper"; *stationary* means "fixed position."
  Please buy some **stationery** so that I can write letters.
  The **stationary** bike at the health club provides a good workout.

Mechanics

CAPITALIZATION:

1. Capitalize the first word of a sentence.
   Please prepare a summary.

2. Capitalize proper nouns and adjectives derived from proper nouns. (A proper noun is the official name of a particular person, place, or thing.)
   Judy Hendrix drove to Albuquerque in her new car, a Pontiac.

DIGITAL Dimension → Language Arts
INTERDISCIPLINARY PROJECT

Link what you will learn about digital input skills in Unit 1 with language arts by analyzing media and advertising. Your project will help you become a savvy consumer, and you will finish with a presentation to the class.

Be a Savvy Consumer

What Influences You to Buy a Product? Everyone wants to feel good and be popular. The media tries to make you believe that if you dress a certain way, listen to certain music, wear a certain cologne, or even eat certain foods you will be part of the in crowd.

Do You Really Need a Product? The main goal of advertisers is to make you feel that you must have a product. Marketing professionals study the buying habits of consumers and then create advertising to influence your thoughts and beliefs. Advertisers use a number of selling techniques, such as humor, promises of increased attractiveness or a youthful appearance, celebrity endorsements, or implications that everyone else already has the product. Being aware of the advertisers' often-subtle techniques can help you be a savvy consumer.

Digital Dimension Activities

Start-Up
Look around your room and at what you are wearing. Create a chart of products you find. Fill in the type of item, the brand name, who bought the item, and why this item was chosen. Instead of your room, you can choose your bathroom or kitchen. You will refer to this chart later in this project.

Look Ahead...
Continue building your project in the following activities:

Chapter 1 .... p. 32
Chapter 2 .... p. 55
Chapter 3 .... p. 80
Chapter 4 .... p. 117
Unit 1 Digital Dimension Lab .... pp. 120–121

Grammar (continued)

6. Subjects joined by and take a plural verb unless the compound subject is preceded by *each*, *every*, or *many a (an)*.
   Every man, woman, and child is included in our survey.

7. Verbs that refer to conditions that are impossible or improbable (that is, verbs in the *subjunctive* mood) require the plural form.
   If the total eclipse were to occur tomorrow, it would be the second one this year.

PRONOUNS:

1. Use nominative pronouns (such as *I*, *he*, *she*, *we*, and *they*) as subjects of a sentence or clause.
   They traveled to Minnesota last week but will not return until next month.

2. Use objective pronouns (such as *me*, *him*, *her*, *us*, and *them*) as objects in a sentence or clause.
   The package has been sent to her.

ADJECTIVES AND ADVERBS:

Use comparative adjectives and adverbs (*-er*, *more*, and *less*) when referring to two nouns; use superlative adjectives and adverbs (*-est*, *most*, and *least*) when referring to more than two.
Of the two movies you have selected, the shorter one is the more interesting.
The highest of the three mountains is Mt. Everest.

WORD USAGE:

Do not confuse the following pairs of words:

- *Accept* means "to agree to"; *except* means "to leave out."
  *We **accept** your offer for developing the new product.*
  *Everyone **except** Sam and Lisa attended the rally.*

- *Affect* is most often used as a verb meaning "to influence"; *effect* is most often used as a noun meaning "result."
  *Mr. Smith's decision will not **affect** our music class.*
  *It will be weeks before we can assess the **effect** of this decision.*

- *Farther* refers to distance; *further* refers to extent or degree.
  *Did we travel **farther** today than yesterday?*
  *We need to discuss our plans **further**.*

- *Personal* means "private"; *personnel* means "employees."
  *The letters were very **personal** and should not have been read.*
  *We hope that all **personnel** will comply with the new rules.*

Digital Communication Tools in Our Lives

CHAPTER 1

Lesson 1.1	New Ways to Enter Text with Digital Tools
Lesson 1.2	Operating Systems, Hardware, and Software
Lesson 1.3	Storage and File Management
Lesson 1.4	Input and Output Devices
Lesson 1.5	External Storage Devices
Lesson 1.6	Digital Skills in the 21st Century Workplace

You Will Learn To:

- **Identify** alternate ways to enter text into a computer other than with a keyboard.
- **Describe** digital technologies.
- **Identify** computer system components.
- **Demonstrate** file management techniques.
- **Explain** the importance of having technology skills.
- **Describe** uses for an e-portfolio.
- **Plan** your own e-portfolio.

When Will You Ever Use a Digital Tool?

Many of you already use digital tools every day. Telephones, cell phones, computers, cameras, electronic games, and pagers are just a few examples of common digital tools. Later, when you start your career, you might use specialized digital tools to assist you in your job. Digital tools seem to be everywhere these days, and their popularity will only continue to grow.

Punctuation (continued)

QUOTATION MARKS:

1. Use quotation marks around the titles of newspaper articles, magazine articles, chapters in a book, reports, conferences, and similar items.
   The best article I found in my research was entitled "Multimedia for Everyone."

2. Use quotation marks around a direct quotation.
   Harrison responded by saying, "This decision will not affect our class."

ITALIC (OR UNDERLINE):

Italicize (or underline) the titles of books, magazines, newspapers, and other complete published works.
I read The Pelican Brief last month. I read The Pelican Brief last month.

Grammar

AGREEMENT:

1. Use singular verbs and pronouns with singular subjects and plural verbs and pronouns with plural subjects.
   I was pleased with the performance of our team.
   Reno and Phoenix were selected as the sites for our next two meetings.

2. Some pronouns (*anybody, each, either, everybody, everyone, much, neither, no one, nobody,* and *one*) are always singular and take a singular verb. Other pronouns (*all, any, more, most, none,* and *some*) may be singular or plural, depending on the noun to which they refer.
   Each employee is responsible for summarizing the day's activities.
   Most of the workers are going to get a substantial pay raise.

3. Disregard any intervening words that come between the subject and verb when establishing agreement.
   The box containing the books and pencils has not been found.

4. If two subjects are joined by *or, either / or, nor, neither / nor,* or *not only / but also,* the verb should agree with the subject nearer to the verb.
   Neither the players nor the coach is in favor of the decision.

5. The subject a number takes a plural verb; *the number* takes a singular verb.
   The number of new students has increased to six.
   We know that a number of students are in sports.

Reference **R10**

Read to Succeed

Key Terms

- software
- operating system
- hardware
- digital
- hard drive
- input
- output
- external storage device
- read-only memory (ROM)
- e-portfolio

Survey Before You Read

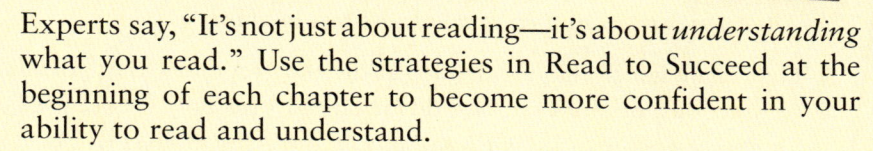

Experts say, "It's not just about reading—it's about *understanding* what you read." Use the strategies in Read to Succeed at the beginning of each chapter to become more confident in your ability to read and understand.

Before reading the chapter, do a quick survey. Read the chapter title and lesson titles to preview the topic. Use a graphic organizer like the one below to organize the chapter by lesson titles and topics. Place each lesson title on one row, and list the main topics beneath. This will make it easy to see and think about the chapter contents.

Example:

Chapter 1 Digital Communication Tools in Our Lives	
Lesson 1.1	New Ways to Enter Text with Digital Tools • Technology Skills Are Important • New Input Tools

DIGITAL CONNECTION

Communicating in a Digital World

How do you communicate with your friends? Years ago, when people were separated by distance, written communication was the only method of exchanging information. For short distances, letters were delivered by a mail carrier on foot. Letters that had to travel longer distances were sent by horse or train. To communicate with a friend today, you can just pick up a cell phone, type a quick e-mail, or send an instant message. No matter where your friend lives, you can make contact in a matter of a few seconds using digital technology.

Writing Activity

Make a list of all the methods people use to send or receive information. How many can you think of? Your list could include telephone, newspapers, and e-mail.

4

Punctuation (continued)

2. Hyphenate compound numbers (between twenty-one and ninety-nine) and fractions that are expressed as words.
   We observed twenty-nine fumbles during the football game.
   All teachers reduced their assignments by one-third.

3. Hyphenate words that are divided at the end of a line. Do not divide one-syllable words, contractions, or abbreviations; divide other words only between syllables.
   To appreciate the full significance of rain forests, you must see the entire documentary showing tomorrow in the library.

APOSTROPHES:

1. Use 's to form the possessive of singular nouns.
   The hurricane caused major damage to Georgia's coastline.

2. Use only an apostrophe to form the possessive of plural nouns that end in *s*.
   The investors' goals were outlined in the annual report.

3. Use 's to form the possessive of indefinite pronouns (such as *someone's* or *anybody's*); do not use an apostrophe with personal pronouns (such as *hers, his, its, ours, theirs,* and *yours*).
   She was instructed to select anybody's paper for a sample.
   Each computer comes carefully packed in its own container.

COLONS:

Use a colon to introduce explanatory material that follows an independent clause. (An independent clause is one that can stand alone as a complete sentence.)
A computer is useful for three reasons: speed, cost, and power.

DASHES:

Use a dash instead of a comma, semicolon, colon, or parenthesis when you want to convey a more forceful separation of words within a sentence. (If your keyboard has a special dash character, use it. Otherwise, form a dash by typing two hyphens, with no space before, between, or after.)
At this year's student council meeting, the speakers—and topics—were superb.

PERIODS:

Use a period to end a sentence that is a polite request. (Consider a sentence a polite request if you expect the reader to respond by doing as you ask rather than by giving a yes-or-no answer.)
Will you please call me.

Lesson 1.1 — New Ways to Enter Text with Digital Tools

You will learn to:
- **Explain** why technology skills are important.
- **Enter** text and save a file.
- **Compare** and contrast the features of different types of digital devices.
- **Identify** handheld computers and Tablet PCs.

Technology Skills Are Important High-tech tools, such as computers, cell phones, PDAs, speech recognition, and Tablet PCs, let us work fast, play electronic games, and communicate instantly with friends. Did you know that having good technology skills can be the key to a successful, well-paying career?

Today, technology is everywhere in the workplace, from offices to factories. In fact, most workplaces depend so much on technology that they cannot function without it. You will need skills using digital tools no matter what career you choose.

New Input Tools In the recent past, people generally used a keyboard and a mouse to communicate with a computer. Three new technologies for communicating with a computer and entering text into documents are:

- **Handwriting on Handheld Computers** Handheld computers are often called personal digital assistants, or PDAs. Using a stylus, which is a penlike device, you can use handwriting to enter text and point to enter commands into a PDA.
- **Speech Recognition** Speech recognition allows a user to enter text and give commands to the computer by speaking into a microphone.
- **Handwriting on Tablet PCs** Using a digital pen, you enter text and commands on the screen of a Tablet PC in your own handwriting!

With the development of these new technologies, computer users can now choose the input method that best suits them: keyboard, mouse, speaking, or handwriting and pointing on a PDA or Tablet PC.

✓ Concept Check

List three new input technologies for entering data into computers. Do you think it is still important to know how to use a keyboard? Why or why not?

DIGI Byte

Speech Recognition
The typical speed of text entry using the keyboard is 40 words per minute. Would you like to enter text at around 120 words per minute? With speech recognition, you can.

Punctuation (continued)

3. Use a comma before and after the year in a complete date.
   We will arrive at the plant on June 2, 2003, for the conference.

4. Use a comma before and after a state or country that follows a city (but not before a ZIP Code).
   Joan moved to Vancouver, British Columbia, in September.
   Send the package to Douglasville, GA 30135, by express mail.

5. Use a comma between each item in a series of three or more.
   There are lions, tigers, bears, and zebras at the zoo.

6. Use a comma before and after a transitional expression (such as therefore and however).
   It is critical, therefore, that we finish the project on time.

7. Use a comma before and after a direct quotation.
   When we left, James said, "Let us return to the same location next year."

8. Use a comma before and after a nonessential expression. (A nonessential expression is a word or group of words that may be omitted without changing the basic meaning of the sentence.)
   Let me say, to begin with, that the report has already been finalized.

9. Use a comma between two adjacent adjectives that modify the same noun.
   We need an intelligent, enthusiastic individual for this project.

SEMICOLONS:

1. Use a semicolon to join two closely related independent clauses that are not connected by a conjunction (such as and, but, or nor).
   Students favored the music; teachers did not.

2. Use a semicolon to separate three or more items in a series if any of the items already contain commas.
   Region 1 sent their reports in March, April, and May; and Region 2 sent their reports in September, October, and November.
   The Home room class sent their reports in 1st, 2nd, 3rd, and 4th hour; the history class sent their reports in 4th, 5th, 6th, and 7th hour.

HYPHENS:

1. Hyphenate compound adjectives that come before a noun (unless the first word is an adverb ending in -ly).
   We reviewed an up-to-date report on Wednesday.
   We attended a highly rated session on multimedia software.

Activity 1

If you need additional help on this topic, open the Help menu; or refer to the documentation for your hardware.

Step-by-Step

Enter Text and Save Files

All computers have a basic word processing program installed on them: WordPad. WordPad will allow you to create simple text documents and save them. You can create folders to save documents in an organized way. Documents are often called files.

1. Look for the **My Documents** folder icon on your desktop and double-click to open it, or click **Start** and look for **My Documents** in the list displayed. The computer saves all files to the My Documents folder unless you change this.

2. To create a new folder, click **File, New, Folder.** A new yellow folder will appear. The words **New Folder** will be highlighted in blue. Enter the words **Digital Tools** to name the folder and then press ENTER. You will save documents to the Digital Tools folder.

3. Click **Start, Programs, Accessories, WordPad.** Enter your name and address in the document.

4. To save the document, click **File, Save As** and enter **Name** in the **File name** box.

5. To choose the folder where you will save the document, click the **Save In** box and choose **My Documents, Digital Tools** from the list of folders.

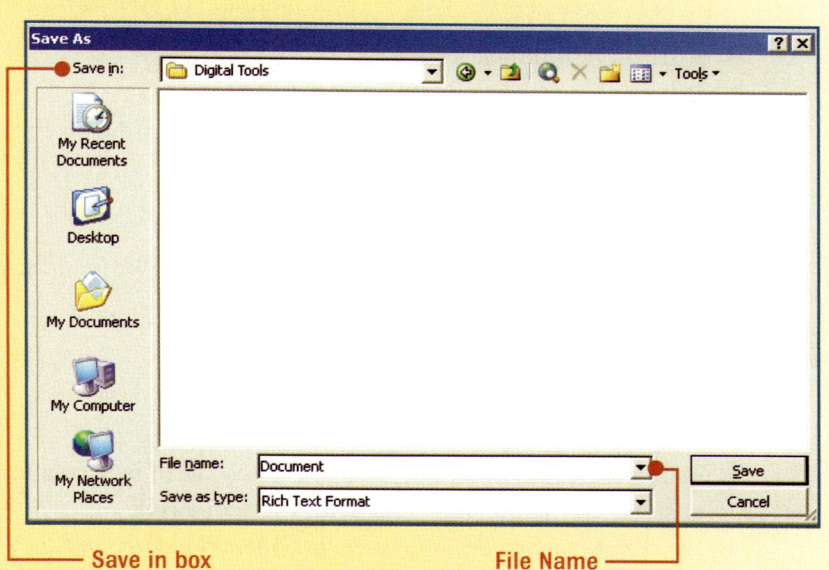

— Save in box   File Name —

6. The first time you save a document you will use **Save As** so you can name your document and choose where to save it.

7. Once a document has a file name, you can choose **File, Save** as you edit and update the document.

Continued

Chapter 1

Punctuation and Usage

ALWAYS SPACE ONCE . . .

- After a comma.
  We ordered two printers, one computer, and three monitors.

- After a semicolon.
  They flew to Dallas, Texas; Reno, Nevada; and Rome, New York.

- After a period following someone's initials.
  Mr. A. Henson, Ms. C. Hovey, and Mr. M. Salisbury will attend the meeting.

- After a period following the abbreviation of a single word.
  We will send the package by 7 p.m. next week. [Note: space once after the final period in the "p.m." abbreviation, but do not space after the first period between the two letters.]

- Before a ZIP code.
  Send the package to 892 Maple Street, Grand Forks, ND 58201.

- Before and after an ampersand.
  We were represented by the law firm of Bassett & Johnson; they were represented by the law firm of Crandall & Magnuson.

- After a period at the end of a sentence.
  Don't forget to vote. Vote for the candidate of your choice.

- After a question mark.
  When will you vote? Did you vote last year?

- After an exclamation point.
  Wow! What a performance! It was fantastic!

- After a colon.
  We will attend on the following days: Monday, Wednesday, and Friday.

Punctuation

COMMAS:

1. Use a comma between independent clauses joined by a conjunction. (An independent clause is one that can stand alone as a complete sentence.)
   We requested Brown Industries to change the date, and they did so within five days.

2. Use a comma after an introductory expression (unless it is a short prepositional phrase).
   Before we can make a decision, we must have all the facts.
   In 1992 our nation elected a president.

Activity 2

Step-by-Step *continued*

Investigate Handheld Computers

In this activity you will explore handheld computers, or PDAs. You will learn to use PDAs in Chapter 5, pages 124–152.

1. Open your Web browser.
2. To research handheld computers in education, use a search engine and enter the keywords "handheld education."
3. Read about handheld computers, or PDAs.
4. Open WordPad and compose a report about the handheld computers you researched on the Internet.
5. Include the following information:
   - Brand name of PDA.
   - Cost.
   - Programs that the PDA integrates with, such as Microsoft Word, Excel, PowerPoint, and so on.
6. Review Activity 1, Steps 4 and 5, page 6, for saving this document. Name the document **PDAs**.

Investigate Speech Recognition

In this activity, you will explore speech recognition software. **Software** is a set of instructions, also called a program or application, that tells a computer how to perform tasks.

You will learn to use speech recognition in Chapter 6, pages 153–189.

1. Open your Web browser.
2. Do a Web search using the keywords "speech recognition software."
3. Compare the applications you find, and take notes.
4. Open a new document in WordPad, and compose a short report about the information you find. Include the following:
   - The name of the software.
   - The cost of the software.
   - The equipment needed to use speech recognition.
   - A brief explanation of how speech recognition works.
5. Review Activity 1, Steps 4 and 5, page 6, for saving this document. Name the document **Speech**.

Activity 3

software A set of instructions, also called a program or application, that tells a computer how to perform tasks.

DIGI*Byte*

Speech Recognition
This technology is not new, but recently it has improved and become useful to the average computer user.

Continued

Boxed Table

Bills Passed for E-Waste or E-Cycling	
State	Bill
Arkansas	SB807, Enacted 4/9/01
California	SP1523, Introduced 2/20/02 SB1619, Introduced 2/21/02
Florida	SB1922, Introduced 2/6/02
Georgia	HB2, Passed the House, in the Senate 2/5/2002
Hawaii	HB1638, Carried over to 2002 Session
Idaho	S1416, Sent to Committee 2/12/2002
Illinois	HB4464, Passed the House, in the Senate 4/10/01
Maryland	HB111, Unfavorable Environmental Committee Report

Contents

CONTENTS

INTR	1
PROBLEM	1
HYPOTHESIS	1
RESEARCH	2
THE EXPERIMENT	5
Materials	5
Procedures	5
Results	7
CONCLUSION	8
BIBLIOGRAPHY	9

Proofreaders' Marks

Proofreaders' Marks	Draft	Final Copy	Proofreaders' Marks	Draft	Final Copy
⌢ Omit space	data base	database	SS Single-space	first line / second line	first line / second line
∨ or ∧ Insert	if he's going, (not)	if he's not going,	ds Double-space	first line / second line	first line / second line
≡ Capitalize	Maple street	Maple Street	⌐ Move right	Please send	Please send
⟋ Delete	a final draft	a draft	⌐ Move left	May I	May I
# Insert space	allready to	all ready to	∼∼ Bold	Column Heading	**Column Heading**
when ∧ Change word	and if you (when)	and when you	ital Italic	ital Time magazine	*Time* magazine
/ Use lowercase letter	our President	our president	u/l Underline	u/l Time magazine	Time magazine readers
¶ Paragraph	¶ Most of the	Most of the	♂ Move as shown	readers will see	will see
⋯ Don't delete	a true story	a true story			
○ Spell out	the only ①	the only one			
∽ Transpose	they all see	they see all			

Activity 4

Suite Smarts
Files from PDAs can be shared with Tablet PCs or desktop computers.

Step-by-Step continued

Investigate Tablet PCs

In this activity you will explore Tablet PCs. When you use a digital pen, your handwriting or printing is recognized by the Tablet PC and can be converted to text. You can also draw sketches, such as pie charts. Your handwriting and sketches will be saved as electronic documents.

Tablet PCs give you the option of entering text, symbols, and commands by touching the soft, or on-screen, keyboard with the digital pen. You will learn to use Tablet PCs in Chapter 7, pages 190–226.

1. Open your Web browser and use a search engine to research Tablet PC features. Use the keywords "Tablet PC" and "Tablet PC question."
2. Review the information at these Web sites, and take notes.
3. Start WordPad. Write a review of the information you find on the Web sites.
4. Include the following in your review:
   - Brand names of Tablet PCs.
   - Cost.
   - Features.
5. Review Activity 1, Steps 4 and 5, page 6, for saving this document. Name the document **Tablet PCs**.

RUBRIC LINK

Go to the Online Learning Center at **digicom.glencoe.com** and click Lesson 1.1, Rubric, to find the application assessment guide(s).

Application 1

Directions Open the file called **Name** that you created in Activity 1, page 6, and saved in the Digital Tools folder. You will recall that the Digital Tools folder is in the My Documents folder.

- Space down several lines under your name on the document and key three or four lines about your favorite sport, hobby, holiday, or subject in school.
- Key a short paragraph describing the technology skills you have learned previously.
- Add a short paragraph describing what you hope to learn about digital communication tools in this class.
- Save your document by going to File, Save.

Close WordPad when you are finished.

Format for Envelopes

A standard large (No. 10) envelope is 9½ by 4⅛ inches. A standard small (No. 6¾) envelope is 6½ by 3⅝ inches. The format shown is recommended by the U.S. Postal Service for mail that will be sorted by an electronic scanning device.

Your Name
4112 Bay View Drive
San Jose, CA 95192

            Mrs. Maria Chavez
            1021 West Palm Blvd.
            San Jose, CA 95192

George Washington High School
6021 Brobeck Street
Flint, MI 48532

            Dr. John Harvey
            Environmental Science Department
            Central College
            1900 W. Innes Blvd.
            Salisbury, NC 28144

How to Fold Letters

To fold a letter for a small envelope:
1. Place the letter *face up* and fold up the bottom half to 0.5 inch from the top edge of the paper.
2. Fold the right third over to the left.
3. Fold the left third over to 0.5 inch from the right edge of the paper.
4. Insert the last crease into the envelope first, with the flap facing up.

To fold a letter for a large envelope:
1. Place the letter *face up* and fold up the bottom third.
2. Fold the top third down to 0.5 inch from the bottom edge of the paper.
3. Insert the last crease into the envelope first, with the flap facing up.

Application 2

Directions Choose one of the following to investigate further:

- Speech recognition software.
- Handwriting recognition using PDAs.
- Handwriting recognition using Tablet PCs.

Find out if there are any special features you need on your computer to use the technology you chose.

Also, find an article on the Internet that discusses the use of that technology in classrooms. Use the keywords "education" or "classroom" in your search.

Do you think more people will use the technology you chose in the future? Write a one-page report of your findings. Explain how you think this technology may affect your daily life and your career. Save your report as **Tech Report** in your folder.

Application 3

Directions Survey a small number of classmates. Discuss your responses to Application 2 above.

- What do you think PDAs, speech recognition software, and Tablet PCs will do for people in the future?
- Why do you feel the way you do?

Prepare a presentation for the class to discuss the pros and cons of these technologies and how they could affect our lives. Support your ideas.

DIGITAL Decisions

Speech Recognition Software

Teamwork Work in a small group. Your group has decided that classroom computers should have speech recognition software available. You came to this decision when you saw that a student with a broken arm was falling behind on his assignments because he could not use a keyboard.

When you share your idea with the principal, however, she rejects it. She is concerned that students will use speech recognition to enter text instead of learning how to use a keyboard correctly.

Critical Thinking Your challenge is to write a short letter to the principal to persuade her to change her mind. Come up with specific reasons why speech recognition software would be helpful in the classroom. Use the student with the broken arm as an example.

What other examples can you think of? Do a Web search to find information about speech recognition software. If you find a good article, print it out and attach it to your letter. Include at least three examples of how speech recognition could be used in the classroom in your letter.

Title Page

↓ Center Vertically

THE STAR-SPANGLED BANNER
↓ 13x

Prepared by
↓ 2x
Hallie Thompson
La Mesa Valley School
↓ 13x

Prepared for
↓ 2x
Ms. Gibson
Social Studies--5th Hour

Today's Date

Multipage Report
(left bound with side headings)

↓ 1 inch

THE STAR-SPANGLED BANNER
By Hallie Thompson

The Story Behind the Flag

During the War of 1812, Americans knew that the British would likely attack the city of Baltimore. In the summer of 1813, Major George Armistead was the commander at Fort McHenry at the Baltimore harbor. He asked Mary Young Pickersgill to make a flag for the fort. Armistead wanted the flag to be so big that the British would be sure to see it from a distance.

Mary's 13-year-old daughter Caroline helped her make the flag. They cut 15 stars. Each star was two feet long from point to point. They also cut eight red stripes and seven white stripes. Each stripe was two feet wide. It took them several weeks to make the flag. When they sewed everything together, the flag measured 30 feet by 42 feet. The flag weighed 200 pounds.

Francis Scott Key's Point of View

Francis Scott Key was 35 years old and he was a well-known and successful lawyer in Georgetown, Maryland. He opposed the War of 1812, but in 1814 he had to get involved. His long-time friend Dr. William Beanes was being held prisoner on a British warship.

On September 3, 1814, Key and a government agent named John S. Skinner boarded a ship that flew a flag of truce. They went to the British warship and negotiated the release of Beanes. On September 7, the British agreed to let Beanes go, but by then Key, Skinner, and Beanes knew too much about the planned attack on the city of

1

Multipage Report continued
(left bound with side headings)

↓ 1 inch

Baltimore. So the British held all three Americans as prisoners on the warship while they attacked Baltimore.

On September 13, the three American prisoners watched from the warship as the British battleships fired upon Fort McHenry. They knew it would be difficult for the American soldiers to fight off the British. The battle continued through the night, and they feared the American soldiers would surrender.

The Story Behind the Song

When the sun rose the next morning, they saw a big American flag flying over the fort. It was the flag Pickersgill had made. The Americans had survived the battle.

Oh! Say, can you see, by the dawn's early light,
What so proudly we hailed at the twilight's last gleaming?
Whose broad stripes and bright stars, through the perilous fight,
O'er the ramparts we watched were so gallantly streaming?
And the rocket's red glare, the bombs bursting in air,
Gave proof through the night that our flag was still there.
Oh! Say, does that Star-Spangled Banner yet wave
O'er the land of the free and the home of the brave.

2

Bibliography

↓ 1 inch

BIBLIOGRAPHY

Armed Forces Collections. "Star-Spangled Banner and the War of 1812." 10 May 2002. <http://www.si.edu/resource.faq/nmah/starflag.htm>.
↓ 2x
Author Unknown. "Francis Scott Key." 10 May 2002. <http://www.usflag.org.francis.scott.key.html>.
↓ 2x
Goertzen, Valerie Woodring. "Star-Spangled Banner." *The World Book Encyclopedia 2002.* Chicago: World Book, Inc., Vol. 18, pp. 853-854.
↓ 2x
"Star-Spangled Banner." *Microsoft Encarta Online Encyclopedia 2002.* <http://encarta.msn.com/encnet/refpages/refarticle.aspx?refid=761575047>.

Lesson 1.2 → Operating Systems, Hardware, and Software

You will learn to:
- **Explore** Control Panel.
- **Navigate** drives, folders, and files.
- **Identify** available software and hardware.
- **Change** mouse settings.

DIGI*Byte*

Software Tasks
A "task" in a software program usually fits into one of these categories: calculating, printing, or storing.

operating system
Software that controls all of the other software programs and allows the computer to perform basic tasks.

hardware All of the physical parts of a computer.

digital Information that is represented as individual pieces of data using the numbers 1 and 0, rather than as a continuous stream.

Software Software is a set of instructions, also called a program or application, that tells a computer how to perform tasks. When a person interacts with a computer, software takes the person's commands and turns them into instructions that the computer will process. Different kinds of software can help with many different tasks, such as writing reports, preparing budgets, designing a newsletter, and browsing the Internet.

Operating System The most important software on a computer is the operating system, or OS. The operating system is software that controls all of the other software programs and allows the computer to perform basic tasks. The OS opens automatically when a computer is turned on, or "booted up."

Handheld computers have their own operating systems, such as Windows CE and Palm OS. Although all operating systems perform similar tasks, not all of their programs are interchangeable. This means that some programs installed on a handheld with the Palm OS may not be able to be installed on a handheld with the Windows CE system.

Hardware Hardware includes all of the physical parts of a computer. You can see some of these parts from the outside, such as the screen, mouse, and keyboard. Other hardware components are hidden inside the computer. These parts include the motherboard, central processing unit (CPU), video card, and memory chips. These parts all work together to perform instructions from the OS software.

Digital Information All of the information on computers and other digital devices is stored in a digital format. Digital information is information that is represented as individual pieces of data using the numbers 1 and 0, rather than as a continuous stream. Every piece of information that a computer handles is broken down into ones and zeroes.

✓ Concept *Check*

What software programs are you familiar with? List the types of tasks you do with these programs. Which of these programs do you consider the most essential to you?

One-Page Report

↓ 1 inch default margin

ALL THE CHOCOLATE YOU CAN EAT!

By Rachelle Cantin

I read the book *Charlie and the Chocolate Factory*. The author of the book is Roald Dahl. The book is 155 pages long, and it was published by Puffin.

Charlie Bucket is a poor boy who lives in a tiny house with his parents. Both sets of grandparents also live with Charlie in that tiny house. Charlie didn't have any money, but he found a dollar bill in the street. He used the money to buy a Wonka candy bar. The Willy Wonka Chocolate Company held a contest. When Charlie opened the candy bar, he found a golden ticket. He was one of five winners.

The other four winners were Augustus Gloop, Violet Beauregarde, Veruca Salt, and Mike Teavee. Charlie went on the tour of the chocolate factory with his Grandpa Joe. The other four winners were there with their parents. They saw lots of amazing things, and they met the Oompa-Loompas. The Oompa-Loompas were the tiny people who lived and worked in the factory.

The other four kids behaved very badly on the tour. When they didn't follow directions, Mr. Wonka punished them. Funny things happened to them that made them disappear. Charlie was kind and polite. Mr. Wonka liked Charlie, and he knew he could trust Charlie. At the end of the tour, Charlie was the only kid left, and Mr. Wonka gave him the chocolate factory. Charlie and his family could live at the factory, and they could have all the chocolate they could eat!

My favorite part about this book was when the other four kids didn't follow directions and Mr. Wonka punished them. Charlie followed directions, and Mr. Wonka rewarded him for that.

Multipage Report
MLA Style

↓ 1 inch

Last Name 1

Your First and Last Name
↓ 1x
Your Teacher's Name
↓ 1x
Class
↓ 1x
Current Date
↓ 1x

King of the Wild Frontier ↓ 1x

ds "Be always sure you are right, then go ahead" (Lofaro 1148d). You're probably wondering what that means. Well, a guy named Davy Crockett used to say that. It is one of his best known quotes. Read on to find out more about this legendary person.

1 inch Actually, his name was David Crockett. He was born in a small cabin in Tennessee on August 17, 1786. (*Davy Crockett*). His family lived in a cabin on the banks of the Nolichucky River. Davy had eight brothers and sisters. Four were older and four were younger. 1 inch

Davy lived with his family in Tennessee until he was 13. He went to school, but he didn't like it. He skipped school a lot. He ran away from home because he knew his dad was going to punish him for playing hooky. He joined a cattle drive to make money. He drove the cattle to Virginia almost 300 miles away. He stayed in Virginia and worked a lot of jobs for over two years. He returned to his family in Tennessee when he was 16 (*Davy Crockett Biography*).

When Davy returned home his dad was in debt. Now Davy was 6 feet tall and he could do a man's work. Davy went to work for Daniel Kennedy. Davy's dad owed Daniel 76 dollars and Davy worked for one year to pay the debt (The Texas State Historical Association).

In 1806 Davy married Polly Finley. They had two sons, John Wesley and William. Then Davy went to fight for the Tennessee Volunteer Militia under Andrew Jackson in the Creek Indian War. When he returned home from the war, he found his wife very ill. She died in 1815 (Davy Crockett Birthplace Association).

Multipage Report continued
MLA Style

↓ 1 inch

Last Name 2

Davy then married Elizabeth Patton in 1817. She was a widow and she had two children of her own, George and Margaret Ann (The Texas State Historical Association).

Davy was well known in Tennessee as a frontiersman. He was a sharpshooter, a famous Indian fighter, and a bear hunter. In 1821, he started his career in politics as a Tennessee legislator. People liked Davy because he had a good humor and they thought he was one of their own. He was re-elected to the Legislature in 1823, but he lost the election in 1825.

In 1827 Davy was elected to Congress. He fought for the land bill. The land bill allowed those who settled the land to buy it at a very low cost. He was re-elected to Congress in 1829 and again in 1833, but he lost in 1836 (Lofaro, 1148d).

1 inch Many Americans had gone to Texas to settle. In 1835, Davy left his kids, his wife, his 1 inch
brothers, and his sisters to go to Texas. He loved Texas. When the Texans were fighting for their independence from Mexico, Davy joined the fight. He was fighting with a group of Tennessee volunteers defending the Alamo in San Antonio on March 6, 1836 (The Texas State Historical Association). He was 49 years old.

Works Cited

Last Name 3

Works Cited

Author Unknown. "Davy Crockett Biography." 6 April 2002.
    <http://www.infoporium.com/heritage/crockbio.shtml>.

ds "Davy Crockett." *Microsoft Encarta Online Encyclopedia 2002*. <http://encarta.msn.com>.

Davy Crockett Birthplace Association. "American West-Davy Crockett." 6 April 2002. <http://www.americanwest.com/pages/davycroc.htm>.

Lofaro, Michael A. "Davy Crockett." *The World Book Encyclopedia 2002*. Chicago: World Book, Inc., Vol. 14, pp. 1148d-1149.

The Texas State Historical Association. *The New Handbook of Texas-Online*. "Davy Crockett (1786-1836)-Biography." 6 April 2002. <http://www.alamo-de-parras.welkin.org/history/bios/crockett/crockett.html>.

TechSIM™
Interactive Tutorials
Ask your teacher about a simulation that is available on this topic.

Activity 5

? If you need additional help on this topic, open the Help menu; or refer to the documentation for your hardware.

Step-by-Step

The most common operating systems are Unix, Macintosh, and Microsoft Windows. In this lesson, you will explore the Microsoft Windows XP operating system. Your screens may vary depending on the computer system and whether your computer is networked.

Many schools and businesses choose to make their Control Panel settings permanent. Users are not allowed to change the settings unless they have special permission. Setting up all computers the same way helps avoid technical problems.

Start the Computer and Open Control Panel

1. Turn on your computer. Watch the information that appears on your screen. These are "boot" instructions that the OS software is sending to the hardware of the computer.

2. Click **Start** in the lower left-hand corner of the screen, and then find and click **Control Panel**. The Control Panel window will open on your screen.

3. The Control Panel displays different categories you can access to change the settings on your computer. Read the categories below.

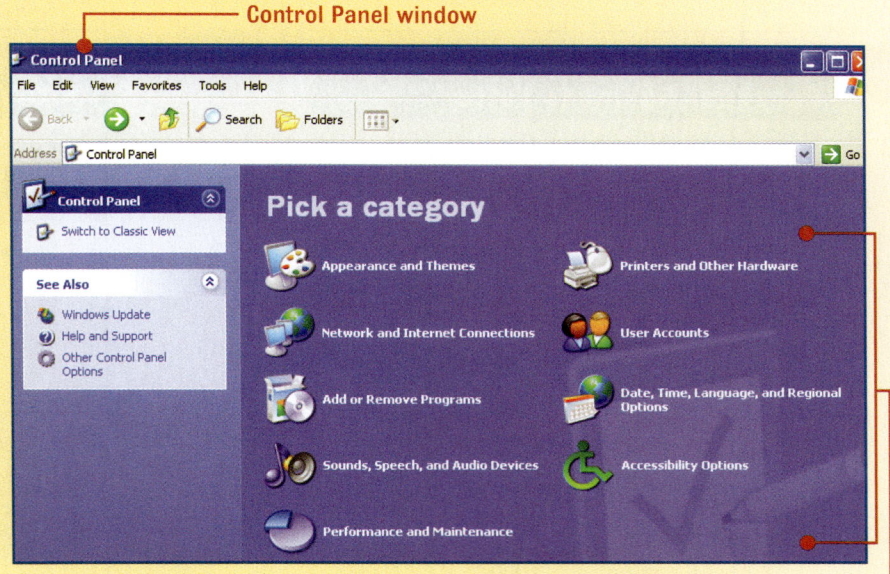

Control Panel window

Categories

4. Click the **Close** button in the upper right-hand corner of the Control Panel to close the window.

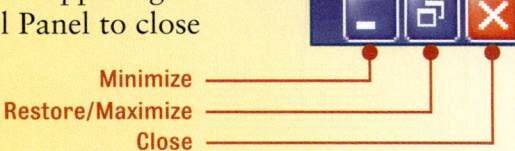

Minimize
Restore/Maximize
Close

Continued

Chapter 1

11

Personal Letter

1719 Lakeview Drive
Fontana, WI 53125
March 21, 20-- ↓ 4x

Ms. Susan Yu, Director
Turtle Creek Animal Shelter
2023 North Lake Shore Drive
Fontana, WI 53125 ↓ 2x

Dear Ms. Yu: ↓ 2x

I am interested in volunteering at your animal shelter. I love animals, and I have pets of my own, including a dog and a cat. I can help you feed, groom, and exercise the animals in your shelter. ↓ 2x

I am available to help you on weekdays after school and on weekends. I would be willing to work up to six hours a week. I can start working next week. ↓ 2x

Please call me at 555-6978 and let me know how I can help. ↓ 2x

Sincerely, ↓ 4x

Your Name

Business Letter

George Washington High School
8021 Brobeck Street • Flint, MI 48532
Phone: 810-555-9001 • Fax: 810-555-9004

Today's Date ↓ 4x

Dr. John Harvey
Environmental Science Dept.
Central College
1900 W. Innes Blvd.
Salisbury, NC 28144 ↓ 2x

Dear Dr. Harvey: ↓ 2x

I would like to invite you to speak to our environmental science class. We are interested in beginning a campus-wide recycling program. ↓ 2x

I understand that Central College recycled many of the materials left over from building your new environmental science building. The environmental science class would like to hear about the process that was used to recycle these materials. ↓ 2x

The class would also like to take a field trip and visit your new environmental science building at Central College. Could you arrange for a tour and some time to discuss recycling with our class? ↓ 2x

Thank you for your assistance. You may call my teacher, Ms. King, at school at 888-555-9023 if you have any questions. ↓ 2x

Sincerely, ↓ 4x

Your Name
Environmental Class Secretary

Personal Business Letter

5419 Mirra Loma Drive
Reno, NV 89502
Today's Date ↓ 4x

Mainstream Music, Inc.
270 Clara Street
San Francisco, CA 94107 ↓ 2x

Ladies and Gentlemen: ↓ 2x

About four weeks ago, I mailed you an order for a DVD movie package. I purchased DVD package #41-809 from page 5 of your catalog. The total cost of the order was $47.35. ↓ 2x

Today I saw a package at my door. I was so excited that my order had finally arrived. But I was really disappointed when I opened it and saw that you sent me the wrong DVDs. ↓ 2x

I am returning the DVD package with this letter and I am enclosing the order return form. I am asking for a full refund of $47.35. And I also want to be reimbursed for the $5.00 it cost me to ship the DVD package back to you. Please send the refund to me at the above address. ↓ 2x

Sincerely, ↓ 4x

Martine Pico ↓ 2x

Enclosure

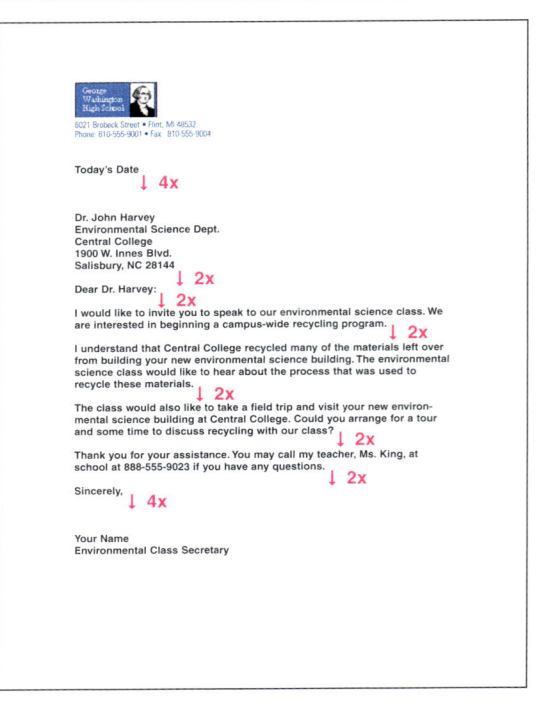

Outline

↓ 1 inch

OCEAN WATER AND LIFE

I. **WAVES AND TIDES**
   A. Waves
      1. How waves move
      2. How waves form
   B. Tides
      1. The gravitational pull of the moon
      2. Spring and neap tides
   C. Life in the intertidal zone

II. **OCEAN CURRENTS**
   A. Definition of currents
   B. Surface currents
   C. Density currents
   D. Upwellings

Step-by-Step *continued*

Activity 6

Explore Drives and Folders

1. Click **Start,** and then click **My Computer.** This window lists the built-in storage devices on the computer, as well as any folders the user has created.

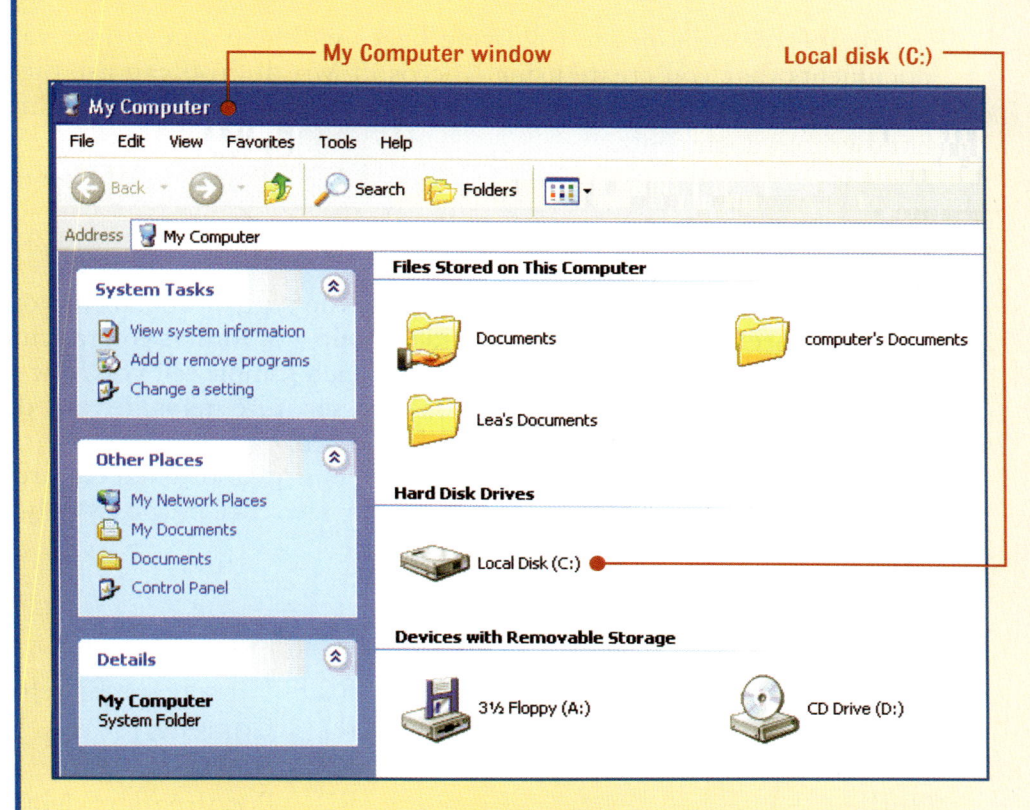

My Computer window — *Local disk (C:)*

2. Locate the **Local Disk (C:)** drive in the My Computer window. This is where the operating system is stored.
3. Double-click the C: drive icon. File folders will appear. These folders contain the files that make your OS run.

Activity 7

Identify Software

There are two basic types of software:

- **Operating system software** manages basic functions, such as starting and shutting down your computer.
- **Application software** performs specific tasks, such as word processing, Web browsing, or sending e-mail. Examples of application software are: Microsoft Internet Explorer®, Microsoft Outlook®, Microsoft Excel®, Microsoft Access®, and Microsoft PowerPoint®.

Continued

Suite Smarts
Software should be installed through the Control Panel.

Reference

Appendix R Reference Section

The information on the following pages will help you format various kinds of documents that you create using Microsoft Word. You can use the proofreaders' marks and the remaining pages to help you edit and proof using any software or handwritten documents.

How to Use the Reference Guide

Use the Contents below to quickly locate the type of document you are creating. Then use the examples shown as a guide to help you format your document properly. The arrows and numbers shown in red on each sample tell you how many times to press Enter or Return on your keyboard to separate items in your document. The letters ds indicate double spacing should be used.

Remember that your work should reflect your own original research and content and that the information provided here is for reference purposes only.

Contents

Personal Letter	R2	Format for Envelopes	R5
Business Letter	R2	Contents	R6
Personal Business Letter	R2	Boxed Table	R6
Outline	R2	Proofreaders' Marks	R6
One-Page Report	R3	Punctuation and Usage	R7
Multipage Report	R3	Grammar	R10
Works Cited	R3	Mechanics	R12
Title Page	R4	Capitalization	R12
Multipage Reports with		Number Expression	R13
Side Headings	R4	Abbreviations	R14
Bibliography	R4	The Web-Safe Color Palette	R15

Step-by-Step *continued*

① To see a list of the software programs on your computer, click **Start,** and then click **All Programs.**

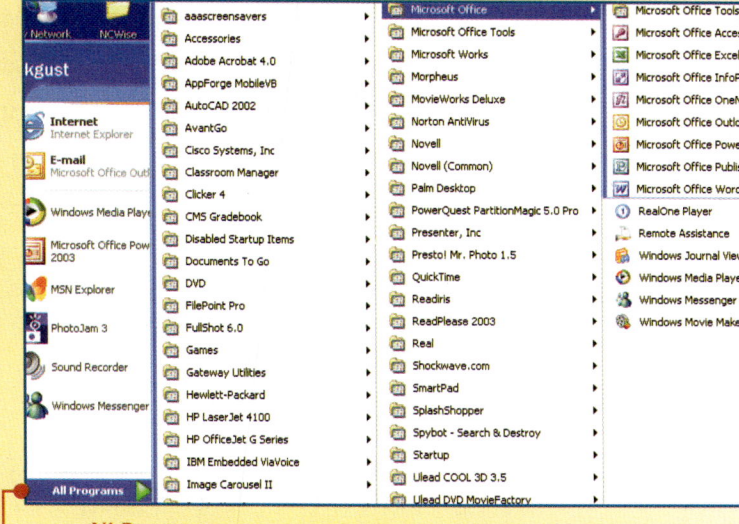

— All Programs

② Read the list to learn which software programs are installed on your computer.

Identify Hardware

A typical computer uses many different hardware devices. Using the Control Panel, a user can view information about the devices and change the way they work.

— Printers and Other Hardware window

① Click **Start,** click **Control Panel,** and click **Printers and Other Hardware.**

② Click **Mouse.** A new window will open. This window displays different options to control how the mouse functions.

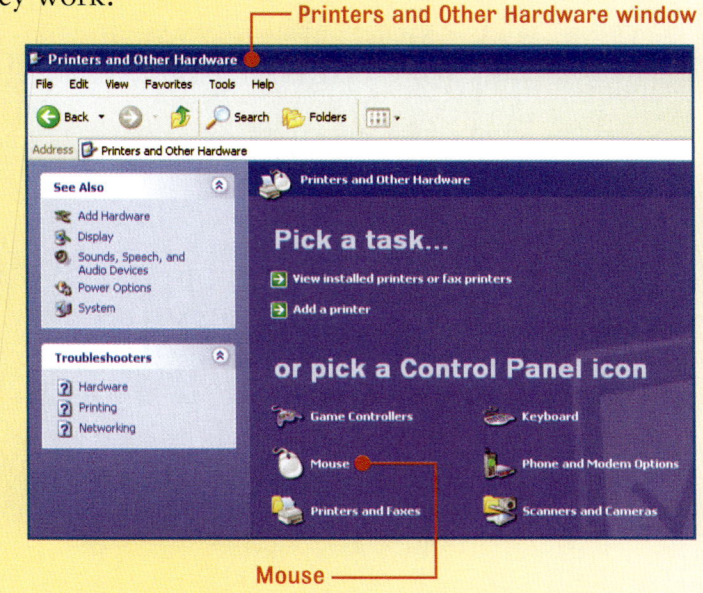

— Mouse

Continued

Activity 8

Suite Smarts

Once hardware is installed, it can be used with any software program.

Chapter 1

13

SKILLBUILDING

H. 1-Minute Alpha-numeric Timings

Take a 1-minute timing on lines 37–39. Note your speed and errors.

```
37      Kim ran the 7.96-mile race last week. Yanni      9
38 ran 14.80 miles. Zeke said the next 5K run will      19
39 be held on August 14 or August 23.                   25
    | 1 | 2 | 3 | 4 | 5 | 6 | 7 | 8 | 9 | 10
```

I. 2-Minute Timings

Take two 2-minute timings on lines 40–45. Note your speed and errors.

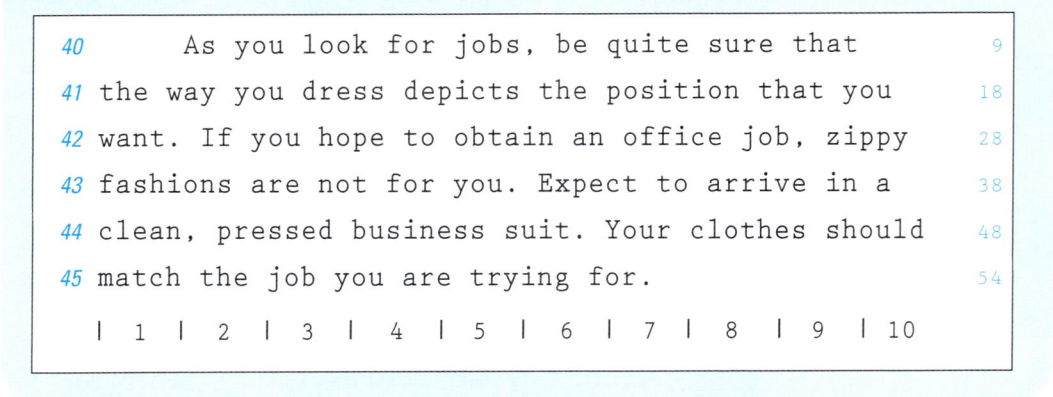

```
40      As you look for jobs, be quite sure that        9
41 the way you dress depicts the position that you      18
42 want. If you hope to obtain an office job, zippy     28
43 fashions are not for you. Expect to arrive in a      38
44 clean, pressed business suit. Your clothes should    48
45 match the job you are trying for.                    54
    | 1 | 2 | 3 | 4 | 5 | 6 | 7 | 8 | 9 | 10
```

DIGI*Byte*

Optical Mouse

An optical mouse is controlled by a beam of light on the bottom of the mouse, rather than a rubber ball.

RUBRIC LINK

Go to the Online Learning Center at **digicom.glencoe.com** and click Lesson 1.2, Rubric, to find the application assessment guide(s).

Step-by-Step *continued*

3. Notice that the double-click speed of the mouse can be adjusted. This refers to how fast you must click twice in order to highlight or open something on-screen. Try adjusting the double-click speed by moving the slider left or right.

4. Click **Apply,** and then click **OK.**

5. Try the new double-click speed setting by opening a folder in My Computer.

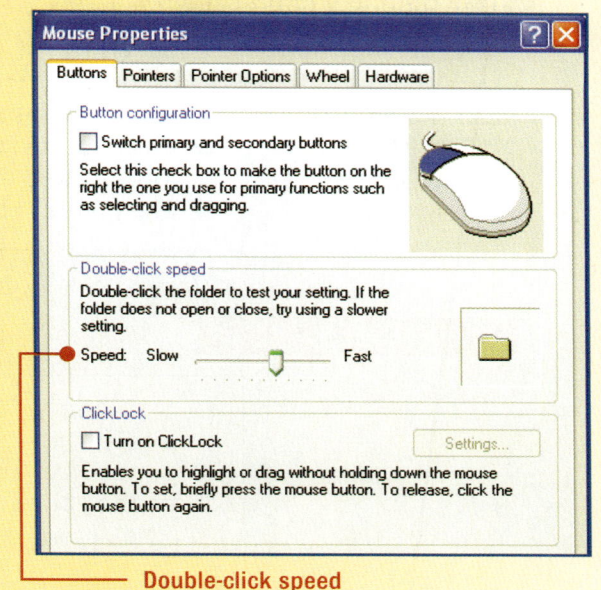

6. If you think the double-click speed should be slower or faster, go back and change it again.

Double-click speed

Application 4

Directions Open Control Panel and click **Sounds, Speech, and Audio Devices.** Then click **Sounds and Audio Devices.** Ask your teacher what adjustments you can make, if any. Are these features hardware or software?

DIGITAL DIALOG

Which Software?

Teamwork In a small group, determine what type of software application you will need to do the following tasks:

- Create a newsletter.
- Design a birthday card.
- Write a letter.
- Prepare a budget for the annual summer picnic.
- Visit a Web site.
- Add up a long list of numbers.

Share your answers with the class.

Critical Thinking Write a paragraph about a software application you would like to buy. Describe what the software does and what it will help you do.

SKILLBUILDING

D. Technique Timings Take two 30-second timings on each line. Focus on the technique at the left.

Sit up straight with your feet flat on the floor.

```
21 Snow leopards are graceful animals with soft fur.
22 They live in the high, rugged mountains of Tibet.
23 These big cats are adept at climbing and leaping.
24 They use their tails to balance on narrow ledges.
```

E. PRETEST Take a 1-minute timing on lines 25–28. Note your speed and errors.

```
25      As a flock, the crows flew to some clumps of    9
26 stalks near the eddy. They seemed to eat the pods   19
27 joyfully as they fed in the field. We like to       28
28 watch them, especially in the morning.              36
   |  1  |  2  |  3  |  4  |  5  |  6  |  7  |  8  |  9  |  10
```

F. PRACTICE

SPEED: If you made 2 or fewer errors on the Pretest, type lines 29–36 two times each.

ACCURACY: If you made more than 2 errors on the Pretest, type lines 29–32 as a group two times. Then type lines 33–36 as a group two times.

Adjacent
Jump
Double
Consecutive

```
29 po pods poem point poise lk hulk silk polka stalk
30 mp jump pump trump clump cr cram crow crawl creed
31 dd odds eddy daddy caddy tt mitt mutt utter ditto
32 un unit punk funny bunch gr grab agree angry grip
```

Alternate
Left/Right
Up/Down
In/Out

```
33 iv give dive drive wives gl glad glee ogled gland
34 fe fear feat ferns fetal jo joys join joker jolly
35 sw swan sway sweat swift k, ark, ask, tick, wick,
36 lu luck blunt fluid lush da dash date sedan panda
```

G. POSTTEST Repeat the Pretest. Then compare your Posttest results with your Pretest results.

Appendix A—Keyboarding Skills

Lesson 1.3 — Storage and File Management

hard drive The built-in storage device inside a computer.

DIGI *Byte*
Organizing Files
To keep your files organized, you can create additional folders within the My Documents folders.

You will learn to:
- **Create** folders to manage your data.
- **Recognize** file extensions.

Storing Information A computer's main storage device is the hard drive. The **hard drive** is a built-in storage device inside a computer. It stores your computer's operating system, as well as the software programs you use. Hard drives today can store huge amounts of information. New documents you create on your computer can also be saved to your hard drive. You usually store applications and the files that you use often on a hard drive.

Managing Data Computer users create many files that they need to store for later use. It is important to keep files well organized so they are easy to find. Computer folders are an on-screen tool for organizing files. Just as people use cardboard folders to organize important papers, they also use on-screen folders to keep their computer files in order. You can create folders on a hard drive, floppy disk, or other storage device of your choice. You will learn more about other storage devices later in Lesson 1.5, pages 23–27.

The storage device is like a file cabinet for the folders. Each folder has its own name, so you can group files by category. For example, you could store essays and reports for your English class in a folder named English.

File Extensions A file extension lets you know what kind of file you are working with. It is usually a three- or four-letter code that appears right after the file name. For example, if you see a file name that ends in .doc, you know it is a word processing document. Graphic files have .jpg, .bmp, or .gif extensions. A Web page file is indicated by the .html extension. Most file extensions have three letters, but some have four.

File extensions also serve as instructions to the computer. When you save a file, the software program you are using automatically assigns the file extension. The next time you open that file, the computer will read the extension and launch the appropriate program. If the computer sees a .doc extension, for example, it will open a word processing program.

✓ Concept *Check*

Why is it important to create folders for storing files? Give two examples of file extensions.

Lesson 19 — New Keys: 5 % 6 ^

OBJECTIVES:
- Learn the 5, %, 6, and ^ keys.
- Refine keyboarding skills.
- Type 27/2'/4e.

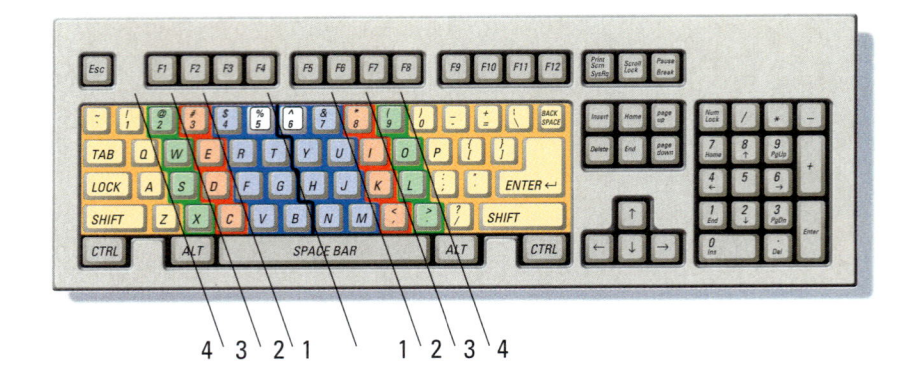

A. Warmup
Type each line 2 times.

Speed
1 Our team at band camp did a new drill for guests.

Accuracy
2 Two jobs require packing five dozen axes monthly.

Numbers
3 Mark read the winning numbers: 190, 874, and 732.

Symbols
4 The shop (J & B) has #10 envelopes* @ $.24 a doz.

NEW KEYS

B. 5 and % Keys
Type each line 2 times. Repeat if time permits.

Use F finger. For 5 and %, anchor A.

5 ftf ft5f f5f 555 f5f 5/55 f5f 55.5 f5f 55,555 f5f
6 55 fins, 55 facts, 55 fields, 55 futures, or 5.55
7 Jo saw 55 bulls, 14 cows, 155 sheep, and 5 goats.
8 I just sold 55 items; his total for today is 555.

% is the shift of 5. The % (percent) is used in statistical data. Do not space between numbers and %

9 ftf ft5 f5f f5%f f%f f%f 5% 55% 555% f%f f5f 555%
10 5%, 5 foes, 55%, 55 fees, 555%, 555 fiddles, 555%
11 The meal is 55% protein, 20% starch, and 25% fat.
12 On June 5, 55% of the students had 5% more skill.

C. 6 and ^ Keys
Type each line 2 times. Repeat if time permits.

Use J finger. For 6 and ^, anchor ;.

13 jyj jy6j j6j 666 j6j 6/66 j6j 66.6 j6j 66,666 j6j
14 66 jaws, 66 jokes, 66 jewels, 66 jackets, or 6.66
15 Her averages were 76.46, 81.66, 86.56, and 96.36.
16 Multiply .66 by .51; the correct answer is .3366.

^ is the shift of 6. The ^ (caret) is used in some programming languages. Do not space between the caret and numbers.

17 jyj jy6 j6j j6^j j^j j^j ^j ^jj ^jjj j^j j6j ^jjj
18 6^, 6 jams, 66^, 66 jets, 666^, 666 jingles, 666^
19 The test problems included these: 75^2, 4^3, 8^6.
20 The ^ (caret) appeared 6 times in a line of code.

Appendix A—Keyboarding Skills A45

Activity 9

If you need additional help on this topic, open the Help menu; or refer to the documentation for your hardware.

Step-by-Step

Explore Folders

1. To see your computer's storage devices, click **Start**, and then click **My Computer**. Another way to do this is to double-click the **My Computer** icon on the desktop.

2. Locate the 3½″ Floppy (A:) drive.

3. Note that the My Computer folder contains a yellow folder named My Documents. This folder is always present on computers that use the Windows XP operating system.

4. Double-click the **C:** drive. This is the hard drive. You will see yellow folders, as well as various files that have been saved to your computer's hard drive.

DIGI Byte

My Documents

When you save a document in Microsoft Word or any other Microsoft program, the file is automatically saved in the My Documents folder unless you instruct the computer to store it somewhere else.

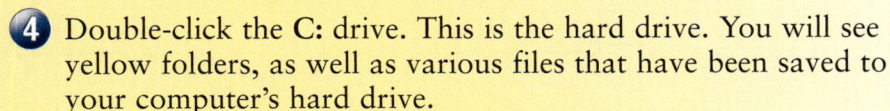

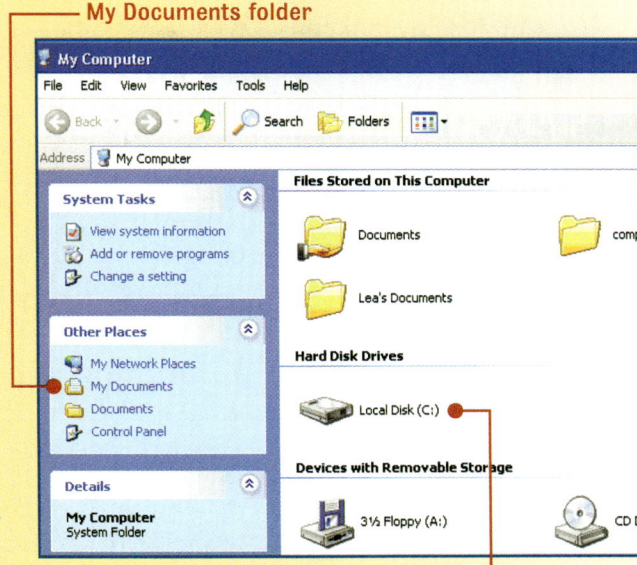

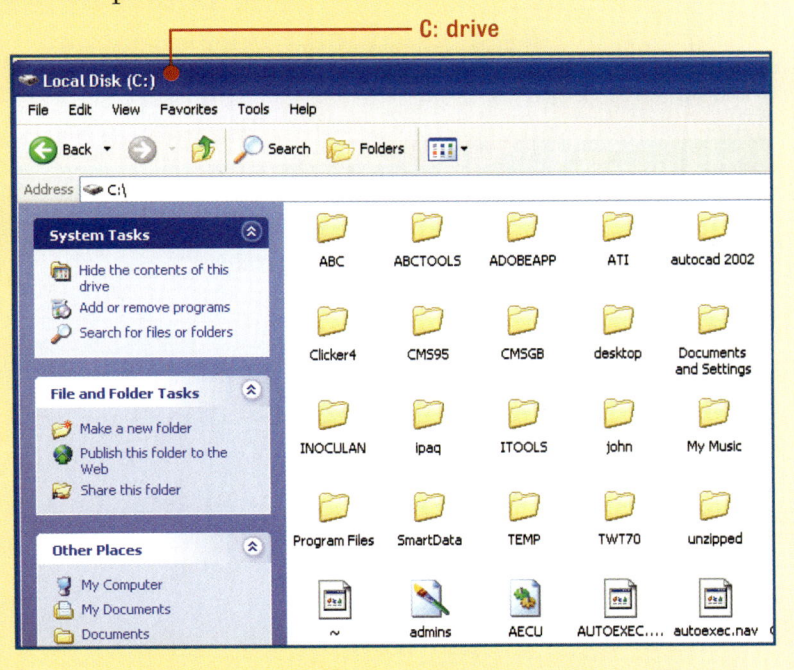

Continued

Chapter 1

16

SKILLBUILDING

H. 1-Minute Alpha-numeric Timings

Take a 1-minute timing on lines 37–39. Note your speed and errors.

```
37      Joy wanted to get a dozen (12) baseball bats     9
38 @ $4.29 from the sports store at 718 Miner Place.    19
39 When I went, only 10 bats were left.                 26
   |  1  |  2  |  3  |  4  |  5  |  6  |  7  |  8  |  9  | 10
```

I. 2-Minute Timings

Goal: 27/2'/4e

Take two 2-minute timings on lines 40–45. Note your speed and errors.

```
40      In the fall of the year, I find pleasure in      9
41 zipping up to the hills to quietly view the trees   19
42 changing colors. Most all aspens turn to shades     29
43 of gold. Oak trees exude tones of red and orange.   39
44 The plants change colors each fall, but all these   49
45 changes are an amazing sight.                       54
   |  1  |  2  |  3  |  4  |  5  |  6  |  7  |  8  |  9  | 10
```

Activity 10

DIGI Byte

Alphabetical Order

When you create a new folder, it appears at the end of the list in the window you are working in. If you close and then reopen the window, the folder will appear in correct alphabetical order.

Activity 11

Suite Smarts

When a software program is installed, it creates its own folder on the hard drive. This is called a program folder.

Step-by-Step *continued*

Create a New Folder on the C: Drive

1 In the File and Folder Tasks menu on the left side of the window, click **Make a new folder**.

Look for the new yellow folder you created. It should appear in the lower right corner of the C: drive window. The folder is temporarily named New Folder, which is highlighted in blue. The highlighted text means you can type over it with a new folder name.

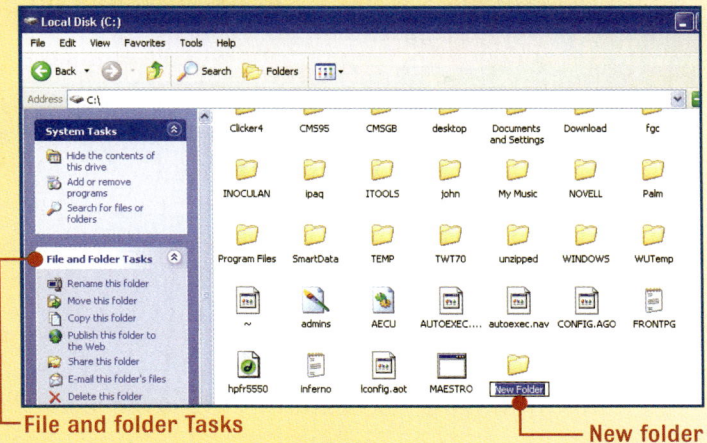

2 Enter **English 1** and press ENTER.

Create a New Folder on a Floppy Disk

1 Insert a floppy disk in the 3½″ disk drive.

2 Double-click the **3½ Floppy (A:)** icon to see the contents of the disk. If the disk has not been used before, it will be blank.

3 In the task window, click **Make a new folder**.

4 Name the folder **English 1**.

5 To get more information about the files on your floppy disk, click **View** in the menu bar and select **Details**. The format of the display will change, and you will be able to see the file types.

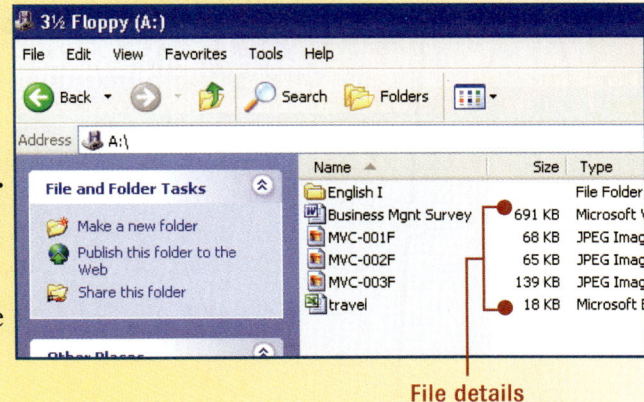

Chapter 1 Digital Communication Tools in Our Lives

Lesson 1.3

17

SKILLBUILDING

D. Technique Checkpoint

Type each line 2 times. Focus on the techniques at the left.

Keep eyes on copy; hold home-key anchors.

```
21  aqa aq1a a1a 111 a1a 1/11 a1a 11.1 a1a 11,111 a1a
22  aqa aq1 a1a a1!a a!a a!a 1! 11! 111! a!a a1a 111!
23  ;p; ;p0; ;0; 000 ;0; 1.00 ;0; 20.0 ;0; 30,000 ;0;
24  p;p p;0 ;0; ;0); ;); ;); );; );;; ;); ;0; );;; ;)
```

E. PRETEST

Take a 1-minute timing on the paragraph. Note your speed and errors.

```
25        Dave and I took our backpacks and started up      9
26  the old mountain trail. Around sunset, we stopped      19
27  to set up camp and have a hot meal. We were very       29
28  tired after such a long day hiking uphill.             37
     | 1 | 2 | 3 | 4 | 5 | 6 | 7 | 8 | 9 | 10
```

F. PRACTICE

SPEED: If you made 2 or fewer errors on the Pretest, type lines 29–36 two times each.

ACCURACY: If you made more than 2 errors on the Pretest, type lines 29–32 as a group two times. Then type lines 33–36 as a group two times.

Up Reaches

```
29  ho shock chose phone shove hover holly homes shot
30  st stair guest stone blast nasty start casts step
31  il lilac filed drill build spill child trail pail
32  de dear redeem warden tide render chide rode dead
```

Down Reaches

```
33  ab squab labor habit cabin cable abate about able
34  ca pecan recap catch carve cable scale scamp camp
35  av ravel gavel avert knave waved paved shave have
36  in ruin invent winner bring shin chin shrink pine
```

G. POSTTEST

Repeat the Pretest. Compare your Posttest results with your Pretest results.

Application 5

Scenario The school year is starting, and you want to keep your computer files organized. You know that you will have a lot of work to keep track of this year.

Directions Open your My Documents folder. Think about how you might want to organize your files and folders.

- Do you need a folder for each class?
- Do you have any special projects to do?
- Add a folder in My Documents for each class you have, and then add another folder of your choice.

Write a paragraph that describes the method that you will use to keep your folders organized.

Application 6

Scenario You see a want ad posted in your local mini-market: "Computer tutor wanted for seventh grader. Needs to learn about computers, hard drives, and managing files and folders. One hour lesson per week, $25." You call the number on the ad and set up the first lesson. This is great—you will use your computer skills to help a younger student. You want to prepare well.

Directions Knowing that younger students have short attention spans, write an outline of your first lesson in 15-minute segments. You plan to cover the topics you have learned in Lesson 1.3, pages 15–17.

Once you have written your lesson, do a walkthrough on your computer to check the steps. Give your lesson to two friends to read, and make any corrections or clarifications that they suggest.

RUBRIC LINK
Go to the Online Learning Center at **digicom.glencoe.com** and click Lesson 1.3, Rubric, to find the application assessment guide(s).

TECH ETHICS: Plagiarism

Scenario As you sit down in the computer lab, you notice the previous user did not log off. This user is a student in your history class. The teacher gave an assignment to write an article about one of the Civil War battles as if you were a newspaper reporter.

Your classmate left the assignment on the computer screen, so you decide to copy and paste the text to a new word processing document. You know you cannot turn in the exact same article, but you think you can change enough of the text to make it sound different. Since you are short on time and have track practice after school, you decide to go ahead and copy the article.

You know that plagiarism means using another's work without giving credit, and you understand that you must put others' words in quotation marks and cite your source(s) even when ideas are paraphrased in your own words.

Critical Thinking Write a paragraph that answers the following questions.

- Is this a fair way to do an assignment?
- Why or why not?
- What ethical issues does this raise?

Discuss your responses with your classmates and teacher.

Lesson 18 → NEW KEYS: 1 ! 0)

OBJECTIVES:
- Learn the 1, !, 0, and) keys.
- Refine keyboarding skills.
- Type 27/2'/4e.

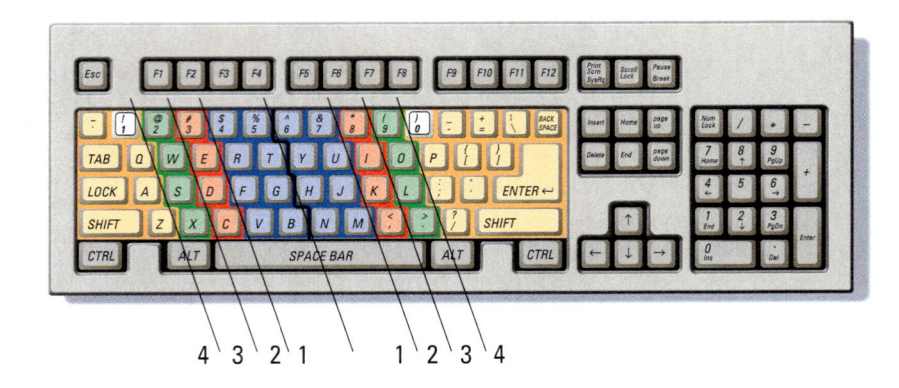

4 3 2 1   1 2 3 4

A. Warmup

Type each line 2 times.

Speed   1 The goal of trade schools is to teach job skills.
Accuracy   2 Jess Mendoza quickly plowed six bright vineyards.
Numbers   3 Nate took this new order: 78, 74, 83, 29, and 23.
Symbol   4 Purchase 32# of grass seed today @ $2.98 a pound.

NEW KEYS

B. 1 and ! Keys

Type each line 2 times. Repeat if time permits.

For 1 and !, anchor F. Use A finger. Do not use the lowercase letter l (el) for 1.

5 aqa aq1a a1a 111 a1a 1/11 a1a 11.1 a1a 11,111 a1a
6 11 arms, 11 areas, 11 adages, 11 animals, or 1.11
7 My 11 aides can type 111 pages within 11 minutes.
8 Joann used 11 gallons of gas to travel 111 miles.

! is the shift of 1. Space once after an exclamation point.

9 aqa aq1 a1a a1!a a!a a!a 1! 11! 111! a!a a1a 111!
10 1!, 1 ant, 11! 11 acres, 111! 111 adverbs, 1 area
11 Listen! There was a cry for help! They need help!
12 Look! It's moving! I'm frightened! Run very fast!

C. 0 and) Keys

Type each line 2 times. Repeat if time permits.

Use Sem finger. For 0 and), anchor J. Do not use the capital letter O for 0.

13 ;p; ;p0; ;0; 000 ;0; 1.00 ;0; 20.0 ;0; 30,000 ;0;
14 300 parts, 700 planks, 800 parades, 900 particles
15 Can you add these: 80, 10, 90, 40, 20, 70, & 130?
16 Some emoticons such as :-(or :(use parentheses.

) is the shift of 0 (zero). Space once after a closing parenthesis except when it's followed by punctuation; do not space before it.

17 ;p; ;p0 ;0; ;0); ;); ;););;);;; ;); ;0;);;; ;)
18 ;0; ;0) ;); ;); 10) 20) 30) 40) 70) 80) 90) 1001)
19 The box (the big red one) is just the right size.
20 My friend (you know which one) is arriving early.

Appendix A—Keyboarding Skills   **A42**

21st Century Connection

Proofread for Successful Communication

If you have strong proofreading skills, your writing will always make a great impression. Proofreading, or reviewing written work for errors or other flaws, is a process for which you may already have learned some guidelines in school. Whatever career you choose, you will need solid proofreading skills.

Spelling and Grammar Checks The spelling and grammar check functions included with word processing and other software programs offer only limited answers. You should never rely on them completely when proofreading your work. Be sure to use other, more thorough tools as well, such as dictionaries and grammar books.

Proofread on Paper By the time you finish writing a first draft, your eyes have probably been fixed on the screen for a long time. You need to rest them at this point. Print a hard copy for rereading later. It is easier to catch errors and content problems when you are reading from a hard copy.

Proofreading Checklist

Basic Tips

- Start your work as early as possible, so you do not have to rush to finish it. This will allow you enough time to polish it and make it stand out.
- Allow some time between writing and proofreading. This refreshes your mind so you can see where you have been in your writing and where you need to go.
- Read slowly and carefully. If you read too fast, you are more likely to miss errors and lose focus.
- Read your work out loud so you can hear how it sounds. If possible, ask someone else to read along silently with you. Does the writing make sense? Are there any mistakes?
- Use a blank sheet of paper to cover unread text. This causes you to slow down and read carefully.

Check Your Facts, Grammar, Spelling, and Punctuation

- Before you begin reading the main text, check the title, headings, page numbers, chart data, graphs, phone numbers, and spellings of names. Make sure these are correct by reviewing them against source documents. If you have any doubt about a spelling, look it up.
- Make sure your thesis agrees with the rest of your text. If the subject and descriptive text do not seem to work together, rewrite your thesis, add more details to paragraphs, or take out unrelated ideas.
- Make sure transitions between sentences and paragraphs are clear.
- Replace passive verbs with active verbs.
- Read the main text word for word to check spelling and punctuation.

Activity Are there a few terms or grammar rules that you forget often? List these, and then look up and list the correct spellings or rules. Keep the list available for future reference.

SKILLBUILDING

H. 1-Minute Alpha-numeric Timing

Take a 1-minute timing on the paragraph. Note your speed and errors.

```
37  The planned ski tour #4 begins at 2:43 p.m.,         9
38  and tour #3 begins at noon. Every tour costs $43,   19
39  and everyone will end at 7:38 p.m.                  26
     | 1 | 2 | 3 | 4 | 5 | 6 | 7 | 8 | 9 | 10
```

I. 2-Minute Timings

Take two 2-minute timings on lines 40–45. Note your speed and errors.

Goal: 27/2'/4e

```
40       It is a joy to end a term with good grades.     9
41  Fall term could be very nice if it were not for     19
42  exams and quizzes. Jan, though, likes to study to   29
43  show how much she has learned. She places great     38
44  value in having high marks. She knows her peers     48
45  admire the grades she achieved.                     54
     | 1 | 2 | 3 | 4 | 5 | 6 | 7 | 8 | 9 | 10
```

Appendix A—Keyboarding Skills

Lesson 1.4 — Input and Output Devices

You will learn to:
- **Define** input and output.
- **Identify** input devices.
- **Identify** output devices.

Digital Devices Digital technology has led to new versions of cameras, phones, and many other common devices. Other recent inventions include MP3 players and scanners. These tools give us new ways to send, receive, and store information. What do all of these devices have in common? They all read digital information. This makes them compatible with computers.

Input and Output Devices The flow of information between you and your computer goes both ways. Sometimes a computer receives information from you, and other times it delivers information to you. **Input** is information that is entered into a computer. The most common input device is the keyboard, which "puts in" the words you key.

Output devices are the opposite: they help the computer deliver (or "put out") information to the user. **Output** is information that a computer produces or processes. The most common output device is the printer.

Specialized Devices Some input and output devices are designed for specific purposes. For example, a joystick is an input device used to play computer games. A Liquid Crystal Diode (LCD) projector is an output device used to give a presentation.

input Information that is entered into a computer.

output Information that a computer produces or processes.

✓ Concept Check

Which kind of information would you enter into a computer? Which kind of information would the computer give you?

DIGITAL DIALOG

Tech Fair

Teamwork Work in a small group. Set up a decorated booth in your classroom for a Tech Fair. Each group should bring in actual technology items or pictures of devices to place on the group's booth. Find photographs and graphics in trade magazines, newspapers, and magazines. Categorize the hardware as input, output, storage devices, or computer hardware. Decide as a class ahead of time what types of devices each group should bring.

Group members should be prepared to demonstrate or explain the items and answer questions about them. As a class, arrange for half the groups to work their booths, while the other group members visit the booths.

SKILLBUILDING

D. Technique Checkpoint

Type each line 2 times. Focus on the technique at the left.

Keep your eyes on the copy when typing numbers and symbols.

```
21  sws sw2s s2s 222 s2s 22.2 s2s 2/22 s2s 22,222 s2s
22  sws sw2 s2s s2@s s@s s@s @2 @22 @222 s@s s2s @222
23  lol lo91 191 999 191 9/99 191 99.9 191 99,999 191
24  lol lo9 191 19(1 1(1 1(1 (9 (99 (999 1(1 191 (999
```

E. PRETEST

Take a 1-minute timing on the paragraph. Note your speed and errors.

```
25       Were you in the biology group that mixed the      9
26  ragweed seeds with some vegetable seeds? Jon and      19
27  Kim sneezed all month because of that. All of us      29
28  agreed that we must be more careful in the lab.       38
    |  1  |  2  |  3  |  4  |  5  |  6  |  7  |  8  |  9  | 10
```

F. PRACTICE

SPEED: *If you made 2 or fewer errors on the Pretest, type lines 29–36 two times each.*

ACCURACY: *If you made more than 2 errors on the Pretest, type lines 29–32 as a group two times. Then type lines 33–36 as a group two times.*

Left Reaches

```
29  tab wards grace serve wears farce beast crate car
30  far weeds tests seeds tread graze vexed vests saw
31  bar crest feast refer cease dated verge bread gas
32  car career grasses bread creases faded vested tad
```

Right Reaches

```
33  you Yukon mummy ninon jolly union minim pylon hum
34  mom nylon milky lumpy puppy holly pulpy plink oil
35  pop oomph jumpy unpin nippy imply hippo pupil nip
36  you union bumpy upon holly hill moon pink ill mop
```

G. POSTTEST

Repeat the Pretest. Compare your Posttest results with your Pretest results.

Appendix A—Keyboarding Skills

Activity 12

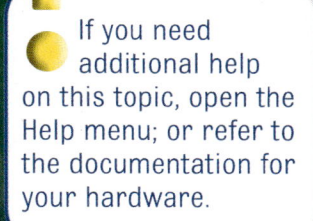

If you need additional help on this topic, open the Help menu; or refer to the documentation for your hardware.

Suite Smarts

Digital input devices such as cameras, touch screens, and trackballs can be used with a variety of software programs.

RUBRIC LINK

Go to the Online Learning Center at **digicom.glencoe.com** and click Lesson 1.4, Rubric, to find the application assessment guide(s).

Step-by-Step

Explore Input and Output Devices

Some of the many input and output devices available are listed below.

1. Choose three devices from the following list.
   - keyboard
   - mouse
   - trackball
   - touchpad
   - joystick
   - graphic tablet
   - optical scanner
   - digital pen
   - monitor
   - touch screen
   - bar code reader
   - optical reader
   - speaker
   - smart card
   - speech recognition
   - digital camera
   - webcam
   - digital camcorder
   - global positioning system
   - handheld computer or PDA
   - LCD projector
   - probe
   - cell phone
   - printer
   - flatbed scanner
   - microphone
   - DVD
   - WORM disk
   - biometric device
   - others you may find

2. Open your Web browser and go to a search engine, such as Google.

3. In the search box, enter the name of one of the devices you chose.

4. On one side of an index card, write down the device you are searching for.

5. On the other side of the card, write a brief description of the device, based on the information you find online.
   Include:
   - The type of device: input, output, or both
   - A detailed description.

6. Repeat Steps 3 through 5 to create a card for each of the three devices you chose.

Application 7

Directions Review the three devices you researched in the previous activity. Would any of them be useful for doing schoolwork or working at a job? Think of examples of how each device could be used and add this information to the back of each card.

Lesson 17

New Keys: 2 @ 9 (

OBJECTIVES:
- Learn the 2, @, 9, and (keys.
- Refine keyboarding skills.
- Type 27/2'/4e.

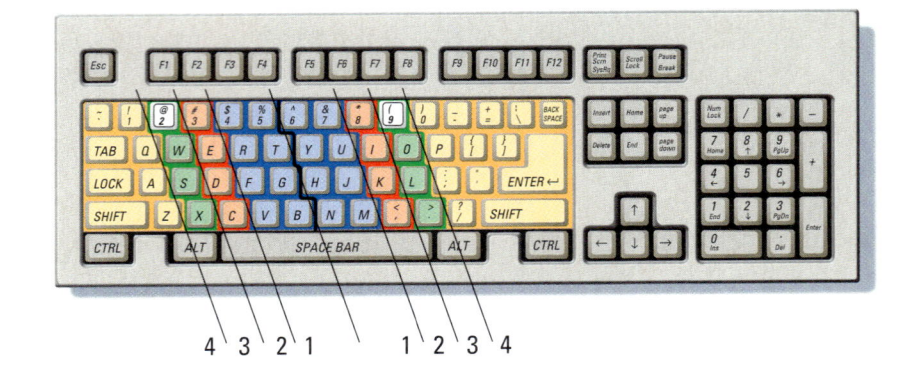

A. Warmup

Type each line 2 times.

Speed 1 The big lake was filled with many ducks and fish.
Accuracy 2 Lazy Jaques picked five boxes of oranges with me.
Numbers 3 The answer is 78 when you add 44 and 34 together.
Symbols 4 Invoices #73 and #48 from C & M Supply were $438.

NEW KEYS

B. 2 and @ Keys

Type each line 2 times. Repeat if time permits.

Use S finger.
For 2 and @, anchor F.

5 sws sw2s s2s 222 s2s 22.2 s2s 2/22 s2s 22,222 s2s
6 22 sips, 22 swings, 22 signals, 22 sites, or 2.22
7 Our class used 22 pens, 23 disks, and 24 ribbons.
8 There were 22 people waiting for Bus 22 on May 2.

@ (at) is the shift of 2. Space once before and after @ except when it is used in an e-mail address.

9 sws sw2 s2s s2@s s@s s@s @2 @22 @222 s@s s2s @222
10 @2, 2 sons, @22, 22 sets, @222, 222 sensors, @222
11 Paul said his e-mail address was smith@acc.co.us.
12 She bought 2 @ 22 and sold 22 @ 223 before 2 p.m.

C. 9 and (Keys

Type each line 2 times. Repeat if time permits.

Use L finger.
For 9 and (, anchor J.

13 lol lo9l 19l 999 19l 9/99 19l 99.9 19l 99,999 19l
14 99 laps, 99 loops, 99 lilies, 99 lifters, or 9.99
15 He said 99 times not to ask for the 99 fair fans.
16 They traveled 999 miles on Route 99 over 9 weeks.

The ((opening parenthesis) is the shift of 9. Space once before an opening parenthesis; do not space after it.

17 lol lo9 19l 19(1 1(1 1(1 (9 (99 (999 1(1 19l (999
18 (9, 9 lots, (99, 99 logs, (999, 999 latches, (999
19 lo9((99((9 lo9(1 lo(9(9(9 (9(9(9 1(lo9(1 9(
20 lo9(19(1 9(91 1((1 (9ol 99 lambs, (999, 999 lads

Appendix A—Keyboarding Skills A39

Application 8

Scenario You want to gain a perspective on computer input and output devices over a period of time. You decide to begin with the early 1950s and end with current innovations.

Directions In a small group, create a timeline that shows when input and output devices were first developed, or first came into popular use. Use the list on the previous page to choose devices to research. Add pictures and graphics to add interest to your timeline.

For recent innovations, look at the Web sites of daily newspapers across the country. They have special technology sections, and new gadgets always get attention. Research newspapers' archives to find information about trends in computer hardware and input and output devices. Include important details of the device's effect on society.

Application 9

Directions Choose two input or output devices. Think about these devices from different points of view. Write a few sentences on the devices from the point of view of:

- A person in advertising.
- A high school student.
- A manufacturer.
- A scientist.
- A person on the street.
- A person who is sick and may be helped by the device.
- A person who is disabled.

DIGITAL Decisions

Gathering Information

Cheng-Li works for a consulting firm. He has been assigned to analyze data from various sources about safety issues at the Pine Valley Rocking Chair Plant. The insurance company thinks the plant has had too many accidents in the past 12 months, so they are requesting an investigation of plant procedures. Cheng-Li wants to give the insurance company a thorough report. He determines that he will need to interview employees, review accident reports, and inspect the facilities at the plant.

Teamwork In a small group, discuss the input and output devices that Cheng-Li can use in his analysis of the plant's safety conditions.

Critical Thinking Describe how each device you decide on will help Cheng-Li as he prepares his report. Write a sentence or two for each device.

DIGITAL DIALOG

Input and Output Flash Cards

Teamwork Trade the cards you filled out in Activity 12, page 21, with a classmate. Sit across from your partner and use the cards like flash cards. Hold up one card with the name of the input or output device facing your partner. Ask your partner to define the device and describe a way to use it. Take turns.

When you have each identified three devices, trade your cards with another team to get six new devices. Repeat the exercise with your partner. See how many each of you can identify.

SKILLBUILDING

H. 1-Minute Alpha-numeric Timing

Take a 1-minute timing on the paragraph. Note your speed and errors.

```
41      B. Warmsly & J. Barnet paid the $847 charges      9
42 for the closing costs of their home at 3487 Cliff     19
43 Road; claim #47* shows the charge.                    26
   | 1 | 2 | 3 | 4 | 5 | 6 | 7 | 8 | 9 | 10
```

I. 2-Minute Timings

Take two 2-minute timings on lines 44–49. Note your speed and errors.

Goal: 27/2'/4e

```
44      We just want to stay all day in the store to      9
45 see the very new shoe styles. Sue quickly saw the     19
46 mix of zany colors. Jo put on a yellow and green      29
47 pair and looked in a mirror. The shoes had wide       39
48 strips on the soles. We were certain of the good      48
49 brand, so I bought two pair.                          54
   | 1 | 2 | 3 | 4 | 5 | 6 | 7 | 8 | 9 | 10
```

Lesson 1.5 → External Storage Devices

You will learn to:
- **Identify** storage devices.
- **Explain** how storage capacity is measured.
- **Save** to a floppy disk, a CD-RW, and a USB flash drive.

external storage device A storage device that can be easily removed from a computer and is often small and portable.

External Storage Devices You already know how to store files on the hard drive. Sometimes, though, you may want to turn off your computer and take your files with you. Also, it is essential to back up your files to an external storage device in case the hard drive on your computer fails.

A variety of external storage devices makes this possible. An **external storage device** is a storage device that can be easily removed from a computer and is often small and portable. Some, such as USB flash drives, are small enough to fit in the palm of your hand. A USB flash drive is a small removable storage device that plugs into a USB, or Universal Serial Bus, port on your computer.

Storage Capacity The storage capacity is the amount of data, usually measured in megabytes or gigabytes, that a storage device can hold. One gigabyte, or GB, equals 1,000 megabytes, or MB. The table below shows how storage capacity differs from device to device.

Device Storage Capacity

Type of Device	Typical Storage Capacity
Floppy disk	1.44 MB
Zip disk	100 MB to 750 MB
Compact disc (CD)	700 MB
USB flash drive	64 MB to 2 GB +
Digital video disc (DVD)	4.7 to 17 GB
External hard drive	20 to 60 GB

SKILLBUILDING

D. PRETEST

Take a 1-minute timing on the paragraph. Note your speed and errors.

```
21       Do you brood when you make errors on papers?     9
22 It would be better to figure out what causes the      19
23 errors and to look for corrective drills to help      29
24 you make fewer errors in the future.                  36
   | 1 | 2 | 3 | 4 | 5 | 6 | 7 | 8 | 9 | 10
```

E. PRACTICE

SPEED: If you made 2 or fewer errors on the Pretest, type lines 25–32 two times.

ACCURACY: If you made more than 2 errors on the Pretest, type lines 25–28 as a group two times. Then type lines 29–32 as a group two times.

Double Reaches

```
25 rr errs hurry error furry berry worry terry carry
26 ll bill allay hills chill stall small shell smell
27 tt attar jetty otter utter putty witty butte Otto
28 ff stuff stiff cliff sniff offer scuff fluff buff
```

Alternate Reaches

```
29 is this list fist wish visit whist island raisins
30 so sons some soap sort soles sound bosses costume
31 go gone goat pogo logo bogus agora pagoda doggone
32 fu fun fume fund full fuel fuss furor furry fuzzy
```

F. POSTTEST

Repeat the Pretest. Compare your Posttest results with your Pretest results.

G. Number and Symbol Practice

Type each line 1 time. Repeat if time permits.

```
33 83 doubts, 38 cubs, 37 shrubs, 33 clubs, 34 stubs
34 87 aims, 83 maids, 88 brains, 73 braids, 84 raids

35 78 drinks, 48 brinks, 43 inks, 83 minks, 33 links
36 88 canes, 78 planes, 73 manes, 34 cans, 84 cranes

37 #7 blue, 4# roast, $3 paint, 77 books,* 3 & 4 & 8
38 7# boxes, 38 lists, $4 horse, #8 tree,* 7 & 3 & 4

39 Seek & Find Research sells this book* for $37.84.
40 The geometry test grades were 88, 87, 84, and 83.
```

read-only memory (ROM) Memory that holds data that can be read by a digital device but cannot be changed or deleted.

Optical Storage Devices The hard drive in your computer stores information magnetically. Compact discs, or CDs, use lasers to store data optically. CDs are small, portable, and capable of storing large amounts of data. Some contain **read-only memory (ROM)**, or memory that holds data that can be read by a digital device but cannot be changed or deleted.

Optical storage devices are available in several different formats:

- A CD-ROM is a compact disc with read-only memory.
- A CD-R is a compact disc that is recordable. Data can be stored on the disc but cannot be erased.
- A CD-RW is a compact disc that is rewritable. Data can be stored, erased, and replaced with new data.
- A DVD is a digital video disc. It can store much more data than a CD and is often used to store movies.

✓ Concept *Check*

Why are external storage devices useful? Name two external storage devices.

D I G I T A L DIALOG

Which USB Flash Drive Is Best?

Teamwork You are a photographer for a magazine. When you are on assignment, you use your digital camera to take hundreds of pictures. You need to find a storage device that will hold all the data.

You have heard of USB flash drives, but you do not know how much they cost or how much data they can hold. Find a classmate to work with and search the Internet, newspapers, and magazines for information about USB flash drives.

Writing Create a list of the names of the flash drives you find. Create a chart similar to the one shown below to gather information. Record the following information:

- Brand name.
- Storage capacity.
- Cost.
- Appearance.

Which USB drive appears to be the best bargain? Share the information you find with the class.

Brand Name	Storage Capacity (MB or GB)	Cost	Appearance

Lesson 16

New Keys: 3 # 8 *

OBJECTIVES:
- Learn the 3, #, 8, and * keys.
- Refine keyboarding skills.
- Type 27/2'/4e.

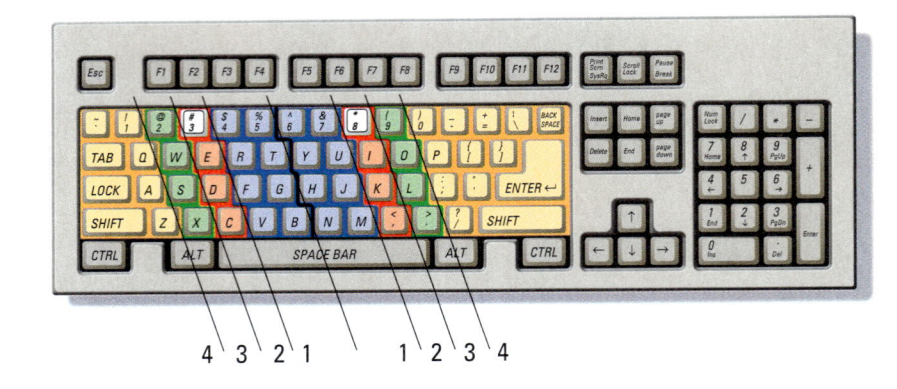

A. Warmup

Type each line 2 times.

Speed   1 The time for Andrew to stop is when the sun sets.
Accuracy   2 Ten foxes quickly jumped high over twelve zebras.
Numbers   3 Lines 47, 77, and 44 were right; line 74 was not.
Symbols   4 Bakes & Deli pays $4, $4.77, and $7.44 for dimes.

NEW KEYS

B. 3 and # Keys

Type each line 2 times. Repeat if time permits.

Use D finger.
Anchor A or F.

5 ded de3d d3d 333 d3d 3/33 d3d 33.3 d3d 33,333 d3d
6 33 dimes, 33 dishes, 33 dots, 33 daisies, or 3.33
7 Draw 33 squares, 3,333 rectangles, and 3 circles.
8 They had 333 dogs in 33 kennels for over 3 weeks.

The # (number or pound sign) is the shift of 3.
Anchor A or F.
Do not space between the number and #.

9 ded de3 d3d d3#d d#d d#d #3 #33 #333 d#d d3d #333
10 #3, 3 dots, #33, 33 dogs, #333, 333 ditches, #333
11 Is Invoice #373 for 344#, 433#, or 343# of fruit?
12 The group used 43# of grade #3 potatoes at lunch.

C. 8 and * Keys

Type each line 2 times. Type smoothly as you use the shift keys. Repeat if time permits.

Use K finger.
Anchor ;.

13 kik ki8k k8k 888 k8k 8/88 k8k 88.8 k8k 88,888 k8k
14 88 kegs, 88 kilns, 88 knocks, 88 kickers, or 8.88
15 Our zoo has 88 zebras, 38 snakes, and 33 monkeys.
16 The house is at 88 Lake Street, 8 blocks farther.

*The * (asterisk) is the shift of 8.*
*Do not space between the word and *.*

17 kik ki8 k8k k8*k k*k k*k *8 *88 *888 k*k k8k *888
18 *8, 88 kits, *88, 88 keys, *888, 88 kimonos, *888
19 This manual* and this report* are in the library.
20 Reports* are due in 8 weeks* and should be typed.

Appendix A—Keyboarding Skills   **A36**

Activity 13

If you need additional help on this topic, open the Help menu; or refer to the documentation for your hardware.

Step-by-Step

Save to a Floppy Disk

The storage capacity of a floppy disk is only 1.44 MB, so it is best used for storing small files, like word processing documents.

1. Place a floppy disk in your computer's A: drive.

2. On your computer desktop, double-click the **My Computer** icon. You will see an icon called **3½ Floppy (A:)**.

3. Start WordPad. Open one of the files you saved in Lesson 1.1, pages 6–8.

4. Click **File**, and then click **Save As.** Do not change the name of the file.

5. Click the box next to **Save in** at the top of the window.

6. Click **3½ Floppy (A:)** in the list. Click the **Save** button. Now your document is saved in two places: on your computer's hard drive and on the floppy disk.

7. Exit WordPad.

Activity 14

Save to a Compact Disc (CD)

1. Create a new folder called **Pictures** on your desktop. Your teacher will tell you where to find some digital pictures on the computer. Open some of the pictures, and use the Save As command to save the pictures into the **Pictures** folder.

2. Double-click the **My Computer** icon on your desktop. You will see your CD-ROM drive in the window. The drive may be named D: or E:.

Continued

Chapter 1

25

SKILLBUILDING

H. 1-Minute Alpha-numeric Timing

Take a 1-minute timing on the paragraph. Note your speed and errors.

36 Luke sent a $47 check to Computers & Such to	9
37 get a disk with 44 games & 4 special programs for	19
38 7 friends. He saw 47 of his friends at 4 p.m.	28
| 1 | 2 | 3 | 4 | 5 | 6 | 7 | 8 | 9 | 10	

I. 2-Minute Timings

Take two 2-minute timings on lines 39–44. Note your speed and errors.

Goal: 27/2'/4e

39 Have you been to our zoo? This is a great	9
40 thing to do in the summer. Bring your lunch to	18
41 eat in the park by the lake. You can watch a bear	28
42 cub perform or just view the zebras. Then explore	2
43 this spot and see the many quail and ducks. Take	48
44 some photos to capture the day.	54
| 1 | 2 | 3 | 4 | 5 | 6 | 7 | 8 | 9 | 10	

Appendix A—Keyboarding Skills   A35

Step-by-Step *continued*

3. In order to save files to this drive, you must have a rewritable disc (CD-RW) and a CD-RW drive. Place a CD in this drive and then double-click on the CD icon.

    You will be able to see the files saved on the disc. Some CDs will open automatically when you place them in the CD drive.

4. Click the **My Computer** icon.
5. From the desktop, open the **Pictures** folder you created.
6. Arrange the two windows so they are side by side.

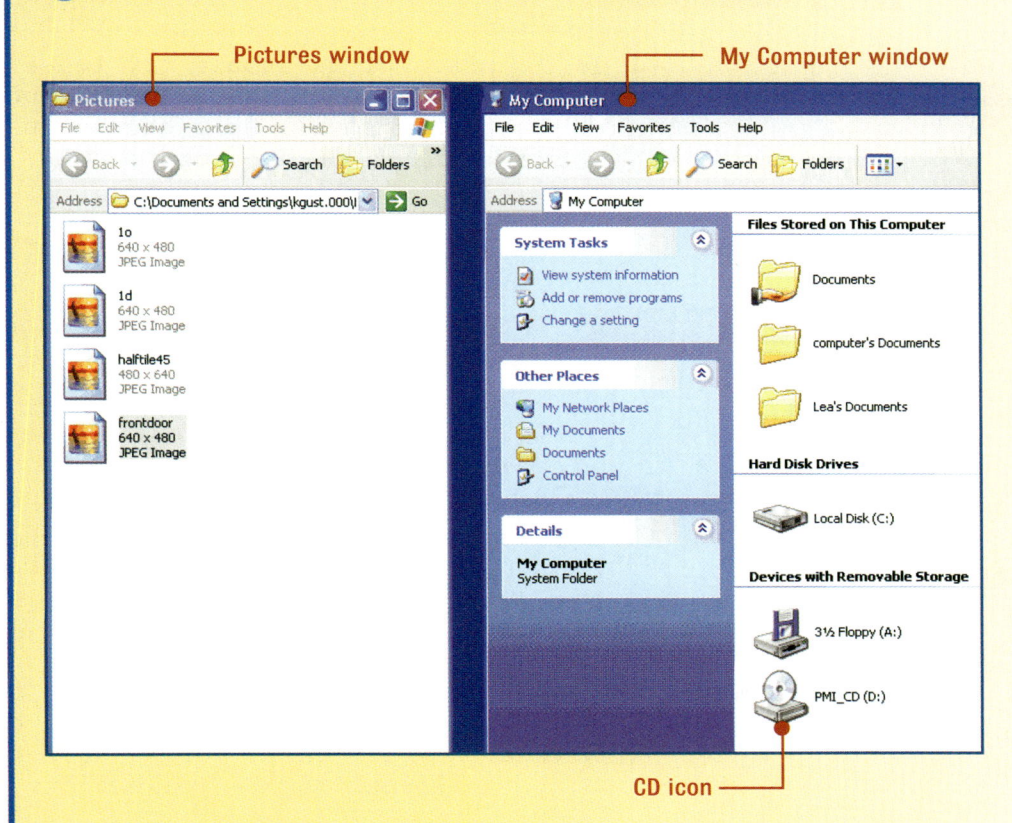

Pictures window — *My Computer window*

CD icon

7. In the Pictures folder, choose a picture you would like to copy to the CD. Click on the picture once and hold down the left mouse button.
8. Drag the picture to the CD icon in the My Computer window. Then let go of the mouse button. The picture will be copied to the CD-RW. This method of moving a file is known as drag and drop.

DIGI Byte

Flash Drive Capabilities

Some flash drives have a built-in MP3 player and a headphone jack for listening to music.

Suite Smarts

USB flash drives can save any type of file from any program. They save files more quickly than a CD-RW drive.

Continued

SKILLBUILDING

D. Technique Timings

Take two 30-second timings on each line. Focus on the techniques at the left.

Lines 20 and 21: Keep your eyes on the copy.
Lines 22 and 23: Space without pausing.

```
20 Kara saw a ship as she was walking over the hill.
21 Ned says he can mend the urn that fell and broke.
22 The five of us had to get to the bus before noon.
23 Lou said he would be at the game to see us later.
```

E. PRETEST

Take a 1-minute timing on the paragraph. Note your speed and errors.

```
24      Each of us should try to eat healthful food,      9
25 get proper rest, and exercise moderately. All of     19
26 these things will help each of us face life with     29
27 more enthusiasm and more energy.                     35
   | 1 | 2 | 3 | 4 | 5 | 6 | 7 | 8 | 9 | 10
```

F. PRACTICE

SPEED: *If you made 2 or fewer errors on the Pretest, type lines 28–35 two times.*

ACCURACY: *If you made more than 2 errors on the Pretest, type lines 28–31 as a group two times. Then type lines 32–35 as a group two times.*

Adjacent Reaches
```
28 tr train tree tried truth troop strum strip stray
29 op open slope opera sloop moped scoop hoped opine
30 er were loner every steer error veers sewer verge
31 po port porter pole pods potter potion pound pout
```

Jump Reaches
```
32 on onion ozone upon honor front spoon phone wrong
33 ex exams exist exact flex exits exalt vexed Texas
34 ve even veers vests verbs leave every verge heave
35 ni nine ninth night nimble nifty nice nickel nighi
```

G. POSTTEST

Repeat the Pretest. Compare your Posttest results with your Pretest results.

Activity 15

Step-by-Step *continued*

Save to a USB Flash Drive

1. Plug a USB flash drive into a USB port. You may see a dialog box similar to the one below.

2. If you wanted to open a file on the flash drive, you would click OK to see a list of files saved on the USB flash drive. Because you want to save something to the USB flash drive instead, click **Cancel**.

3. Start **WordPad**, and then open one of the files you saved in Lesson 1.1, pages 6–8.

4. When the file is open, click **File**, then **Save As**.

5. Click the **Save in** box.

6. Click once on **UDISK20 (E:)**, or the name displayed in your list. The name may vary, depending on the manufacturer of the USB flash drive.

7. Do not change the name of the file. Click **Save**.

8. Before removing the USB flash drive, you may first need to click the **Unplug or Eject Hardware** icon on the Task bar to make sure removal is safe. Then take the USB flash drive out of the USB port.

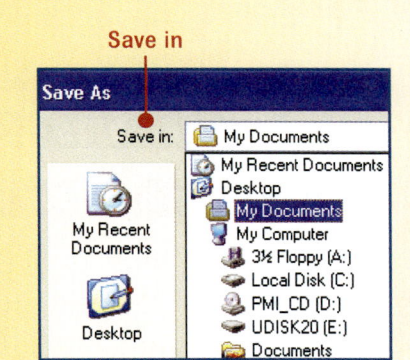

RUBRIC LINK

Go to the Online Learning Center at **digicom.glencoe.com** and click Lesson 1.5, Rubric, to find the application assessment guide(s).

Application 10

Directions Use a digital camera to take several pictures around your classroom or school. Save the pictures to the **Pictures** folder you created in this lesson.

Next, save the pictures to a floppy disk and a USB flash drive. If you have a CD-RW drive, copy the pictures to that drive as well, using the drag-and-drop method.

Lesson 15 — New Keys: 4 $ 7 &

OBJECTIVES:
- Learn the 4, $, 7, and & keys.
- Refine keyboarding techniques.
- Type 27/2'/4e.

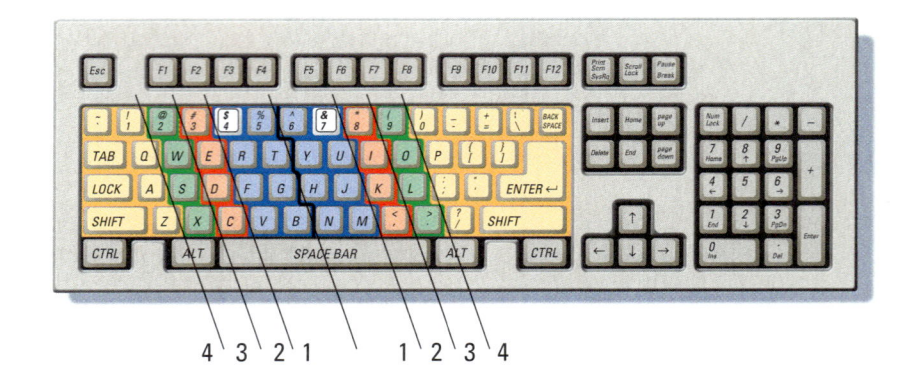

A. Warmup
Type each line 2 times.

Words   1 shot idea jobs corn quip give flex whey maze elks
Speed   2 It is not a good idea to play ball in the street.
Accuracy 3 My joke expert amazed five huge clowns in Quebec.

NEW KEYS

B. 4 and $ Keys
Type each line 2 times. Repeat if time permits.

Use F finger. Anchor A.

4 frf fr4f f4f 444 f4f 4/44 f4f 44.4 f4f 44,444 f4f
5 44 films, 44 foes, 44 flukes, 44 folders, or 4.44
6 I saw 44 ducks, 4 geese, and 4 swans on the lake.
7 Today, our team had 4 runs, 4 hits, and 4 errors.

$ is the shift of 4. Do not space between the $ and the number.

8 frf fr4 f4f f4$f f$f f$f $4 $44 $444 f$f f4f $444
9 $444, 44 fish, 4 fans, $44, 444 fellows, $4, $444
10 Jo paid $44 for the oranges and $4 for the pears.
11 They had $444 and spent $44 of it for 4 presents.

C. 7 and & Keys
Type each line 2 times. Repeat if time permits.

Use J finger. Anchor ;.

12 juj ju7j j7j 777 j7j 7/77 j7j 77.7 j7j 77,777 j7j
13 77 jokers, 77 joggers, 77 jets, or 7.77, 77 jumps
14 Hank will perform July 4 and 7, not June 4 and 7.
15 On July 4, we celebrated; on August 7, we rested.

Use J finger and left shift. Anchor ;. Space before and after the ampersand.

16 juj ju7 j7j j7&j j&j j&j j& &j& ju7& j&j j7j ju7&
17 7 jugs & 7 jars & 7 jewels & 7 jurors & 7 jungles
18 He thinks he paid $44 & $77 instead of $47 & $74.
19 B & C ordered 744 from Dixon & Sons on January 7.

Appendix A—Keyboarding Skills

Lesson 1.6 — Digital Skills in the 21st Century Workplace

You will learn to:
- **Describe** the purpose of an e-portfolio.
- **Plan** your own e-portfolio.

High-Tech Job Hunting Computers and the Internet have become valuable tools for people as they look for work or apply to colleges. You can apply for jobs online, e-mail your résumé to potential employers, and search the Web for jobs that match your interests and skills. Now, many job-seekers create an electronic portfolio to boost their chances of getting a well paying job.

E-Portfolios You will use the skills you learn in this course to build an electronic portfolio, or e-portfolio. An e-portfolio is a collection of the best documents and projects you have created and saved in an electronic format. You use your e-portfolio to showcase your best skills and talents. For example, you could include your best reports in word processing documents, spreadsheets, databases, and electronic slide presentations. You will learn about spreadsheets in Chapter 10, pages 313–358, and databases in Chapter 11, pages 359–392. You could also include your best handwritten reports in Journal on the Tablet PC, digital photographs of you working in teams, and video clips of you giving presentations.

The format is electronic, so you can share your e-portfolio files with others in a variety of ways: e-mail them, publish them on your own Web site, or save them on a CD to show to a prospective employer.

Your e-portfolio showcases your best work, so be sure that you:

- Organize your e-portfolio and include a table of contents.
- Proofread and edit all your work for correct use of punctuation and grammar. See the 21st Century Connection, *Proofreading*, page 19, for more on proofreading.
- Be creative and professional.

You will build your own e-portfolio as you work through this course. In each chapter Digital Dimension and unit lab, you will save to your e-portfolio. For example, take a look at the E-Portfolio Activity in Chapter 1 Digital Dimension—Language Arts, page 32, and the Unit 1 Lab, pages 120–121.

✓ Concept Check

What would you put in your e-portfolio to showcase your talents? Illustrations? Writing samples? Computer projects?

---

e-portfolio A collection of the best documents and projects you have created and saved in an electronic format.

DIGIByte

A Web Site of Your Own

Many people create personal Web sites to promote themselves when they are looking for work. Often, they name the sites after themselves, for example, www.YourName.com.

SKILLBUILDING

E. Technique Timings

Take two 30-second timings on each line. Focus on the techniques at the left.

Sit up straight, keep your elbows in, and keep your feet flat on the floor.

16 Steward and Phon drove a car down to the shelter.
17 Ten people helped serve meals to thirty children.
18 They said it was hard work. Jung felt happy then.
19 This might help solve these problems in our city.
   | 1 | 2 | 3 | 4 | 5 | 6 | 7 | 8 | 9 | 10

F. PRETEST

Take a 1-minute timing on the paragraph. Note your speed and errors.

20   Look up in the western sky and see how it is   9
21 filled with magnificent pinks and reds as the sun   19
22 begins to set. As the sun sinks below the clouds,   29
23 you will see an amazing display of great colors.   39
   | 1 | 2 | 3 | 4 | 5 | 6 | 7 | 8 | 9 | 10

G. PRACTICE

SPEED: *If you made 2 or fewer errors on the Pretest, type lines 24–31 two times each.*

ACCURACY: *If you made more than 2 errors on the Pretest, type lines 24–27 as a group two times. Then, type lines 28–31 as a group two times.*

Left and right reaches are a sequence of at least three letters typed by fingers on either the left or the right hand. (lease, think)

24 was raged wheat serve force carts bears cages age
25 tag exact vases rests crank enter greet moves ear
26 was raged wheat serve force carts bears cages age
27 tag exact vases rests crank enter greet moves ear

28 get table stage hired diets gears wages warts rat
29 hop mouth union input polka alone moors tunic joy
30 him looms pumps nouns joked pound allow pours hip
31 lip mopes loose equip moods unite fills alike mop

H. POSTTEST

Repeat the Pretest. Compare your Posttest results with your Pretest results.

I. 1-Minute Timings

Take two 1-minute timings on the paragraph. Note your speed and errors.

Goal: 25/1'/2e

32   We saw where gray lava flowed down a path.   9
33 At the exit, Justin saw trees with no bark and a   19
34 quiet, fuzzy duck looking at me.   25
   | 1 | 2 | 3 | 4 | 5 | 6 | 7 | 8 | 9 | 10

Appendix A—Keyboarding Skills

Activity 16

? If you need additional help on this topic, open the Help menu; or refer to the documentation for your hardware.

Suite Smarts
With software such as PowerPoint, you can deliver your e-portfolio as a multimedia presentation in front of an audience.

RUBRIC LINK
Go to the Online Learning Center at **digicom.glencoe.com** and click Lesson 1.6, Rubric, to find the application assessment guide(s).

Step-by-Step

Plan an E-Portfolio

In this exercise, you will look at some sample e-portfolios and set up folders to begin to plan your own.

1. First, view some e-portfolios online to get an idea of what the possibilities are. Do a Web search using the keywords "Penn State University personal e-portfolio." Click on the **Gallery** link to view some examples.

   Another way to find examples is to do a Web search using the keywords "student e-portfolio."

2. On a piece of paper list the names of some of the folders that you find in the example e-portfolios.

3. Create a new folder on your desktop named **My e-portfolio**.

4. You will create new folders in the My e-portfolio folder. Look at the folder names you listed in Step 2 and think about the names of folders that will be useful to you. For example, you might include folders named: Personal Information, Academic Skills, Digital Devices, Handheld Computer, Speech Recognition, Tablet PC, Word Processing, Spreadsheet, Database, and Presentation.

   You may not have skills yet in many of these areas, and that is okay. Choose which folders you want to add and create them in your My e-portfolio folder. You will use the folder names as the table of contents for your e-portfolio.

Application 11

Directions In Activity 16 above, you began planning a personal information folder for your e-portfolio. Now you will add a file to that folder.

Open WordPad and start a new document. Enter the following information about yourself:

- Your name.
- Grade.
- Extra-curricular activities.
- Hobbies.
- Academic interests, for example classes you want to take.
- Goal for the future when you graduate from high school.

Save your document and name it **Personal**. Save it in the Personal Information folder you created in your My e-portfolio folder.

Lesson 14

New Keys: - _

OBJECTIVES:
- Learn the hyphen (-) and underscore (_) keys.
- Type 25/1'/2e.

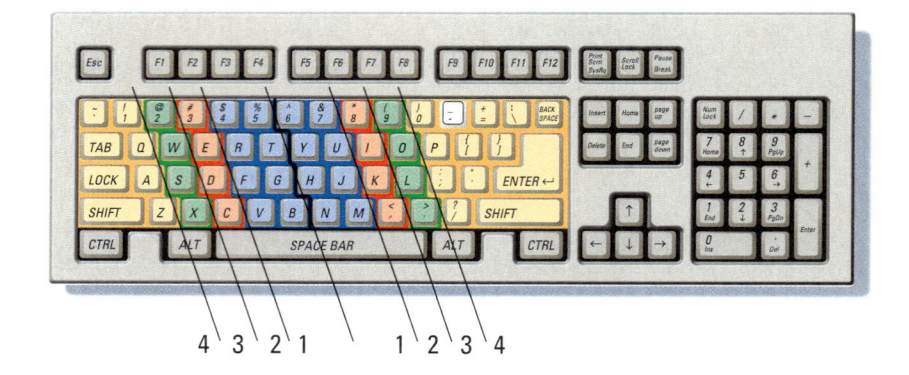

4 3 2 1   1 2 3 4

A. Warmup
Type each line 2 times.

1 rave jinx tact safe mind glib quit yelp hawk doze
2 We all must be good friends to have good friends.
3 We have quickly gained sixty prizes for best jam.

NEW KEYS

B. - Keys
Type each line 2 times. Repeat if time permits.

Use Sem finger.
Anchor J.
Do not space before or after hyphens.

4 ;;; ;p; ;p-; ;-; -;- ;;; ;p; ;p-; ;-; -;- ;;; ;-;
5 ;p- ;-; self-made ;-; one-third ;p- one-sixth ;-;
6 ;p- ;-; part-time ;-; one-tenth ;p- two-party ;-;
7 Self-made Jim stopped at an out-of-the-way place.

C. _ Keys (UNDER-
Type each line 2 times. Repeat if time permits.

Use the Sem finger and the left shift key.
Anchor J.

8 ;p; ;p- ;-; ;-_; ;-_; ;p-_ _;_ ;p-_ ;-_; ;_; ;p-_
9 ;;; ;p; ;p_; ;_; _;_ ;;; ;p; ;p_; ;_; _;_ ;;; ;_;
10 Quick, create this seven-character line: _____.
11 Be sure to use her e-mail name, jennifer_cochran.

SKILLBUILDING

D. Technique Checkpoint
Type each line 2 times. Repeat if time permits. Focus on the techniques at the left.

Keep your feet on the floor, back straight, elbows in.

12 ;;; ;p; ;p-; ;-; -;- ;;; ;p; ;p-; ;-; -;- ;;; ;-;
13 ;;; ;p; ;p_; ;_; _;_ ;;; ;p; ;p_; ;_; _;_ ;;; ;_;
14 Are you an easy-going person who gets along well?
15 The new name he now uses for e-mail is jute_rope.

Appendix A—Keyboarding Skills

21ST CENTURY CONNECTION

Career Clusters

Career clusters are groups of related occupations and industries. The 16 career clusters were developed by the U.S. Department of Education. Learning about the career clusters can help you to focus on careers that you will enjoy.

In the Digital Career Perspective feature in this book, you will learn about 16 careers, one from each cluster. You will also learn more about career clusters in Lesson 13.2, pages 436–438.

Activity Choose a Digital Career Perspective from one of the pages listed below. Research the career featured, and write a paragraph about it.

The 16 Career Clusters

Agriculture and Natural Resources — 389
Farmer, park ranger, animal trainer, food inspector, logger, ecologist, veterinarian

Architecture and Construction — 176
Building inspector, surveyor, architect, bricklayer, electrician, civil engineer

Arts, Audio/Video Technology, and Communication — 288
Actor, computer animator, musician, telecommunication specialist/technician

Business and Administration — 79
Business analyst, human resources specialist, accountant

Education and Training — 407
Teacher, coach, corporate trainer, librarian, school administrator

Finance — 309
Stockbroker, financial planner, financial analyst, banker, insurance agent, auditor

Government and Public Administration — 254
City manager, postal worker, armed services officer, sanitation worker

Health Science — 149
Physician, nurse, pharmacist, paramedic, physical therapist, dentist, dietician

Hospitality and Tourism — 223
Caterer, hotel manager, lifeguard, chef, housekeeper, travel agent, concierge

Human Services — 160
Social worker, hair stylist, counselor, child care worker, product safety tester

Information Technology — 416
Computer programmer, Web designer, systems analyst, network administrator

Law and Public Safety — 355
Police officer, firefighter, parole officer, lawyer, paralegal, judge, dispatcher

Manufacturing — 116
Production manager, robotic engineer, baker, machinist, industrial designer

Retail/Wholesale Sales and Service — 69
Market researcher, salesperson, inventory clerk, real estate broker

Scientific Research and Engineering — 31
Mathematician, physicist, chemist, oceanographer

Transportation, Distribution, and Logistics — 54
Truck driver, cargo agent, shipping clerk, air traffic controller

SKILLBUILDING

E. Technique Timings

Take two 30-second timings on each line. Focus on the technique at the left.

Hold those anchors. Quickly return your fingers to home-key position.

```
16  Tony was a better friend than Hope was to Salena.
17  Lu and I were at Camp Piney Forest in early fall.
18  I rode the Rocky Ford train to San Juan in March.
19  Maya, Sue, and Grace were there. It was exciting.
    | 1 | 2 | 3 | 4 | 5 | 6 | 7 | 8 | 9 | 10
```

F. PRETEST

Take a 1-minute timing on the paragraph. Note your speed and errors.

```
20       The blind slats are broken. Can you fix the      9
21  broken ones? My WILY dog jumped out of the window    19
22  which is how this happened. There should be some    29
23  way to stop him. For a young dog, he is AMAZING.    38
    | 1 | 2 | 3 | 4 | 5 | 6 | 7 | 8 | 9 | 10
```

G. PRACTICE

Type each line 2 times.

```
24  slat slit skit suit quit quid quip quiz whiz fizz
25  LASS bass BASE bake CAKE cage PAGE sage SAGA sags
26  maze mare more move wove cove core cure pure pore
27  mix; fix; fin; kin; kind wind wild wily will well

28  cape cane vane sane same sale pale pals pats bats
29  jump pump bump lump limp limb lamb jamb jams hams
30  slow BLOW blot SLOT plot PLOP flop FLIP blip BLOB
31  mite more wire tire hire hide hive jive give five
```

H. POSTTEST

Repeat the Pretest. Compare your Posttest results with your Pretest results.

I. Composing at the Keyboard

Language Link

Answer the following questions with a single word.

Keep your eyes on the screen as you type

32 What day of the week is today?
33 What is your favorite animal?
34 What is your favorite food?
35 What is your favorite ice cream flavor?
36 What month is your birthday?

DIGITAL Career Perspective

Scientific Research and Engineering

John Gust
Environmental Consultant

Working with the Government to Protect the Environment

"Technology has become a welcomed partner in the environmental world. It allows scientists and consultants to see the world in ways never before realized," says environmental consultant John Gust. Gust works for an international consulting firm. His responsibilities require him to travel all over the country as he meets with various governmental agencies.

"Government agencies hire my firm to analyze their systems and make sure they comply with federal regulations and laws set forth by Congress," explains Gust.

He continues, "I usually work with a team of other consultants in the firm, so it is imperative that we keep in touch with each other. We design surveys that gather data about information systems that manage the Department of Defense. As part of this survey, Geographical Information Systems, or GIS, is used as a tool to give us a visual representation of an area so we can interpret air, land, and water regulations."

Gust's firm uses an in-house instant messaging program to provide quick communication access for team members between offices, floors, and buildings.

"My laptop allows me to work from multiple places and stay plugged into the fast-paced consulting world," Gust adds. "I use the laptop for a number of things, including document retrieval off the World Wide Web, e-mailing, creating documents and presentations, displaying graphical interpretations of complex systems, and spatial mapping," states Gust. During his travels, Gust takes his laptop with him to access files while he flies from one location to another.

Why does Gust like being an environmental consultant? "I like having the ability to blend technology and environmental policy to aid in the development of our changing world. There are many ways technology can help the environment. I am excited to be part of that accomplishment."

Training
Experience is the most important factor. It can be gained through previous jobs, education, or on-the-job training. A bachelor's degree in a relevant field, such as science or engineering, is required, although a master's degree will increase the likelihood of advancement.

Salary Range
Typical earnings are in the range of $60,000 to $80,000 a year.

Skills and Talents
Environmental consultants need to be:

- Excellent verbal and written communicators
- Strong, logical thinkers
- Willing to continue their education, both on the job and for advanced degrees
- Willing to work long hours when necessary
- Highly ethical
- Able to manage their time effectively to meet deadlines
- Willing to travel when necessary

Career Activity
There are many areas of scientific research and engineering. Research an area of science in which you are interested and write a paragraph about it.

Lesson 13 — New Keys: ? Caps Lock

OBJECTIVES:
- Learn the ? key.
- Use the caps lock key to type all-capital letters.
- Compose at the keyboard.

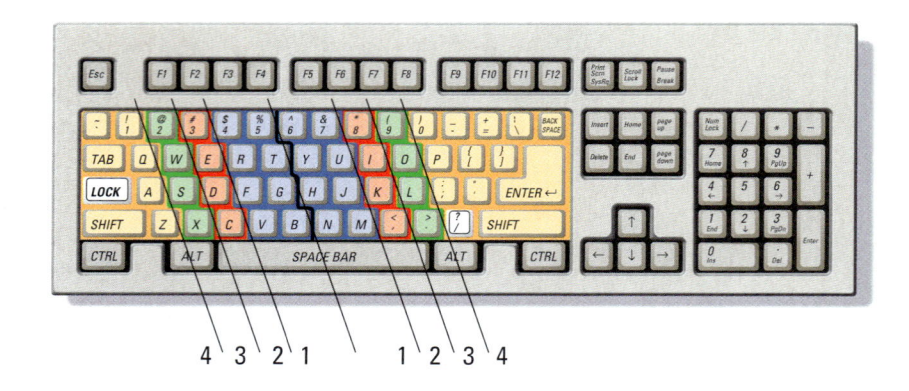

A. Warmup

Hold those anchors.

Type each line 2 times.

1 herbs jinx gawk miff vest zinc ploy quad best zoo
2 Dozy oryx have quit jumping over the huge flocks.
3 Lax folks quickly judged the lazy dogs unfit now.

NEW KEYS

B. ? Keys

Shift of /.
Use Sem finger and left shift key.
Anchor J.
Space once after a question mark.

Type each line 2 times. Repeat if time permits.

4 ;;; ;/; ;/? ;?; ;?; ;;; ;/; ;/? ;?; ;?; ;;; ;/ ;?
5 ;/; ;?; now? now? ;?; how? how? ;?; who? who? ;?;
6 Who? What? Why? Where? When? Next? How many? Now?
7 How can Joe get there? Which way are the outlets?

C. Caps Lock Key

Use A finger.

Use the caps lock key to type letters or words in all-capital letters (all caps). You must press the shift key to type symbols appearing on the top half of the number keys.

Type each line 2 times. Repeat if time permits.

8 A COMPUTER rapidly scanned most AIRMAIL packages.
9 Another START/STOP safety lever was stuck lately.
10 Was JOSE elected CLASS PRESIDENT today or sooner?
11 You should not answer my door WHEN YOU ARE ALONE.

SKILLBUILDING

D. Technique Checkpoint

Quickly return fingers to home keys after reaching to other keys.

Type each line 2 times. Repeat if time permits.

12 ;;; ;/; ;/? ;?; ;?; ;;; ;/; ;/? ;?; ;?; ;;; ;/ ;?
13 Did you see HELEN? Did you learn about her crash?
14 Her auto was hit by a TRAIN. She broke BOTH arms.
15 HOW will she manage while both arms are in casts?

Appendix A—Keyboarding Skills

Chapter 1

DIGITAL Dimension — Language Arts
INTERDISCIPLINARY PROJECT

Be a Savvy Consumer

Using the knowledge and skills you have learned in this chapter, describe how the use of technology online can assist in advertising, compared with ads in hardcopy publications.

Compare Hardcopy and Online Publications Find four magazines and one newspaper. Check the publications and find their Web sites. Go to the Web site listed and compare the articles on the Web site to the information in the hardcopy publication. Look at the advertisements. Do you notice any difference in the online ads compared with the hardcopy ads? Consider how the use of technology online can assist in advertising.

On a separate sheet of paper make a list of the advertisements you find in the online version and compare them to the ads you find in the hard copy. Include the following information:

- Name of the advertising company.
- Graphics or special effects that are used online.
- Size of the advertisement, for example, full page, half page, small corner ad, and so on.

Describe how the use of technology online can assist in advertising.

Organize and Analyze If you can, cut out at least two ads from each magazine and two from the newspaper. Include the text with the advertisement. Organize the ones you cut out into a notebook or folder. Tape or glue the cut-out ads to a sheet of paper.

On a separate sheet of paper, write several statements about how the ad makes you feel. Answer the questions:

- Does the product look good?
- Do you want to buy it?
- Is the ad intended for your age group?
- What technique is the advertiser using?
- Is the ad appealing to your taste?

Incorporate your answers in a paragraph or two.

E-Portfolio Activity

Showcase Your Skills

Key your description of how the use of technology online can assist in advertising. Save the file to the e-portfolio folder you created in Activity 16, page 29.

Self-Evaluation Criteria

Have you:

- Reviewed ads in four magazines and one newspaper?
- Looked up the Web site for these publications?
- Made a list to compare online and hardcopy ads?
- Described how the use of technology online can assist in advertising?
- Made a notebook of cut-out ads from the magazines and newspaper?
- Recorded your comments about the ads?

SKILLBUILDING

E. Technique Timings

Take two 30-second timings on each line. Focus on the technique at the left.

Keep your elbows in by your sides.

```
15 Type fast to reach the end of the line.
16 Keep your eyes on the copy as you type.
17 Tests are easy if you know the answers.
18 If they go to the zoo, invite them too.
   | 1 | 2 | 3 | 4 | 5 | 6 | 7 | 8
```

F. PRETEST

Take a 1-minute timing on the paragraph. Note your speed and errors.

Remember to press ENTER only at the end of the paragraph.

```
19      As Inez roamed the ship, she told        7
20 fond tales. She slipped on that waxy         14
21 rung and fell to the deck. She hurt her     22
22 face and was dazed, but felt no pain.       30
   | 1 | 2 | 3 | 4 | 5 | 6 | 7 | 8
```

G. PRACTICE

Type each line 2 times.

Check your posture.

```
23 waxy wavy wave save rave raze razz jazz
24 ship whip whop shop stop atop atoms At:
25 rung rang sang sing ring ping zing zinc
26 cure pure sure lure lyre byre bytes By:

27 tale kale Kate mate late lace face faze
28 fond pond bond binds bins inns Inez In:
29 gaze game fame same sale dale daze haze
30 roam loam loom zoom boom books took To:
```

H. POSTTEST

Repeat the Pretest. Compare your Posttest results with your Pretest results.

Chapter 1 Review

Read to Succeed PRACTICE

After You Read

Summarize and Assess The skills and topics you learn about in this textbook are important to your life now, and in the future. Create a graphic organizer similar to the illustration. Rate each on its importance to you now and in the future by placing an X on the line between low and high.

Example:

Skill or Topic	Current Importance	Future Importance
What digital tools are	Low ←—X—→ High	Low ←———X→ High
How to manage files	Low ←—X—→ High	Low ←———X→ High

Using Key Terms

Every day it seems technology changes the workplace. You are learning cutting-edge technology skills. See how ready you are to apply them by matching the correct statement with the following key terms.

- software (7)
- operating system (10)
- hardware (10)
- digital (10)
- hard drive (15)
- input (20)
- output (20)
- external storage device (23)
- read-only memory (ROM) (24)
- e-portfolio (28)

1. All of the physical parts of a computer.
2. Information that is entered into a computer.
3. Memory that holds data that can be read by a digital device but cannot be changed or deleted.
4. Information that is represented as individual pieces of data using the numbers 1 and 0, rather than as a continuous stream.
5. A set of instructions, also called a program or application, that tells a computer how to perform tasks.
6. The built-in storage device inside a computer.
7. Software that controls all of the other software programs and allows the computer to perform basic tasks.
8. A collection of the best documents and projects you have created and saved in an electronic format.
9. Information that a computer produces or processes.
10. A storage device that can be easily removed from a computer and is often small and portable.

Lesson 12 — New Keys: Z Colon (:)

OBJECTIVE:
- Learn the Z and colon keys.

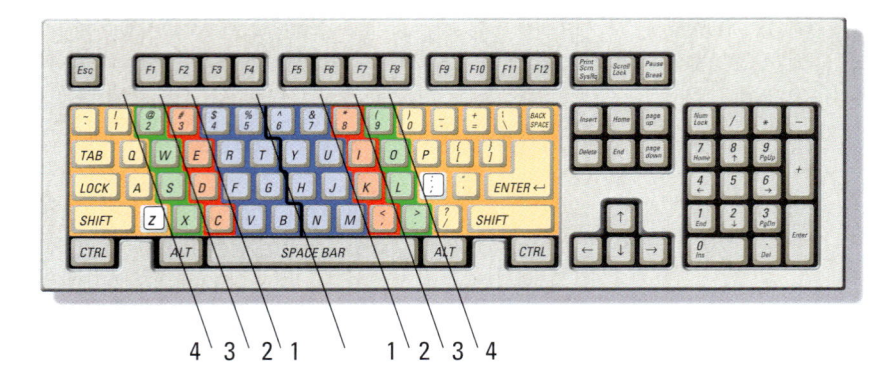

A. Warmup

Type each line 2 times.

1 bake chin jogs wave quip dome onyx left
2 His soft big lynx quickly jumped waves.

NEW KEYS

B. Z Key

Type each line 2 times. Repeat if time permits.

Use A finger.
Anchor F.

3 aaa aza aza zaz aza aaa aza aza zaz aza
4 aza zip zip aza zoo zoo aza zap zap aza
5 aza dozing zebu, he zags, dazed zebras,
6 Zachary ate frozen pizza in the gazebo.

C. : Key

Type each line 2 times. Repeat if time permits.

Shift of ;
Use left shift key.
Anchor J.
Space once after a colon.

7 ;;; ;:; ;:; ;:; ;:; ;;; ;:; ;:; ;:; ;;;
8 Dr. Webb: Mr. Que: Mrs. Downs: Ms. Lia:
9 Mr. Dode: Mrs. Chin: Ms. Finn: Dr. Mai:
10 To: From: Date: Subject: Attention: To:

SKILLBUILDING

D. Technique Checkpoint

Type each line 2 times. Repeat if time permits. Focus on the technique at the left.

Keep your elbows close to your body.

11 aaa aza aza zaz aza aaa aza aza zaz aza
12 ;;; ;:; ;:; ;:; ;:; ;;; ;:; ;:; ;:; ;;;
13 Zach and zany Hazel visited local zoos.
14 They saw: lazy zebras, apes, and lions.

Appendix A—Keyboarding Skills

CHAPTER 1 Self-Assessment

Take a moment to review what you have learned in this chapter. Rank your understanding of the topics below.

4 means, "I understand all of this."
3 means, "I understand some of this."
2 means, "I understand very little of this."
1 means, "I don't remember this."

To use a printout of this chart, go to **digicom.glencoe.com** and click on **Chapter 1, Self-Assessment.**
Or:
Ask your teacher for a personal copy.

Rank Your Understanding

Lesson	Topic	4	3	2	1
1.1	• List three ways of entering text without using a keyboard				
	• Compare and contrast different digital devices				
	• Explain the uses of handheld computers and Tablet PCs				
	• Describe what speech recognition software does				
	• Enter text and save a file				
1.2	• Explore the Control Panel				
	• Navigate drives, folders, and files				
	• Identify the software and hardware used by a computer				
	• Change how a mouse works				
1.3	• Create a new folder				
	• Explain the importance of keeping files organized				
	• Explain how computers use file extensions				
	• List three file extensions				
1.4	• Define digital				
	• Identify digital devices				
	• Explain input and output				
	• Identify four external input and output devices				
1.5	• List and describe the differences between two kinds of CDs				
	• Explain how megabytes (MB) relate to gigabytes (GB)				
	• Describe how to save a file to a USB flash drive and a CD-RW				
1.6	• Explain the importance of having technology skills				
	• Describe what e-portfolios are used for				

If you ranked all topics 4, congratulations! Consider doing a quick review.
If you ranked yourself 3 or lower in any topic, consider reviewing these topics first.

digicom.glencoe.com

SKILLBUILDING

D. Technique Checkpoint

Type each line 2 times. Focus on the technique at the left. Repeat if time permits.

Keep your eyes on the copy.

```
11 jjj jyj jyj yjy jyj jjj jyj jyj yjy jyj
12 I saw yards of yellow fabric every day.
13 They happily played in the lonely yard.
14 Yes, the daily reports are ready today.
```

E. Technique Timings

Take two 30-second timings on each line. Focus on the technique at the left.

Keep your eyes on the copy as you take each timing.

```
15 Push your fingers to find the keys now.
16 You will see your typing speed improve.
17 Have a goal to type faster than before.
18 Try every day to achieve that new goal.
   | 1 | 2 | 3 | 4 | 5 | 6 | 7 | 8
```

F. PRETEST

Take a 1-minute timing on lines 19–22. Note your speed and errors.

Remember: Press ENTER only at the end of the paragraph (line 22).

```
19      A jury will meet next January to       7
20 get a verdict. People stole costly fuel    15
21 from the boys. We found bags of cards      22
22 next to the mops in the broom closet.      30
   | 1 | 2 | 3 | 4 | 5 | 6 | 7 | 8
```

G. PRACTICE

Type each line 2 times.

```
23 fuel duel duet suet suit quit quip quid
24 gape nape cape cave wave wage wags bags
25 mops pops maps hops tops toys joys boys
26 rope lope lops laps lips lids kids kiss

27 card cart curt hurt hurl furl fury jury
28 cost most lost lest best test text next
29 slab flab flap flaw flay slay clay play
30 pan, fan, tan, man, can, ran, Dan, Jan,
```

H. POSTTEST

Repeat the Pretest. Compare your Posttest results with your Pretest results.

CHAPTER 2: Communicating with Digital Technology

Lesson 2.1   Digital Image Technology

Lesson 2.2   Communicating Online

Lesson 2.3   Internet Security and Acceptable Use

Lesson 2.4   Ethics and Netiquette

You Will Learn To:
- **Describe** the uses of digital cameras and camcorders.
- **Explore** different webcams on the Internet.
- **Describe** weblogs and begin planning your own.
- **Explain** the uses of e-mail.
- **Describe** an acceptable use policy.
- **Identify** security risks of computer use.
- **Explain** ethical uses of technology and netiquette.
- **Define** ethics and netiquette.

When Will You Ever Use Digital Technologies?

For many people, using digital technology is just another part of everyday life. Have you used a cell phone? How about a computer or an MP3 player? If so, you have already developed some impressive digital technology skills. It is important to keep learning new skills, because digital technology is constantly changing in new and exciting ways.

Taking this course is one way you will keep up with the new digital technology skills that are important for the 21st century student. You will learn to use new technologies, to include PDAs, Tablet PCs, and speech recognition.

Enjoy your digital journey!

Lesson 11 — New Keys: Y Tab

OBJECTIVE:
- Learn the Y and the tab keys.

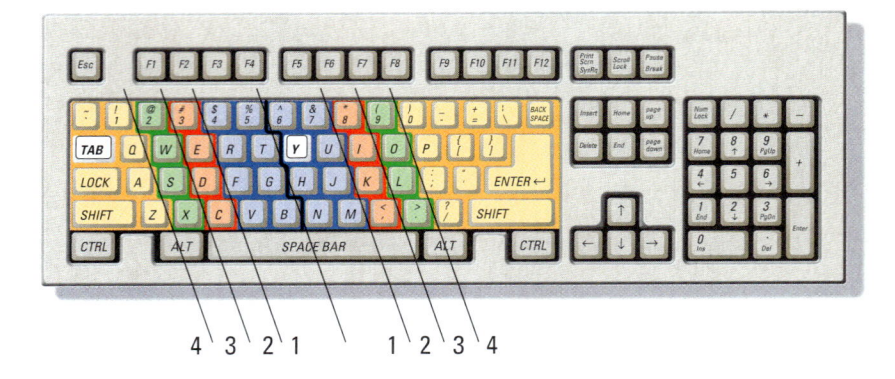

A. Warmup

Type each line 2 times.

1 jibe wing more vase deft lack hex; quid
2 Max just put a pale slab over the gate.

NEW KEYS

B. Y Key

Type each line 2 times. Repeat if time permits.

Use J finger.
Anchor ; L, and K.

3 jjj jyj jyj yjy jyj jjj jyj jyj yjy jyj
4 jyj yes yes jyj joy joy jyj aye aye jyj
5 jyj yard of yarn, July joy, yellow yam,
6 Shelley yearns to yodel but only yells.

C. Tab Key

The tab key is used to indent paragraphs. The tab key is located to the left of the Q key. Reach to the tab key with the A finger. Keep your other fingers on home keys as you quickly press the tab key.

Pressing the tab key will move the cursor 0.5 inch (the default setting) to the right.

Word wrap automatically moves a word that does not fit on one line down to the next line.

Type each paragraph 2 times. Press Enter only at the end of a paragraph. Repeat if time permits.

7   If you are happy, you will be able
8 to set goals. You will also smile more.
9   The jury was out and no one could
10 leave the room. We all had to stay put.

Appendix A—Keyboarding Skills

Read to Succeed

Key Terms

- resolution
- pixel
- weblog
- webcam
- firewall
- ethics
- netiquette

Use the Glossary

Throughout this book, look for the key terms highlighted in a lesson. You will see that the key terms are highlighted in the margin, too. The definition is included in the margin. This definition is also listed in the glossary at the back of the book. If you need to quickly find the definition of a key term, you can look it up in the Glossary, pages G1–G4.

As you read through Chapter 2, write each key term on the front of a 3″ by 5″ card. On the back of the card write the definition. You can find the definition in the margin, or look it up in the glossary. You will use these cards in the reading activity Trade-a-Question at the end of Chapter 2, page 56.

Front of card: pixel

Back of card: A single point in a digital image.

DIGITAL CONNECTION

Everyday Technology

Human beings have always invented new kinds of technology. The wheel was a technological invention. So were the sundial, the printing press, and the refrigerator. It may be hard to imagine, but those things were pretty high-tech for their time. Today, we use an amazing variety of technological tools. Stereo systems, radios, telephones, televisions, microwave ovens, CD players, security systems, and computers are all part of our daily lives. Computers are especially useful because they can be used for many different purposes. You can use them to communicate instantly with friends and family, find information, listen to music and watch movies, play video games, take pictures, and create home movies.

Writing Activity

Make a list of some of the technological tools that you use every day. Your list could include everything from an alarm clock to a microwave oven to a cell phone. Now imagine how your life would be different without technology. Write a short paragraph describing what you would need to do differently if these tools did not exist.

SKILLBUILDING

E. Technique Timings

Take two 30-second timings on each line. Press ENTER at the end of each sentence. Focus on the technique at the left.

Keep your rhythm steady as you reach to the ENTER key and back to home position.

```
15  Pull on the tabs.↵  The box will open.↵
16  Speed is good.↵  Errors are not good.↵
17  Glue the picture.↵  The book is done.↵
18  Get the clothes.↵  Bring me their caps.↵
     | 1 | 2 | 3 | 4 | 5 | 6 | 7 | 8
```

F. PRETEST

Take a 1-minute timing on lines 19–20. Note your speed and errors.

```
19  slag chop gate plop tops bows veal dart      8
20  apex slab gave quit fix, hoax text jell     16
     | 1 | 2 | 3 | 4 | 5 | 6 | 7 | 8
```

G. PRACTICE

Type each line 2 times.

To type faster:
- *Read copy before typing.*
- *Type with smooth strokes.*

```
21  slag flag flap flax flux flex Alex apex
22  chop clop clap clan claw slaw slap slab
23  gate gale pale page pave have cave gave
24  plop flop flip slip ship whip quip quit

25  tops tips sips sits sit, six, mix, fix,
26  bows bowl jowl howl cowl coal coax hoax
27  veal real seal meal meat neat next text
28  dart part park bark balk ball bell jell
```

H. POSTTEST

Repeat the Pretest. Compare your Posttest results with your Pretest results.

Lesson 2.1 Digital Image Technology

You will learn to:
- **Describe** how digital cameras, digital camcorders, and DVD recorders are used.
- **Describe** how resolution affects digital images.
- **Create** a scrapbook using digital photographs.
- **Print** digital images from a folder.

resolution The amount of detail in a digital image, measured in pixels.

pixel A single point in a digital image.

Digital Cameras With a digital camera, you can take as many photos as you want and simply delete the ones you do not like. Digital cameras connect directly to computers, so you can edit and e-mail photos. With some cameras, you can even make paper prints right from the camera.

Resolution One big difference among digital cameras is resolution. The **resolution** is the amount of detail in a digital image, measured in pixels. A **pixel** is a single point in a digital image. More pixels mean better resolution, which means better picture quality.

Digital Camcorders Digital camcorders are used to record media clips and movies. You can then edit the video and add special effects on a computer. A digital camcorder and computer may be all you need to get your start as a movie director!

DVD Recorders DVD recorders let you record and save television shows, copy home movies from VHS tapes, play music CDs, store digital pictures, and much more.

✓ Concept *Check*

What can you do with a digital camera that is not possible with a regular camera?

DIGITAL DIALOG

Digital Camcorders and DVD Recorders

Teamwork You have been hired by the technology department of your school district. The department wants to purchase a DVD recorder and digital camcorder for each school. Work in a group to search the Internet for the best price on each device.

Writing Create a list that includes the price, model name, and features found on each device. Compare your list to those of the other groups. Which model is best? Explain why you chose that model.

Lesson 10 — **New Keys: P X**

OBJECTIVE:
- Learn the P and X keys.

4 3 2 1   1 2 3 4

A. Warmup

Type each line 2 times.

1 fade cave what swim quad blot king jars
2 Black liquids vanish from the jug I saw.

NEW KEYS

B. P Key

Type each line 2 times. Repeat if time permits.

Use Sem finger.
Anchor J and K.

3 ;;; ;p; ;p; p;p ;p; ;;; ;p; ;p; p;p ;p;
4 ;p; nap nap ;p; pen pen ;p; ape ape ;p;
5 ;p; perfect plot, a pale page, pen pal,
6 Pam pulled a pouting pup past a puddle.

C. X Key

Type each line 2 times. Repeat if time permits.

Use S finger.
Anchor A or F.

7 sss sxs sxs xsx sxs sss sxs sxs xsx sxs
8 sxs tax tax sxs mix mix sxs axe axe sxs
9 sxs lax taxes, vexed vixen, six Texans,
10 Fix the next six boxes on next weekend.

SKILLBUILDING

D. Technique Checkpoint

Type each line 2 times. Focus on the techniques at the left.

Remember to keep:
- Wrists up.
- Fingers curved.
- Feet flat on the floor.

11 ;;; ;p; ;p; p;p ;p; ;;; ;p; ;p; p;p ;p;
12 sss sxs sxs xsx sxs sss sxs sxs xsx sxs
13 Phil will fix ripped carpets alone now.
14 Go see that duplex before next weekend.

Appendix A—Keyboarding Skills

Step-by-Step

Activity 1

If you need additional help on this topic, open the Help menu; or refer to the documentation for your hardware.

Take and Upload Digital Photographs

1. Use a digital camera to take several pictures around your school. For example, you could take pictures of the auditorium, classrooms, school mascot, or special features on your campus. Take notes about each picture.

2. Upload or copy your pictures to the Pictures folder that you created on the desktop in Lesson 1.5, Activity 14, page 25.

   If your digital camera has a USB port, connect the camera directly to your computer's USB port. Open the icon that appears, and select the pictures you want to upload.

   If your camera uses a removable storage device, such as a memory card, place the card in the appropriate slot of your computer to copy the pictures to your folder. Open the icon that appears, and select the pictures you want to upload.

Activity 2

View and Rename Photos

1. Open the **Pictures** folder on your desktop where you saved the pictures.

2. To see small versions, or thumbnails, of your pictures without opening them, click **View**, and then click **Filmstrip**. You can also click **View**, and then click **Thumbnail**.

3. Choose the picture you want to use for your scrapbook page, and click on it.

4. Click **Rename this file** from the menu at the left side of the window.

5. Enter a descriptive name for the picture in the blue highlighted area under the picture.

6. Click **Edit** in the Menu bar, and then click **Copy**.

Rename This File

Photo name

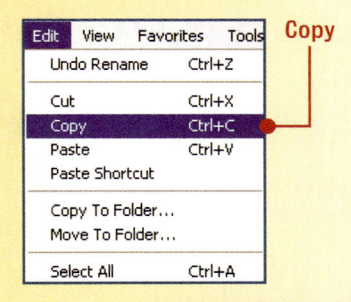

Copy

Continued

Chapter 2

38

SKILLBUILDING

E. Technique Timings

Take two 30-second timings on each line. Focus on the technique at the left.

Keep your eyes on the copy.

```
15  Robb's clothes and image don't "match."
16  Mr. Quill said, "Wait." Lee did not go.
17  Jane's visit was "quick"; she ran back.
18  I haven't enough time to "quibble" now.
    | 1 | 2 | 3 | 4 | 5 | 6 | 7 | 8
```

F. 12-Second Sprints

Take three 12-second timings on each line. Try to increase your speed on each timing.

```
19  Go to the cabin and get us the dog now.
20  Now is the time to call all men for me.
21  She made a face when she lost the race.
22  Ask them if the vase is safe with them.
    |  |  | 5 |  |  |  | 10 |  |  | 15 |  |  | 20 |  |  | 25 |  |  | 30 |  |  | 35 |  |  | 40
```

G. PRETEST

Take a 1-minute timing on lines 23–24. Note your speed and errors.

```
23  We can't "remember" how Bo got bruised.         8
24  Burt's dad "asked" Kurt to assist Ross.        16
    | 1 | 2 | 3 | 4 | 5 | 6 | 7 | 8
```

H. PRACTICE

Type each line 2 times.

```
25  made fade face race lace lice nice mice
26  Burt Nora Will Mame Ross Kurt Olaf Elle
27  he's I've don't can't won't we've she's
28  Bo's dogs Lu's cows Mo's cats Di's rats

29  "mat" "bat" "west" "east" "gone" "tone"
30  He "quit"; she "tried." I hit a "wall."
31  sand/land vane/cane robe/lobe quit/suit
32  asks bask base vase case cast mast last
```

I. POSTTEST

Repeat the Pretest. Compare your Posttest results with your Pretest results.

Step-by-Step continued

Activity 3 — Create a Scrapbook Page

1. Start the WordPad application.
2. Click at the beginning of the WordPad document. Click **Edit**, and then click **Paste**. Your picture will appear in your document.
3. Click in the area to the right of your picture. Press ENTER.
4. Find the information about this photo that you wrote in your notebook. Key that information, and anything else you would like to add, in the space next to the photo.
5. Save your document, and close WordPad.

Activity 4 — Print Pictures

1. Open the **Pictures** folder.
2. On the command list at the left, choose **Print Pictures**. The Welcome to the Photo Printing Wizard window will open.

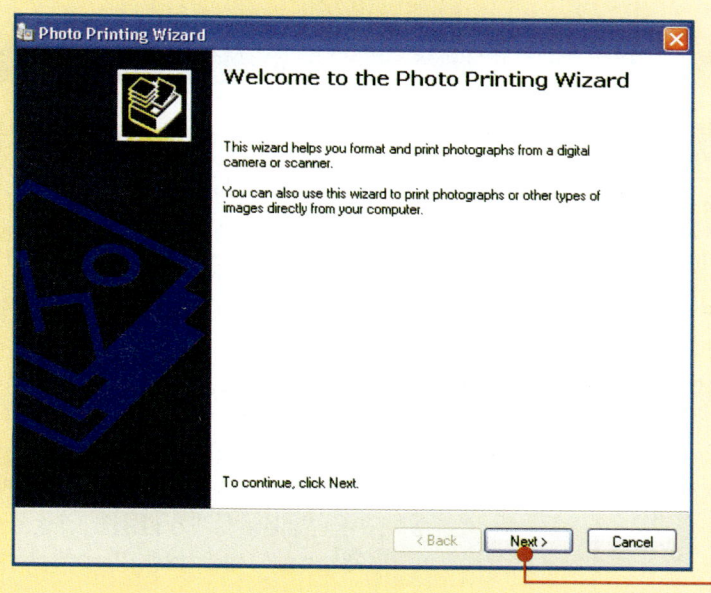

—Next

3. Click **Next**. You can print all the pictures by selecting all of them, or you can choose the ones you want to print.

Suite Smarts

Digital pictures usually have a file extension of .jpg or .jpeg, pronounced "dot jay-peg."

RUBRIC LINK

Go to the Online Learning Center at **digicom.glencoe.com** and click Lesson 2.1, Rubric, to find the application assessment guide(s).

Application 1

Directions Copy each of your other digital pictures into a new WordPad document. Key a description below each picture. Save each file with a name you choose. Print one of the pictures.

Chapter 2 Communicating with Digital Technology — digicom.glencoe.com — Lesson 2.1

Lesson 9 — New Keys: ' "

OBJECTIVES:
- Learn the apostrophe (') and the quotation mark (") keys.
- Improve speed and accuracy.

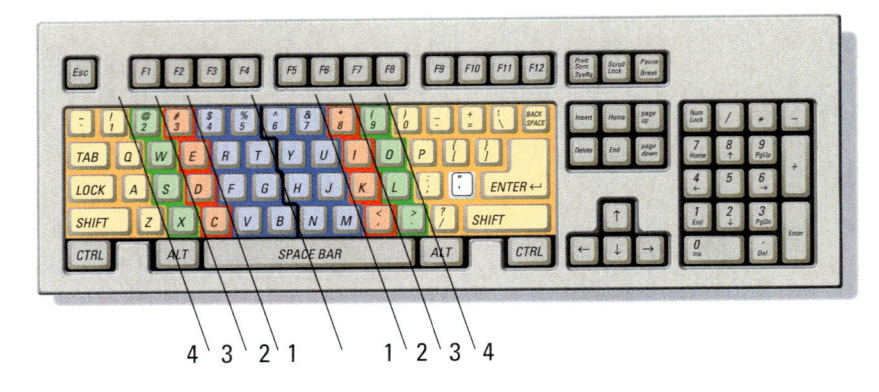

A. Warmup

Type each line 2 times.

1 quill wagon cabin valued helms, and/or;
2 Jake is quite good in math but not Val.

NEW KEYS

B. ' Keys

Type each line 2 times. Repeat if time permits.

Use Sem finger.
Anchor J.
Do not space before or after an apostrophe within a word.

3 ;;; ;'; ;'; ';' ;'; ;;; ;'; ;'; ';' ;';
4 ;'; he's he's ;'; where's ;'; it's it's
5 ;'; ';' Kit's barn ;'; Ed's car ;'; ';'
6 Bill's car isn't running; it's at Li's.

C. " Keys

Type each line 2 times. Repeat if time permits.

Shift of apostrophe.
Use Sem finger.
Anchor J.

7 ;;; ;"; ;"; ";" ;"; ;;; ;"; ;"; ";" ;";
8 ;"; "win" "win" ;"; "big" "big" ;"; ";"
9 ;"; "mew" "oink" "woof" "moo" "baa" ";"
10 "Green" means "go"; "red" means "wait."

SKILLBUILDING

D. Technique Checkpoint

Type each line 2 times. Repeat if time permits. Focus on the technique at the left.

Keep your eyes on the copy.

11 ;;; ;'; ;'; ';' ;'; ;;; ;'; ;'; ';' ;';
12 ;;; ;"; ;"; ";" ;"; ;;; ;"; ;"; ";" ;";
13 He said "no thanks," but it was "lame."
14 Rita "forgot," but Milo added "favors."

Appendix A—Keyboarding Skills

Lesson 2.2 — Communicating Online

You will learn to:
- **Define** weblogs.
- **Explain** how to send e-mail and add attachments.
- **Describe** how webcams are used.

weblog An online journal that is updated almost every day by its author.

webcam A small camera that connects to a computer and sends an image over the Web.

The Weblog Revolution
In its early years, the Internet was a high-tech tool used mainly by the government. Now it is so user-friendly that anyone can use the Web to share opinions and information with others. Millions of people have created weblogs, or blogs. A **weblog** is an online journal that is updated almost every day by its author. Some weblogs are read only by a small circle of online friends. Others have thousands of readers.

E-Mail
E-mail, which stands for electronic mail, makes it possible for computer users to communicate all over the world with just the click of a button! Messages can be sent to just one person or many people at once. A file, such as a newsletter, digital photo, or spreadsheet budget, can be attached to an e-mail message.

Listservs
Listservs are large e-mail groups that are created because the members all share a common interest. Anyone can join a listserv. The members of the listserv e-mail information and opinions on the listserv's topic.

Webcams
Thanks to webcams, you can turn on your computer and see what is happening all over the world. A **webcam** is a small camera that connects to a computer and sends an image over the Web. Some people use webcams for one-on-one video chats. Others use them for educational purposes, such as putting a webcam in the elephant enclosure at a zoo. Webcams are also used for entertainment, such as broadcasting concerts over the Web. Anyone with Internet access can view an enormous variety of webcam sites, including sites like these:

- Alamo Cam.
- Bonaire Reef Cam.
- Fall Foliage Cams.
- Mount St. Helens Volcano Cam.
- Philadelphia International Airport.

✅ Concept Check
How has the Internet changed the way we communicate? What is the purpose of a weblog? Give uses for a webcam.

DIGI Byte

Voice Chats
More and more people are using their computers to have voice chats with other computer users. Some experts think that in the future, computers could replace telephones completely.

SKILLBUILDING

D. Technique Checkpoint

Type each line 2 times. Repeat if time permits. Focus on the technique at the left.

Keep fingers curved and wrists level.

```
11  aaa aqa aqa qaq aqa aaa aqa aqa qaq aqa
12  ;;; ;/; ;/; /;/ ;/; ;;; ;/; ;/; /;/ ;/;
13  The quick squash squad requested quiet.
14  He/she said that we could do either/or.
```

E. Technique Timings

Take two 30-second timings on each line. Focus on the technique at the left.

Keep your eyes on the copy.

```
15  Louise will lead if she makes the team.
16  Their bands will march at the quadrant.
17  Brad just had time to finish his goals.
18  I was quiet as he glided over the wave.
    | 1 | 2 | 3 | 4 | 5 | 6 | 7 | 8
```

F. PRETEST

Take a 1-minute timing on lines 19–20. Note your speed and errors.

Hold those anchors.

```
19  find/seek boat fate jail cube brad swat      8
20  walk shut quid mile vane aqua slot quit     16
    | 1 | 2 | 3 | 4 | 5 | 6 | 7 | 8
```

G. PRACTICE

Type each line 2 times. Repeat if time permits.

To build skill:
- Type each line two times.
- Speed up the second time you type the line.

```
21  find/lose cats/dogs hike/bike walk/ride
22  seek/hide soft/hard mice/rats shut/ajar
23  boat goat moat mode rode rude ruin quid
24  fate face race rice nice Nile vile mile

25  jail fail fall gall mall male vale vane
26  cube Cuba tuba tube lube luau quad aqua
27  brad brat brag quag flag flat slat slot
28  swat swam swim slim slid slit suit quit
```

I. POSTTEST

Repeat the Pretest. Compare your Posttest results with your Pretest results.

Activity 5

If you need additional help on this topic, open the Help menu; or refer to the documentation for your hardware.

TechSIM™

Interactive Tutorials
Ask your teacher about a simulation that is available on this topic.

Step-by-Step

Create an E-Mail

1. Start the Microsoft Outlook or Microsoft Outlook Express application.

   You can read messages, send messages, and delete messages from the main window.

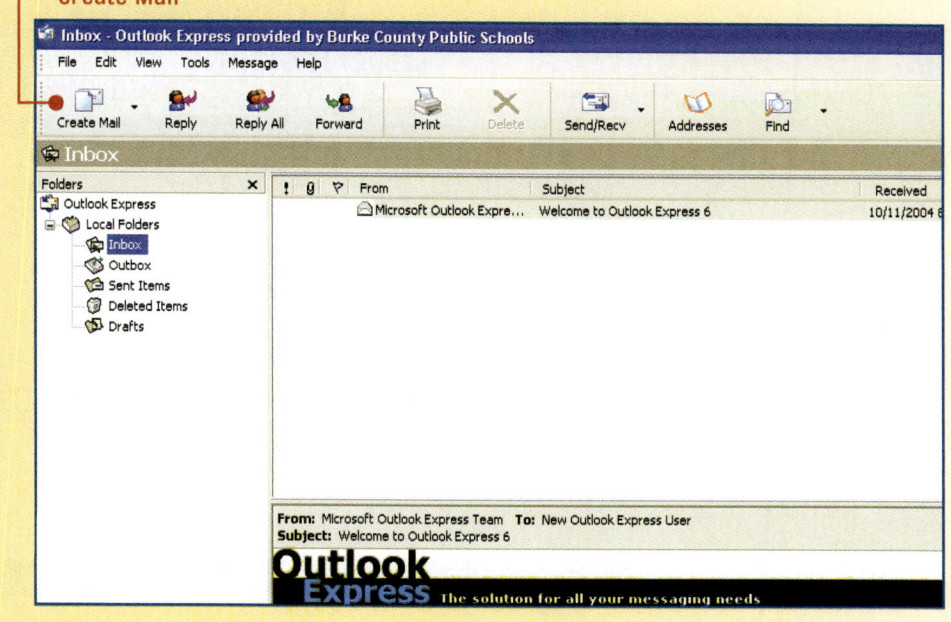

2. Click **Create Mail**. The New Message window will open.

3. Enter an e-mail address on the **To:** line. If you are sending the e-mail to more than one person, separate the e-mail addresses with a semicolon.

4. To send a copy of the message to someone else who does not need to reply to the message, enter his or her e-mail address on the **Cc:** line. This stands for carbon copy.

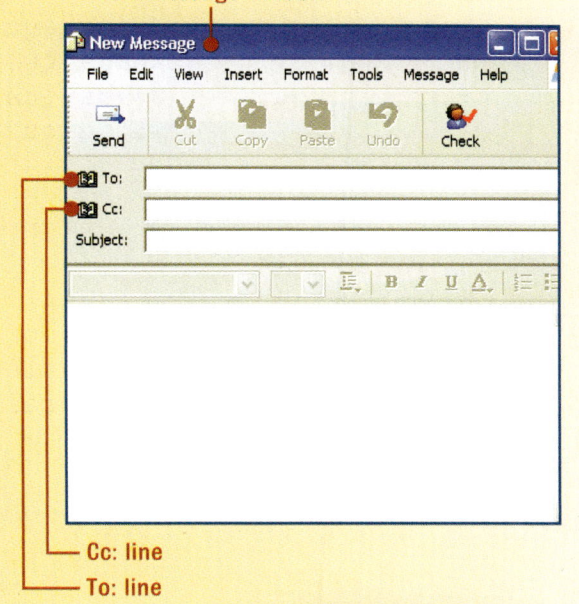

Continued

Chapter 2

41

Lesson 8 — New Keys: Q /

OBJECTIVE:
- Learn the Q and / (slash or diagonal) keys.

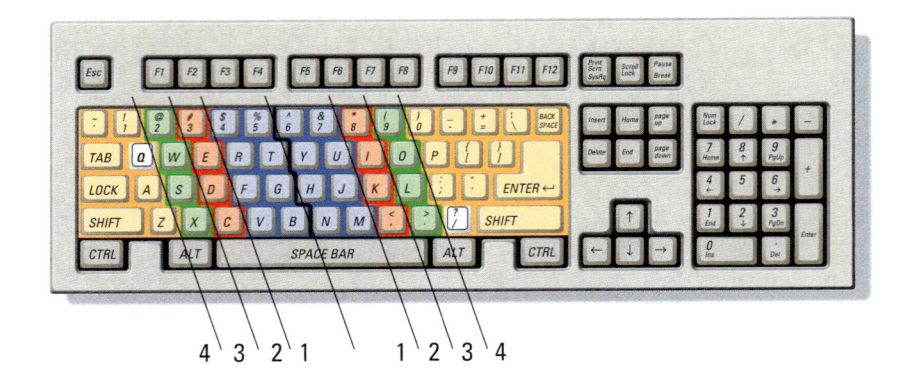

A. Warmup

Type each line 2 times.

1 club face when silk mold brag java blue
2 Jana went biking, and Cila waved flags.

NEW KEYS

B. Q Key

Type each line 2 times. Repeat if time permits.

Use A finger.
Anchor F.

3 aaa aqa aqa qaq aqa aaa aqa aqa qaq aqa
4 aqa quo quo aqa qui qui aqa que que aqa
5 aqa quail, quit quick quid, half quest,
6 The quints squabbled on a square quilt.

C. / Key

Type each line 2 times. Repeat if time permits.

Use Sem finger.
Anchor J.
Do not space before or after a slash (diagonal).

7 ;;; ;/; ;/; /;/ ;/; ;;; ;/; ;/; /;/ ;/;
8 ;/; her/him ;/; us/them ;/; his/her ;/;
9 ;/; slow/fast, walk/ride, debit/credit,
10 The fall/winter catalog has new colors.

Appendix A—Keyboarding Skills A19

Step-by-Step *continued*

5 On the **Subject** line, enter the reason for your e-mail. This will let the recipients of your e-mail know what you are writing about right away.

6 In the main body of the e-mail, enter the text of your message.

Activity 6

Set Priority

If you want the recipient to know that a message is urgent, you can set the priority level to high.

1 Click **Message, Set Priority, High**.

2 Priority can also be set from the toolbar.

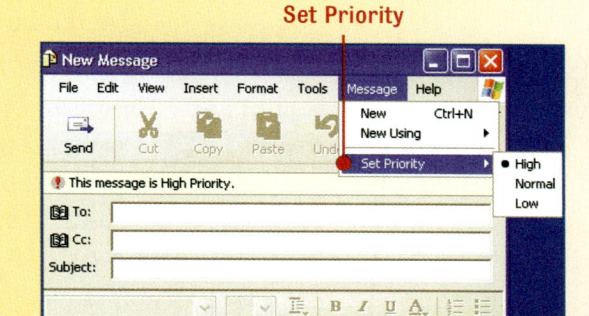

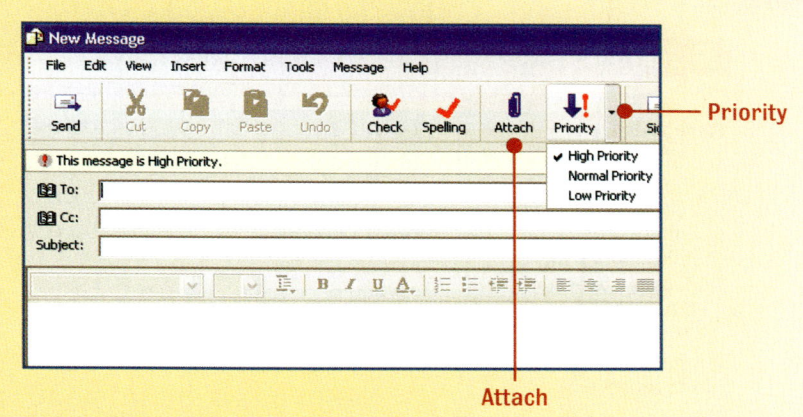

Activity 7

Insert an Attachment

1 A file, such as a letter or picture, can be attached to an e-mail message. Click the **Attach** button on the toolbar or click **Insert, File Attachment**.

2 From the **Insert Attachment** window that opens, locate the file you want to attach.

Continued

Chapter 2

42

SKILLBUILDING

E. Technique Checkpoint

Type each line 2 times. Repeat if time permits. Focus on the technique at the left.

Keep F or J anchored when shifting.

```
15  fff fbf fbf bfb fbf fff fbf fbf bfb fbf
16  jjj juj juj uju juj jjj juj juj uju juj
17  aaa Kaa Kaa aaa Jaa Jaa aaa Laa Laa aaa
18  Jo told Mike and Nel that she would go.
```

F. PRETEST

Take a 1-minute timing on lines 19–20. Note your speed and errors.

```
19  bran gist vast blot sun, bout just beef       8
20  craw just rest bran sum, dole hunk bear      16
    |  1  |  2  |  3  |  4  |  5  |  6  |  7  |  8
```

G. PRACTICE

Type each line 2 times. Repeat if time permits.

Place your feet:
- In front of the chair.
- Firmly on the floor, square, flat.
- Apart, with 6 or 7 inches between the ankles.
- One foot a little ahead of the other.

```
21  bran brad bred brew brow crow crew craw
22  gist list mist must gust dust rust just
23  vast vest jest lest best west nest rest
24  blot blob blow blew bled bred brad bran

25  sun, nun, run, bun, gun, gum, hum, sum,
26  bout boat boot blot bold boll doll dole
27  just dust dusk dunk bunk bulk hulk hunk
28  beef been bean bead beak beam beat bear
```

I. POSTTEST

Repeat the Pretest. Compare your Posttest results with your Pretest results.

Step-by-Step *continued*

3 Click the file name to select it. Click **Attach**. The file will be attached to your e-mail.

4 After you compose your message in the message area, click **Send** to deliver your e-mail.

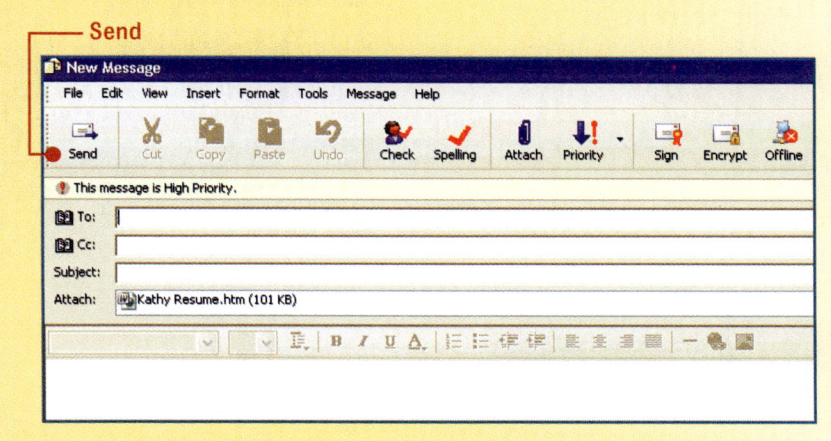

Activity 8

Delete an E-Mail Message

1 Select the message in the window.

2 Press DELETE or click the **Delete** button on your e-mail toolbar.

E-mail messages that are created and sent by you are kept in the **Sent** folder. Deleted messages are kept in the **Deleted** folder.

Activity 9

Read and Reply to an E-Mail Message

1 When you get a message, it will appear in your Inbox. Double-click on the message to read it.

2 To reply to an e-mail, select the message. Click the **Reply** button to send the reply to the person who sent you the message. Click the **Reply to All** button to send the reply to everyone in the To: and Cc: lines.

Continued

Chapter 2
43

Lesson 7

New Keys: B U Left Shift

OBJECTIVE:
- Learn the B, U, and Left Shift keys.

A. Warmup

Type each line 2 times.

1 dim logo wags jive foal corn them wags,
2 Wanda mailed the jewels that Carl made.

NEW KEYS

B. B Key

Type each line 2 times. Repeat if time permits.

Use F finger.
Anchor A and S.

3 fff fbf fbf bfb fbf fff fbf fbf bfb fbf
4 fbf rob rob fbf ebb ebb fbf bag bag fbf
5 fbf a bent bin, a back bend, a big bag,
6 That boat had been in a babbling brook.

C. U Key

Type each line 2 times. Repeat if time permits.

Use J finger.
Anchor ; L and K.

7 jjj juj juj uju juj jjj juj juj uju juj
8 juj jug jug juj urn urn juj flu flu juj
9 juj jungle bugs, just a job, jumbo jets
10 Students show unusual business success.

D. Left Shift Key

Type each line 2 times. Repeat if time permits.

Use A finger.
Anchor F.

11 aaa Kaa Kaa aaa Jaa Jaa aaa Laa Laa aaa
12 aaa Kim Kim aaa Lee Lee aaa Joe Joe aaa
13 aaa Jan left; Nora ran; Uncle Lee fell;
14 Mari and Ula went to Kansas in October.

Step-by-Step continued

3. A window will open with the e-mail address information already entered. The original message is also in the reply. Enter the reply message, and then click **Send**.

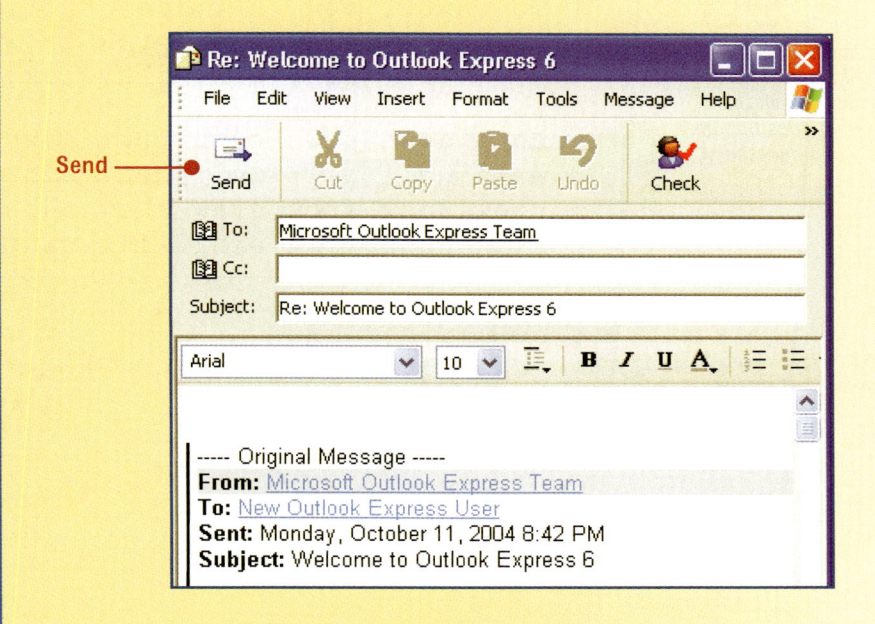

Activity 10

Forward an E-Mail Message

If you receive a message and you want to share it with other friends, you can forward an e-mail.

1. Click the message in your **Inbox** to select it.
2. Click the **Forward** button.

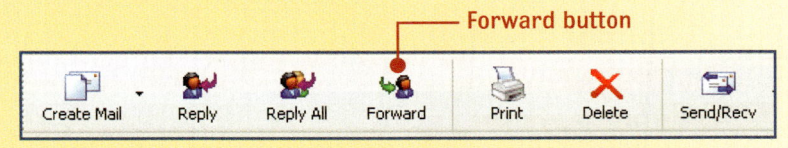

3. Enter the e-mail addresses of the people you want to receive the forwarded message in the **To:** line.
4. Click **Send**.

Activity 11

Explore Weblogs

1. Go to a search engine, and key the words "education weblogs."
2. Choose a weblog to visit. You may need to try several weblogs before you find one that interests you.

Continued

Chapter 2

SKILLBUILDING

H. PRETEST

Type each line 2 times. Repeat if time permits.

Focus on:
- Wrists up; do not rest palms on keyboard.
- Fingers curved; move from the home position only when necessary.

```
24  sag, sag, wag, wag, rag, rag, hag, hag,
25  mow, mow, how, how, hot, hot, jot, jot,
26  crew crew grew grew grow grow glow glow
27  elf, elf, elk, elk, ilk, ilk, ink, ink,

28  down down gown gown town town tows tows
29  well well welt welt went went west west
30  scow scow stow stow show show snow snow
31  king king sing sing wing wing ring ring
```

I. POSTTEST

Repeat the Pretest. Compare your Posttest results with your Pretest results.

Appendix A—Keyboarding Skills

DIGI*Byte*

Blogs
Weblogs are also known as blogs.

Activity 12

RUBRIC LINK
Go to the Online Learning Center at **digicom.glencoe.com** and click Lesson 2.2, Rubric, to find the application assessment guide(s).

Step-by-Step *continued*

3. Start WordPad. Key a summary of what you found on the weblog you visited.
   - Would you read this weblog again? Why or why not?
   - Describe the most interesting entry in the education weblog.
   - Explain how the information on the weblog could be useful to you or to others.
   - Identify the intended audience for this weblog.

Explore Webcams

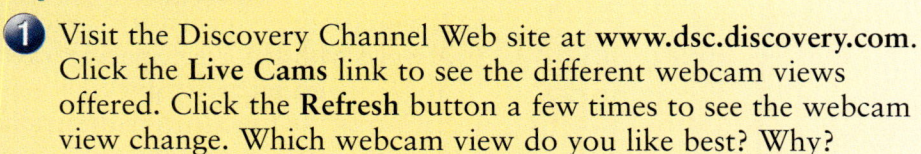

1. Visit the Discovery Channel Web site at **www.dsc.discovery.com**. Click the **Live Cams** link to see the different webcam views offered. Click the **Refresh** button a few times to see the webcam view change. Which webcam view do you like best? Why?

2. Visit the San Diego Zoo Web site at **www.sandiegozoo.org**. Click the **Panda Cam** link to watch the pandas, or a similar link.

Application 2

Directions Think about how e-mail, weblogs, and webcams may be especially useful to scientists and students. Create an e-mail or a WordPad document to write a summary of your thoughts. Give specific examples of how scientists could benefit from each of these technologies. For example, if you were a scientist, on what kind of project could you use a webcam? How, and in what settings, would students be able to make similar uses of these technologies? Use all of the e-mail skills you have learned in this lesson to share your document with the class.

DIGITAL DIALOG

Weblog Planning

Teamwork Work in a small group. Your school has asked you to create a school weblog to post calendar events and keep students informed. With your group, brainstorm ideas for the weblog. What kinds of information could be included? What kind of news would people want to know about?

Writing Each person in the group will write a sample post on any subject for the new weblog. Include a headline, and it can be one or two paragraphs long.

SKILLBUILDING

E. Technique Checkpoint

Technique Checkpoint Type each line 2 times. Repeat if time permits. Focus on the techniques at the left.

Hold anchor keys.
Keep elbows in.

```
15  sss sws sws wsw sws sss sws sws wsw sws
16  kkk k,k k,k ,k, k,k kkk k,k k,k ,k, k,k
17  fff fgf fgf gfg fgf fff fgf fgf gfg fgf
18  Wanda watched the team jog to the glen.
```

F. Counting Errors

Count 1 error for each word, even if it contains several errors. Count as an error:

1. A word with an incorrect character.
2. A word with incorrect spacing after it.
3. A word with incorrect punctuation after it.
4. Each mistake in following directions for spacing or indenting.
5. A word with a space.
6. An omitted word.
7. A repeated word.
8. Transposed (switched in order) words.

Compare these incorrect lines with the correct lines (19–21). Each error is highlighted in color.

```
Frank sold sold Dave old an washing mshcone .
     Carl joked with Al ice, Fran, Edith.
Wamda wore redsocks; Sadie wore green,
```

Type each line 2 times. Proofread carefully and note your errors.

```
19  Frank sold Dave an old washing machine.
20  Carl joked with Alice, Fran, and Edith.
21  Wanda wore red socks; Sadie wore green.
```

G. PRETEST

Take a 1-minute timing on lines 22–23. Note your speed and errors. Keep your eyes on the copy.

```
22  sag, mow, crew elf, down well scow king     8
23  hag, jot, glow ink, tows west snow ring    16
    |  1  |  2  |  3  |  4  |  5  |  6  |  7  |  8
```

Appendix A—Keyboarding Skills A15

21st Century Connection

Digital Cameras

Do you want to e-mail a photo to a friend, include it in a presentation, or post it on a Web site? This is easy to do when you use a digital camera. Digital cameras store pictures as digital files that you can transfer to your computer and then immediately use in a presentation or e-mail.

What Happens When You Take a Photo? When you take a photo using a digital camera, a small computer processor inside the camera records the image in an electronic form. An image is made up of pinpoint-sized dots of color called pixels. The image is stored on a memory device, such as a memory card, instead of on film. The image is usually saved with the file extension .jpg, pronounced "dot jay-peg".

Some cameras connect to a desktop computer to upload the pictures. Memory cards can be inserted directly into some printers so images can be printed without uploading.

Video Many digital cameras are capable of taking short video clips, which can be posted on Web sites or shared through e-mail.

Power or Batteries? Camera batteries are rechargeable. As with cell phones, be sure to fully charge the battery before using the camera.

Activity What are four benefits of taking pictures with a digital camera rather than with a film camera?

Resolution

What to Look For When You Shop The resolution of the digital camera is the most important consideration, as higher resolution means higher image quality. Resolution is measured in megapixels.

What Do You Want to Do with Your Photos?	Resolution Needed	Advantages	Disadvantages
E-mail to friends or post on a Web site	Low resolution—about 1 megapixel	• Camera is less expensive • Images take less storage space • Images take less time to transfer to your computer	• Images are not print quality
Print or include in presentations	Medium resolution—about 2 megapixels	• Choice of printing, compressing and sending by e-mail, or posting on a Web site	• Camera is more expensive • Images take up more storage space
Print enlarged photos, such as 8 by 10 inches	High resolution—3.5 megapixels or more	• Very high image quality	• Camera is much more expensive, up to several thousand dollars

Lesson 6 — New Keys: W Comma (,) G

OBJECTIVES:
- Learn the W, comma, and G keys.
- Learn the spacing with a comma.

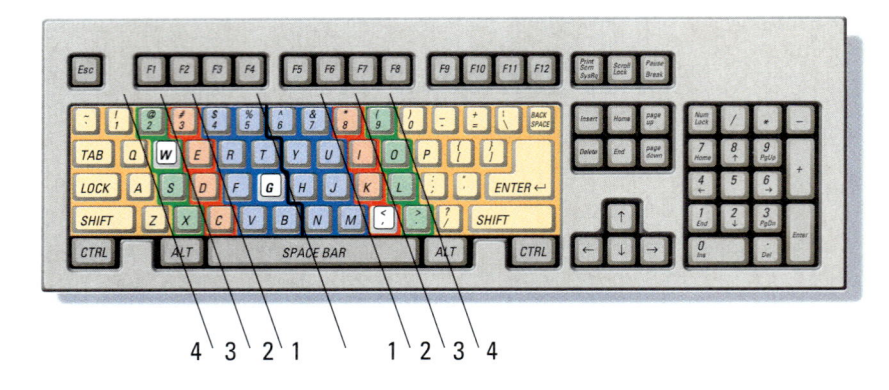

A. Warmup

Type each line 2 times.

1 fail not; jest mist chin Rev. card sake
2 Rick did not join; Val loves that fame.

NEW KEYS

B. W Key

Type each line 2 times. Repeat if time permits.

Use S finger.
Anchor F.

3 sss sws sws wsw sws sss sws sws wsw sws
4 sws was was sws own own sws saw saw sws
5 sws white swans swim; sow winter wheat;
6 We watched some whales while we walked.

C. , Key

Type each line 2 times. Repeat if time permits.

Use K finger.
Anchor ;.
Space once after a comma.

7 kkk k,k k,k ,k, k,k kkk k,k k,k ,k, k,k
8 k,k it, it, k,k or, or, k,k an, an, k,k
9 k,k if it is, two, or three, as soon as
10 Vic, his friend, lives in Rich, Alaska.

D. G Key

Type each line 2 times. Repeat if time permits.

Use F finger.
Anchor A S D.

11 fff fgf fgf gfg fgf fff fgf fgf gfg fgf
12 fgf leg leg fgf egg egg fgf get get fgf
13 fgf give a dog, saw a log, sing a song,
14 Gen gets a large sagging gift of games.

Lesson 2.3 → Internet Security and Acceptable Use

You will learn to:
- **Identify** security risks.
- **Define** acceptable use policies.

DIGI*Byte*

Passwords

When you create a password, make sure no one can guess what it is, and never share it with anyone.

What Could Happen

If someone gained access to your password, they could use it to do illegal or unethical things online. If the activities are traced to the password, you could be blamed for the actions.

firewall A software program that acts as a barrier between a computer and the Internet to protect the computer from unauthorized access.

Security Risks The Internet gives your computer access to the world—but it also gives the world access to your computer! There are security risks involved in using the Internet. Some dishonest people use the Internet to gain access to other people's computers without permission. This is called unauthorized access. They want to steal information, or damage files, hard drives, or Web sites. For example, they may try to steal a credit card number. Businesses store financial information on computers, so they must take special care to protect that information.

Denial of Service Another security risk is denial of service (DoS) attacks. In a DoS attack, someone uses various computer tricks to overload and eventually shut down a company's Web site.

The Culprits Two of the worst culprits of security risk are:
- **Crackers** Crackers steal information or break down software's security protections. Their goal is often to destroy computer systems.
- **Hackers** Hackers are specialists in gaining unauthorized access to computer systems. They are usually motivated by curiosity or the challenge of breaking in, rather than the desire to do real damage. Hackers often steal usernames and passwords.

Firewall Protection To guard against hackers, spyware, and other security risks, many people use firewalls. A **firewall** is a software program that acts as a barrier between a computer and the Internet to protect the computer from unauthorized access.

Acceptable Use Many organizations, such as businesses and schools, require people to sign an acceptable use policy (AUP) before using their computer systems. An AUP contains rules for using the Internet and computer network. The AUP is like a contract: if a user violates one of the rules, he or she could lose certain privileges. Organizations have the right to make these rules because they own the computers, and they do not want their computers used for unethical or illegal purposes.

✓ Concept *Check*

Do you think schools and businesses have the right to make people sign an AUP? Do they have the right to take away computer privileges from someone who violates the AUP? Explain your answer.

SKILLBUILDING

H. PRETEST

Type lines 23–24 for 1 minute. Repeat if time permits. Note your speed. Keep your eyes on the copy.

```
23  fold hide fast came hold ride mast fame        8
24  hone rice mask fade none vice task jade       16
```

I. PRACTICE

Type each line 2 times. Repeat if time permits.

Build speed on repeated word patterns.

```
25  fold fold hold hold sold sold told told
26  hide hide ride ride rice rice vice vice
27  fast fast mast mast mask mask task task
28  came came fame fame fade fade jade jade

29  last last vast vast cast cast case case
30  mats mats mars mars cars cars jars jars
31  fell fell jell jell sell sell seal seal
32  dive dive five five live live love love
```

J. POSTTEST

Repeat the Pretest. Compare your Posttest results with your Pretest results.

Appendix A—Keyboarding Skills

Activity 13

Step-by-Step

Investigate Security and Firewalls

Investigate security and firewalls through the following Web sites.

1. Go to **HowStuffWorks.com**. In the site's What are you searching for? search box, key the word "firewall." Choose "How Firewalls Work" and read the information.
2. What does a firewall protect you from?

Explore Acceptable Use Policies

1. Go to a search engine and key the words: "acceptable use policy." Use the quotation marks.
2. Click on one organization's or school district's Web site to view their AUP.
3. Find three unacceptable uses that are listed in the AUP.
4. List the three unacceptable uses, and explain whether you agree or disagree with them.
5. Repeat Steps 1 through 4, choosing a different organization or school district.
6. Find three unacceptable uses that are listed in their AUP.
7. Are the unacceptable uses between the two organizations the same or different?

Activity 14

? If you need additional help on this topic, open the Help menu; or refer to the documentation for your hardware.

RUBRIC LINK

Go to the Online Learning Center at **digicom.glencoe.com** and click Lesson 2.3, Rubric, to find the application assessment guide(s).

DIGI Byte

Internet Tricks

Advertisements on the Web may be a trick to get you to reveal your username and password. Be suspicious of any advertisement or e-mail that asks for personal information.

Application 3

Directions Create a small group to discuss the use of firewalls. One thing firewalls can do is block access to certain Web sites. Do you think schools should block access to Web sites? Why or why not?

Write three paragraphs about your group's responses to this question.
- In the first paragraph, describe the risks that can occur when using shared computers and computer resources.
- In the second paragraph, summarize the various responses in your group.
- In the third paragraph, describe how to avoid computer security risks at school and at home.

Post the responses in your classroom where other students can see them. How do the other students in your class feel about the use of firewalls in schools to block access to some Web sites?

D. . Key

Use L finger.
Anchor ; or J.

Type each line 2 times. Repeat if time permits.

```
11  111 1.1 1.1 .1. 1.1 111 1.1 1.1 .1. 111
12  1.1 Fr. Fr. 1.1 Sr. Sr. 1.1 Dr. Dr. 1.1
13  1.1 std. ctn. div. Ave. Rd. St. Co. vs.
14  Calif. Conn. Tenn. Colo. Fla. Del. Ark.
```

SKILLBUILDING

E. Spacing After Punctuation

Space once after:
- *A period at the end of a sentence.*
- *A period used with an abbreviation.*
- *A semicolon.*

Type each line 2 times. Repeat if time permits.

```
15  The draft is too cold. Close this door.
16  Ask Vera to start a fire. Find a match.
17  Dr. T. Vincent sees me; he made a cast.
18  Ash Rd. is ahead; East Ave. veers left.
```

F. Technique Checkpoint

Hold anchor keys.
Eyes on copy.

Type each line 2 times. Focus on the technique at the left.

```
19  fff fvf fvf vfv fvf fff fvf fvf vfv fvf
20  ;;; T;; T;; ;;; C;; C;; ;;; S;; S;; ;;;
21  111 1.1 1.1 .1. 1.1 111 1.1 1.1 .1. 111
22  Dee voted for vivid vases on her visit.
```

G. Figuring Speed

Typing speed is measured in words a minute (wam). To determine your typing speed:

- Type for 1 minute.
- Determine the number of words you typed. Every 5 strokes (characters and spaces) count as 1 word. Therefore, a 40-stroke line equals 8 words. Two 40-stroke lines equal 16 words.
- Use the cumulative word count at the end of lines to determine the number of words in a complete line.

To determine the number of words in an incomplete line:
- Use the word scale below the last line (below line 24 on this page).
- The number over which you stopped typing is the number of words for that line. For example, if you typed line 23 and completed up to the word vice in line 24, you have typed 14 words a minute (8 + 6 = 14).

```
23  fold hide fast came hold ride mast fame      8
24  hone rice mask fade none vice task jade     16
    |  1  |  2  |  3  |  4  |  5  |  6  |  7  |  8
```

Appendix A—Keyboarding Skills   **A12**

Application 4

Scenario The company you work for has decided that it needs a new acceptable use policy (AUP). Your boss asks you to work with a committee to create a new AUP for all the employees to sign. The committee's job is to create the rules of the AUP.

Directions Form a small committee with your classmates. On index cards or sheets of paper, write down some possible rules for your AUP. For example, "Do not use company computers to download music from the Internet."

For some helpful ideas, search on the Web for the key words "acceptable use policy." Include the quotation marks. Make notes about what you find. You can also check with your school to see if it has an AUP. If it does, ask for a copy.

Share your ideas with the rest of the class. Then, welcome your classmates to give you feedback and contribute new ideas.

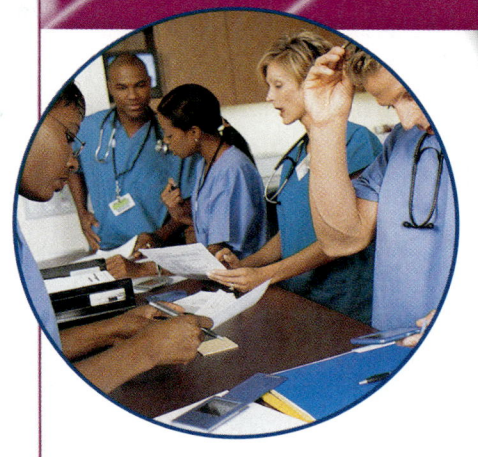

Look Who's DIGITAL

Digital ER Team

There's been an accident; a hiker tumbled into a ravine and needs help. His buddies get GPS coordinates from their PDA and call 911. Once at the site, an EMT (Emergency Medical Technician) checks the hiker, records the data on a PDA, snaps images of the injuries, and phones the hospital.

The ER team accesses the patient's medical records and uploads the photo and data on their Tablet PC. A nurse searches PDA files to review the right treatments to use. The patient arrives, and they get to work. An assistant holds the Tablet PC and records the event via speech recognition. Once the patient is stabilized, the doctor reviews and signs the electronic chart.

Activity In emergency medicine, sharing information quickly can save a life. What are other ways these technologies would be valuable to doctors, dentists, school nurses, or team coaches? Work with a partner to create a story that describes such a situation.

Lesson 5 — New Keys: V Right Shift Period (.)

OBJECTIVES:
- Learn the V, right shift, and period keys.
- Learn spacing with the period.
- Figure speed (typing rate in words a minute).

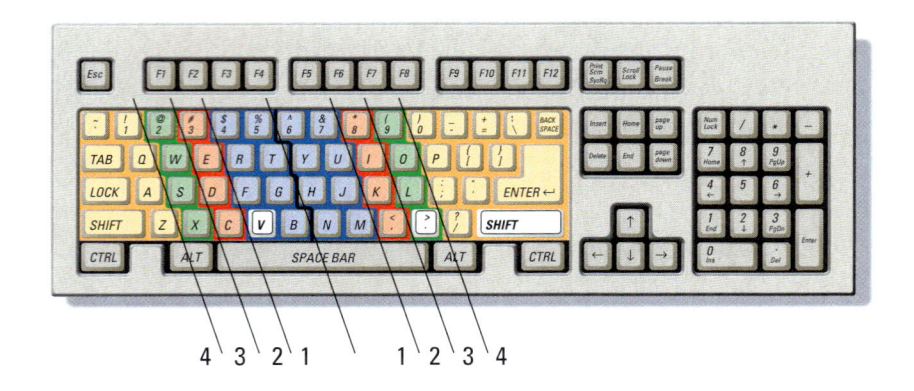

A. Warmup

Type each line 2 times.

1 asdf jkl; jh de lo jm fr ki ft jn dc ;;
2 cash free dine jolt milk iron trim star

NEW KEYS

B. V Key

Type each line 2 times. Repeat if time permits.

Use F finger.
Anchor A S D.

3 fff fvf fvf vfv fvf fff fvf fvf vfv fvf
4 fvf vie vie fvf eve eve fvf via via fvf
5 fvf vie for love; move over; via a van;
6 vote to move; even vitamins have flavor

C. Right Shift Key

Type each line 2 times. Repeat if time permits.

Use Sem finger.
Anchor J.

7 ;;; T;; T;; ;;; C;; C;; ;;; S;; S;; ;;;
8 ;;; Ted Ted ;;; Cal Cal ;;; Sam Sam ;;;
9 ;;; Ed likes Flint; Rick ran; save Tom;
10 Vera loved Florida; Aaron and Sam moved

Appendix A—Keyboarding Skills

Lesson 2.4 — Ethics and Netiquette

You will learn to:
- **Explore** computer ethics.
- **Investigate** netiquette.

ethics Society's rules for what is right and what is wrong.

The Importance of Ethics Ethics are society's rules for what is right and what is wrong. For example, it is ethical to tell the truth. It is unethical to cheat on a test or to spread rumors about someone. When you use computers, you should apply the same good standard of ethics that you use in other areas of your life. On the Internet, computer users cannot see each other face to face, but it is still important to be respectful of others.

The choices you make demonstrate your sense of ethics. For example:

- If someone accidentally leaves personal information on a classroom or workplace computer, do you:
  - Consider that information "fair game" for you to look at? Or:
  - Alert the person who left it there?
- If you are using an online discussion forum, do you:
  - Use threatening language against others because you know they cannot see or hurt you? Or:
  - Use the same manners you would in a face-to-face conversation?
- Someone offers you a free copy of software. You know the software costs money, and users are not supposed to copy it. Do you:
  - Accept the copy? Or:
  - Buy your own?

Computers and the Internet have had a huge positive impact on people's lives and on businesses. In return, computer users need to use ethical standards to make using computers safe for everyone.

netiquette The manners people use in electronic communications.

Netiquette A thoughtful computer user pays attention to netiquette. Netiquette refers to the manners people use in electronic communications. When you communicate using a computer, and the person on the other end cannot hear your voice or see your face, it is important to prevent people from misunderstanding you, especially in a business setting. For example:

- Do not use obscene or angry language. This is called flaming and can offend others.
- Do not type in all capital letters. You may think you are simply emphasizing your words, but it comes across as shouting.

✓ Concept Check

A friend tells you about a Web site that lets you download research papers. All you need to do is put your name on it and pretend you wrote it. Is this ethical? Explain your answer.

D. C Key

Type each line 2 times. Repeat if time permits.

Use D finger.
Anchor A.

```
11  ddd dcd dcd cdc dcd ddd dcd dcd cdc dcd
12  dcd ace ace dcd can can dcd arc arc dcd
13  dcd on a deck; in each car; cannot act;
14  act at once; call to cancel the tickets
```

SKILLBUILDING

E. Technique Checkpoint

Type each line 2 times. Repeat if time permits. Focus on the techniques at the left.

Hold anchor keys.
Eyes on copy.

```
15  fff ftf ftf tft ftf fff ftf ftf tft ftf
16  jjj jnj jnj njn jnj jjj jnj jnj njn jnj
17  ddd dcd dcd cdc dcd ddd dcd dcd cdc dcd
18  the carton of jam is here on this dock;
```

F. PRETEST

Type lines 19–20 for 1 minute. Repeat if time permits. Keep your eyes on the copy.

```
19  sail farm jets kick this none care ink;
20  rain hand jots tick then tone came sink
```

G. PRACTICE

Type each line 2 times. Repeat if time permits.

To increase skill:
- Keep eyes on copy.
- Maintain good posture.
- Speed up on the second typing.

```
21  sail sail said said raid raid rain rain
22  farm farm harm harm hard hard hand hand
23  jets jets lets lets lots lots jots jots
24  kick kick sick sick lick lick tick tick

25  this this thin thin than than then then
26  none none lone lone done done tone tone
27  care care cake cake cane cane came came
28  ink; ink; link link rink rink sink sink
```

H. POSTTEST

Type lines 19–20 for 1 minute. Repeat if time permits. Compare your Posttest results with your Pretest results.

Appendix A—Keyboarding Skills   A10

Activity 15

Step-by-Step

Explore Ethics

1. Go to a search engine and key the words "ethics in computing." Use the quotation marks.
2. Click on an educational site. You will recognize an educational site as a domain name with the domain extension **.edu**.
3. View a list of computer ethics that you find.
4. Choose one rule and explain what it means in your own words.
5. Give an example of how someone may violate this rule. Give an example of how someone would comply with this rule.

Activity 16

Explore Netiquette

1. Go to a search engine, such as Google or Yahoo!, and key the word "netiquette."
2. Choose a reliable site. You may choose to click on an educational site by looking for the domain extension **.edu**.
3. View a list of netiquette that you find.
4. Write a summary of three important rules of netiquette that you find.

? If you need additional help on this topic, open the Help menu; or refer to the documentation for your hardware.

Application 5

Directions Work with classmates to create a poster for your classroom that lists ethical guidelines for using computers. Each person should contribute at least one idea. For example: "Ask permission before viewing computer files that belong to someone else."

Make sure all information is correct and factual. Include at least one image and a minimum of six guidelines on your poster.

RUBRIC LINK

Go to the Online Learning Center at **digicom.glencoe.com** and click Lesson 2.4, Rubric, to find the application assessment guide(s).

Application 6

Directions Many companies have a code of conduct that outlines ethical behavior. Do a Web search for "code of conduct" and choose a code of conduct to examine.

Form a small group with other students and compare your code of conduct with the ones they found. What are the similarities? What are the differences? Make a list and share your findings with the class.

Lesson 4 — New Keys: T N C

OBJECTIVE:
- Learn the T, N, and C keys.

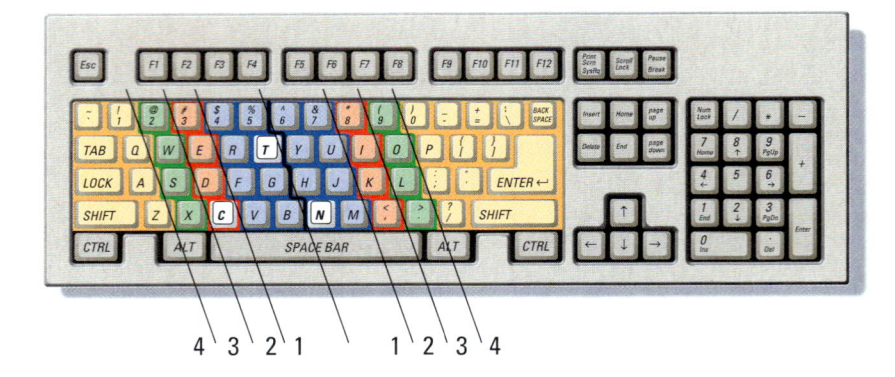

4 3 2 1   1 2 3 4

A. Warmup

Type each line 2 times. Leave 1 blank line after each set of lines.

1 asdf jkl; heo; mri; asdf jkl; heo; mri;
2 herd herd mild mild safe safe joke joke

NEW KEYS

B. T Key

Type each line 2 times. Repeat if time permits.

Use F finger.
Anchor A S D.

3 fff ftf ftf tft ftf fff ftf ftf tft ftf
4 ftf kit kit ftf toe toe ftf ate ate ftf
5 ftf it is the; to them; for the; at it;
6 that hat is flat; it ate at least three

C. N Key

Type each line 2 times. Repeat if time permits.

Use J finger.
Anchor ; L K.

7 jjj jnj jnj njn jnj jjj jnj jnj njn jnj
8 jnj ten ten jnj not not jnj and and jnj
9 jnj nine tones; none inside; on and on;
10 nine kind lines; ten done in an instant

Application 7

Directions It is important that you are aware of the school and classroom rules for using computers and other digital tools. Work with two other students to discuss why using a computer and technology at school is a privilege. Consider what you think are appropriate rules for acceptable use of this equipment. Be sure to include rules regarding the use of the Internet in school.

Key the following rules in WordPad. Add at least five more rules to this list from your group's discussion.

```
1. I will not damage the computers
   or networks in any way.
2. I will not view or use other
   people's folders, files, or work
   without their permission.
3. I will not waste computer
   resources, such as paper, ink,
   or disk space.
4. I will not change any of the
   school's computer settings
   unless my teacher has asked me
   to do so in class.
5. I will not check my personal
   e-mail account while in class.
```

With your teacher's permission, post the rules. Note any rules that were not included on your group's list.

Application 8

Directions Interview a classmate. Create a list of four or five questions about how technology affects everyone. Include such questions as:

- What are some examples of technology that you use every day?
- How would your life be different if you did not have this technology?
- How do you think your life would be different if you were born fifty years in the future?

After your interview, key the interview responses in a WordPad document. Include what you learned in the interview.

DIGITAL Decisions

Children's Internet Protection Act

Form a small group to work with. You have just been hired as an assistant technology director for a school district. Your supervisor, Maya Ventullo, tells you that she wants to be sure the district upholds the Children's Internet Protection Act (CIPA).

Internet Investigation Your first assignment is to research the law and provide an outline of everything it covers. Ventullo needs to include this information in a report for the school board. Be thorough in your research. Use WordPad to record the information and to create the outline. Be sure the outline includes the following information:

- The date on which CIPA became law.
- Background information on the creation of CIPA.
- A summary of the general requirements of CIPA.
- Benefits to schools complying with CIPA.
- Problems with implementation of CIPA.
- Recent news articles relating to CIPA.

Share the information that you and your group have gathered with your class. Discuss how CIPA protects you and your classmates.

SKILLBUILDING

E. Technique Checkpoint

Type each line 2 times. Repeat if time permits. Focus on the techniques at the left.

Focus on these techniques:
- Fingertips touching home keys.
- Wrists up, off keyboard.

```
15 jjj jmj jmj mjm jmj jjj jmj jmj mjm jmj
16 fff frf frf rfr frf fff frf frf rfr frf
17 kkk kik kik iki kik kkk kik kik iki kik
18 he did; his firm red desk lid is a joke
```

F. PRETEST

Type lines 19–20 for 1 minute. Repeat if time permits. Keep your eyes on the copy.

```
19 joke ride sale same roam aims sire more
20 jars aide dark lame foal elms hire mare
```

G. PRACTICE

Type each line 2 times. Repeat if time permits.

Keep eyes on copy. It will be easier to keep your eyes on the copy if you:
- Review the charts for key positions and anchors.
- Maintain an even pace.
- Resist looking up from your copy.

```
21 joke joke jade jade jams jams jars jars
22 ride ride hide hide side side aide aide
23 sale sale dale dale dare dare dark dark
24 same same fame fame dame dame lame lame

25 roam roam loam loam foam foam foal foal
26 aims aims arms arms alms alms elms elms
27 sire sire dire dire fire fire hire hire
28 more more mire mire mere mere mare mare
```

H. POSTTEST

Type lines 19–20 for 1 minute. Repeat if time permits. Keep your eyes on the copy. Compare your Posttest results with your Pretest results.

Appendix A—Keyboarding Skills

21ST CENTURY CONNECTION

All Aspects of Industry

Every successful business needs to balance these eight aspects. Every employee of a business also needs to know how to implement the aspects according to the vision of the business. Being familiar with these aspects will help you in any career you choose.

Activity After reading the chart below, discuss with a small group the steps that you think a company needs to take to recruit employees. Which of the eight Aspects of Industry will be involved in these steps?

All Aspects of Industry

Planning
How an organization determines its goals and meets the needs of its customers

Management
How an organization accomplishes its goals using facilities, staff, and resources

Finance
How an organization gets and manages funds for operation

Technical and Production Skills
Basic academic and computer skills, interpersonal skills, and job-specific skills needed by employees

Underlying Principles of Technology
How employees use and continue to learn about new technology

Labor Issues
Employees' rights, wages, benefits, and working conditions

Community Issues
How the organization and the community affect each other

Health, Safety, and Environmental Issues
Practices and laws protecting the employee and the environment

53

Lesson 3 — New Keys: M R I

OBJECTIVE:
- Learn the M, R, and I keys.

4 3 2 1   1 2 3 4

A. Warmup

Type each line 2 times. Leave 1 blank line after each set of lines.

1 asdf jkl; heo; asdf jkl; heo; asdf jkl;
2 jade jade fake fake held held lose lose

NEW KEYS

B. M Key

Type each line 2 times. Repeat if time permits.

Use J finger.
For M anchor ; L K.

3 jjj jmj jmj mjm jmj jjj jmj jmj mjm jmj
4 jmj mom mom jmj mad mad jmj ham ham jmj
5 jmj make a jam; fold a hem; less flame;
6 messes make some moms mad; half a dome;

C. R Key

Type each line 2 times. Repeat if time permits.

Use F finger.
For R anchor A S D.

7 fff frf frf rfr frf fff frf frf rfr frf
8 frf far far frf for for frf err err frf
9 frf more rooms; for her marks; from me;
10 he reads ahead; more doors are far ajar

D. I Key

Type each line 2 times. Repeat if time permits.

Use K finger.
For I anchor ;.

11 kkk kik kik iki kik kkk kik kik iki kik
12 kik dim dim kik lid lid kik rim rim kik
13 kik if she did; for his risk; old mill;
14 more mirrors; his middle silo is filled

DIGITAL Career Perspective

Transportation, Distribution, and Logistics

Javier Lopez
Light Rail Train
Maintenance Specialist

Keeping People Moving

"Millions of people take trains to and from work every day, so they count on me to keep the trains running safely and efficiently," says Javier Lopez, a maintenance specialist for a major transportation organization. Lopez began his career as a train yard laborer. He later took classes at a local college, enrolled in electronics technology courses at work, earned his associate's degree, and finished a maintenance apprenticeship. He is now a journey-level, or senior-level, technician.

"I work in this industry because I've always enjoyed watching trains. I also love the challenge of mechanical projects, and this career keeps me well-equipped to meet those challenges," says Lopez. His job is to identify all vehicle system problems and their causes and repair the vehicle and its related systems.

Once repairs are made, Lopez and his crew test the rail cars to be sure they are working properly. When new or rebuilt trains are delivered to the tracks, Lopez sees that they are fully operational before they go into service.

"Most workdays are routine, but I need to be prepared for system emergencies and equipment failures," says Lopez. "To be sure I am ready, there are strict company guidelines for training, sensible conduct, and proper equipment operation." Lopez is also responsible for promoting and keeping a safe work environment. "In emergency situations, it is important to demonstrate sensitivity and interest in customer concerns and needs," says Lopez.

Lopez brings a PDA and a walkie-talkie mobile phone with him every day. "When I need to order new parts, I can enter a detailed request into my PDA and send it to headquarters quickly and accurately. I also communicate with coworkers on my walkie-talkie phone; in urgent situations, I can reach someone more easily. With digital technology, the development of detailed specifications for the purchase of technical and specialized equipment is completed without unnecessary paperwork."

Training

Employers require that entry-level applicants have a high school diploma. Employees wishing to advance can enroll in additional training programs at work or take applicable electronics technology courses in accredited colleges. Employees are required to pass certification exams. They also take periodic classes and tests to ensure that their training is up to date.

Salary Range

Typical earnings are in the range of $21.00 to $23.00 per hour.

Skills and Talents

Transportation maintenance specialists need to be:

- In good health. Good vision and hearing are very important
- Physically fit
- Good at performing intricate repairs
- Experts in how machines work
- Good communicators
- Able to make responsible judgments quickly

Career Activity

Why do transportation maintenance specialists need to maintain current training for their occupations?

SKILLBUILDING

E. Technique Checkpoint

Type each line 2 times. Repeat if time permits. Focus on the technique at the left.

Focus on this technique: Press and release each key quickly.

```
15  ddd ded ded ede ded ddd ded ded ede ded
16  lll lol lol olo lol lll lol lol olo lol
17  jjj jhj jhj hjh jhj jjj jhj jhj hjh jhj
18  she has old jokes; he has half a salad;
```

F. PRETEST

Type lines 19–20 for 1 minute. Repeat if time permits. Keep your eyes on the copy.

Hold anchor keys.

```
19  heed jade hoof elf; hash folk head hole
20  seed lake look jell sash hold dead half
```

G. PRACTICE

Type each line 2 times. Repeat if time permits.

When you repeat a line:
- Speed up as you type the line.
- Type it more smoothly.
- Leave a blank line after the second line (press ENTER 2 times).

```
21  heed heed feed feed deed deed seed seed
22  jade jade fade fade fake fake lake lake
23  hoof hoof hood hood hook hook look look
24  elf; elf; self self sell sell jell jell

25  hash hash lash lash dash dash sash sash
26  folk folk fold fold sold sold hold hold
27  head head heal heal deal deal dead dead
28  hole hold hale hale hall hall half half
```

H. POSTTEST

Type lines 19–20 for 1 minute. Repeat if time permits. Keep your eyes on the copy. Compare your Posttest results with your Pretest results.

Appendix A—Keyboarding Skills

Chapter 2

Digital Dimension → Language Arts
INTERDISCIPLINARY PROJECT

Be a Savvy Consumer

Using the digital input skills you have learned in this chapter, create a digital video clip with your digital camera, webcam, or digital camcorder to make a commercial.

Make a Commercial Think about a product that you use and like very much. Why do you like this product? Consider what you can tell your friends to influence them to buy this product.

Search the Internet for information about your product. Before a product can be advertised effectively, you will need to gather all the information you can about the product.

Find out:

- Where the product can be purchased
- How much the product costs
- What ingredients the product contains
- Where the product is manufactured
- Interesting details in the history of the product

Write the Script In Notepad, write a commercial about your product. Write a paragraph or two that would sell your product. Tell why someone should buy it. Include the information you gathered from the Internet. If you need script ideas, think about other commercials you have liked.

Get Ready to Record Think about and gather any props or visual aids that would enhance your commercial.

Here are some ideas to get you started:

- Backgrounds
- Competitors' products
- Costumes
- Large charts or graphs
- Special effects

Rehearse your script, and then record it with a digital camera, webcam, or camcorder.

E-Portfolio Activity

Digital Skills

To showcase the new digital input skills you have learned in this chapter, use Notepad to create a list of your skills with a digital camera, webcam, or camcorder. Include skills such as resizing pictures, capturing video clips, and capturing a picture with the webcam. Give a title to your document, and include a graphic or photo. Save the document to your e-portfolio.

Self-Evaluation Criteria

Have you included:

- The name of your product?
- Details about your product?
- The cost of your product?
- Where you can buy your product?
- The reason you like the product?

Lesson 2 → New Keys: H E O

OBJECTIVE:
- Learn the H, E, and O keys.

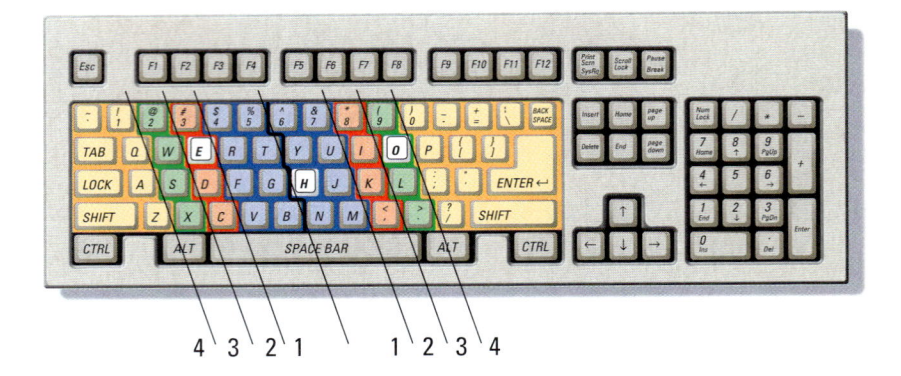

4 3 2 1   1 2 3 4

A. Warmup

Type each line 2 times. Leave 1 blank line after each set of lines.

Hold anchor keys
For H anchor ; L K
For E anchor A
For O anchor J or ;

1 ff jj dd kk ss ll aa ;; f j d k s l a ;
2 adds adds fads fads asks asks lads lads

NEW KEYS

B. H Key

Type each line 2 times. Repeat if time permits.

Use J finger.
For H anchor ; L K.

3 jjj jhj jhj hjh jhj jjj jhj jhj hjh jhj
4 jhj ash ash jhj has has jhj had had jhj
5 jhj a lass has; adds a half; a lad had;
6 has a slash; half a sash dad shall dash

C. E Key

Type each line 2 times. Repeat if time permits.

Use D finger.
For D anchor A S.

7   ddd ded ded ede ded ddd ded ded ede ded
8   ded led led ded she she ded he; ded he;
9   ded he led; she fell; he slashes sales;
10 he sees sheds ahead; she sealed a lease

D. O Key

Type each line 2 times. Repeat if time permits.

Use L finger.
For L anchor J K.

11 lll lol lol olo lol lll lol lol olo lol
12 lol odd odd lol hoe hoe lol foe foe lol
13 load sod; hold a foe; old oak hoes; lol
14 she sold odd hooks; he folded old hoses

Appendix A—Keyboarding Skills

CHAPTER 2 Review

Read to Succeed PRACTICE

After You Read

Trade-a-Question In the Read-to-Succeed activity, Use the Glossary, at the beginning of Chapter 2, you started collecting 3" by 5" cards of the key terms and their definitions. Make sure you have completed a card for each key term in this chapter.

Team up with a partner to participate in Trade-a-Question. Trade a card with your partner and ask, "What is the definition of this key term?" Your partner should attempt to answer the question before looking at the back of the card. You can use this activity for each chapter in this book.

When you have completed Trade-a-Question, place all the cards on a table with the definition side of the card facing up. Point to a definition and ask your partner, "What is the key term that is defined?" Your partner should attempt to identify each key term as quickly as possible without picking up the card and looking at the back of the card.

Using Key Terms

In today's technological world, people often use the following terms. See how many of these terms you know by writing each statement on a sheet of paper and filling in the blanks.

- resolution (37)
- pixel (37)
- weblog (40)
- webcam (40)
- firewall (47)
- ethics (50)
- netiquette (50)

1. A(n) _____ is a small camera that connects to a computer and sends an image over the Web.

2. A(n) _____ is a software program that acts as a barrier between a computer and the Internet to protect the computer from unauthorized access.

3. The manners people use in electronic communications are called _____.

4. An online journal that is updated almost every day by its author is called a(n) _____.

5. _____ is/are society's rules for what is right and what is wrong.

6. A single point in a digital image is called a(n) _____.

7. The _____ is/are the amount of detail in a digital image, measured in pixels.

SKILLBUILDING

I. Technique Checkpoint

Technique Checkpoints enable you to practice new keys. They also give you and your teacher a chance to evaluate your keyboarding techniques. Focus on the techniques listed in the margin, such as:

- Use correct fingers.
- Keep eyes on copy.
- Press ENTER without pausing.
- Maintain correct posture.
- Maintain correct arm, hand, and finger position.

Focus on these techniques:
- Keep eyes on copy.
- Keep fingers on home keys.

Type lines 9 and 10 one time.

```
 9 ff jj dd kk ss ll aa ;; f j d k s l a ;
10 ff jj dd kk ss ll aa ;; f j d k s l a ;
```

J. PRETEST

Type lines 11–12 for 1 minute. Repeat if time permits. Keep your eyes on the copy.

Hold anchor keys.

```
11 sad sad fad fad ask ask lad lad dad dad
12 as; as; fall fall alas alas flask flask
```

K. PRACTICE

Type lines 13–14 one time. Repeat if time permits

Leave a blank line after each set of lines (13–14, 15–16, and so on) by pressing ENTER 2 times.

```
13 aaa ddd sad sad aaa sss lll lll all all
14 aaa ddd sad sad aaa sss lll lll all all

15 aaa sss kkk ask ask fff aaa ddd fad fad
16 aaa sss kkk ask ask fff aaa ddd fad fad

17 aaa ddd ddd add add lll aaa ddd lad lad
18 aaa ddd ddd add add lll aaa ddd lad lad

19 aaa sss ;;; as; as; ddd aaa ddd dad dad
20 aaa sss ;;; as; as; ddd aaa ddd dad dad

21 f fl fla flas flask; l la las lass lass
22 f fl fla flas flask; l la las lass lass

23 f fa fal fall falls; a al ala alas alas
24 f fa fal fall falls; a al ala alas alas
```

L. POSTTEST

Type lines 11–12 for 1 minute. Repeat if time permits. Keep your eyes on the copy. Compare your Posttest results with your Pretest results.

M. End-of-Class Procedure

To keep hardware in good working order, treat it carefully. Your teacher will tell you what should be done at the end of each class period.

Chapter 2 Self-Assessment

Take a moment to review what you have learned in this chapter. Rank your understanding of the topics below.

4 means, "I understand all of this."
3 means, "I understand some of this."
2 means, "I understand very little of this."
1 means, "I don't remember this."

To use a printout of this chart, go to **digicom.glencoe.com** and click on **Chapter 2, Self-Assessment.**
Or:
Ask your teacher for a personal copy.

Rank Your Understanding

Lesson	Topic	4	3	2	1
2.1	• Describe how people use digital cameras, camcorders, and DVD recorders				
	• Define resolution				
	• Explain the connection between pixels and resolution				
	• Create a flyer using digital pictures				
	• Print digital photographs				
2.2	• Explain how communication has evolved over the years				
	• Describe how to send an e-mail				
	• Explain how e-mail attachments work				
	• Describe the purpose of weblogs				
	• Explore online webcams and describe how webcams could be used in science or education				
2.3	• Identify security risks				
	• Describe acceptable use policies				
	• List three unacceptable uses that could be in an AUP				
2.4	• Define ethics and explain the importance of ethical computer use				
	• Identify poor netiquette practices				

If you ranked all topics 4, congratulations! Consider doing a quick review.
If you ranked yourself 3 or lower in any topic, consider reviewing these topics first.

4 3 2 1   1 2 3 4

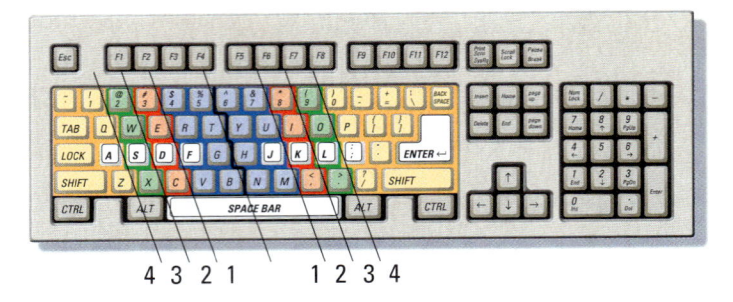

Use the thumb of your writing hand (left or right) to press the space bar.

1. With your fingers on the home keys, type the letters a s d f. Then press the space bar once.
2. Type j k l ;. Press the space bar once.
3. Type a s d f. Press the space bar once; then type j k l ;.
4. Repeat Steps 1–3.

D. Enter Key

The ENTER key moves the insertion point to the beginning of a new line. Reach to the ENTER key with the Sem finger. Lightly press the enter key. Return the Sem finger to home position.

Practice using the enter key. Type each line 1 time, pressing the space bar where you see a space and pressing the Enter key at the end of a line.

```
asdf jkl; asdf jkl; asdf jkl;↵
asdf jkl; asdf jkl; asdf jkl;↵
asdf jkl; asdf jkl; asdf jkl;↵
asdf jkl; asdf jkl; asdf jkl;↵
```

E. F J Keys

Use F and J fingers.

Type each line 1 time.

```
1 fff jjj fff jjj fff jjj ff jj ff jj f j↵
2 fff jjj fff jjj fff jjj ff jj ff jj f j↵
```

F. D K Keys

Use D and K fingers.

Type each line 1 time.

```
3 ddd kkk ddd kkk ddd kkk dd kk dd kk d k↵
4 ddd kkk ddd kkk ddd kkk dd kk dd kk d k↵
```

G. S L Keys

Use S and L fingers.

Type each line 1 time.

```
5 sss lll sss lll sss lll ss ll ss ll s l↵
6 sss lll sss lll sss lll ss ll ss ll s l↵
```

H. A ; Keys

Use A and Sem fingers.

Type each line 1 time.

```
7 aaa ;;; aaa ;;; aaa ;;; aa ;; aa ;; a ;↵
8 aaa ;;; aaa ;;; aaa ;;; aa ;; aa ;; a ;↵
```

Appendix A—Keyboarding Skills

CHAPTER 3
Using the Internet

Lesson 3.1	Browse the Internet
Lesson 3.2	Customize Your Browser and Search Online
Lesson 3.3	Security and Viruses
Lesson 3.4	Copyright, Fair Use, and Plagiarism
Lesson 3.5	Computer Crimes

You Will Learn To:

- **Explore** the Internet by using a browser and hyperlinks.
- **Save** Web sites as Favorites.
- **Recognize** a Web site by its Web address.
- **Apply** advanced search techniques.
- **Choose** a home page for your browser.
- **Print** Web pages.
- **Describe** anti-virus software.
- **Describe** security for the Internet.
- **Explain** the importance of copyright, fair use, and avoiding plagiarism.
- **Describe** computer crimes and fraud.

When Will You Ever Use the Internet?

For many people, the Internet is an essential part of everyday life. The Internet entertains us, lets us communicate anywhere in the world instantaneously, teaches us new things, and gives us the ability to go shopping without leaving our homes. The Internet is the world's largest computer network. It has revolutionized the way we think, work, and search for information.

Lesson 1 — New Keys: A S D F J K L ; Space Bar Enter

OBJECTIVE:
- Learn the home keys, the space bar, and the enter key.

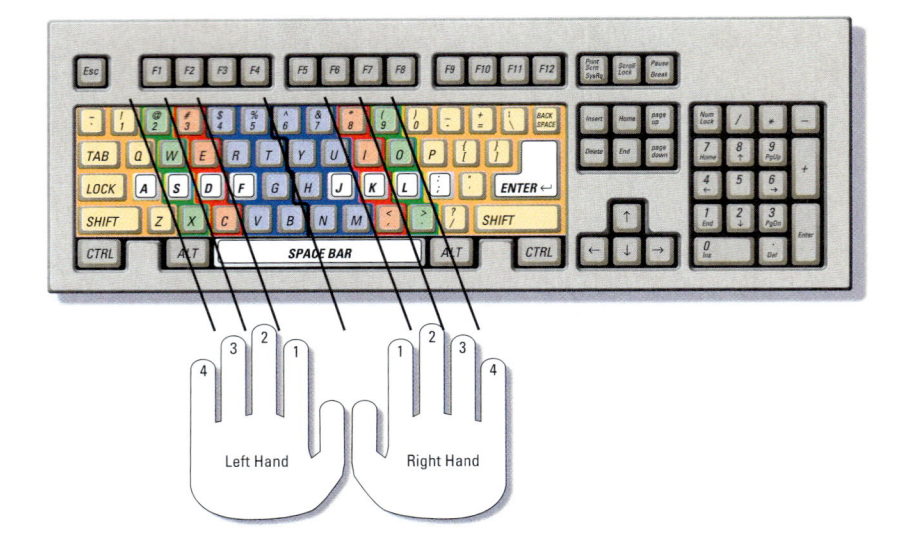

NEW KEYS

A. Home-Key Position

The A S D F J K L ; keys are called the home keys. Each finger controls a specific key and is named for its home key: A finger, S finger, D finger, and so on, ending with the Sem finger on the ; (semicolon) key.

1. Place the fingers of your left hand on A S D and F. Use the illustration as a guide.
2. Place the fingers of your right hand on J K L ;. Again, use the illustration as a guide.

You will feel a raised marker on the F and J keys. These markers will help you keep your fingers on the home keys.

3. Curve your fingers.
4. Using the correct fingers, type each letter as you say it to yourself: a s d f j k l ;.
5. Remove your fingers from the keyboard and replace them on the home keys.
6. Type each letter again as you say it: a s d f j k l ;.

B. Using Anchors

An anchor is a home key that helps you return each finger to its home-key position after reaching for another key. Try to hold the anchors listed, but be sure to hold the first one, which is most important.

C. Space Bar

The space bar, located at the bottom of the keyboard, is used to insert spaces between letters and words, and after punctuation.

Appendix A—Keyboarding Skills

Read to Succeed

Key Terms

- browser
- hyperlink
- uniform resource locator (URL)
- domain name
- search engine
- Boolean operator
- virus
- copyright
- fair use
- plagiarism

Read with a Purpose

How can you use what you learn in this chapter in your own life? Having a purpose or reason for reading each chapter will help you apply what you learn to your life now and in the future and will help you learn more easily. Create a graphic organizer similar to the illustration.

Before reading each lesson in the chapter, ask yourself these questions: How does this subject relate to my life? How might I be able to use what I learn in my life?

Example:

Browse the Internet
- **Relates to my life:** I use the Internet every day, and I need to know how to use search engines efficiently.
- **I can use this:** I can learn how to set my own home page and how to print Web pages.
- **Other purpose:** I can learn how to protect my computer from viruses and protect myself from cyberbullies.

DIGITAL CONNECTION

Quick, Set Up Your Internet

A company in New Zealand needs your expertise. The company asks you to set up your room at home as an office—all expenses paid. You are due to start telecommuting next week with the help of communication tools such as cell phones, computers, fax machines, and the Internet.

You decide you had better learn how to use the Internet efficiently. You will be writing reports, so you want to make sure to be current on copyright issues and plagiarism. Also, it will be important to protect your computer and your work from viruses and unauthorized use.

Writing Activity

What is in the recent news about the Internet that you will need to know to protect your computer? Write a paragraph about news and trends you know about, such as Internet browsers, security, and computer crime.

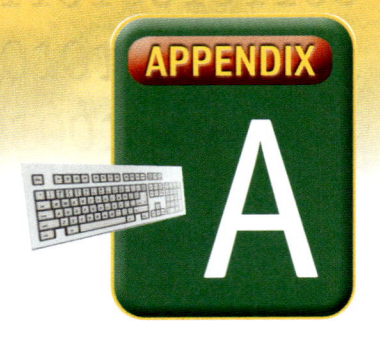

Keyboarding Skills

APPENDIX A

Table of Contents

Lesson	Page
1 New Keys: A S D F J K L ; Space Bar Enter	A2
2 New Keys: H E O	A5
3 New Keys: M R I	A7
4 New Keys: T N C	A9
5 New Keys: V Right Shift Period (.)	A11
6 New Keys: W Comma (,) G	A14
7 New Keys: B U Left Shift	A17
8 New Keys: Q /	A19
9 New Keys: ' "	A21
10 New Keys: P X	A23

Lesson	Page
11 New Keys: Y Tab	A25
12 New Keys: Z Colon (:)	A27
13 New Keys: ? Caps Lock	A29
14 New Keys: - _	A31
15 New Keys: 4 $ 7 &	A33
16 New Keys: 3 # 8 *	A36
17 New Keys: 2 @ 9 (	A39
18 New Keys: 1 ! 0)	A42
19 New Keys: 5 % 6 ^	A45

Lesson 3.1 — Browse the Internet

You will learn to:
- **Use** hyperlinks to explore the Internet.
- **Add** and organize Web sites in Favorites.

Browsers To visit Web sites on the Internet, you need a special software program called a browser. A **browser** is an application that reads the contents of Web pages and displays them on your computer screen.

Hyperlinks You navigate from page to page on the Internet by clicking hyperlinks. A **hyperlink** is any text or graphic on a Web page that will take you to a new location when clicked. Hyperlinks are often underlined.

URLs Every Web page is identified by a uniform resource locator, or URL. A **uniform resource locator (URL)** is a unique Web address that directs a browser to a Web page. A URL has several parts. Some of the parts are common to all Web sites, but the URL as a whole belongs only to one site.

Here is what each part means:

- Hypertext transfer protocol, or http, refers to the set of rules used by Web sites to send information over the Internet.
- www refers to World Wide Web.
- The domain name is glencoe.com. A **domain name** is the part of a URL that usually identifies the owner or the subject of a Web site.
- The domain extension is .com. A domain extension is a two- to four-letter code that identifies the category of a Web site.

URL

http://www.glencoe.com

- hypertext transfer protocol
- World Wide Web
- domain name
- domain extension

Sidebar definitions:

browser An application that reads the contents of Web pages and displays them on your computer screen.

hyperlink Any text or graphic on a Web page that will take you to a new location when clicked.

uniform resource locator (URL) A unique Web address that directs a browser to a Web page.

domain name The part of a URL that usually identifies the owner or the subject of a Web site.

Common Domain Extensions

Extension	Description of Site	Example
.com	Originally intended for commercial or for-profit businesses; now anyone can use .com	www.glencoe.com
.edu	Educational institution	www.ucla.edu
.gov	U.S. government organization	www.senate.gov
.mil	U.S. military organization	www.usmc.mil
.org	Professional or nonprofit organization	www.fbla-pbl.org

✓ Concept Check

What can you tell about a Web site by looking at the URL?

UNIT 4 LAB *continued*

Step 3 Often people find that their total budgeted expenditures for the month are larger than their income. Review your expenses to see where you can reduce them. In the short term, it is usually easier to lower variable expenses than to lower fixed expenses. If your fixed expenses are too high, however, it may be necessary to reconsider rent and car expenses..

Step 4 Investigate tips on budgeting and ways to reduce expenses. Look in books and magazines or search online using keywords such as budgeting tips, how to budget, and so on. Take notes, if possible using a PDA or Tablet PC.

Step 5 Choose your top ten budgeting tips.

Step 6 Work in a small group to share your budgeting tips and create a list of the group's top ten tips, ready to give a presentation on budgeting tips.

Step 7 Your handy budgeting tips will be helpful to many people, not just your classmates. If possible, work with your teacher and your class to invite family members and members of the community to the presentations.

Step 8 With your group, create a five- to eight-minute presentation with the following goals:

- To summarize what you have learned about budgeting.
- To share your group's top ten budgeting tips with your audience.

Step 9 Before giving your presentation to the live audience, practice your presentation. Ensure that the presentation will end within the time limits.

Step 10 On the day of the event, provide handouts of the presentation to your audience.

ONLINE Resources

To find Web sites on this topic, visit the Digital Communication Tools Online Learning Center at **digicom.glencoe.com**. Click on Chapter 15, Resources.

Self-Evaluation Criteria

Have you included:
- Ideas for ways to reduce expenses?
- Additional budgeting tips?

digicom.glencoe.com

Step-by-Step

Activity 1

If you need additional help on this topic, open the Help menu; or refer to the documentation for your hardware.

Use Hyperlinks

1. Open your browser.
2. In the Address bar at the top of the screen, enter **digicom.glencoe.com**. Press ENTER.
3. Move your pointer around the page. Whenever the pointer looks like a hand, it is on a hyperlink.
4. Click one of the hyperlinks. You will be taken to a new page.
5. After reading the new page, click the Back button in the toolbar. You will return to the previous page.
6. Click on another hyperlink.

Activity 2

Add Favorites to Your Browser

You can save Web sites that you want to be able to return to easily as Favorites. Instead of entering the URL, you simply click on the Favorite to get to the site. Favorites are also known as bookmarks.

1. Open your Web browser and go to the Digital Communication Tools Web site at **digicom.glencoe.com**.
2. When the Web site appears, click **Favorites** in the Menu Bar at the top of the screen. Do not click the yellow star icon in the Toolbar.
3. Click **Add to Favorites**. Click **OK**.

DIGI*Byte*

Browser Choices
There are many browsers to choose from, including Internet Explorer, Netscape, Mozilla, and Safari.

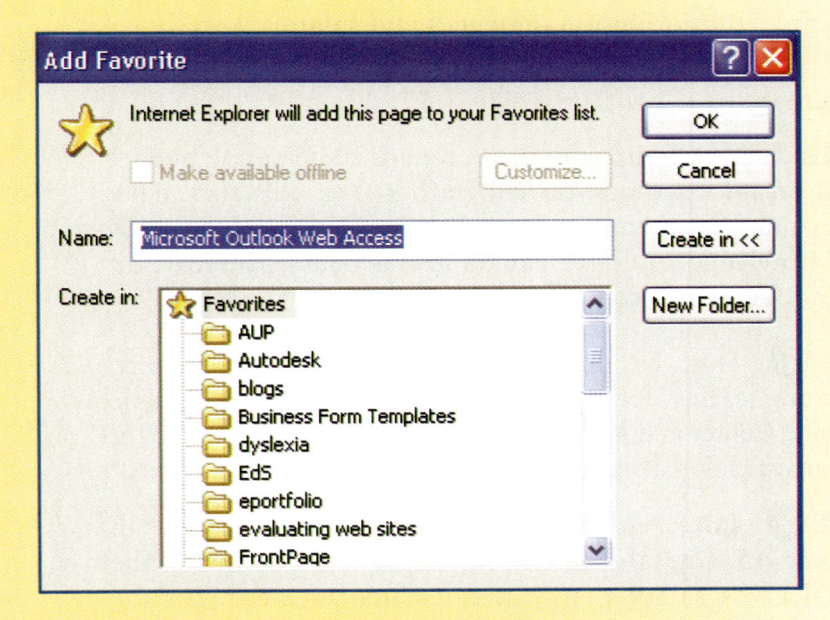

4. Close your browser.

Continued

Chapter 3

61

UNIT 4 LAB

DIGITAL Dimension → Math
INTERDISCIPLINARY PROJECT

Managing Your Money

Background In Unit 4 you have been learning about money management:

- In Chapter 13 Digital Dimension, page 444, you evaluated credit card expenses.
- In Chapter 14 Digital Dimension, page 465, you calculated the costs of car ownership.
- In Chapter 15 Digital Dimension, page 493, you compared information on wages and salaries.

Presentation on Budgeting Tips

In this Unit 4 Lab, you will prepare a personal budget based on information you will gather. You will also make a presentation on handy budgeting tips. Apply all the skills and knowledge you have gained in this course to create an outstanding presentation.

Step 1 Start by doing the Unit 4 Lab Prep Activity. Ask your teacher for the prep activity, or go to the Online Learning Center at **digicom.glencoe.com** and click Chapter 15, Resources, Unit 4 Lab Prep Activity.

Step 2 Look at the spreadsheet you created in Chapter 15 Digital Dimension, page 493. Compare the annual gross salary that you calculated in Step 5 of the Unit 4 Lab Prep Activity with the annual gross salaries of the jobs in your spreadsheet.

E-Portfolio Activity

Showcase Your Work

To showcase your spreadsheet and presentation skills, save your budget and presentation in your e-portfolio.

Continued

digicom.glencoe.com

Step-by-Step *continued*

5. Open the browser again. Click the **Favorites** icon with the yellow star in the toolbar.

6. You will see the Favorite that you just created. Click the Favorite to go to the Digital Communication Tools Web site.

Organize Favorites

You can create folders in Favorites to group Web sites by category.

1. Click **Favorites** in the Menu Bar, and then click **Organize**.

2. Click **Create Folder**. Enter the name **Digital Communication Tools** for this folder. Press ENTER.

3. Move the Digital Communication Tools favorite to this folder by clicking and dragging the item into the folder.

Activity 3

Suite Smarts

Creating folders in Favorites is similar to the steps used for creating folders in My Computer. Folders are created to help you organize data and make it easier to find information.

RUBRIC LINK

Go to the Online Learning Center at **digicom.glencoe.com** and click Lesson 3.1, Rubric, to find the application assessment guide(s).

Application 1

Directions Bookmark the Web sites for the U.S. Senate at **www.senate.gov** and the U.S. House of Representatives at **www.house.gov**. Create a new folder in your Favorites called **Legislative Branch,** and move these two favorites into the new folder.

DIGITAL DIALOG

Creating a Domain Name

Teamwork Work in a small group. You work for a new recording company. Decide what type of music your company is going to promote, and then think of an original name for the company. Think of four unique domain names that will help identify your company. The domain names should include the name or part of the name of your company.

Communication Share your domain names with the rest of the class. Make a list of all the domain names created by the groups. Ask the class to vote on the one they like best.

CHAPTER 15

Self-Assessment

Take a moment to review what you have learned in this chapter. Rank your understanding of the topics below.

4 means, "I understand all of this."
3 means, "I understand some of this."
2 means, "I understand very little of this."
1 means, "I don't remember this."

To use a printout of this chart, go to **digicom.glencoe.com** and click on **Chapter 15, Self-Assessment.**
Or:
Ask your teacher for a personal copy.

Rank Your Understanding

Lesson	Topic	4	3	2	1
15.1	Name two résumé formats				
	Explain what makes a résumé effective				
	List important factors about electronic résumés				
	Describe the parts of a cover letter				
	List reasons to use a skills résumé				
15.2	How to search for jobs				
	How to find jobs online				
	Describe the information given in online job postings				
15.3	How to fill out job application forms				
	Explain what a reference is				
	Give three examples of legal information on job application forms				
	List examples of disclaimers				
15.4	How to organize your job search				
	List the information you would collect about potential employers				
	How to research companies to prepare for an interview				
15.5	List four characteristics or skills you already have that are useful for interviews				
	List what you need to do to prepare for an interview				
	Give examples of interview techniques				
	Explain the parts of a thank-you letter to send after an interview				

If you ranked all topics 4, congratulations! Consider doing a quick review.
If you ranked yourself 3 or lower in any topic, consider reviewing these topics first.

Lesson 3.2 — Customize Your Browser and Search Online

You will learn to:
- **Choose** a home page for your browser.
- **Select** options for printing Web pages.
- **Search** for information on the Internet.

search engine A Web site that finds Web pages that contain words you tell it to search for.

Boolean operator A word that can make an Internet search more specific and effective.

DIGI Byte
Successful Searching
The key to a successful search is to use descriptive words that apply to your topic and to eliminate unnecessary words.

Home Pages You may find that you have a favorite Web site that you keep going back to. It might be the first site you visit when you get on the Internet every day. Why not make that site your home page? A browser can be customized to go to the site that you specify as your home page each time it is opened.

Search Engines The Internet is home to vast amounts of information. Luckily, there are Web sites to help you find the specific information you need. A search engine is a Web site that finds Web pages that contain words you tell it to search for. Some major search engines include Google at www.google.com, AltaVista at www.altavista.com, and Ask Jeeves at www.ask.com.

Boolean Search Boolean operators are words that can make an Internet search more specific and effective. The basic Boolean operators are AND, OR, and NOT. See the table below for examples of how Boolean operators can affect your search results.

Boolean Search Comparison

Key Words	Possible Number of Results	Explanation of Results
computers OR laptops	112,000,000 results	These Web pages contain either the word "computers" OR the word "laptops."
computers NOT laptops	102,000,000 results	These Web pages contain the word "computers" but NOT the word "laptops."
computers AND laptops	5,150,000 results	These Web pages contain both the word "computers" AND the word "laptops."

✓ Concept Check

What is an advantage to setting a home page in your browser?

CHAPTER 15 Review

Read to Succeed PRACTICE

After You Read

Recall After you have read each chapter and completed the activities, recall one factor about each main topic of the chapter. Use Memo Pad on your PDA, Journal on your Tablet PC, or a piece of paper to list the main idea. Recall one factor about each topic.

For this chapter, recall the main idea about each of the chapter topics in a graphic organizer.

Example:

	Main Idea
Résumés	A résumé is the first impression you give to a prospective employer.
Interviews	Prepare before the interview and practice responses to interview questions

Using Key Terms

The following terms will help you understand and communicate effectively as you prepare to get the job you want. In Memo Pad, Journal, or on a separate piece of paper, rewrite the sentences below and match with the correct key term.

- résumé (470)
- job interview (470)
- cover letter (473)
- reference (482)
- employment-at-will (482)

1. A short letter you send along with a résumé, specifying the job you are interested in and highlighting specific skills or accomplishments you feel make you a good candidate for the job.

2. The name of a person an employer can contact to find out about a job applicant.

3. A summary or your skills, abilities, education, work history, and achievements.

4. Employment that can be terminated with or without cause by the employer.

5. A formal meeting between an employer and a job applicant.

Activity 4

If you need additional help on this topic, open the Help menu; or refer to the documentation for your hardware.

Step-by-Step

Set a Home Page

In this exercise, you will use your Web browser's Menu bar and toolbar.

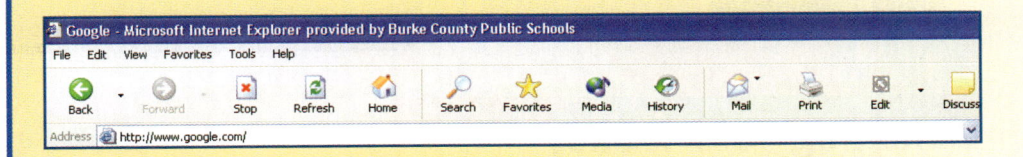

1. Click **Tools** in the Menu bar. Then click **Internet Options**.

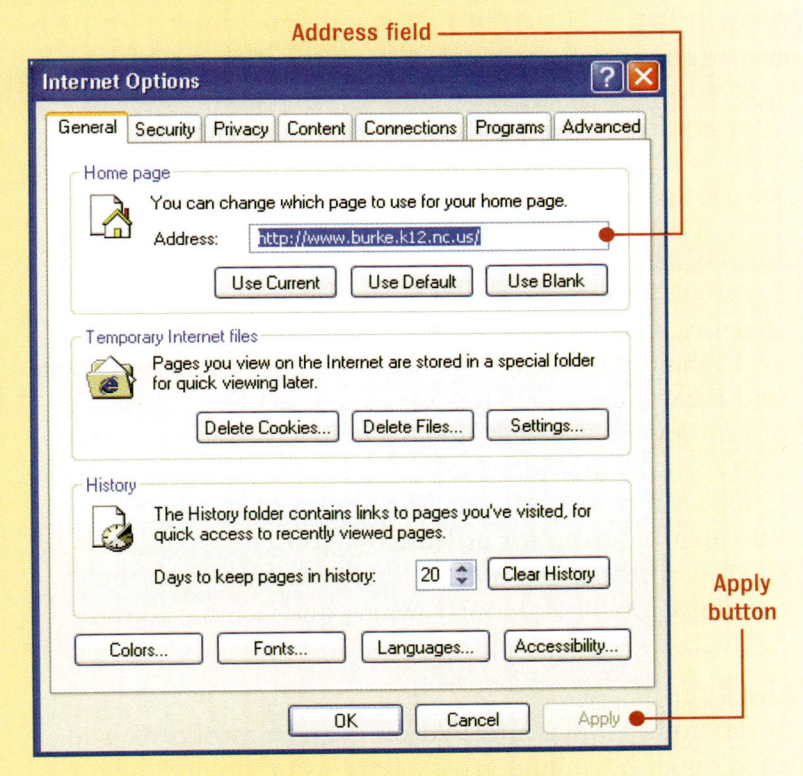

2. In the **Address:** field near the top of the window, enter `digicom.glencoe.com`.

3. Click **Apply** to set the new home page.

4. Close the browser.

5. Open the browser again. It will open at the home page you have just set. While browsing, you can also go to the home page at any time by clicking the Home button in the toolbar.

DIGI Byte

Customizing a Browser

There is a lot you can do to change the way a browser looks and the way it works. You can change the size of the text you see on the screen, the font style, and much more.

Continued

Chapter 3

64

CHAPTER 15

DIGITAL Dimension → Math
INTERDISCIPLINARY PROJECT

Managing Your Money

Using the skills and knowledge you have gained in this course, you will create a spreadsheet and compare salaries.

Comparing Salaries When you look at job compensation, pay is sometimes given in terms of hourly wages and sometimes in terms of an annual salary. You need to know how to convert hourly wages to annual salaries so you can compare the compensation you will receive from similar jobs.

Investigate

1. Read the information sheet Your Pay, Taxes, and Benefits. Ask your teacher for the information sheet, or go to the Online Learning Center at **digicom.glencoe.com** and click Chapter 15, Resources, Digital Dimension.
2. Visit a few of the Web sites you have looked at in this unit for jobs that interest you.
3. Create a spreadsheet to collect the salary or wage information given for at least five jobs.
4. In your spreadsheet, include three columns with the headings Job Title, Hourly Wage, and Annual Gross Salary.

Calculate

1. If the hourly wage information is given on the Web site, determine the annual gross salary using the formula:

    Annual gross salary =
    hourly wage × hours per week × weeks per year

    If the annual gross salary is given, determine the hourly wage using the formula:

    Hourly wage =
    gross salary / (hours per week × weeks per year)

    Assume a 40-hour workweek and pay for all 52 weeks of the year.

2. Fill in the information in the appropriate columns in your spreadsheet. Give a title to your spreadsheet.

E-Portfolio Activity

Showcase Your Spreadsheet Skills

To showcase your spreadsheet skills, set up a file that provides the information requested for a number of different jobs.

Save the document as an Excel spreadsheet file in your e-portfolio.

Self-Evaluation Criteria

For each job in your spreadsheet, did you include:

- Job title?
- Annual gross salary?
- Hourly wage?

---

digicom.glencoe.com

Activity 5

Step-by-Step continued

Print Web Pages

It is easy to print a Web page, but you will want to preview it first. That way, you will make sure you are printing only the pages you need.

1. Go to the Web site of the Future Business Leaders of America at **www.fbla-pbl.org**.

2. Click **File**, then **Print Preview**.

   The Print Preview screen tells you how many pieces of paper are needed to print the Web page.

3. Look at the Toolbar. Note that you can change the page size percentage and view more than one page at a time.

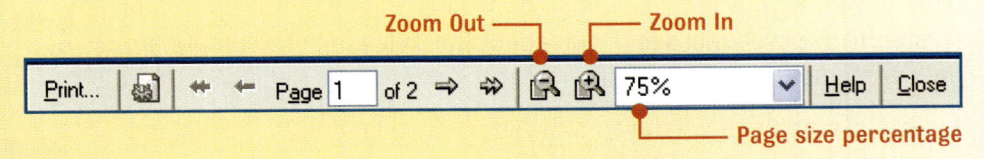

4. Click the arrow where 75% is displayed. Select 50%. Now try 100%. Note what happens in the Print Preview window when you change the page size percentage.

5. Click the **Zoom Out** button. Click the **Zoom Out** button again.

6. Now click the **Zoom In** button.

7. To get ready to print the pages, click **Print...**.

8. In the dialog box, click **Print**.

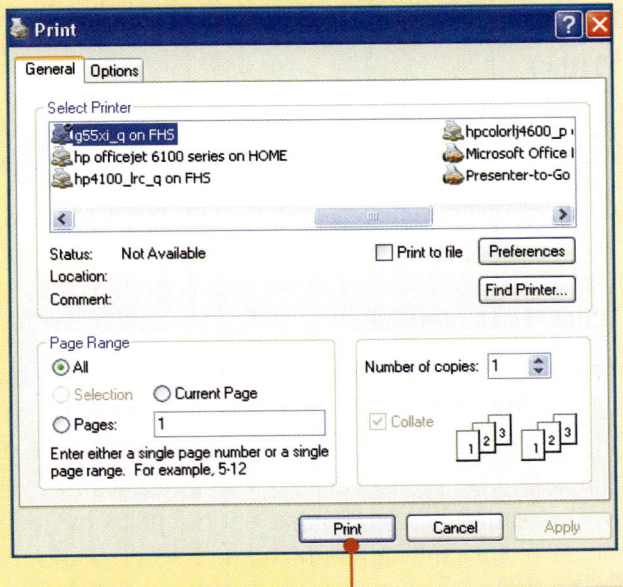

DIGI Byte

Cookies

A cookie is a small file that is entered into your browser when you access certain Web sites. If a Web site places a cookie in your browser, the site will "remember" you when you return. If the site requires a username and password, for example, you may not have to enter that information each time you visit the site.

Continued

PERSONAL Career Perspective

Part 4 of 4

Your Career Path

This is the final activity of a four-part project where you explored your career path. You may have already completed Parts 1, 2, and 3, pages 443, 464, and 491. Now, you will present your personal career story.

Leadership

You are currently working in your career of choice, and your supervisor has asked you to evaluate your leadership expertise and your ability to work as a team member in your daily life. To truly evaluate yourself, you must be willing to recognize both your strengths and weaknesses. Think about the following statements and then write down your thoughts.

1. Give examples of when you showed positive leadership skills and when you felt you could improve your leadership qualities in your home, at school, in a youth group, sports team, or among your friends. Be specific about the situations.

2. Give examples of when you showed excellent team-building characteristics and where you could improve in this area.

3. Consider three things that would help you become a more successful leader and team player in all areas of your life. Make them short-term goals for your future. Write down these goals.

Using your notes, write a brief report on your leadership skills, if possible using a Tablet PC.

Panel Presentation

Your class will give a panel presentation. Be prepared to answer questions on the topic of "You in the 21st Century Workplace." Work together as a class team and, with your teacher, gather local community representatives, such as parents, industry leaders, and so on to be your audience. Ahead of time, e-mail audience members an outline of the activities you have explored in this unit, and ask them to come with at least five questions.

Create a rotating panel presentation. Everyone in the class takes a five-minute turn on the panel. When you join the panel, state your name and the career path you have chosen. Have your audience members ask questions on topics such as leadership skills, mentors, 21st century workplace skills, workplace ethics, personal budgets, working collaboratively, resolving conflicts, and so on. Answer each question briefly, stating the main points and giving examples.

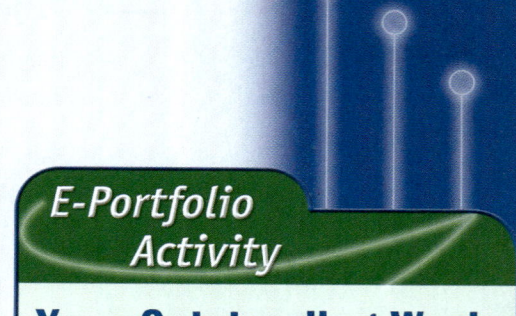

E-Portfolio Activity

Your Outstanding Work

Arrange for the panel presentation to be recorded. Save the recording to your e-portfolio.

Self-Evaluation Criteria

Did you:

- Give examples of your leadership skills and then write a report?
- Assist in arranging the "You in the 21st Century Workplace" panel presentation?
- State your name and your career of choice when you were on the panel?
- Answer the questions briefly and to the point?

Step-by-Step *continued*

Do a Yahoo! Search

Activity 6

If you need additional help on this topic, open the Help menu; or refer to the documentation for your hardware.

1. Open your browser. Go to the Yahoo! home page at www.yahoo.com.
2. Enter the words "white tiger" in the search box. Press ENTER or click the **Yahoo! Search** button.
3. Note the number of results found near the top of the page. There are likely several million Web pages that contain your search words.

— Number of results
— Advanced Search link

Do an Advanced Search

Activity 7

1. On the results page from the last Activity, click the **Advanced Search** link near the top of the page.
2. Enter the word "habitat" in the box labeled "any of these words."
3. Click the **Yahoo! Search** button.

   This time, there are not as many results because you limited your search by adding "habitat."

— Yahoo! Search button

DIGI*Byte*

Hits or Results

When talking about Web searches, "hits" is another way of saying "results."

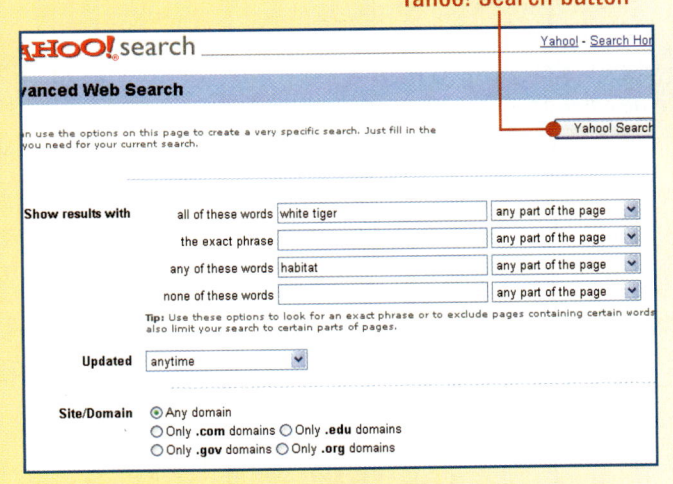

Continued

Chapter 3

66

PERSONAL Career Perspective

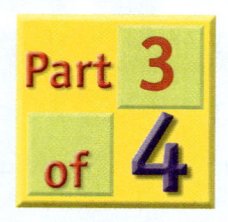

Part 3 of 4

Your Career Path

This is the third of a four-part project that gives you the chance to explore and present your own career story. You may have already completed Parts 1 and 2 on pages 443 and 464. In this chapter, you have prepared a résumé and a cover letter, and practiced for an interview. What is the next step? Get ready for the face-to-face interview!

Marketing Yourself Is Not Bragging

The interview is your chance to market yourself as the best person for the job. When you give facts and examples about your skills and abilities, you are being sincere and truthful. You are not bragging.

To prepare to market yourself:

1. List honors, awards, and words of praise that you have received at school, work, or in teams and clubs. If possible, use your PDA.
2. Write a sentence or two for each item you listed in Step 1. Explain why you received each.
3. Think of four or more times you have been proud of your skills and abilities in getting a task done. Write a sentence or two describing each task.
4. Include how you used your skills, such as problem-solving and communication skills, and your abilities to complete each task.
5. Read aloud the examples you have written.
6. Visualize saying these examples in an interview.

Body Language

Body language, or the way you act, tells an interviewer a great deal about you. What do you think when you are talking to a person and they are making eye contact with you, nodding their head, and sitting up straight? You probably think that they are interested in what you are saying and paying close attention.

Role-play an interview situation with a classmate where you market yourself using the examples you practiced in Steps 1 through 6 above.

- Take notes on the body language that you observe.
- Switch roles and again take notes on body language.
- Share your observations in a class discussion.

Look Ahead...

Continue your project in the final activity: Personal Career Perspective Part 4, Chapter 15, page 492.

E-Portfolio Activity

Your Outstanding Work

Gather examples of your outstanding work from school, extracurricular activities, and work. Scan those examples that are documents, if possible. Take digital photographs of other examples. Save the scanned files and photograph files to your e-portfolio.

Self-Evaluation Criteria

To prepare to market yourself did you:

- List honors, awards, and words of praise that you have received and explain why you received them?
- Write about four or more times you have been proud of your skills and abilities in getting a task done?
- Read aloud the examples you wrote and visualize using them in an interview?

For the role play, did you:

- Take notes on the body language that you observed?
- Switch roles and again take notes on body language?
- Share your observations with the class?

Activity 8

DIGI Byte

Varying Results

Search results may vary from day to day. Web sites are constantly being added to and removed from the Internet.

Suite Smarts

Microsoft Office has an online search feature built into the Help menu of all of their programs. You can quickly access Help resources online when you choose this option.

Activity 9

Step-by-Step continued

Do a Google Search

1. Go to the Google Web site at **www.google.com**.
2. Key the words "white tiger" in the search box, and click the **Google Search** button. Notice the number of results.
3. Use the scrollbar on the right side of the screen to move down the page. At the bottom, you will see links to additional pages.
4. Click **Next** to move to the next page of search results. Search engines try to show the best results on the first page, but the following pages will also have useful information.

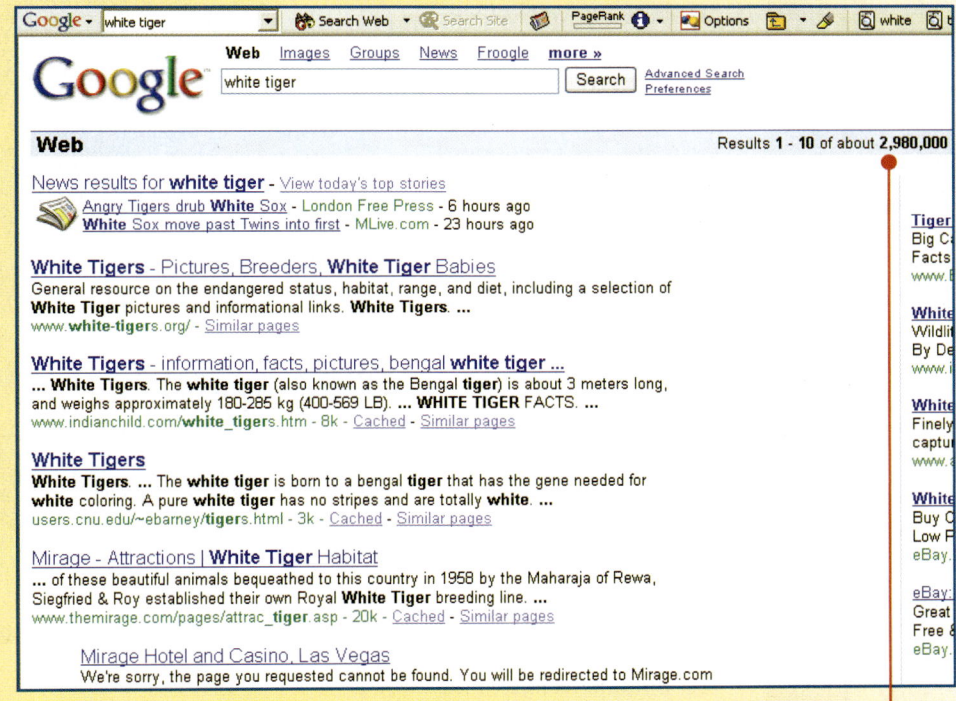

Number of results

Do a Boolean Search

1. Draw a table with two columns and three rows.
2. In the left column, write "camera OR scanner" in the first row. Write "camera AND scanner" in the second row. Write "camera NOT scanner" in the third row.
3. Go to the Yahoo! Web site at **www.yahoo.com**.
4. Search with each phrase in the table you created.
5. Fill in the second column of your table with the number of search results you receive.

Chapter 3 Using the Internet

Lesson 3.2

67

21ST CENTURY CONNECTION

Safety in the Workplace: Keep Yourself Healthy

When you begin a new job, your employer will distribute policies and procedures for protecting yourself from personal injury in the workplace. The items will also be posted in workplace common areas. It will be up to you to learn and review them periodically for additions or changes to the information. Some jobs require special equipment and procedures, such as jobs requiring welding or working with hazardous chemicals.

For most of us, though, simply knowing how to work safely with computers is especially important. Whether you are a warehouse manager, a registered nurse, or a graphic artist, you will use at least one computer program specifically created for your occupation or organization.

Is Your Workstation Suitable For Your Work?

It is critical to your good health and job performance that the conditions in your workspace include comfortable furnishings with good ergonomics and good lighting.

Ergonomics is the science of designing and arranging things people use so that people work efficiently and safely. If your workstation is not designed according to ergonomic guidelines, you could be at serious risk for carpal tunnel syndrome, back problems, eyestrain, or other painful physical conditions. Most businesses will provide ergonomically designed computer stations, chairs, and assistance or instructions to adjust them to fit you.

Protect Your Eyes with the Right Light

Computer screens can sometimes transmit excessive glare, a bright, unpleasant light that reflects off your screen and can harm your vision. Be sure to have a properly fitted glare screen protector if one is not already built into it. If you wear glasses, glare-resistant lenses are a good form of additional protection. They are available from your eye doctor. Make sure there is sufficient light in your work area to help you avoid eye strain.

Stretch to Stay Clear and Focused

Sitting for a long time at your desk can make you feel inflexible and tired. Remember to take periodic, short stretch breaks at your workstation to help increase your circulation and keep your mind clear and focused.

If you have any doubts about your ability to do your job safely and effectively, talk to your manager or safety officer. Take the initiative to ensure your health and productivity.

Activity Write a paragraph describing four preventive measures you can take at work or at home to avoid workspace-related personal injuries.

Application 2

Directions Change your browser's home page to your school's Web page. Print preview the new home page.

Application 3

Directions Choose one of the following questions to answer:

1. How many volcanoes are there in the United States?
2. What is the population for the state you live in?
3. How many people voted in the 2004 Presidential election?
4. How many books are published worldwide every year, on average?

Open your browser. Use a search engine to do searches that will help you answer the question you chose. If your initial results are not helpful, try an advanced search or Boolean search.

RUBRIC LINK

Go to the Online Learning Center at **digicom.glencoe.com** and click Lesson 3.2, Rubric, to find the application assessment guide(s).

DIGITAL Decisions

Search for Poetry

Internet Investigation You have been assigned a poetry project for your English class. Gather the following information about an American poet:

- Date of birth.
- Where he or she lived as a child.
- Educational background.
- Most famous poems.

Use three different search engines to gather information about the poet you choose. If you do not know the names of any American poets, begin your search on the Internet with the key words "American poets." Write a paragraph comparing the information you find in the different search engines.

DIGITAL DIALOG

Search Engine Keywords

Teamwork Work with another student. Create a diagram similar to the one shown. In the center circle, write one of the topics that appears in the list below. In the outer circles, add descriptive words you could use with a search engine to search for the topic on the Internet.

Topics:
- Baseball
- Cars
- College
- Science
- Music
- Sports

Choose three topics from the list, and create one diagram for each topic.

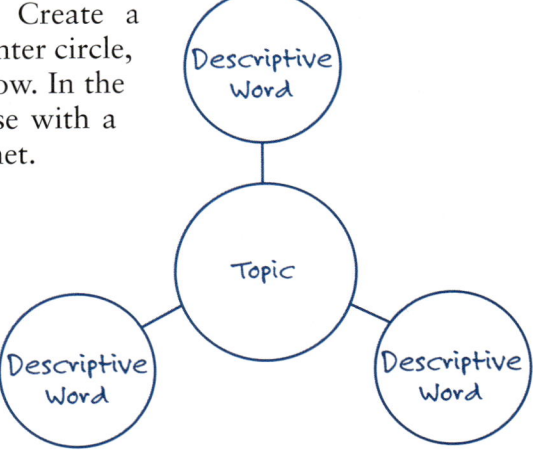

Activity 9

If you need additional help on this topic, open the Help menu; or refer to the documentation for your hardware.

Activity 10

DIGI Byte

Salary Questions

In general, do not bring up the topic of salary. If the interviewer asks you the salary you want, you can choose to say you are sure the company pays fair compensation for this type of job.

RUBRIC LINK

Go to the Online Learning Center at **digicom.glencoe.com** and click Lesson 15.5, Rubric, to find the application assessment guide(s).

Step-by-Step

Practice Interview Responses

1. Choose a student as your partner. Choose a job in which you are interested. Tell your partner which job you chose.

2. Decide which of you will be the interviewer first and which will be the job candidate. If you are the interviewer, choose two questions from the list in the 21st Century Connection on page 488. Ask your partner the questions, and let him or her know how well you think he or she answered. If you are the candidate, answer the questions as best you can. Look at the questions first, and think of some possible answers.

3. Trade roles and repeat Step 2 above.

Follow-Up after an Interview

1. Open your word processing software, and format the page for a business letter.

2. Write a short, one-page thank-you letter. First, thank the interviewer for taking the time to meet with you.

3. Mention again any particular strengths of yours that you feel are important to your success at this job. Also, if the interviewer made any specific comments about a weakness in your qualifications or mentioned any objections to your ability to do the job, this is your chance to try to overcome them.

4. End your letter with a closing. Leave several lines blank so there is room for your signature. Then enter your name.

5. Proofread a printed copy, and if possible, have someone else proofread it as well.

6. Name the file *Your Name* **Thank You,** and save the file.

7. Print the thank-you letter on good-quality bond paper.

Application 7

Scenario In Chapter 15, Application 1, page 476, you prepared a résumé for a position of camp counselor at Camp Evergreen. You have just found out that the camp director wants to interview you for the job.

Directions What questions is the camp director likely to ask? What questions do you have for her? Make a list of how you will prepare, if possible using your PDA or Tablet PC.

DIGITAL Career Perspective

Retail/Wholesale Sales and Service

Yvette Herrera
Marketing Professional

Building Relationships Between Corporations and the Community

As a marketing professional, Yvette Herrera develops ideas that will promote a client's product and figures out how to get those ideas to potential customers. She then works with others to create materials that get those ideas across, such as newspaper ads, billboards, and postcards.

After working for large companies for several years, she decided to start her own business. "I wanted to focus my efforts on encouraging corporations to support the community through non-profit organizations," she says. She also wanted the flexibility to be able to develop her own projects and work with clients who have similar goals and interests.

"A typical work day for me often includes meeting with a client to discuss their resource needs. We brainstorm ideas that would help them develop funding relationships with corporate donors. For example, a client who provides non-profit child and health care services might ask me how they can strengthen their efforts to fund a parenting class program," she says.

Herrera takes her miniature USB flash drive with her on appointments. She says, "It holds a lot more data than a floppy, is easy to travel with, and downloads quickly." She can plug it into a client's computer port and download documents for planning materials and designing new Web sites.

"On days when I am in the office, I develop the ideas we discussed in our meeting or work with outside communications contractors to turn our plans into real materials," she says. She and her business partner also handle the paperwork. She says, "My partner and I scan our documents, e-mails, and contacts to organize and store them in Access files. This frees space in the office for active information and helps us keep client and vendor contacts within easy reach."

"Preparation, planning, and persistence guide my partner and me in our effort to meet our clients' goals. The rewards are definitely worth all the hard work," says Herrera.

Training
Employers prefer candidates with liberal arts degrees and related experience, but requirements vary by job. For some marketing or sales management positions, employers prefer degrees in business administration and marketing.

Salary Range
Typical earnings are in the range of $45,000 to $65,000 a year.

Skills and Talents
Marketing professionals need to be:

- Able to form and maintain effective professional relationships
- Strong, persuasive verbal and written communicators
- Creative
- Highly motivated
- Flexible
- Effective, decisive leaders
- Able to exercise good judgment
- Able to come up with new ideas quickly
- Prepared with backup ideas for clients
- Familiar with all kinds of media

Career Activity
Find out about another career in retail or wholesale sales and service. What training, skills, and talents are needed?

21st Century Connection

Interview Success = Preparation + Practice

The more you prepare for meetings with potential employers, the better your chances for having successful interviews and receiving job offers will be.

Step One: Their Questions and Your Questions Interview questions fall into two categories: factual questions about your education and experience, and questions that require more thought on your part. You should also come to an interview ready to ask a few questions of your own. Create two lists before the interview. In one, list and answer questions an interviewer will probably ask you, such as:

- Tell me about yourself.
- Are you more comfortable working in a group or on your own?
- Tell me about the accomplishment that you are most proud of.
- Can you tell me about a difficult situation you have been in at work and how you resolved it?
- Why do you think you are the best person for this job?

Take some time to think about your answer to each question.

In the other list, include questions you might ask the interviewer, such as:

- How would you describe the ideal person for this job?
- What kinds of projects will I be doing?
- What goals would you expect me to fulfill during my first few weeks here? What goals would I have during my first several months?
- How do you measure employee performance and give feedback?
- What growth opportunities exist?

Step Two: Practice Once you have thought about your questions and answers, ask a parent, another adult, or a friend to be the employer in a practice interview with you. Remember that you are also deciding if the job would be right for you, so your questions for the interviewer are equally important.

- Before beginning to practice, visualize yourself in a real interview. Picture yourself in conversation with the interviewer. You are confident, professional, and relaxed. You are listening well and articulating your responses and questions clearly.
- As you speak and listen, maintain eye contact. Show your interviewer you are genuinely interested in what he or she has to say.
- If you can, record the practice run on video or audiotape, so you can hear or see it yourself afterward.
- After the first practice, ask your parent or friend to critique your appearance, tone, and answers. Use their feedback with your own thoughts to adjust your approach for a second practice interview.

Activity Choose one of the questions above that an interviewer might ask you. Write down the first answer that comes to mind. Now take a minute to think about and revise the answer. Which version of your answer do you think would be more impressive to a potential employer?

Lesson 3.3 — Security and Viruses

You will learn to:
- **Describe** types of security used on the Internet.
- **Identify** viruses and how they work.

Internet Security Internet security has become a major issue for schools and organizations that use the Internet in daily operations. Businesses that sell products over the Internet have a responsibility to keep their Web sites safe for customers. Many people use credit cards to make online purchases. Hackers penetrate computer systems and try to steal credit card numbers. Security is important at home, too.

Firewalls Computers need to be protected so that no one can gain unauthorized access to them. One way to protect a computer at home or at work is with a firewall, which you learned about in Chapter 2, pages 47–49. A firewall is software that acts as a barrier between your computer and the Internet.

Anti-Virus Software A virus is a small computer program that can copy itself repeatedly in a computer system and cause a variety of problems. Viruses are often hidden inside harmless-looking e-mails or files. When you download the e-mail or file, you also download the virus without realizing it. Computers can be protected with anti-virus software. This software will alert you if a file you are trying to open contains a virus. The software will also clean an infected file and get rid of the virus for you. You need to update the anti-virus software often because harmful new viruses are constantly being released on the Internet.

virus A small computer program that can copy itself repeatedly in a computer system and cause a variety of problems.

✓ Concept Check

Why is it important to have anti-virus software on all business and personal computer systems?

DIGITAL DIALOG

Anti-Virus Software

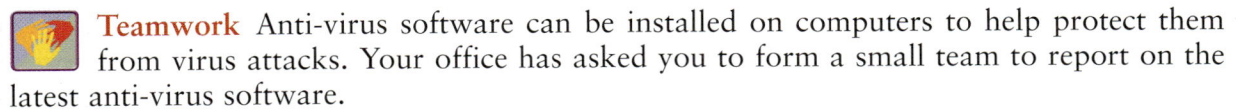

Teamwork Anti-virus software can be installed on computers to help protect them from virus attacks. Your office has asked you to form a small team to report on the latest anti-virus software.

Writing Search the Internet for information about anti-virus programs. Prepare a written report about the information you find. Explain in your report the reasons for installing this type of software. Share this information with the rest of the class.

Lesson 15.5 → Succeed in Interviews

You will learn to:
- **List** what you need to do to prepare for an interview.
- **Develop** interview techniques.
- **Write** thank-you letters to follow up with contacts after an interview.

You Already Have Some Interview Skills Have you ever asked your neighbors if they would hire you to mow the lawn or your parents to allow you to go somewhere with friends? To make these requests, you had to be confident, know your capabilities, and let your neighbors or parents know they could trust you. These experiences will help you prepare for an interview for a job.

Preparing for an Interview Your goal is to show the employer that you have the skills to match the job requirements, are the right person for the job, and will contribute to the employer's success.

Before the Interview Prepare to answer common interview questions, such as, "Can you give an example of how you work in a team?" Also, prepare to ask questions about the position, for example, "What task or project will I help with first?" Research the company so you can show background knowledge of the company and its products. Practice being in an interview situation. The next 21st Century Connection, Practicing for an Interview, page 488, will give you a chance to practice.

On the Day of the Interview To help you be successful in an interview, on the day:
- Arrive about 15 minutes before your appointment.
- Be neatly dressed in a businesslike manner.
- Bring several copies of your résumé, a copy of your references, and a pad, PDA, or Tablet PC to take notes.
- When the interviewer arrives to take you to the meeting room, stand and greet him or her and shake hands.
- Leave a positive impression at the end of the interview by again standing, shaking hands, looking the interviewer in the eye, thanking him or her, and saying, "I look forward to hearing from you."

Follow-Up after an Interview Send a thank-you letter to the person who interviewed you as soon as possible after the interview. Use this opportunity to remind the interviewer of your qualifications for the job.

✓ Concept Check

How will preparing for interviews help you?

DIGIByte

Practice Your Responses

You can use your word processing software or PDA to work on answers to the questions you are likely to be asked during the interview. Sometimes writing answers down helps you remember them better than simply repeating them aloud.

Activity 10

Step-by-Step

Security and Viruses

Microsoft's Web site is one of many locations where you can learn about Internet security and viruses. Microsoft offers frequent updates to keep your computer secure and protect your privacy. Your computer can be set up to let you know when a new update is available.

1. Go to the Microsoft Web site, **www.microsoft.com**, and click the **Windows Update** link.
2. Click the **Frequently Asked Questions** link.
3. What are the three basic types of updates you can get?

Activity 11

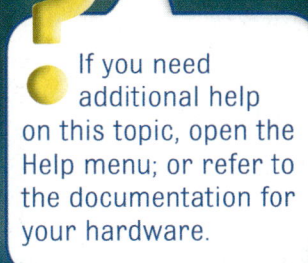

Learn About Viruses

Your computer can spread a virus to another computer without your even realizing it. That is one good reason why knowing about viruses is so important.

1. Go to the Google Web site or the search engine of your choice.
2. Search for information about viruses, worms, and Trojan horses.
3. Create a table similar to the one below. Fill in the information from the Web sites you found.

Type of Risk	Description
Virus	
Worm	
Trojan horse	

> If you need additional help on this topic, open the Help menu; or refer to the documentation for your hardware.

RUBRIC LINK

Go to the Online Learning Center at **digicom.glencoe.com** and click Lesson 3.3, Rubric, to find the application assessment guide(s).

Application 4

Scenario A friend is buying her first computer and wants to know about the security risks.

Directions On a sheet of paper, write three things your friend should know. For example: You can use firewall software to create a barrier between your computer and the Internet.

Activity 7

> If you need additional help on this topic, open the Help menu; or refer to the documentation for your hardware.

Step-by-Step

Job Lead Contact List

1. Open a new file in your database software.
2. Choose one of your existing files that contains names and contact information or job leads.
3. In the new file, make a new entry for each job lead. Enter all the information you have about the lead: company name, address, telephone number, contact name, e-mail address of contact, a short description of the position, and any other information you think might be helpful.
4. Add another column or entry field for comments about how you think these people may be able to help you in your job search. Do they know managers or human resources personnel at potential employers' offices? Do they have experience in a field of interest to you?
5. Add another column or entry field to keep track of what you did or will do to follow up on the lead and when.

Activity 8

Research Companies to Prepare for an Interview

1. Enter the name of a company you are interested in into a search engine.
2. From the results, select the company's official site. Find two pieces of information that give you a better idea of what it is like to work for the company.
3. Some of the results from the search in Activity 8, Step 1, will be from sources other than the company, such as news articles. Find one piece of information from an article that could be useful in an interview with the company.

RUBRIC LINK

Go to the Online Learning Center at **digicom.glencoe.com** and click Lesson 15.4, Rubric, to find the application assessment guide(s).

Application 6

Directions Think of a job in which you are interested. Using a search engine, find two companies that might offer that job.

Enter the companies in your job lead database.

Then, enter the contact information you will need. Enter any information you find that tells you what it is like to work there or that could be used in an interview. You can find this information on the company's official Web site or on other sites.

21ST CENTURY ⟷ CONNECTION

Cyberbullies

Ask a friend if he or she has ever received a threatening e-mail, and your friend is apt to reply, "Yes, it feels bad, but haven't we all?" According to ChannelOne.com, more than a million students are harassed and bullied each week in cyberspace. Cyberbullies are not found just in the United States. A recent study in Great Britain reported that one in four children had been bullied either on the Internet or through cell phones.

Hurtful Messages

Cyberbullies use any communication medium they can to get their destructive point across, including text messaging on cell phones, instant message programs, chat rooms, e-mail, and Web sites containing embarrassing photos and hurtful comments. Some cyberbullies send vicious messages about people they know to a network of friends. Other cyberbullies send threatening messages directly to individuals as a form of harassment. Students have reported receiving messages that threaten their life or physical safety.

What Should You Do?

If you receive mild harassment online, it is best to ignore the message. If you are on the receiving end of severely harassing or threatening messages, let someone know. The Center for Safe and Responsible Internet Use recommends that students speak out when they become aware of cyberbullying. Enlist the help of a teacher, coach, counselor, or parent. Cyberbullies do get caught, and are punished.

Giving the Wrong Message

Are you ever a cyberbully without realizing it? It is easy for cybercommunications to be misunderstood. If you are feeling angry when you are writing a message, think before you send it. Take a break from the computer, and do something relaxing before going back to reread your message. Remember that messages may not be private and may be sent to the wrong person. Would you want your family or other people you know to read what you have written? Have a look at the results of a search online using your own first and last name as the keywords.

Technology Lifeline

Today's technology is a social and career lifeline. It is how we all keep in touch. It is often a vital tool to look for and find jobs. It is important that you know how to protect yourself from online harassment.

> **Activity** To help protect yourself from cyberbullies, enter the keywords "Center for Safe and Responsible Internet Use" in a search engine. Look up Internet use guidelines.
>
> If you have not tried this already, try searching online for your own name.

Lesson 15.4 — Manage Your Job Search

You will learn to:
- **Set up** and **organize** files to help you keep track of job leads and your progress.
- **Collect** and **manage** information about potential employers.

Suite Smarts

Spreadsheet software provides many of the functions you need to develop a job-lead database. You can enter names, contact information, and follow-up information in different columns, and sort and search the information. You also can make mailing labels and e-mail lists from the spreadsheet file of contacts.

Organizing Your Job Search

Some experts say that looking for a job is a full-time job itself! To get the most out of your efforts, you will need a system for keeping track of leads and when and how you follow up.

You can set up a job lead tracking system on your computer's word processing, spreadsheet, or database software or in the Address Book and To Do applications of your PDA. For each job lead, set up a new entry in your software. **Figure 15.4**, below, is an example of such an entry. You can also track referrals and networking contacts with this system.

Research Potential Employers

Researching the company to which you are applying will give you a better idea of what working there is actually like. This research will also be helpful if you are interviewed for the job. It makes a good impression on an interviewer when you can ask questions that indicate that you have done some "homework" on the company. Simply enter the name of the company in the keyword box of a search engine. Another way to find out more about the company is to visit it.

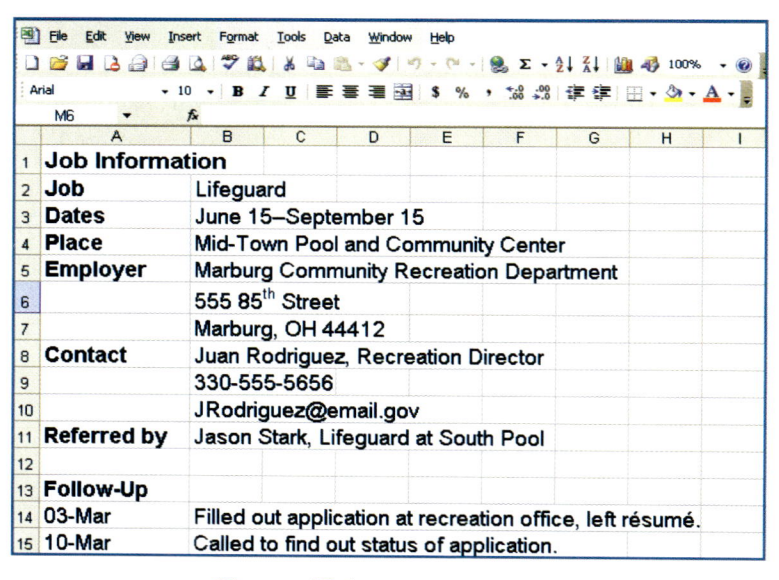

	A	B
1	Job Information	
2	Job	Lifeguard
3	Dates	June 15–September 15
4	Place	Mid-Town Pool and Community Center
5	Employer	Marburg Community Recreation Department
6		555 85th Street
7		Marburg, OH 44412
8	Contact	Juan Rodriguez, Recreation Director
9		330-555-5656
10		JRodriguez@email.gov
11	Referred by	Jason Stark, Lifeguard at South Pool
12		
13	Follow-Up	
14	03-Mar	Filled out application at recreation office, left résumé.
15	10-Mar	Called to find out status of application.

Figure 15.4 *Job Information*

✓ Concept Check

Why is it a good idea to keep all job lead information in one place?

Lesson 3.4 — Copyright, Fair Use, and Plagiarism

You will learn to:
- **Define** copyright, fair use, and plagiarism.
- **List** examples of copyright, fair use, and plagiarism.

copyright A type of legal protection for works that are created or owned by a person or company.

fair use Describes when you may use a copyrighted work without permission.

plagiarism Using other people's ideas and words without acknowledging that the information came from them.

Computer Ethics In Chapter 2, pages 50–52, you learned that ethics are the rules we use to define someone's behavior as right or wrong. In all areas of life, you are responsible for your actions. On the Internet, for example, a tremendous amount of information is available. How you use that information is determined by your sense of ethics.

Copyright A copyright is a type of legal protection for works that are created or owned by a person or company. A copyrighted work could be a song, a novel, a play, a software program, or an invention, to name a few. If you find a copyrighted work online that you want to copy parts of for your own use, the law says that you need to ask permission first. If you are not sure whether a work is copyrighted or not, you should assume that it is and therefore not copy it.

Fair Use Is it always illegal to borrow from a copyrighted work without permission? Definitely not. Fair use describes when you may use a copyrighted work without permission. How can you tell if fair use applies to your situation? It depends on how you plan to use the copyrighted material. Consider these examples:

- A teacher copies part of a book to hand out to students.
- A student uses a small amount of copyrighted material in a report and gives credit to the original author.
- A book critic quotes a paragraph from a novel in a book review.

These uses would all be considered fair use, because the material is being used for purposes of criticism, news reporting, education, or research.

Plagiarism Plagiarism is using other people's ideas and words without acknowledging that the information came from them. It is okay to use a phrase or sentence written by another author, but you must always acknowledge the author's work. Plagiarism is a problem in schools. Students often get in trouble for copying entire papers from the Internet. Software now helps teachers scan student papers for plagiarism.

✓ Concept Check

Give reasons why you think authors and composers are careful to copyright their work.

21ST CENTURY CONNECTION

Job Application Essentials

Employers will look at a job application as a sample of your abilities. Applications should be completed with the same care you would use for your résumé, and the information reported in them must be consistent with your résumé.

Basic Guidelines

- Read through the entire application carefully before writing anything on it. If you have any questions about the form, ask someone in the organization for help.
- Type your information onto the application, if possible. Otherwise, write as neatly as you can in dark blue or black ink.
- If you are completing the application at home, make a copy of the application to fill out as a practice. Then do a final copy for the employer.
- If you are completing the application at a business, be sure to bring the right documents along to help you. These include your driver's license or state identification card, your Social Security card, the names and phone numbers of at least three references, and a copy of your résumé for reference.
- Get permission from your references before sharing their names and phone numbers with anyone.
- Be honest on your application. Your information will be verified with past employers, your references, and academic institutions.
- Include any relevant activities or occasional jobs you have had, such as school newspaper editor, scorekeeper, or dog walker. You have learned many transferable skills, or skills you can use in almost any job, through these activities.
- List any academic honors, awards, or community service.

Final Steps

- Proofread the application carefully, making sure your ID numbers, dates, and contact data are current and accurate. Also check for spelling and grammar mistakes. Ask a family member or friend to read the application as well, if possible.
- Make a copy of the completed application for your personal files. It will be useful to you in preparing for a job interview or in filling out other applications.
- If you are mailing the application, do not fold it. Send it in a large envelope. A form that is not creased is easier to read, scan, and photocopy.

Activity Trade one of the job applications that you filled out for practice earlier in this lesson with another student. Look at the application. Find something on the application that makes you want to hire this person. Why does it make you want to hire him or her?

484

Activity 12

If you need additional help on this topic, open the Help menu; or refer to the documentation for your hardware.

Step-by-Step

Learn About Fair Use and Copyright

The Internet has posed a big challenge to copyright laws. For example, many people download and share music files without realizing they are breaking the law.

1. Open your Web browser and go to the Google search page, www.google.com.
2. Do a Web search for the keywords "fair use," including the quotation marks. What kind of information do you find?
3. Do a search for the keyword "copyright." Explore the links that interest you. Write brief notes to summarize what you find.
4. Go to the Microsoft Web site, www.microsoft.com. Scroll down the page and click the **Legal** link at the bottom.
5. What legal information is listed on this page?

RUBRIC LINK

Go to the Online Learning Center at **digicom.glencoe.com** and click Lesson 3.4, Rubric, to find the application assessment guide(s).

Application 5

Directions Think of occasions in school when you have written a report or prepared a presentation.

- Did you give credit to authors whose writing you used?
- Did you ever copy and paste a graphic from the Internet?
- Based on what you have learned in this chapter, would you do anything different now?

Research on the Internet to find out about plagiarism. Write a short paragraph explaining what you know about copyright, fair use, and plagiarism.

DIGITAL DIALOG

Hidden Viruses and Spyware

Critical Thinking Research on the following keywords: "hidden viruses" and "spyware." Think about these types of programs and computer ethics.

- How are hidden viruses and spyware unethical?
- How can you protect yourself from these types of programs?
- How do you think such threats will be handled in five years?

Write a paragraph or two that summarizes your thoughts.

Activity 5

If you need additional help on this topic, open the Help menu; or refer to the documentation for your hardware.

Step-by-Step

Fill Out a Job Application Form

1. Open one of the résumés you developed in Lesson 15.1, page 475: *Your Name* Resume 1 or *Your Name* Resume 2.
2. In a Web search engine, key "job application forms" in the search box. Include the quotation marks as shown, to ensure that your search results are actual forms. You can also search to see if your state or city government has a job application form at its Web site.
3. Fill out the job application form, using your résumé for help.
4. Compare the job application form with your résumé.
   - Is there any information requested on the application form that is missing from your résumé? Should this information be included in your résumé?
   - Is there information on your résumé that is not required on the form? If so, do you still feel it is important to include in your résumé?

Activity 6

Evaluate Disclaimers

1. Look at the job application form that you used in Activity 5 above, or find any other application form.
2. Identify each disclaimer on the form.
3. Read each disclaimer and write a brief description of what you think it means.

RUBRIC LINK

Go to the Online Learning Center at **digicom.glencoe.com** and click Lesson 15.3, Rubric, to find the application assessment guide(s).

Application 5

Directions On the Internet, search for "job application form," including the quotation marks, and the name of your city. Download, or fill in online, the general application form used by your city for employment. Alternatively, find and fill in another city's form.

- What is your opinion of such general purpose application forms? Do the forms give the employer an accurate and complete view of the candidate's skills and capabilities?
- What are the advantages and disadvantages of a general purpose form?
- What disclaimers are on the form and how do they relate to the job?

Write a paragraph that answers the questions above.

Lesson 3.5 — Computer Crimes

You will learn to:
- **Identify** computer crimes.
- **Learn** about the government's role in solving computer crime.

Cybercrime Cybercrime is the term given to illegal acts committed on the Internet. The Internet has proven to be a haven for many criminals. Criminal activity on the Internet includes fraud, embezzlement, unauthorized use of computers, creating and releasing viruses, harassment, and stalking.

Investigation The FBI, or Federal Bureau of Investigation, has a special team dedicated to investigating computer crimes: the National Computer Crime Squad (NCCS). The crimes they investigate often involve suspects in other countries.

Computers may be used to commit crimes, but they are also used to solve crimes. For example, the National Center for Missing and Exploited Children at www.missingkids.com uses the power of the Web to track down runaways and kidnapping victims.

Protecting Your Computer In Lesson 3.3, you learned ways to keep your computer secure. Securing your computer will help prevent criminal activity.

✓ Concept Check

Why do you think some people choose the Internet as a place to commit crimes?

Suite Smarts

All software programs have a copyright license that states how many computers you can install the software on. If you install the software on more computers than allowed, you could be fined by the software developer.

DIGITAL DIALOG

Computer Ethics

Teamwork Create a small group with other students. Work together to create a list of ideas to share with the other students about ethics, copyright, fair use, and plagiarism. Include your thoughts on downloading music, movies, and other documents illegally.

Critical Thinking Compare your group's list to the other lists created in your class. What are the similarities? What are the differences? Why would a reader be more likely to trust a paper that cites sources versus a paper that does not? Have a discussion with your class.

Lesson 15.3

Fill Out Job Application Forms

You will learn to:
- **Fill out** a job application form.
- **Recognize** legal information included by the employer on the job application.

reference The name of a person an employer can contact to find out about a job applicant.

Job Application Forms When you apply for a job, most employers require you to fill out a job application form. You can use information on your résumé to help fill out an application form. As a job applicant, you are usually required to fill in a section on the application form asking for references. A **reference** is the name of a person an employer can contact to find out about a job applicant. Obtain permission from at least three people that you know, such as a teacher, coach, or your internship supervisor, to be references for you.

It is important to fill in the job application carefully and neatly to make a good impression. Use correct spelling and punctuation and check that you have filled in every answer.

Legal Information on the Application Form Job applicants may be told on application forms that they must provide proof of identity and legal work authorization before they are hired. Employers may include work rules, notification about preemployment drug screening, and other disclaimers that generally protect the employer. The application form may include various disclaimers and statements:

- **Disclaimers** Statements that clarify a policy or protect the employer in case of a misunderstanding are termed disclaimers. For example, the prospective employer might ask you to sign a release form that gives the employer permission to check your references.
- **Employment-at-Will** The form may state that you, if hired, will be subject to employment will. **Employment-at-will** is employment that can be terminated with or without cause by the employer.
- **Statement of Confidentiality** There may be a statement of confidentiality, meaning that all information on the form will be kept private. By signing this document, you promise not to reveal any company secrets such as upcoming projects.

employment-at-will Employment that can be terminated with or without cause by the employer.

✓ Concept *Check*

Why do you think it is helpful to first develop a résumé, even if the jobs you are investigating all require that you fill out an application form?

Activity 13

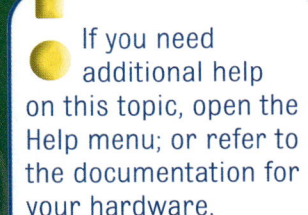

? If you need additional help on this topic, open the Help menu; or refer to the documentation for your hardware.

Step-by-Step

Explore the Anti-Cybercrime Effort

In this lesson, you will learn what the U.S. government is doing to keep the public informed about cybercrime.

1. Open your browser and go to the Department of Justice's Cybercrime Web site, **www.cybercrime.gov**.
2. You will find many links on this page. Click the ones that interest you. Write notes on what you find.
3. This site has special pages for students. Find these pages and write two paragraphs to summarize the information you find.

Application 6

Scenario The government has asked you to design an advertisement that warns the public about cybercrime. The advertisement will appear on the government's Web site and in government pamphlets, brochures, and other publications.

Directions Before you begin, you must determine the target audience and the message that you think will effectively provide the public enough information about cybercrime. Think of a headline for your ad. For example, "Don't Let Cybercrime Happen to You." Then come up with a few ideas for photographs or designs that could accompany the headline. Make a sketch of your ad.

RUBRIC LINK

Go to the Online Learning Center at **digicom.glencoe.com** and click Lesson 3.5, Rubric, to find the application assessment guide(s).

DIGITAL DIALOG

Wireless Computing

Critical Thinking Wireless Internet connectivity is becoming more and more popular. Many desktop computers, laptops, PDAs, Tablet PCs, and cell phones can connect to the Internet without needing a wire plugged into a telephone jack or cable modem. Wireless access points are found in many hotels, airports, and cafés. Some cities even offer wireless Internet access in outdoor spaces.

Some wireless networks cannot be accessed unless the user enters the correct password. Sometimes, networks are password-protected for security reasons. Other times, it is because some companies charge a fee to access their networks.

Writing The mayor of your city is considering a plan to provide free wireless Internet access for the whole city. Write a short letter to the editor of your local newspaper. Express your opinion about the mayor's plan. Support your opinion with specific reasons.

Application 3

Directions Consider the self-assessment you developed in Chapter 13, Activity 6, page 437. What companies in your area could you apply to for positions like this? Check the newspaper and phone book, and talk with people you know about the types of job you are interested in. Make a list of five or more local companies that may have a position for this type of job.

Application 4

Directions Search online for a newspaper in your state or a state you might want to move to. Use the name of the state and the key words: online newspaper. Find the link to the classified section of the paper and search for three types of jobs you are interested in. In your PDA or on a Tablet PC, if possible, write the job title, company name, brief description, qualifications, salary, and how to apply for each position. What type of skills do you need to develop so you can apply for this job?

RUBRIC LINK
Go to the Online Learning Center at **digicom.glencoe.com** and click Lesson 15.2, Rubric, to find the application assessment guide(s).

DIGITAL Decisions

Summer Intern

Scenario Your guidance counselor has talked to your class about internships. Many companies hire interns for the summer so students can learn more about a career they are interested in. Some internships are paid positions, but many of them are volunteer positions. You have decided to find an internship, but you do not know what types of internships are available.

Internet Investigation Search the Internet to find several Web sites that list internships. Research these sites to find four internships that sound interesting to you.

Teamwork In a small group, evaluate the internships you have each found. Prepare a presentation to the class on four of the internships your group has discussed.

DIGITAL DIALOG

Compare Jobs

Critical Thinking Check your local newspaper to see what types of jobs are advertised. Try to look at jobs in two different newspapers so you can compare the positions.

Writing Create a Word table of at least 15 jobs you find in the newspaper. In your table include the name of the job, the name of the company, skills required, education level required, and salary, if listed. Compare your list with the lists created by two other students in your class.

Application 7

Scenario You work for a company as a computer specialist. One of your responsibilities is to stay informed on computer security risks that employees should know about. You are preparing a report on Internet fraud to share with your department.

Directions Investigate the Web site of the Internet Fraud Complaint Center, **www.ifccfbi.gov**. Prepare a brief outline describing the purpose of this Web site. Include an explanation of two of the most recent warnings posted on the site.

Choose two Internet fraud tips to include in the report that you will be sending to all employees.

Answer the following questions in your report:

- Who should check this site?
- How often should this site be checked?
- What Internet fraud statistics do you think the employees should know? List three statistics that you feel are significant and will be persuasive to employees.

Edit and proofread your report before sharing it.

TECH ETHICS

Term Paper for Sale

Scenario You have a term paper due for English in three days. You have had to work extra hours lately, so you have not had time to start the paper. While searching for your topic online, you come across a Web site that boasts complete term papers on any subject. Out of curiosity, you click the link. There are five different term papers on your topic —and they are all for sale!

You ponder this opportunity. Why not, you think? After all, you are exhausted from the extra hours at your job. You figure that if you buy a term paper and make a few changes, your teacher will never know.

? Critical Thinking In a small group, discuss what should be done in this situation. Discuss the questions below and write down your responses. Defend your answers with good reasons.

- What would be wrong with buying a term paper online?
- By changing some of the information, does that give you copyright ownership of the term paper? If so, how much do you think needs to be changed?
- Do you think there is anything illegal about this situation? Explain your answer.
- What would you tell a friend to do in this situation?

Share your group's responses with your class and teacher.

Activity 3

If you need additional help on this topic, open the Help menu; or refer to the documentation for your hardware.

Step-by-Step

Search a Local Newspaper

Use a Sunday newspaper you have brought to class or one your teacher has given you.

1. Turn to the Classifieds section, and locate the Employment listings.
2. Choose four jobs you would like to find out more information about.
3. Create a table with seven columns in Word.
4. Enter the following headings in the columns: Position Name, Company Name, Address, Experience Required, Salary, Education Level, and How to Apply.
5. Fill each cell in the table with the information from the classified ad. If any of this information is not available, enter "Not Available" in the appropriate table cell.
6. Save the file as *Your Name* **Newspaper.**
7. Open another Word document. Enter the name of the first position you listed in your table. Below the position title enter the following text: `Steps I must complete to apply for this job.`
8. List the steps you believe will be necessary for you to apply for the job you have listed. Think about what type of education and experience you will need. You may have only a few items to list or you may have many items in your list. Be thorough.
9. For ideas to list, check the Department of Education 16 Career Clusters you studied in Chapter 13, pages 436–437.
10. Save the file as *Your Name* **Apply.**

Activity 4

Find a Job Online

1. Go to a search engine and search for the state employment office of your state. Use the name of your state and the key words "employment office."
2. Navigate through the site to the Search Jobs section. Look for jobs in your state or your local area.
3. Make a list of four jobs that sound interesting to you.
4. Repeat Steps 1–5 in Activity 3, above.
5. Did the online jobs list more or less information about the positions advertised? Answer this question in the file, and save the file as *Your Name* **Online Job.**

21ST CENTURY CONNECTION

Treasure Hunting with Technology

A Global Positioning System, or GPS, device is a small unit that receives satellite signals. The signals allow the GPS device to calculate the exact position of the device anywhere on the planet, within a few yards.

About the Global Positioning System

The Global Positioning System uses 24 satellites. The first one was launched in 1978, and the last one was launched in 1994. The satellites circle the earth twice a day at an altitude of 12,600 miles. It costs about $400 million per year to maintain the system. The U.S. Department of Defense designed and controls the Global Positioning System, but anyone can use the system.

A New Sport Some people are using handheld GPS devices, which range in price from $100 to $1000, to embark on a new sport known as geocaching. *Geo* stands for geography, and *caching* is the process of hiding a cache, or small treasure.

Geocaching is like an Easter egg hunt on a huge scale. The hunt can take you anyplace on Earth. The latitudes and longitudes of the hidden caches are published on Web sites by geocachers. Caches are hidden in parks, cities, and the wilderness of over 200 countries. A geocacher enters the location of a cache into a GPS device. The device then tells the geocacher how far away the cache is and how to get there.

When the geocacher reaches the coordinates of the cache, he or she still has to find the treasure itself. Sometimes the caches are hidden in places that are difficult to reach, like the side of a cliff or underwater. The discoverer is free to take whatever treasure he or she finds but is expected to replace the treasure with a new find for the next treasure hunter. The treasures are usually small items, such as maps, coins, books, or CDs. A cache also contains a logbook. Each person who finds the cache can enter the date when he or she found the cache and any other helpful information he or she wants to share.

Any geocacher can also start a new cache by choosing a location, hiding a treasure, and publishing the coordinates. There are over 120,000 active caches, and more are being created every day!

> **Activity** Find a geocaching Web site, and enter your ZIP code to see how many geocaches are nearby.

Lesson 15.2 — Find Job Openings

You will learn to:
- **Search** for jobs through local publications.
- **Search** for jobs online.

Help Wanted! Many people read the Help Wanted section of classified ads in a newspaper when they are looking for jobs. The jobs listed here are divided into categories based on the type of job being advertised. Some common categories you may find in all newspapers are General, Professional, Education, and Technical.

Job Banks Local, state and federal government positions are listed in a job bank with local employment offices. Jobs from private industry employers are also listed with your local employment office. All states have an employment office which assists people who are looking for jobs by helping them locate available positions within their community.

Job Fairs A job fair is an event where you can meet representatives of many different companies and talk to them face to face about employment opportunities. Job fairs are often held at local community buildings.

Job Recruiters Many large companies have recruiters who visit community colleges, universities, and local job fairs. A job recruiter seeks future employees for large companies by matching a person's skills to the needs of open positions.

Jobs Online The Internet has become a very useful tool for both employers and prospective employees. Many people post their résumés online or fill out a preliminary screening application when looking for employment. There are many reputable free Web sites dedicated to helping people find jobs and allowing them to post their résumés to apply for jobs online.

Searching online for a job makes it quick and easy to find out about jobs in other parts of the country or even the world. Each state's employment office maintains a searchable Web site. Many newspapers also place their job posting classified section on a Web site.

✓ Concept Check

Think about the type of job you might want to apply for after high school or college.

- Where might you begin looking for openings for this type of job?
- Are there jobs in your area in this field? How can you find out?

Write a short paragraph about what you will do if the only opportunities in the career you are interested in are not in your local area.

DIGIByte

Jobs Online

Jobs posted online give you a description of the position and the skills needed for the position.

DIGITAL Career Perspective

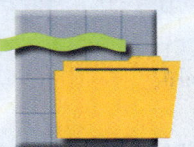

Business Management and Administration

Olivia Acacio
Project Manager

Driving Successful Publication Projects

Olivia Acacio, a desktop publishing manager for a utility company, says, "I love my computers!" Acacio juggles many responsibilities and uses her PC or Mac for nearly everything she does at work. She oversees a wide range of activities, supports multiple departments within the firm, supervises employees, and assigns and tracks projects.

Acacio works with a team to design and write newsletters, advertisements, internal magazines, and other documents. She stays prepared by keeping her digital camera phone handy whenever she is away from the office. "You never know when you're going to see an image you can use in a newsletter," she says.

Acacio and her staff also develop presentations for management with Microsoft Publisher, CorelDRAW®, Microsoft PowerPoint, and many other publishing software packages. PowerPoint is especially important for creating presentations.

"Depending on an internal client's needs, we sometimes create presentations from scratch, using the software to create our own graphics and add photos that we take. Most managers depend on PowerPoint for their discussion materials. They like the attractive, clear, readable messages on the slides."

What does Acacio like most about her job? "Along with the opportunity to have so much creative fun with technology every day, the best part of my job is that I get to meet and work with many wonderful people in every area of our business. I am happy when my team is pleased with their work, and when we meet and exceed client expectations."

"Desktop publishing is an exciting career due to new technology tools. Personal computers now enable us to perform publishing tasks that would otherwise require complicated equipment and human effort. Instead of receiving simple typed text from customers, we receive the material over the Internet or on a computer disk and put our creativity hats on," says Acacio.

Training

Although formal training is not always required, those with certificates or degrees will have the best job opportunities. Students interested in a career in desktop publishing may obtain an associate's degree in applied science, or a bachelor's degree in graphic arts or graphic communications. Internships and part-time work are good ways to gain experience early in a career.

Salary Range

Typical earnings are in the range of $31,000 to $74,000 a year.

Skills and Talents

A project manager needs to have:

- Strong verbal and written communication skills
- Excellent leadership skills
- Excellent project management skills
- A flexible attitude
- Knowledge in and enthusiasm for their project area

Career Activity

Look for other careers in business management and administration. What training and abilities are needed?

21ST CENTURY CONNECTION

Résumé Dos and Don'ts

An employer will get many résumés in response to a job opening, and it only takes one mistake for yours to be removed from consideration. An effective résumé makes the information that is relevant to the job easy for the employer to find. Follow these tips for a clear, concise résumé that will get more interviews!

🟢 Dos

- Match your experience and capabilities with the job's requirements as much as possible. This is one reason preparing several different résumés with different emphases can be very useful.
- Do keep sentences, paragraphs, and bullets short.
- Do use a plain 10- or 12-point font for the body of the résumé. Headings can be slightly larger. Use bold or italic formatting sparingly.
- Do make sure any electronic version of your résumé contains key words that are relevant to the job you are interested in. Prospective employers use key words to find résumés online.
- Do include a cover letter. No résumé should go anywhere without one.
- Do proofread your résumé and cover letter carefully. Reread it yourself several times, and have one or two other people read it as well.
- Do print any hard copies of your résumé on good quality, opaque white bond paper.
- Examine any hard copies to ensure that they have printed clearly.

🔴 Don'ts

- Don't lie on your résumé.
- Don't include anything negative or critical about yourself or a previous employer, such as poor grades or getting fired.
- Don't include personal interests or hobbies unless they are somehow related to skills needed for the job. For example, playing golf is important if you want a summer job as a caddy, and interest in horseback riding or crafts may be helpful if you want a job as a camp counselor.
- Don't include references. If needed, they will be requested at the interview.
- Don't use pronouns in your descriptions. Start sentences with a verb instead.
- Don't include family information.
- Don't include jokes, humor, or slang.
- Don't include anything about salary, either previous or desired.

Activity Think of three key words that you could include in an electronic résumé that relate to one of the jobs in which you are interested. Write a paragraph to explain why you chose these three.

CHAPTER 3

DIGITAL Dimension → Language Arts
INTERDISCIPLINARY PROJECT

Be a Savvy Consumer

Use the awareness and skills you have gained in this chapter on using the Internet to examine advertisements that are placed on Web sites.

Advertisers Want Your Attention Advertisers want you to buy their product. The Internet gives advertisers a new way to reach out to millions of consumers. How often do you open a Web site on your computer and get all kinds of pop-up, banner, or screen ads?

Consumer Beware! Anyone can create a Web site. There is no agency that monitors what is posted to the Internet. How can you be sure the item you are purchasing online is from a reputable seller? Check out the company before you buy, and only use a secure site when you make a purchase.

Examine Advertising Online. Go to a search engine and look up five of the products you listed in the chart you created in the Digital Dimension activity at the start of Unit 1, page 2. Make a list in WordPad that includes:

1. The product name, brief description, and price, if available.
2. An example of a pop-up, banner, or screen cyber ad that appeared.
3. The type of advertising technique:
   - Humor
   - Appeal to your emotions
   - Celebrity endorsement
   - Everyone has the product
   - Other
4. Include your opinion of the advertising technique and whether it would persuade you to buy the product.

Give a title to your document. Save your document as **Savvy Internet Consumer**.

E-Portfolio Activity

Internet Savvy

Proofread and edit the document you saved as **Savvy Internet Consumer**.

Save the document to your e-portfolio to showcase:

- The skills you have learned in this chapter.
- Your ability to think critically and be a savvy consumer.

Self-Evaluation Criteria

In your document, did you include:

- Five products, with a brief description of each?
- An example of a cyber ad?
- Your description, opinion, and analysis of the advertising techniques?

Application 2

Scenario You are a college junior working on a degree in architecture. You want to save some money to help pay for your education. You have agreed to work for the summer at your aunt's resort, as you did last year.

Your aunt has promised that this year you will be in charge of scheduling and supervising the golf caddies, lifeguards, and tennis instructors.

Then one of your professors tells you that a summer internship is available at an important architectural firm in a city about 100 miles away. The firm will provide a small apartment for the summer and a small salary to cover basic expenses. What will you do?

Directions Consider the skills you would gain in each experience—resort supervisor and architectural intern. For each option:

- Make a list of the skills and accomplishments you predict you would gain, if possible using Memo Pad on your PDA.
- List what in your opinion are the long-term advantages and disadvantages.
- List what in your opinion are the short-term advantages and disadvantages.
- Create a résumé to show how each experience will be used on your future résumé. Decide which type of résumé is more appropriate: chronological or skills.

Once you have created both résumés, decide what to do. Will you go with the resort supervisor position, or be the architectural intern? Think how you will explain your decision to your parents, your aunt, and your professor. Present your answers to your class.

TECH ETHICS

Opportunity Decision

Scenario A friend of yours, Joel, who has the same skills and experience as you have in your chosen field, is out of work. You currently have a job, and you are interested in advancement opportunities in your field. You are excited about taking the next step in your career.

You are a member of a professional association and regularly attend meetings. At the association meeting, an employee of CDS Company tells you the company is looking for someone with your qualifications for a challenging position in the near future. Joel, to save money, dropped his membership in the professional association as soon as he lost his job.

Critical Thinking Should you tell your friend about the potential opportunities at CDS Company or keep the information to yourself?

Teamwork Work in small groups and discuss the questions below.

Analyze

- Should your friend have left the organization?
- What will you do?
- What will you tell your friend? Why?

Present your answers to the class, and explain the reasons for your answers.

Chapter 3 Review

Read to Succeed PRACTICE

After You Read

Self-Esteem The more confident you are in your reading abilities, the better you will read and learn. It is important to believe in yourself and have confidence in your own abilities. Do you remember learning to ride a bike? As with many things, when you practice you improve your skills.

Create a chart similar to the illustration. Write an original sentence using the term.

New Term	Original Sentence Using Term
virus	It is important to install virus protection software on my computer.

Using Key Terms

It is important to understand key terms to help you use the World Wide Web as a communication and information tool. Match the sentences below with the correct key term.

- browser (60)
- hyperlink (60)
- uniform resource locator (URL) (60)
- domain name (60)
- search engine (63)
- Boolean operator (63)
- virus (70)
- copyright (73)
- fair use (73)
- plagiarism (73)

1. A small computer program that can copy itself repeatedly in a computer system and cause a variety of problems.
2. Using other people's ideas and words without acknowledging that the information came from them.
3. Any text or graphic on a Web page that will take you to a new location when clicked.
4. Describes when you may use a copyrighted work without permission.
5. An application that reads the contents of Web pages and displays them on your computer screen.
6. A type of legal protection for works that are created or owned by a person or company.
7. A Web site that finds Web pages that contain words you tell it to search for.
8. The part of a URL that usually identifies the owner or the subject of a Web site.
9. A unique Web address that directs a browser to a Web page.
10. A word that can make an Internet search more specific and effective.

Activity 2

Step-by-Step continued

Write a Cover Letter

Choose one of the résumés you developed in Activity 1, page 475. You will write a cover letter to go with this résumé.

1. Refer to Figure 15.3, page 474, as a general guide for your cover letter.
2. Start the cover letter with your address, the date, the employer's name and address, and the greeting. Format the letter as in Figure 15.3, page 474.
3. Introduce yourself in a brief opening statement. Include at least two characteristics that are qualities the employer is likely to value.
4. For the next paragraph, replace the details in the sample with the specifics from the job description you found.
5. Include your qualifications in the third paragraph.
6. Write a final paragraph that asks for an interview, lets the employer know how to contact you, and thanks the employer for his or her consideration.
7. End with a formal closing.
8. Proofread your cover letter, and make any necessary changes.
9. Save your document as *Your Name* **Cover Letter**.

RUBRIC LINK

Go to the Online Learning Center at **digicom.glencoe.com** and click Lesson 15.1, Rubric, to find the application assessment guide(s).

Application 1

Scenario You find out that a summer job you always dreamed of is available—a camp counselor at Camp Evergreen, a summer camp for children with disabilities. You grab your PDA and start a list of your relevant experience:

- Every week, you volunteer a few hours at the local children's hospital, and you went to camp yourself when you were younger.
- Last semester, you were a mentor in an after-school program for elementary school students.
- For the past three years, you babysat your neighbor's child, who has disabilities. In fact, you are your neighbor's favorite babysitter. The only other paying job you had was as a cashier in a local grocery store.

Directions Create and format your résumé to convince the camp director that you are the best person for the camp counselor job. Create a cover letter to go with your résumé. Save your résumé as *Your Name* **Camp Counselor**.

CHAPTER 3

Self-Assessment

Take a moment to review what you have learned in this chapter. Rank your understanding of the topics below.

4 means, "I understand all of this."
3 means, "I understand some of this."
2 means, "I understand very little of this."
1 means, "I don't remember this."

To use a printout of this chart, go to **digicom.glencoe.com** and click on **Chapter 3, Self-Assessment.**
Or:
Ask your teacher for a personal copy.

Rank Your Understanding

Lesson	Topic	4	3	2	1
3.1	• Define uniform resource locator (URL)				
	• Explain the different parts of a URL				
	• Add and organize Favorites in your browser				
3.2	• Create a home page in your browser				
	• Print a Web page				
	• Explain the purpose of search engines				
	• Define Boolean operators				
	• Do searches in Yahoo! and Google				
3.3	• Explain the importance of Internet security				
	• Describe viruses and how they work				
	• Describe worms and Trojan horses				
3.4	• Explain the purpose of copyright protection				
	• Identify situations that are considered fair use				
	• Define plagiarism and explain why it is wrong				
3.5	• Describe cybercrime				
	• Identify examples of cybercrime				

If you ranked all topics 4, congratulations! Consider doing a quick review.
If you ranked yourself 3 or lower in any topic, consider reviewing these topics first.

Activity 1

If you need additional help on this topic, open the Help menu; or refer to the documentation for your hardware.

Step-by-Step

Develop Two Résumés

Using material you have created in Chapters 13 and 14, you will create a list of your experience, choose two types of jobs, and then develop a résumé for each job.

1. Review the documents *Your Name* Skills, *Your Name* Interests, *Your Name* Aptitudes, and *Your Name* Values that you developed in your self-assessment in Chapter 13, pages 433–434.

2. Look through your notes on career clusters and careers that interest you from Chapter 13, Lesson 13.2, page 437.

3. Refer also to your career planning information in Chapter 14, pages 457.

4. Select two job types of particular interest to you. You will develop a résumé for each type of job.

5. Find the job duties and qualifications required for each of the two jobs by searching an online job board, such as **Monster.com**.

   Use the information on duties and required qualifications to choose the experience you will emphasize in the résumé you create. Create a list of key words based on the required skills and position responsibilities for each job.

6. For each of the jobs, make a list of your experience that is relevant to the job, including your:
   - Work experience.
   - Education.
   - Skills and abilities.
   - Achievements.

7. Read back over your list, and make sure each section contains the appropriate key words from the list you created in Step 5.

   Use the lists you created in Step 6 to develop résumés customized to each of the jobs you chose in Step 4, above. Refer to the sample résumés on pages 471 and 472 to help you organize your information. Choose the most logical résumé format for each job: chronological or skills.

8. Edit, proofread, and correct your résumés as needed.

9. Save your résumés as *Your Name* Resume 1 and *Your Name* Resume 2. You will use these documents later in Chapter 15.

Continued

CHAPTER 4
Create a Web Page with HTML

Lesson 4.1   Get Started with HTML

Lesson 4.2   Format Text

Lesson 4.3   Insert Images and Backgrounds

Lesson 4.4   Text and Graphic Alignment

Lesson 4.5   Bulleted and Numbered Lists

Lesson 4.6   Link to Other Pages and Web Sites

You Will Learn To:

- **Read** basic HTML code.
- **Create** a simple Web page using HTML code.
- **Add** background color and graphics to a Web page.
- **Change** font sizes, colors, and styles on a Web page.
- **Add** paragraphs and line breaks.
- **Add** images to a Web page.
- **Align** text, graphics, and clip art on a Web page.
- **Create** a bulleted or numbered list on a Web page.
- **Add** external and internal hyperlinks to a Web page.
- **Insert** a link to an e-mail address.

When Will You Ever Need to Create a Web Page?

There are countless reasons to create a Web page. Many people create personal Web pages to share information with people who have similar interests. Teachers create Web pages to post information for students. Businesses create Web pages to advertise and sell their products. Some companies even set up Web pages that let customers pay their bills online.

444 Southeast Balmy Avenue
Marburg, OH 44406
May 17, 2004

Mr. J. T. Adams
Workplace USA
555 34th Street
Marburg, OH 44411

Dear Mr. Adams:

I would like to apply for the position of Computer Help Desk Assistant at your company, as advertised in a recent issue of the electronic newsletter Jobs Unlimited.

Specific qualifications that meet your requirements include my three years of experience as a part-time help desk clerk for XYZ Company while I attended high school, plus full-time work in this field over the summer breaks.

I am available for an interview at your convenience. You may contact me by telephone at (303) 555-1515 or e-mail at jreading@marburgschools.edu. Thank you for your consideration.

Sincerely,

Jennifer Reading

Figure 15.3 *Cover Letter*

Read to Succeed

Key Terms

- **Hypertext Markup Language (HTML)**
- **tag**
- **attribute**
- **font**
- **horizontal rule**
- **clip art**
- **ordered list**
- **unordered list**

Exploring HTML Visually

Creating a concept map can be a great way to organize information, especially when that information consists of many related items. In addition to organizing information, a concept map draws attention to the information in a visual way, making the items more distinct from one another and easier to remember.

In this chapter, you will learn many HTML tags. Create a graphic organizer similar to the illustration. As you read and practice a new tag, add the tag to your organizer.

Example:

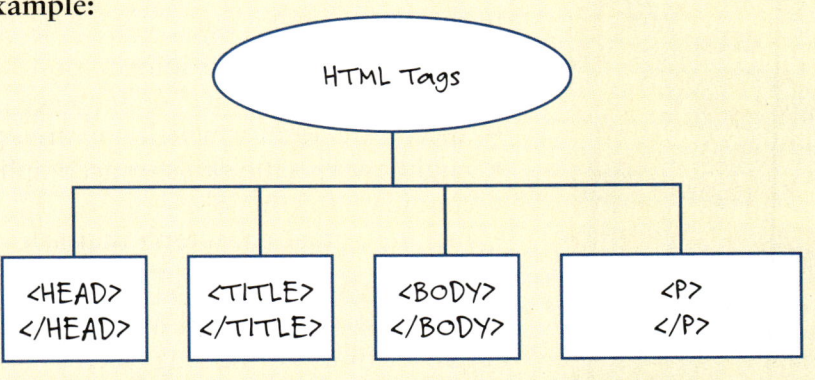

```
                    HTML Tags
        ┌──────┬──────┴──────┬──────┐
     <HEAD>  <TITLE>      <BODY>   <P>
     </HEAD> </TITLE>     </BODY>  </P>
```

DIGITAL CONNECTION

How Do You Use Web Sites?

Do you use Web sites to help with your schoolwork? Many students have favorite Web sites they visit frequently. Some people shop for items online that cannot be purchased locally. Students do research on the Internet. What kinds of Web sites do you visit?

Writing Activity

Write four Web sites on an index card. In one sentence, describe what activities or information can be found at each Web site. Form a small group, and trade your card with another student. Take turns reading the information on the card you have. Decide as a group what category the Web site would be: educational, informational, or entertainment. Write the group's decision of category next to the site's address on the card.

Electronic Résumés Many employers today accept—and many prefer—résumés submitted by e-mail. Career counselors advise that you use simple formatting for résumés you will send electronically, as formatting such as bold text and underline may be lost. Many companies use a software program that scans applicants' résumés for key words that are associated with the job position. Think carefully about what key words to use that match the requirements of the job you are applying for. For example, the job description might contain a sentence like this: "Candidate must have excellent **technical writing** skills, **attention to detail**, and the ability to **multitask**." Make sure that the bold words can be found in your résumé. These keywords will catch the attention of the prospective employer.

Cover Letters The cover letter introduces you and helps the employer find the most important information in your résumé. A ==cover letter== is a short letter you send along with a résumé, specifying the job you are interested in and highlighting specific skills or accomplishments you feel make you a good candidate for the job. See **Figure 15.3**, page 474.

In the cover letter, you:
- Include your contact information.
- Introduce yourself in one sentence.
- Specify the job for which you are applying.
- Highlight specific skills or accomplishments you feel make you a good candidate for the job.
- Include a final paragraph that asks for an interview.

E-Mail Cover Letters Including your cover letter and your résumé in the body of an e-mail may get you noticed as it saves the prospective employer time. Formatting, such as bold text, may be lost in the e-mail so you should also include your cover letter and résumé as attachments.

> ==cover letter== A short letter you send along with a résumé, specifying the job you are interested in and highlighting specific skills or accomplishments you feel make you a good candidate for the job.

✅ Concept *Check*

When would you find it useful to develop more than one résumé for your job search? Give examples. What is the purpose of a cover letter?

DIGITAL DIALOG

Employability Survey

Communication Survey friends, relatives, and managers of local companies to find which skills, abilities, and characteristics employers consider most important in their employees.

Writing Chart the results of your survey. Give an oral report to emphasize the skills, abilities, and characteristics job seekers should include in their résumés.

Lesson 4.1 — Get Started with HTML

You will learn to:
- **Add** basic HTML code to create a Web page.
- **Create** a Web page with text and background color.
- **Save** and preview a Web page.

> **Hypertext Markup Language (HTML)** A set of codes used to create documents for the Web.
>
> **tag** A piece of HTML code that tells a Web browser how to display a particular section of text, a graphic, or other Web page element.

Understanding HTML Tags

Hypertext Markup Language (HTML) is a set of codes used to create documents for the Web. Web browsers, such as Microsoft Internet Explorer and Netscape Navigator, read the coded document and turn it into a Web page that people can view.

HTML consists of many different tags, or elements, that surround the text that will appear on a Web page. A **tag** is a piece of HTML code that tells a Web browser how to display a particular section of text, a graphic, or other Web page element. You will usually enter an opening tag before and a closing tag after the content you want the tag to affect. Opening tags are enclosed in brackets like these: < >. Closing tags are enclosed in the same brackets, but they also have a slash after the first bracket: </ >.

Learning to read HTML code is like learning to read someone's shorthand. Many of the codes are abbreviations for what the codes will do. Decoding HTML tags is not difficult. For example, the <B> tag makes the text between the tags **bold**. The <I> tag makes the text between the tags *italic*. Once you learn the basic HTML tags, you will be able to create and publish your own Web page. Activities in this chapter will prepare you to do just that.

Creating an HTML Document

A Web page can be created in any word processing program. Attention to detail is very important when keying HTML code. If just one character in an HTML tag is wrong, the tag will not work. A Web browser will not recognize the spaces you insert between tags and between lines. You insert these spaces to make the HTML document easier for you to read.

Web Editors

More advanced software programs, such as Dreamweaver® and FrontPage®, will enter HTML codes for you as you design a Web page. These Web editors allow you to enter text and graphics much like you would in a word processor. Since it is important for you to learn how HTML works, you will use HTML in the basic WordPad and Notepad programs that come with all versions of Microsoft Windows.

DIGI Byte

Plain Text

The WordPad toolbar has features to enhance text, but these should not be used in HTML documents. Use plain text when creating an HTML document.

✓ Concept Check

What does an HTML tag do? Write how opening and closing tags appear in enclosed brackets.

Jennifer Reading

444 Southeast Balmy Avenue • Marburg, OH 44406 • (303) 555-1515
jreading@marburgschools.edu

Job Objective   A position in computer/software/hardware training or customer assistance

Skills and Accomplishments

Computer and Training Skills
- Provided help desk support to employees of XYZ Company, Marburg, OH
- Provided one-on-one training to new employees on general software applications
- Responded to employee requests for assistance regarding computer malfunctions

Teamwork and Leadership
All experience has involved working with colleagues as part of a team to organize the group to enable it to better respond to in-house client needs.

Responsibility and Dedication
I have worked part-time throughout my last two years in high school and full-time during summers. I have worked overtime when requested. I have met all deadlines imposed, and provided additional assistance when requested. I particularly enjoy the challenge of helping new employees learn about the applications used by the company and how their functionality can be used to its fullest.

Education   Marburg High School, June 2004

Awards and Activities   National Honor Society, 2003–2004
Marburg Elementary School Mentoring Project, 2002–2004

Figure 15.2 *Skills Résumé*

Activity 1

If you need additional help on this topic, open the Help menu; or refer to the documentation for your hardware.

Step-by-Step

Create a New HTML Document

In this exercise, you will use the WordPad program to begin building a Web page from scratch. HTML documents have a head section and a body section.

1. Create a new folder on your desktop. Name this folder **FBLA Web Page.** You will use this folder to save all graphics and documents related to your Web page.

2. Start WordPad by clicking **Start, Programs, Accessories,** and then **WordPad.**

3. On the first line of the new WordPad document, key `<HTML>`

    This opening tag tells your browser that this document is a Web page.

4. Press ENTER ten times to move the insertion point down the page. Key `</HTML>`

    This closing tag will tell your browser that it has reached the end of the Web page. You will key the rest of your code between these two tags.

5. Move the insertion point to the line under the opening `<HTML>` tag. Key the following:

```
<HEAD><TITLE>FBLA Home Page</TITLE></HEAD>
```

    The head section contains the title of the Web page, and can contain other information. The closing </head> tag tells the browser that it has reached the end of the head section.

    When the Web page is viewed in a browser, text between the opening and closing title tags will appear at the top of the screen in the browser's title bar.

6. Press ENTER twice.

7. Move the insertion point to the line under the `<HEAD>` tag and enter the next tag: `<BODY>`

    This is the opening of the body section of the HTML document.

8. Move the insertion point to the line above the `</HTML>` tag and enter the closing tag: `</BODY>`

    The body section must have an opening tag and a closing tag.

    You will enter all text and graphics appearing on your Web page between the opening and closing body tags.

Continued

Chapter 4

Jennifer Reading
444 Southeast Balmy Avenue
Marburg, OH 44406
(303) 555-1515
jreading@marburgschools.edu

Job Objective   A position in computer/software/hardware training or customer assistance

Experience
June 2002–June 2004   Help Desk Assistant, XYZ Company, Marburg, OH
- Provided one-on-one training to new employees on general software applications
- Responded to employee requests for assistance regarding computer malfunctions
- Worked part-time (15 hours/week) during school year and full-time during summer

June–August 2001   Cashier, Best Foods, Marburg, OH

Education   Marburg High School, June 2004

Awards and Activities National Honor Society, 2003–2004
Marburg Elementary School Mentoring Project, 2002–2004

Figure 15.1 *Chronological Résumé*

Step-by-Step continued

9. Move the insertion point to the line below the opening <BODY> tag, and enter the following text:

```
Welcome to Our High School FBLA Web Page
```

The words you just keyed will appear on your browser screen when this document is opened in your browser. Any text you enter between the tags in the body of an HTML document shows on the browser screen.

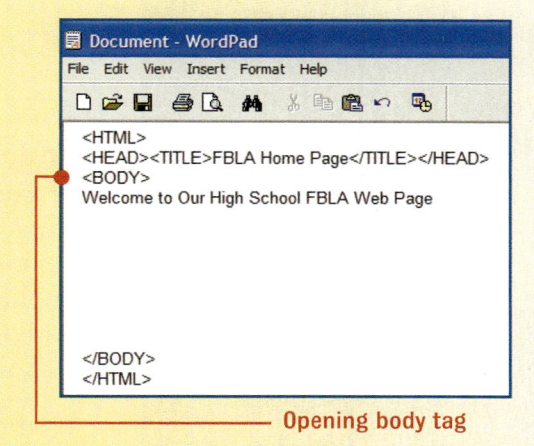

```
<HTML>
<HEAD><TITLE>FBLA Home Page</TITLE></HEAD>
<BODY>
Welcome to Our High School FBLA Web Page



</BODY>
</HTML>
```

Opening body tag

Save and Preview a Web Page

1. Click **File, Save As**. In the **Save in:** box, navigate to the FBLA Web Page folder. In the **File name:** box, enter `FBLA.html`

   In the **Save as type:** box, choose **Text Document**.

2. Click **Save**.

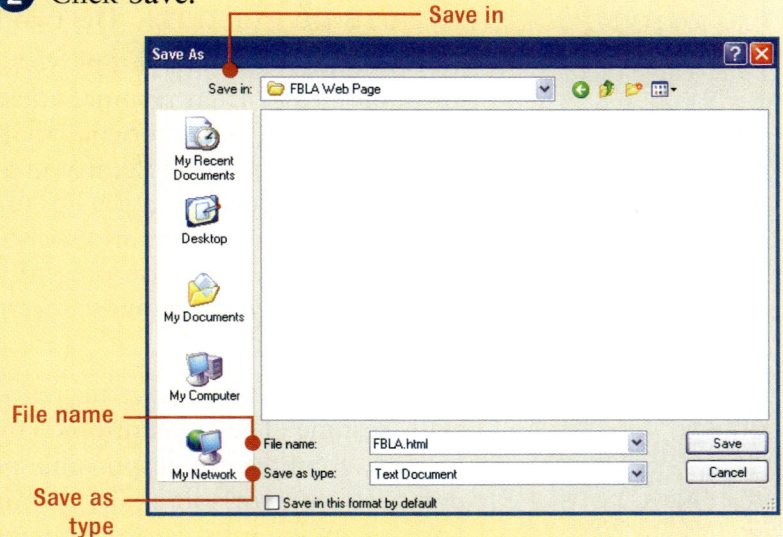

   Save in — FBLA Web Page
   File name — FBLA.html
   Save as type — Text Document

3. Minimize all windows on your desktop.

4. Open the FBLA Web Page folder on your desktop by double-clicking the folder icon.

Activity 2

DIGI Byte

Backup

You should always save a backup copy of your HTML documents before you experiment with adding new code. That way, you can go back to the original if the experiment does not work.

Continued

Chapter 4

87

Lesson 15.1

Write Your Résumé

You will learn to:
- **Write** two types of résumés.
- **Write** cover letters.
- **Determine** which style of résumé works best, depending on your qualifications and on the job you are applying for.

résumé A summary of your skills, abilities, education, work history, and achievements.

job interview A formal meeting between an employer and a job applicant.

Your Résumé: Your First Contact Your résumé is a summary of your skills, abilities, education, work history, and achievements. You want your résumé to help convince the employer that you are the best person for the job. Employers do not have time to interview everyone who applies for a job. A job interview is a formal meeting between an employer and a job applicant. Employers look at the résumés they receive to choose a short list of applicants to interview.

What Makes a Résumé Effective? Create an effective résumé by clearly stating information about your capabilities in a way that is easy to read, and free of grammatical, spelling, and formatting errors. Experts recommend limiting a résumé to one page and making sure the information that is most relevant to the job is easy to find. Keep sentences short, and use bullet points whenever possible.

Two Résumé Formats The two most common résumé formats are:

- **Chronological Résumé** Employers generally expect to see the chronological format résumé, and this is the better format to use when you have job experience that is relevant to the job you are applying for. For example, if you are applying for a job as an assistant Web developer, job-specific skills to list could be: helped develop school Web site and working knowledge of HTML. In the chronological résumé you present the information in reverse time order—that is, with the most recent work experiences first. See **Figure 15.1,** page 471.

- **Skills Résumé** If you have no relevant on-the-job experience, use the skills résumé. The skills résumé emphasizes overall accomplishments in paid, volunteer, or responsible positions you have held that demonstrate your soft skills, or transferable skills. For example, if you are applying for a job in customer service, but have no related experience, you could list soft skills that you know are needed in all jobs. Provide examples to show how you excel in soft skills such as, excellent communication and listening skills, flexibility, positive attitude, working well in a group to achieve results, and so on. See **Figure 15.2,** page 472.

Chapter 15 Get the Job You Want

DIGI*Byte*

Work in Two Windows

As you work on a Web page, you will frequently switch back and forth between the WordPad document and the browser window.

Address Line

The address line in your browser displays the location of this file.

Step-by-Step *continued*

5. The FBLA document will have an Internet icon. This indicates an HTML document. Double-click to open this document.

   The Web browser will now open and display your new Web page! Note that your page is not actually on the World Wide Web. Your browser is simply showing you what the page would look like if you chose to publish it on the Web.

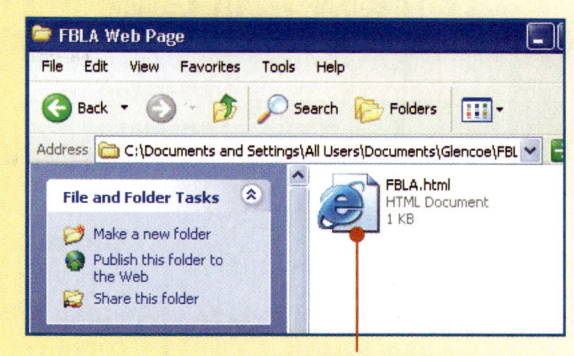

   HTML document icon

   Title of the Web page

Add Background Color

Now you will return to the original document and change the background color of your Web page. To do this you will add the background color attribute, bgcolor=, to the body tag.

An **attribute** is a specific instruction that tells the browser how to display the text or graphics enclosed by the HTML tags. Here the attribute bgcolor specifies the background color for the Web page. You choose the value of the attribute—in this activity you will give "YELLOW" as the value of the background color.

1. Minimize the browser window.
2. From the Taskbar, open the FBLA.html WordPad document.

Activity 3

attribute A specific instruction that tells the browser how to display the text or graphics enclosed by the HTML tags.

Continued

Read to Succeed

Key Terms
- résumé
- job interview
- cover letter
- reference
- employment-at-will

Time Your Reading

How long does it take you to read and understand a lesson introduction, for example, Lesson 15.2, page 479? The answer will vary depending on several factors. The more difficult and unfamiliar the subject matter, the longer it will take you to complete the reading and understand the material.

Use time to your advantage. To help maintain focus on your reading, time how long it takes you to read each lesson introduction in this chapter. Attempt to better your time for each lesson, but make sure you also understand the material. Can you beat your own time, and still understand the material, when you get to Lesson 15.5?

Example:

Lesson Number	Reading Time
15.1	10 minutes
15.2	8 minutes
15.3	6 minutes
15.4	6 minutes
15.5	5 minutes

DIGITAL CONNECTION

Keeping Track of Business

You have been a member of the school swim team for the past two years and have completed the Red Cross First Aid, Lifesaving, and Lifeguard courses. You decide to apply to all the community recreation departments and private swim clubs within a 25-mile radius of your home for a summer job as a lifeguard.

Writing Activity

Write a paragraph on how you can use your PDA to start and manage the job application process for the job as a lifeguard.

Step-by-Step *continued*

③ Place the insertion point immediately after the <BODY> tag. Insert one space, and then key the background color attribute and value:

```
<BODY bgcolor="YELLOW">
```

④ Click **File, Save**.

⑤ On the Taskbar, click the FBLA Home Page document. The browser window will open. The page needs to be reloaded, or refreshed, to display the change you made. Click the **Refresh** button on the Menu bar.

⑥ Notice the background color. Did the background color of the page turn yellow? If so, congratulations! If not, go back to the WordPad FBLA.html document and make sure you entered the tags, attribute, and value correctly.

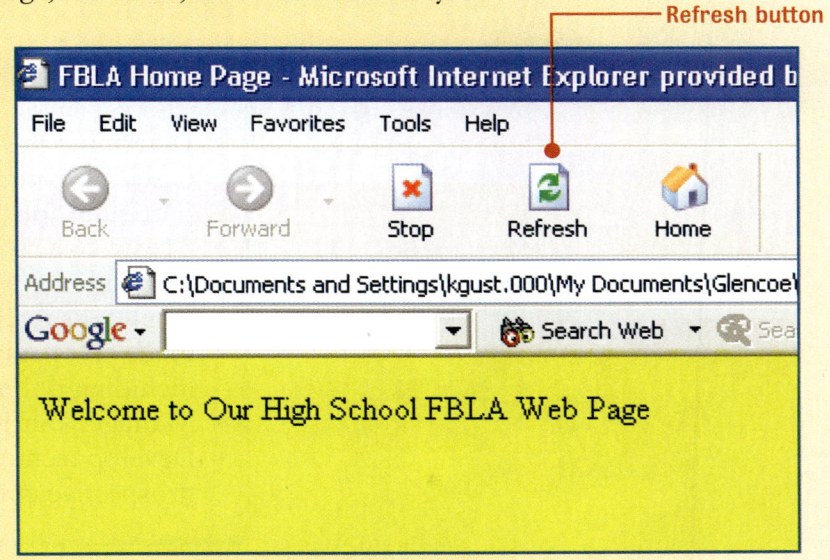

Use Color Codes to Change Background Color

Color code values are made up of six letters and/or numbers.

① Open the FBLA document from the Taskbar.

② Delete the word YELLOW in the BGCOLOR attribute and enter the value: "#99CC99"

③ Click **File, Save**.

④ Open the browser window that is minimized on the Taskbar. Click the **Refresh** button. The background will now be greenish-blue.

DIGI*Byte*

When Color Codes Are Needed

To create a background color in an HTML document, sometimes all you need is the bgcolor attribute and the name of the color, such as bgcolor="YELLOW". For more unusual colors, like aqua or burnt orange, you may need to use color codes instead of names.

Activity 4

Continued

Chapter 4

89

CHAPTER 15
Get the Job You Want

Lesson 15.1   Write Your Résumé

Lesson 15.2   Find Job Openings

Lesson 15.3   Fill Out Job Application Forms

Lesson 15.4   Manage Your Job Search

Lesson 15.5   Succeed in Interviews

You Will Learn To:

- **Determine** the most effective résumé style to use, based on your experience and work history for a particular job.
- **Create** two styles of résumés.
- **Write** brief cover letters that encourage prospective employers to consider you for the job.
- **Find** job openings through many different means, including online and in publications.
- **Manage** your job search by using digital tools.
- **Develop** techniques to use in an interview with a prospective employer.

How Do You Get the Job You Want?

If you want a job, you first need to catch the attention of the employer. You can do this best by showing that you meet the requirements of the job.

Your first communication with a prospective employer is often with your résumé and cover letter. In the résumé and cover letter, you summarize your qualifications for the job. Your PDA and computer skills will prove to be very useful throughout the job-seeking process.

Step-by-Step continued

5. To close the WordPad application, click **File, Exit**.
6. To close the Web browser, click **File, Close**.

Open the FBLA Document as a Web Page and as a Notepad Document

1. Double-click the FBLA folder you created on your desktop.
2. Double-click the FBLA document file in the folder. Your Web browser will open and display the Web page you have started.
3. In the Menu bar of the browser, click **View, Source**. A window will open, displaying the HTML you have entered. The title bar of this window will read "FBLA.html—Notepad."

Although you started your Web page in WordPad, you will edit and save your HTML Web page in Notepad from now on. Notepad is the default Web editor for your browser.

Activity 5

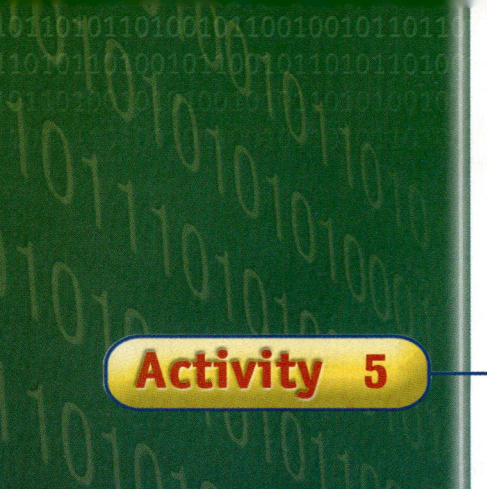

Suite Smarts
Web pages can also be created in Word, Excel, and PowerPoint applications.

RUBRIC LINK
Go to the Online Learning Center at **digicom.glencoe.com** and click Lesson 4.1, Rubric, to find the application assessment guide(s).

Application 1

Directions Open the FBLA Web page you have been working on. Open the FBLA Notepad document using Step 3 in Activity 5 above. Try changing the background to at least three different colors. Remember to save the HTML document and refresh the browser window to see each change.

Make a note of which colors require you to use a code, rather than just the name of the color. Refer to the Web-Safe Color Palette in the Reference Section at the back of this book, page R15, and try at least three of the color codes listed there.

- Which of the six colors you tried make the text easier to read?
- Did any colors make the text harder to read? If so, which ones?

Summarize your answers in a table for future reference.

DIGITAL DIALOG

The Effect of Colors in a Web Page

Teamwork In a small group, discuss the following questions:
- How did you feel about each background color you tried?
- How did your text look? Was it easy to read?
- What should you do to the text if you want to use a dark background?

Writing Have one person record the group's responses. Share your responses with the rest of the class.

Chapter 14 Self-Assessment

Take a moment to review what you have learned in this chapter. Rank your understanding of the topics below.

4 means, "I understand all of this."
3 means, "I understand some of this."
2 means, "I understand very little of this."
1 means, "I don't remember this."

To use a printout of this chart, go to **digicom.glencoe.com** and click on **Chapter 14, Self-Assessment.**
Or:
Ask your teacher for a personal copy.

Rank Your Understanding

Lesson	Topic	4	3	2	1
14.1	• Analyze job descriptions for jobs that interest you				
	• Rate your interest in each job-related task				
14.2	• Identify degrees and certifications required for your career				
	• List institutions and organizations that offer the degrees and certifications you need for a job				
	• Develop an education and training plan				
14.3	• Develop your career plan				
	• List short-, medium-, and long-term career goals				
	• Determine a timeline for achieving your goals				
	• Identify needs for education, training, and experience and ways to achieve them				
14.4	• List the steps to follow when making career decisions				
	• Determine options for achieving career goals				
	• Create an action plan for implementing a career decision				
	• Create a backup plan for a career decision				

If you ranked all topics 4, congratulations! Consider doing a quick review.
If you ranked yourself 3 or lower in any topic, consider reviewing these topics first.

Lesson 4.2 Format Text

You will learn to:
- **Change** font colors and sizes.
- **Make** text bold or italic.
- **Add** paragraph codes.
- **Add** line breaks.
- **Add** a horizontal rule.

font A combination of specific visual characteristics of text, including size, typeface style, bold, and italics.

horizontal rule A line that runs across the width of a page.

Applying Font Changes Text can be styled in a variety of different fonts. A **font** is a combination of specific visual characteristics of text, including size, typeface style, bold, and italics. These characteristics affect the overall appearance of a Web page. Font styles, colors, sizes, and attributes like bold and italics are very important when designing a Web page. They help to emphasize text and catch the attention of people viewing your page.

As you design a Web page, you may decide to change the font to make the page look better. Use contrasting colors: light text with dark backgrounds and dark text with light backgrounds. In this lesson, you will learn the HTML codes needed to make these changes.

Creating Paragraphs and New Lines Once you have learned how to change the way text looks on a Web page, it is time to learn how to control the layout of text on the page.

In text documents created with word processors, breaks between paragraphs are created by pressing the Enter key. Line breaks are made differently in HTML. The paragraph tag, <P>, adds two blank lines between paragraphs, like pressing the Enter key twice. The break tag, <BR>, inserts a single space after a line of text, like pressing the Enter key once.

Another common tag used to separate text is the horizontal rule tag: <HR>. A **horizontal rule** is a line that runs across the width of a page.

Viewing Source Code You can look at the HTML code of any Web site by clicking View and then Source in the browser's Menu bar.

✓ Concept Check

If your Web page has small, dark text against a dark background, are people likely to read the text? Why or why not? Describe what you can do as a Web designer to make the text on your Web page appealing. Is text easier to read when it is broken into paragraphs? Explain your answers.

CHAPTER 14 Review

Read to Succeed PRACTICE

After You Read

Create a Quiz Studies show that a good way to reinforce what you have learned is to write yourself a quiz after reading a chapter. For topics to set the questions on, look at each lesson's "You will learn to" section, key terms, and the lesson paragraph headings. Create the following types of questions: true or false, multiple choice, matching, or fill-in-the-blanks.

Sample question:

A job description includes the job _____, location, and _____ required.

Sample answer: title, skills

For this chapter, use a separate page, or your Tablet PC, and create quiz questions of your own.

Example:

Questions	Answers
What characteristics are needed to write good goals?	Specific, achievable, and flexible
Why is it important to have a backup plan in the career decision-making process?	You have already figured out what to do if you are not able to achieve a goal?

Using Key Terms

Planning a career requires a lot of careful decisions, but the rewards are great. Knowing the following key terms will help you to plan more effectively. To test your knowledge, write each sentence on a sheet of paper, Tablet PC, or PDA and fill in the blanks.

- job description (449)
- chronological order (456)
- prioritize (459)
- backup plan (459)

① To _____ is to put in order from most important to least important.

② An alternative course of action is a _____ _____.

③ The order in which events happen is _____.

④ A _____ is an explanation of the major areas of a job or position necessary for an employee to perform the job successfully.

466

Step-by-Step

In Activities 6–9, you will apply font color and style codes and change font size. In Activities 10–12, you will apply paragraph and line spacing codes.

Apply Font Color Codes

In a similar way to when you added a background color, you will add the font color attribute, <FONT color=>, to the document. Again, you choose the value of the attribute—in this activity you will give "WHITE" as the value of the font color.

Activity 6

1. Open the FBLA Web page and the FBLA Notepad document as you did in Activity 5, page 90.
2. Place the insertion point in front of the word *Welcome* in the Notepad document.
3. Press ENTER twice to move this line of text down the page.
4. Move the insertion point to the line above the word *Welcome*. This is where you will enter the font code.
5. Enter `<FONT color="WHITE">`
6. Move the insertion point to the line above the </BODY> tag. Enter the closing font tag: `</FONT>`
7. Click **File, Save.** Expand the Web browser, which is on the Taskbar. Click the **Refresh** button to reload the Web page.
8. Open the Notepad document from the Taskbar. Change WHITE to "#FFFFFF"
9. Click **File, Save.** Switch to the Web browser and click the **Refresh** button.

Apply Header Codes

There are six header tags, with <H1> the largest font size and <H6> the smallest font size.

Activity 7

1. Open the FBLA Notepad document. Place the insertion point after the color code. Now press ENTER. On the new line, enter `<H2>`
2. Click **File, Save.** Open the Web browser and click the **Refresh** button to view the changes you made.

Apply Bold and Italics

Activity 8

Another way to emphasize text is to use the <B> and <I> tags to bold the text, italicize the text, or do both.

Continued

> **?** If you need additional help on this topic, open the Help menu; or refer to the documentation for your hardware.

> **DIGI*Byte***
> **Alternate Tags**
> An alternate tag for:
> • Bold, <B>, is <STRONG>.
> • Italics, <I>, is <EM> for emphasis.

Chapter 4

92

CHAPTER 14

DIGITAL Dimension → Math
INTERDISCIPLINARY PROJECT

Managing Your Money

Using your spreadsheet skills, you will estimate the costs of car ownership by preparing a budget.

Calculating Costs of Owning a Car This activity will help you decide if the costs of owning a car outweigh the convenience.

Investigate Talk to someone you know who has a car and ask them how many miles they drive in a year. Then ask what they pay each year for:

- Gasoline and regular maintenance.
- Repairs and tires.
- Insurance—keep in mind that drivers under age 25 pay more.
- Registration, parking, and cleaning.

Calculate Create a spreadsheet to perform each of the following calculations:

- Add all of the above expenses to determine the total annual operating expense.
- To find the average cost per mile of operating a car, take the total annual operating expense and divide it by the number of miles driven in a year.
  Remember, this cost does not include paying for the car itself! Most car owners make a monthly car payment.
- If the monthly car payment is $275, what is the total annual cost of owning and operating the car? To get the total cost, multiply the monthly car payment by 12 and add the annual operating costs to it.
- What is the total cost per mile of owning and operating the car for a year?

Decision Making Now that you have learned about the real-world costs of car ownership in this project, decide whether you consider car ownership a necessity or an unnecessary expense. Write a summary paragraph defending your decision.

E-Portfolio Activity

Decision-Making Skills

To showcase your spreadsheet and decision-making skills, give a title to your document. Save the documents to your e-portfolio as Excel and Word files.

Self-Evaluation Criteria

Have you included:

- The annual cost of each operating expense?
- The total annual cost of all operating expenses?
- The annual cost per mile of the operating expenses?
- The total annual cost of all operating and owning expenses?
- The annual cost per mile of operating and owning expenses?
- A paragraph defending your portfolio regarding car ownership.

Step-by-Step *continued*

1. Open the FBLA Notepad document. Place the insertion point just after the <H2> code and press ENTER.

2. On the new line, key <B>

   This is the HTML code for bold. Press ENTER.

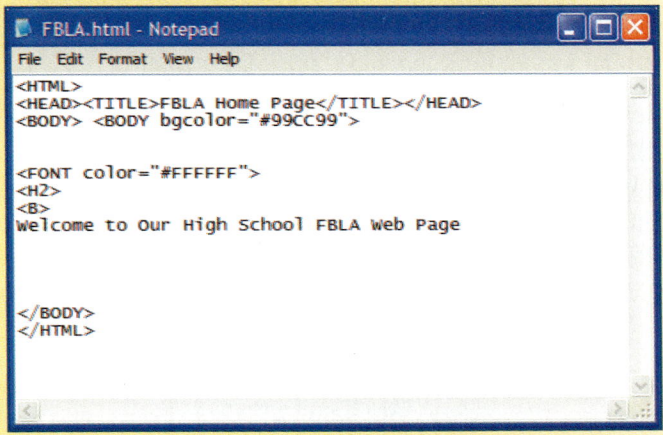

3. On the next line, key the HTML code for italics, <I>

4. Click **File, Save**. Expand the Web browser from the Taskbar, and click the **Refresh** button to see the changes.

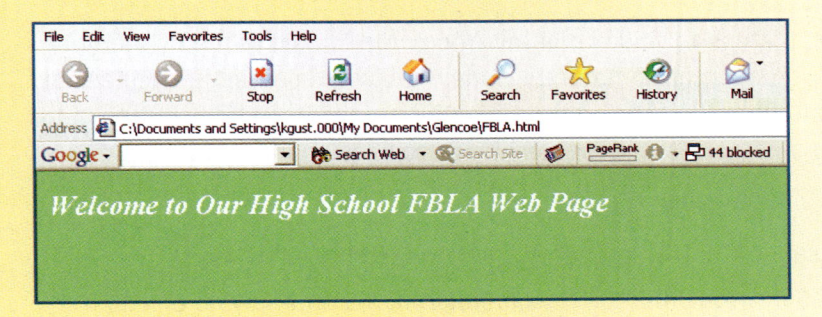

Add Closing Tags

You have created opening tags for font size, bold, and italics. Now you need to create the closing tags for these attributes: </H2>, </B>, and </I>. If you do not enter closing tags for font attributes, the attributes will apply to all of the text after the starting tag. The closing tags must be placed after the text you are formatting.

1. The italics tag was the last one you entered before the text, so enter the closing italics tag first, </I>, placing it on the line right below the text. Press ENTER.

Activity 9

Continued

PERSONAL Career Perspective

Part 2 of 4

Your Career Path

This is the second part of a four-part project that lets you explore your own career story. You may have already completed Part 1, page 443. Now you will prepare for lifelong learning and find a mentor.

Mentors: The Key to Lifelong Learning

Careers in the 21st century change so rapidly that success is more likely for lifelong learners. Lifelong learning means continuing to learn after you finish high school. It includes reading, on-the-job training, volunteering in the community, apprenticeships, taking night classes at adult school, or taking courses at a university or community college.

The teachers in many lifelong learning settings are called mentors. A mentor is someone who you respect and admire, and a person who can teach you new skills. The best mentors are enthusiastic, committed, and effective listeners. They should understand your strengths and weaknesses so they can help you grow toward your goals. A mentor can help you select the best college or training.

Find a Mentor

Take the steps to find a mentor. You could choose a family member, coach, teacher, or community leader. You could contact a business or two and find out if they have a mentor program. Another option is to locate a company or university that provides online mentors.

When you have found a mentor:

1. Arrange to meet in person, over the phone, or online.
2. Find out your mentor's accomplishments and how he or she achieved them.
3. Share the career plans you made in Lesson 14.2, page 453.
4. Get your mentor's feedback and make changes to your plan if you feel they are appropriate.

After the meeting or online exchange, summarize your experience with the mentor and present the summary to your class. Explain the qualities of your mentor and why you chose him or her as a mentor.

Look Ahead...

Continue your project in the following activities:

Part 3, Chapter 15..........p. 491
Part 4, Chapter 15..........p. 492

E-Portfolio Activity

Presentation Skills

If you created an electronic presentation, save your presentation to your e-portfolio. Alternately, arrange to have a digital recording of your presentation, and save that to your e-portfolio.

Self-Evaluation Criteria

For this activity, did you:

- Select a mentor?
- Meet with the mentor in person, over the phone, or online?
- Review your career plans?
- Summarize the highlights of your meeting?
- Present the summary to your class?
- Explain your mentor's qualities?

Step-by-Step *continued*

2. Enter the closing bold tag: </B>
   Press ENTER.

3. Enter the </H2> closing tag.

```
FBLA.html - Notepad
File Edit Format View Help
<HTML>
<HEAD><TITLE>FBLA Home Page</TITLE></HEAD>
<BODY> <BODY bgcolor="#99CC99">

<FONT color="#FFFFFF">
<H2>
<B>
<I>
Welcome to Our High School FBLA Web Page
</I>
</B>
</H2>
</FONT>
```

4. Save your Notepad file and exit the application.

Use Paragraph and Break Tags

As with any written document, the text on a Web site should not run together in one long paragraph. You will apply the paragraph and break tags to separate sentences into paragraphs and sections.

1. Open the FBLA Notepad document.

2. In the FBLA Notepad document, place the insertion point after the closing </FONT> tag. Press ENTER.

3. Key the text below exactly as it appears, as two paragraphs with <P>, the paragraph tag, on the first line and </P> at the end of the second pargraph.

```
<P>

Future Business Leaders of America is an
organization for high school students interested
in a business or business-related career.

To join, simply pay annual national membership
dues to your local high school chapter. You will
enjoy all that FBLA has to offer.

</P>
```

Continued

Suite Smarts

Microsoft Word, Excel, and PowerPoint documents can easily be saved as HTML for use as Web pages.

Activity 10

21ST CENTURY CONNECTION

The Power of Negotiation

The process of two or more people or groups resolving conflicts and moving toward a mutually beneficial agreement is known as negotiation. Negotiation requires strong, polished skills and the willingness to work fairly without harm to either side. Negotiations range from simple decisions, such as which film two friends might go to see, to issues of great seriousness and impact, such as the terms of a peace treaty between two or more nations.

Negotiation at Work Your ability to negotiate well is one of the most valuable life skills you will use in the world of work. You should be prepared to negotiate different issues in your career, such as:

- Wages or salary.
- A promotion or job change.
- A special project in which you would like to participate.
- Adjustments to your work schedule.

Preparing to Enter Negotiations The more prepared you are, the more confident you will be as you negotiate. Follow these tips to prepare:

- List what you will ask for and provide evidence of your need. If negotiating salary or wages, research what a company might pay for similar jobs, calculate your own financial needs, and compare the two. Reach a comfortable figure that might benefit both you and your employer.
- Describe what you can offer your employer in return. Ask your employer about other needs the company might have.
- Practice your negotiation session with family or friends ahead of time.
- Visualize positive negotiations in which both you and your employer profit from the outcome.
- Focus on interests, not positions. Try to find a solution that meets the needs or interests of all parties.

During Negotiations Use the following techniques as you negotiate:

- Enter with a positive, flexible attitude.
- Speak in a confident but respectful tone.
- Present the most important request and supporting facts first, if you have more than one request. Be sure you can show that your need is valid and that there would be mutual gain.
- Thank your employer, leave the room politely, and remain optimistic about your future even if you do not end up achieving what you had hoped. A negative reaction on your part could give your employer a bad impression and make it harder for you to negotiate in the future.

Activity Think of a time when you negotiated for something. Which of the tips and techniques listed above did you use? Write a paragraph about how you used them.

Step-by-Step continued

4. Click **File, Save**.

5. In the browser, click the **Refresh** button to view the text you added.

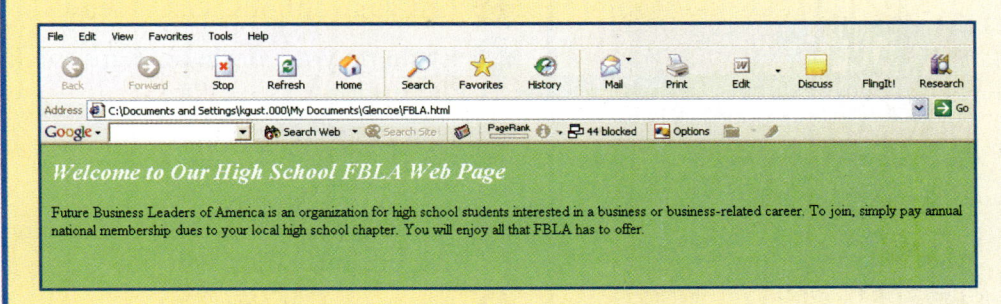

The text on the Web page appears as one paragraph, even though the text you keyed in Notepad was two paragraphs. This is because no paragraph or break tag was used in the text.

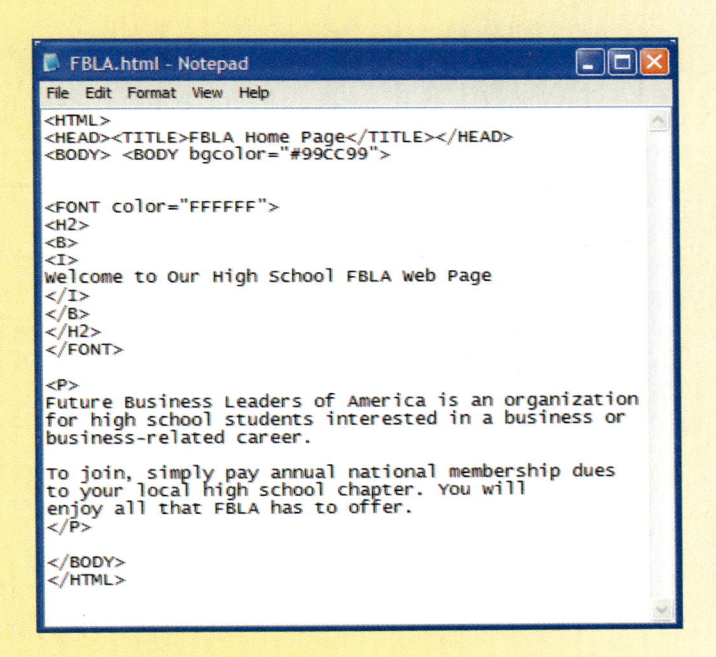

6. In the Notepad document, key `<BR>` on the blank line between paragraphs. Click **File, Save**.

7. Open the FBLA Web page and click the **Refresh** button. The sentence beginning "To join…" now begins on a new line, however, there is no space between the paragraphs. You will now fix that.

Suite Smarts

Spell check is not available in Notepad, so be sure to proofread your Web pages before they are published.

Continued

Step-by-Step continued

2. **Review your resources.** Use books, Web sites, people you know, and the resources you have gathered in Chapters 13 and 14 so far, pages 430–458, to help you choose.
3. **List your options.** List at least three options.
4. **Evaluate your options.** With the knowledge you have, evaluate the options.
5. **Pick the best option.** Choose the option you think would work the best.
6. **Set up a plan of action.** Set up an action plan for implementing this option. What will you do to achieve this goal?
7. **Check your decision.** Are you comfortable with this decision? Do you feel this decision makes the best sense for you?

Set a Backup Plan

1. Set up a backup plan for the option. What will you do if your action plan does not work?
2. Add the backup plan to your **Career Decisions** file, and then save the document.

Activity 6

RUBRIC LINK
Go to the Online Learning Center at **digicom.glencoe.com** and click Lesson 14.4, Rubric, to find the application assessment guide(s).

Application 5

Directions Using the process you followed in Activities 4, 5, and 6, pages 461–462, make decisions for your medium-term and long-term goals.

DIGITAL DIALOG

Career Decisions

Teamwork Work in a group. Take turns asking each other the following questions about your career goals from Lesson 14.3, pages 457. Enter responses in your PDA.

1. How did you make your career decision?
2. What obstacles did you have to overcome in making this decision? How did you deal with these obstacles?
3. What would you do differently the next time?
4. What did you learn in the process of making this decision?
5. Do you need to modify your career plan based on what you learned during this exercise? If so, how would you change it?

Examine the answers you received. Do they inspire you to adjust your own career plan?

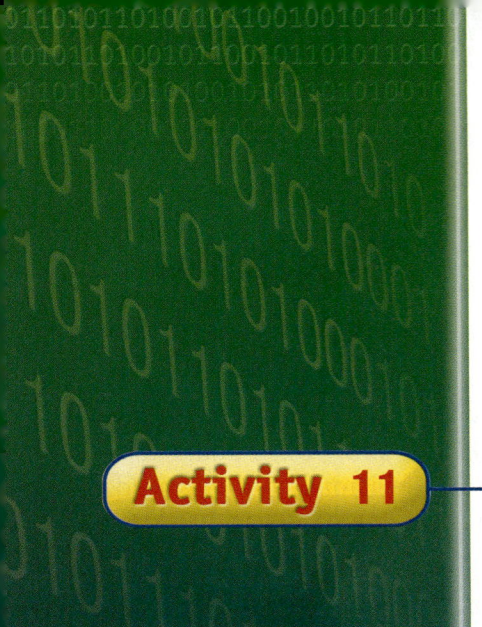

Step-by-Step *continued*

8 Go back to the Notepad document and change the `<BR>` tag to `</P>`. Press ENTER. Key `<P>`.

9 Click **File, Save**.

10 Open the FBLA Web page and click the **Refresh** button. The text should now appear as two separate paragraphs.

Control Line Length with Break Tags

The text on a Web page is easier to read if the line does not stretch all the way across the window. You can use break tags to control the length of the lines.

1 Open the FBLA Notepad document.

2 Enter the break tag, `<BR>`, after the word *interested* in the first paragraph.

3 Enter `<BR>` before the word *You* in the second paragraph.

```
<P>
Future Business Leaders of America is an organization
for high school students interested <BR>
in a business or business-related career.
</P>
<P>
To join, simply pay annual national membership dues
to your local high school chapter. <BR>You will
enjoy all that FBLA has to offer.
</P>
```

4 Save the Notepad document.

5 Open the FBLA Web page and click the **Refresh** button.

Add a Horizontal Rule

You can use the horizontal rule tag, `<HR>`, to add emphasis or improve the appearance of your Web page.

1 Open the FBLA Notepad document.

2 Place the insertion point after the word *offer*. Press ENTER.

3 Enter the horizontal rule tag: `<HR>`

4 Click **File, Save**.

Continued

Activity 4

If you need additional help on this topic, open the Help menu; or refer to the documentation for your hardware.

DIGI Byte

Seven-Step Decision-Making Process

Do you have a big decision to make in your life? Follow these essential seven steps:
1. Focus on one decision at a time.
2. Review your resources.
3. List your options.
4. Evaluate your options.
5. Pick the best option.
6. Set up a plan of action.
7. Check your decision.

Step-by-Step

Make Decisions

1. Open a file in a word processing or spreadsheet program.
2. Save the file as **Career Decisions**. Create a table similar to the one below.
3. Refer to the file **Career Plan** that you saved in Lesson 14.3, Activity 3, page 457. Look at the column of short-term goals and list in your **Career Decisions** file all of the decisions you will need to make in order to achieve the short-term goals.
4. Arrange the list of decisions in chronological order. What decisions do you need to make first, next, after that?

Career Decisions

Career Goals	Circle One: Short-Term Goals / Medium-Term Goals / Long-Term Goals				
Decisions in Chronological Order (Circle One)	1.	2.	3.	4.	5.
Options (at least 3) — 1.					
2.					
3.					
Best Option					
Action Plan					
Backup Plan					

Activity 5

Decision-Making Steps

In this activity, you will use the seven-step decision-making process. You can adapt this process to all big decisions you may need to make in your life, including deciding on your career path.

1. **Focus on one decision at a time.** Select one of the decisions you just listed in Activity 4, Step 4, to focus on. Circle that decision in your table.

Continued

Step-by-Step *continued*

5. Open the FBLA Web page and click the **Refresh** button to see the new horizontal rule.

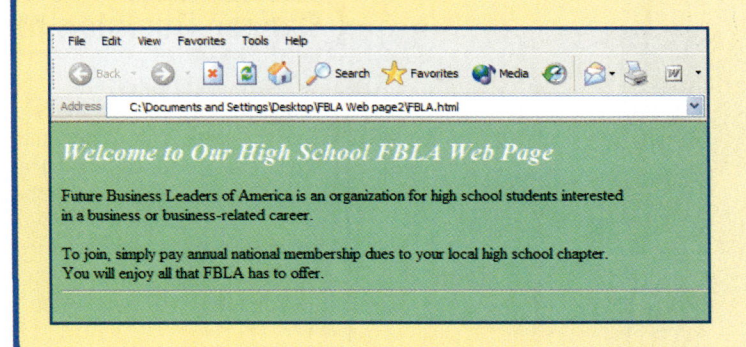

RUBRIC LINK

Go to the Online Learning Center at **digicom.glencoe.com** and click Lesson 4.2, Rubric, to find the application assessment guide(s).

Application 2

Directions Open the FBLA Notepad document. Change the background to a color of your choice. Then change the font color and size. Choose how you want to use bold and italics on the text. Change the text to show the name of your high school.

Save the document and close the file when you are done. View the Web page you created.

Application 3

Directions Write a paragraph that contains reasons to use bold, italics, different font sizes, and different font and background colors. View some of your favorite Web sites to see how they apply these font styles. Find three examples of text that you think is hard to read, and explain why.

DIGITAL DIALOG

Formatting Text

Teamwork In a small group, discuss reasons to use bold, italics, different font sizes, and font and background colors. View some of your favorite Web sites to see how they apply these font styles.

Writing Find three examples of text that you think is hard to read, and explain why. Find three examples of text that you think is easy to read, and explain why. Write your group's list on the board.

Steps in the Career Decision-Making Process The following steps will help you to make wise career decisions. In fact, these steps can be used to help you make any kind of decision.

Career Decision-Making Steps

Step 1 — **Focus on one career decision at a time.**
List and prioritize the decisions you need to make about your career based on your career plan. List them in chronological order. Which one needs to be made first, next, after that?

Step 2 — **Review your career resources.**
Who or what can help you choose your options? Think of books, Web sites, and people you know.

Step 3 — **List your options.**
Using your resources, consider all possibilities. Try to come up with a minimum of three choices.

Step 4 — **Evaluate your options.**
Research each option so that you can make an informed decision. Consider the advantages and disadvantages of each one.
- What is its value?
- Why is it important to you?
- Which will be most helpful in achieving your career goals?
- What impact do your skills, interests, aptitudes, and values have on your decision?
- What about your personality and learning style?

Step 5 — **Pick the best option.**
As you go through the career decision-making process, the best option becomes clearer. As you research and evaluate each one, it becomes more obvious which one suits your needs.

Step 6 — **Set up a plan of action.**
Now that you have made your decision, set some goals and figure out how you are going to accomplish them.

Step 7 — **Check your decision.**
Evaluate your choice to make sure you are on the right track. What have been the benefits of your decision? What have been the drawbacks? Be honest with yourself. If there have been drawbacks, identify them and determine how you can correct them.

✓ Concept *Check*

Why is it important to follow a step-by-step process when making decisions about your career?

Application 4

Directions Add the text below to the FBLA Web page. Use the <P> and <BR> codes as necessary to create two paragraphs. Add a horizontal rule as well, using the <HR> tag.

```
FBLA is a student organization
that will give you a competitive
edge through career exploration,
self-improvement, and community
service opportunities. As a member,
you will build your resume, meet
business leaders in your community
and beyond, experience the rewards
of volunteerism, and enjoy travel
and special activities.

You will also be able to
participate in special events,
such as member, adviser, and
chapter recognition programs.
There are over 250,000 members and
advisers in FBLA.

If your school does not have
a chapter, find out how to
begin one by searching online
using the keywords: start FBLA
chapter. Your chapter must be
officially chartered by your state
organization to be recognized by
FBLA.
```

Digital Decisions

Investigate New Tags

Scenario You have just started a new job as a designer at an up-and-coming clothing design company. The company does not have a Web site, but you have convinced the owner that a Web site would increase sales. Since you are familiar with what the customers like, you decide to create the Web site yourself. You begin to assemble a basic reference document for HTML tags.

Internet Investigation Use a search engine to find HTML tutorials. Choose one tutorial, and look for more tags that can be applied to text in a Web page.

Critical Thinking On an index card, list and describe each new tag. Describe how it could be applied to the Web page you have been working on in this chapter.

Tape your index card to the designated area your teacher has organized on a bulletin board.

Digital Dialog

Discuss Tags

Teamwork In a small group, discuss the tags you have learned about so far. How do these tags affect the text on a Web page? Each member of the group will choose two tags to discuss.

Writing Write a sentence or two that explains what each tag does and why it is important in designing a Web page.

Lesson 14.4 → The Career Decision-Making Process

You will learn to:
- **List** steps to follow when making career decisions.
- **Determine** options for achieving career goals.
- **Create** an action plan for implementing a career decision.

Why a Career Decision-Making Process Is Important No one is responsible for your future except you. Knowing how to make responsible career decisions will give you confidence and a sense of freedom. You will be in control of the process.

As you pursue your career, you may change your mind, modify your path, and adjust your plan accordingly. You can increase your chances for success by following the career decision-making process shown on page 460.

How to Use the Career Decision-Making Process Evaluating your decisions is an ongoing process. In order to decide which decision to make first, you will **prioritize** them, or put in order from most important to least important.

prioritize To put in order from most important to least important.

You will consider several options and learn from each decision you make. This will help you make better decisions the next time. As you follow the career decision-making steps, you will become more skilled and more confident in your decision-making ability.

It is not always easy to make good decisions. To make a good decision, you have gather information, evaluate your choices, and predict what the outcome of your decision will be. Data used to make decisions should be as current as possible and should relate directly to your needs. When you have all the information you need, it is easier to evaluate your choices. You become empowered to predict the outcome of your decision and will feel comfortable about the decision you have made.

backup plan An alternative course of action.

Backup Plans Remember to always have a **backup plan**, or an alternative course of action, in place. If you have a backup plan and you are not able to achieve one of your goals, you have already figured out what to do instead. You may simply need to adjust or revise one of your goals. You may decide to start over with a new plan and choose a different career. Always be prepared and willing to adjust your decisions along the way. This will keep you moving along your career path.

21ST CENTURY CONNECTION

Are You Self-Directed? Become an Entrepreneur!

Take the following quiz to see how self-directed you are:

- Have you ever babysat, mowed lawns, planned birthday parties, tutored, helped someone learn a computer program, assisted someone with homework, or taken photographs for your family or the school newspaper?
- Did you do these jobs well?
- Did you enjoy the independence of doing these jobs on your own?

If so, you may have what it takes to be your own boss by becoming an entrepreneur. An entrepreneur creates his or her own business and takes on the financial risk of starting, operating, and managing the business.

How to Get Started If starting your own business appeals to you, here are some tips for getting started:

1. List the reasons you want to start your own business. Some of the most common reasons are the desire for independence, financial freedom, creativity, and flexible working hours.
2. Consider the type of business you would like to start. What are your talents, hobbies, personal interests, and work experience? Whenever you get ideas for businesses, record them on your PDA.
3. Determine the market for your business. Is there a need for the type of products and services in your area?
4. Determine your financial needs. What equipment, tools, and materials will you need? Do you have personal funds to use, or will you need to borrow money to get started?
5. Identify people who can and will help you. Some good candidates are relatives, friends, previous employers, community leaders, coaches, and neighbors.
6. Develop your business plan. Include your business goals and objectives and measures for success.
7. Develop a marketing plan. How will you let people know about your products and services?

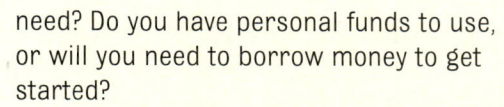

Advantages of Youth Do you need to wait until you have real world experience before you start a business? Not necessarily. Milton Hershey opened his first candy store when he was 18.

Starting a business while you are young can be to your advantage. You are not likely to have much competition among your peers. People will naturally want to help you succeed, including your teachers, parents, neighbors, relatives, local business owners, nonprofit organizations, and community leaders. Your achievements may even help you get into the college of your choice.

Being young, you have energy and creativity just waiting to be unleashed. You have a fresh perspective and are not set in your ways, so you see opportunities where others see obstacles.

> **Activity** What are four or more benefits of starting your own business compared to working for someone else?

Application 4

Directions Now that you know how to categorize goals, you can fill out a career plan more thoroughly. Once you clearly and specifically commit to a target, your mind helps you adjust your behavior to get there. Commit to specific goals with a timeline, and you will become empowered to reach the goals. The most workable career plan is the one that you design for yourself.

1. Set more short-, medium-, and long-term goals in your Career Plan document that you developed in Activity 3. Enter them into the appropriate boxes, or cells.
2. Choose a time frame for completing each goal. Fill in as much as you can now, and add a To Do List entry in your PDA for those items you still need to research. To back up your data, be sure to synchronize your PDA with your personal computer.
3. Identify what technology-related skills and knowledge are required for ultimate career goal. Use research you have already done for this chapter as well as information from the *Occupational Outlook Handbook*, trade journals, newspapers, magazines, and the Internet. Use the information you gather to help you further refine your career goals.
4. Assess your plan.
   - Are your goals specific, achievable, and flexible?
   - Can you complete them in the time frame specified?
5. Evaluate how well your goals match your values, skills, and interests. In what ways do your goals identify who you are? What goals, if any, do not match and why?
6. Make adjustments, if needed, to your career plan to better match your values, skills, and interests.

RUBRIC LINK

Go to the Online Learning Center at **digicom.glencoe.com** and click Lesson 14.3, Rubric, to find the application assessment guide(s).

Digital Decisions

Career Day

Teamwork Work in a small group. One of you will play the role of a television news anchor who will be leading a discussion on careers. The discussion will be broadcast as part of a special National Career Day program. The other two or three students in the group will play the roles of experts in the careers they indicated as their ultimate career goal on the plan completed in Activity 3, page 457.

Writing Work together to write a script of the discussion that includes the questions the moderator will ask and the responses from each of the experts. Items you may want to address include:

- Description of the job.
- Qualifications, training, and education needed for the job.
- Salary range.
- What you like most about the job.
- What the most challenging aspect of your job is.
- Suggested goals for people interested in pursuing that career.
- Career related lifelong learning opportunities.

Present your discussion to the rest of the class.

Lesson 4.3 Insert Images and Backgrounds

You will learn to:
- **Save** clip art and insert it into a HTML document.
- **Add** digital pictures to an HTML document.
- **Save** and insert background clip art.

clip art Electronic illustrations that can be inserted into a document.

Adding Images to Web Pages Nothing helps to get a point across better than a picture. When you are designing a Web page, you can use clip art, digital pictures, and backgrounds as part of your design.

Clip Art Clip art is electronic illustrations that can be inserted into a document. There are countless Web sites that offer many different kinds of clip art. Many of these sites allow you to download their clip art for free, as long as you plan to use it for a noncommercial purpose. Noncommercial Web pages include those created for a nonprofit organization, for educational use, or simply for personal interest.

Always check the user policy on a clip art Web site, however, to be sure it is permissible to download the clip art. A clip art file usually has the extension .png, .gif, or .jpg. You may want to review Lesson 1.3, pages 15–18, about file extensions.

Animated Clip Art You can use animated clip art on your Web pages to make the pages come alive. Animated clip art moves when you view it in a Web browser.

Digital Pictures You can create your own Web page images. With a digital camera, you can take pictures and download them to your computer for use on a Web page. If you take photos of people, be sure to get their permission before placing the image on a Web site.

Backgrounds You can also use clip art as the background image for your Web page, instead of the solid-color background you learned about in Lesson 4.1, pages 85–90. When you use an image as a background, be sure that your text is still easy to read. You may have to change the color or adjust the formatting of the text.

✓ Concept *Check*

What kinds of clip art could you add to the Web site you have been working on? What kinds of digital photos could you take to add to the site? Describe the photos.

Activity 3

If you need additional help on this topic, open the Help menu; or refer to the documentation for your hardware.

DIGI*Byte*

Adapt Your Career Plan

You are likely to change jobs several times, so career planning is a lifelong process. That means that your career plan must adapt to fit your ever-changing needs. As you develop your values and expand your interests and skills, you may need to change your plan to fit your new directions.

Step-by-Step

Categorize Career Goals

1. Open a file in a word processing or spreadsheet program.
2. Create a Career Plan table similar to the one below.
3. Save the file as **Career Plan**.
4. Enter your top career choice in the top row.
5. Think of five goals that will help you toward that career. Enter them below the table in your document.
6. Categorize each goal as a short-term goal, medium-term goal, or long-term goal.
7. Categorize each goal as education/training, experience/jobs, or career research.
8. Move each of your goals to the appropriate cell in your table.

Career Plan

Ultimate Career Goal: _____

	Short-Term Goal	Medium-Term Goal	Long-Term Goal
Time Frame for Completion	Start within a year	Start within 2 or 3 years	Start within 5 years
Education/Training What courses, degrees, or certificates do I need?			
Experience/Jobs What can I do to gain experience?			
Career Research What do I want to know? How can I find it out?			

Chapter 14 Plan Your Career — *Lesson 14.3*

Step-by-Step

Activity 13

If you need additional help on this topic, open the Help menu; or refer to the documentation for your hardware.

DIGI Byte

Accurate Keying
When adding clip art to an HTML document, it is important to key the file name exactly as it has been saved and to use the correct extension.

Animated Clip Art
To see an animated clip art actually move, you need to view it in a Web browser.

Activity 14

Add Clip Art to Your Web Page

There are two steps to adding an image to a Web page:
- The image file itself must be placed in the same folder as your Notepad document.
- A special <IMG src=> image source tag must be added to the FBLA Notepad document.

1. Go to a Web site that offers free clip art. To find sites, do a Web search with the keywords "free clip art."
2. Find an image that you like. Right-click the image and click **Copy**.
3. Open the folder where you have been saving the FBLA Notepad document. Click **Edit, Paste**.
4. Rename the clip art so it will be easy to remember what the picture is. Right-click on the clip art file, click **Rename**, and enter a short, descriptive name. Make sure you keep the original file extension.
5. Open the FBLA Notepad document from the Taskbar. Place the insertion point after the <HR> tag. Press ENTER.
6. Key the image tag and file name:

```
<IMG src="clip_art_file_name.gif">
```

   Instead of *clip art file name.gif*, key the actual file name and extension of your clip art file. Be sure to key the file name in quotes.

7. Press ENTER.
8. Click **File, Save**. Open the FBLA Web page and click the **Refresh** button to see the clip art on your page.

Add a Digital Photo to Your Web Page

A digital photograph is added to a Web page the same way as any other image. Remember that the photo must be stored in the folder with your Web page.

1. Use a digital camera to take a picture of a classmate.
2. Follow the directions for your camera to transfer the picture from the camera to your computer.
3. Move the picture to your Web page folder.
4. Open the FBLA Notepad document.

Continued

Chapter 4

101

Lesson 14.3 ▶ Develop Your Career Plan

You will learn to:
- **List** short-, medium-, and long-term career goals.
- **Set** a timeline for achieving your goals.

Set Career Goals A good career plan includes short-, medium-, and long-term goals that will lead you to your chosen career. Short-term goals are activities you can start right away and might be able to complete quickly. For example, you could get a part-time job that is related to your career goal. Medium-term goals are more challenging and take more time to accomplish, like getting a degree. Long-term goals take even more time to complete. Your ultimate career goal is a long-term goal. Arrange your goals in **chronological order**, or the order in which events happen.

chronological order The order in which events happen.

Putting your career plan in print makes your plan more real and tangible. You know what you want to do with your life rather than just drifting from job to job.

Types of Career Goals

Education and Training—You are already working on the first part of your career plan. The courses you are taking now and will take after high school are the first steps in preparing you for your career.

Experiences and Jobs—A career plan can include internships, volunteering, job shadowing, temporary positions, part-time jobs, and full-time jobs. These activities will help you to prepare for your chosen career.

Career Research—The career planning process never ends. You will continue to learn about new opportunities and changes in any field. Resources include the library, the Internet, guidance counselors, and coworkers.

How to Set Good Goals When you set goals, be sure they are:
- **Specific**—If you know exactly what you are planning to do, it will be easier for you to achieve your goals.
- **Achievable** within the time frame you have set up.
- **Flexible** so that you can change them as your needs change.

DIGI Byte

Average Number of Jobs
The average person in the U.S. will hold about 9 jobs from age 18 to 34.
Source: Bureau of Labor Statistics

✓ Concept Check

Why is it important to write out your career plan? Why is it important both to plan for your career and to be flexible?

Step-by-Step continued

5. Move the insertion point to where you want to place the picture, and key the image tag and file name:

```
<IMG src="digital_picture_file_name.gif">
```

Instead of *digital_photo_file_name.gif*, key the actual file name and extension of the digital photo.

Use Background Images

In this activity, you will add a <background => tag to use a background image. If you add a background image after you have entered text, you may need to go back and change the font color of your text so it will show up against the background image.

1. Go to a Web site that offers free background images. To find these sites, do a Web search with the keywords "free background images."

2. Find a background image that you like. Right-click on the image and click **Copy**.

3. Go to the folder where the FBLA Notepad document is saved and click **Edit, Paste**.

4. Open the FBLA Notepad document.

5. Move the insertion point to the beginning of the document where you entered the <BODY bgcolor="#99CC99"> code in Lesson 4.1.

6. Change the tag to read:

```
<BODY background="background_file_name.gif">
```

Instead of *background_file_name.gif*, key the actual file name and extension of the background.

7. Click **File, Save**. Open the FBLA Web page and click the **Refresh** button.

   If you do not like the way the background looks, repeat Steps 2–7 and select another background.

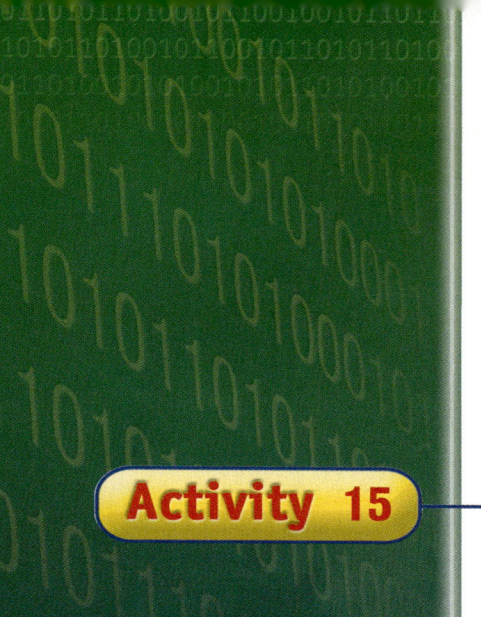

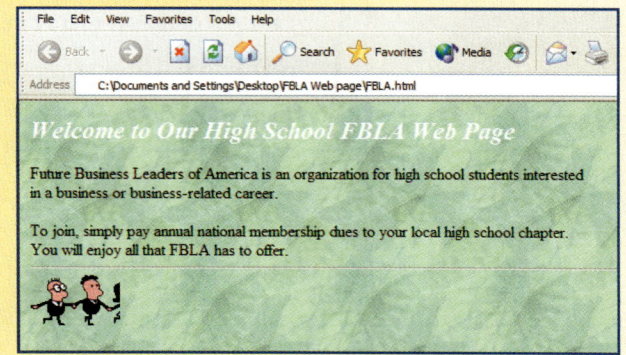

Activity 15

DIGI Byte

File Naming
File names should not have blank spaces. Use the underscore character (_) for blank spaces.

Suite Smarts

The background and clip art files you are viewing on the Microsoft Web site can also be added to Word, Excel, Access, and PowerPoint documents.

21ST CENTURY CONNECTION

Overcome Career Obstacles

As you pursue your career, you may face obstacles, or things that stand in the way of progress or achievements. Obstacles may tempt you to give up or change direction. To avoid changing direction too early or giving up, look for ways to overcome obstacles.

You Can Do It The chart below shows some common obstacles and possible ways to overcome them. Remember, you can always find a way to do it.

> **Activity** Think of one obstacle you are facing right now. Read the chart below and write a paragraph that includes two ways to overcome the obstacle.

Career Obstacles—What Is Holding You Back?

Obstacle	Ways to Overcome
Money "I cannot afford it."	There is always a way to finance what you want to do. Look into scholarships, grants, assistantships, internships, loans, and part-time work. Think of the expense as an investment in yourself.
Fear "I am afraid I will fail." "I have never done that before." "I always get nervous." "What if it does not work?"	If you do not try, you will not fail, but you will not succeed either. Learn from your mistakes. A failure is truly a failure only if you do not learn anything from it. Talk to others about how they deal with fears and nervousness and see if they have some ideas that would help you. Take small steps toward achieving your goal rather than trying to do it all at once. Gain confidence along the way.
Negativity "I like the old way of doing it." "I might not do it correctly." "Jill is better at that than I am." "I am too old to do that." "I am not qualified to do that."	Sometimes we are our own worst enemies. We get in the way of our own success. Rather than telling yourself what you cannot do, tell yourself what you can do. Give yourself positive messages. Make an "I Can Do" list in your PDA. Writing things down reinforces them in your mind.
Other People "You cannot do that." "That is a stupid thing to do." "You should…" "If I do that, my friends may not like me."	Stand your ground. Be true to the person that you are and do not let others pressure you. If it is the right thing for you to do, then do it. Be confident in your decisions.
Procrastination "Tomorrow I will…" "After I finish this, I will…"	If you do nothing, then someone will make the decision for you or you may miss out on opportunities. Use your PDA to set a schedule and stick to it.

Application 5

Directions Find two clip art pictures and a new background to add to your Web page. Add a digital photo as well. Be sure to save the pictures in the FBLA folder and add the correct tags to the HTML document.

Application 6

Directions You will create a new FBLA Web page. Create a new Web page. Go to the FBLA Web site at **www.fbla-pbl.org** to find out about the upcoming state and national conferences that are available. Include the name, dates, location, intended audience, and purpose of each conference on your page.

Format the text using all of the features and tags you have learned about in Lessons 4.1 through 4.3, pages 85–103. Add graphics as appropriate and place them to balance the text. Use a background color or image that draws attention to the information but does not conflict with it or overwhelm it.

Name this Web page **conferences.htm,** and save it in the same FBLA folder you have been working on.

RUBRIC LINK
Go to the Online Learning Center at **digicom.glencoe.com** and click Lesson 4.3, Rubric, to find the application assessment guide(s).

DIGITAL Decisions

Web Page Images

Critical Thinking Work with a partner and trade the Web pages that you created in Application 5 on this page. You can trade printouts of the Web pages, or you can sit at each others' computers. By looking at a page other than your own, you will get a fresh perspective on the use of images on a Web page.

As you look at the images on your partner's Web page, write down your responses to the following questions:

- When is clip art more effective than photos?
- When are photos more effective than clip art?
- What kinds of backgrounds are most effective?
- How should images and text work together?

Discuss your conclusions with your partner.

DIGITAL DIALOG

Web Page Information

Teamwork Work in a small group. Discuss the FBLA Web page you have each started. What other information do you think should be included on this page? You can visit the national FBLA Web site at **www.fbla-pbl.org** for inspiration.

Writing Each student should list on a sheet of paper two ideas for other information to include on the Web page. After everyone has recorded their ideas, place them front-side down in the middle of the table. Assign one person in the group to record the ideas. As a group, decide on the impact each idea would have on the user of the Web site and prioritize your ideas. Share your group's top three ideas with the rest of the class.

Application 2

Scenario You check your schedule in your PDA and realize that in two weeks you will be meeting with your school's career counselor to discuss your education and training options. To prepare for the meeting, you decide to research in detail the education and training requirements for one of the occupations that interest you.

Directions Look up a career that interests you in the *Occupational Outlook Handbook* at **stats.bls.gov/oco**. You can also use the *Occupational Outlook Handbook* from the library. Record your answers to the following questions related to the education and training requirements in a Tablet PC or PDA.

1. What degrees or certifications are required for the occupation?
2. What kinds of educational institutions or agencies offer the training or certifications needed?
3. Once a person is hired, what additional training or certifications are required?
4. What is the job outlook for the position?
5. What is the potential income range for the position?

Application 3

Directions Prepare a presentation on an education or training option in your community. Research the organization on the Internet, and request information about their programs. You may even want to visit the organization and take some pictures with a digital camera to include in your presentation. Include details that will help your classmates determine if the option would be good for them.

RUBRIC LINK
Go to the Online Learning Center at **digicom.glencoe.com** and click Lesson 14.2, Rubric, to find the application assessment guide(s).

Digital Decisions

Diversity in the Workplace

The number of people working continues to grow and shows no signs of stopping. As the workforce grows, it also changes. Changes mean more opportunities for people of all backgrounds.

Critical Thinking Talk to an adult who has a different background than you. Ask how he or she has overcome any unique challenges in the workforce related to his or her background. If possible, take notes on your PDA or Tablet PC.

Writing Think about any unique challenges related to your background that you have overcome in school. Write a one-page report comparing your experience to the experience of the adult you talked to. How do you think the workforce has been, and will be, enhanced by both experiences? Also, describe ways you can contribute to a positive and productive work environment when working with a diverse group of people.

Lesson 4.4 → Text and Graphic Alignment

You will learn to:
- **Center** text and graphics on a Web page.
- **Align** text and graphics on the right or left side of a Web page.

Page Layout When designing a Web page, it is important to consider how the text and graphics are laid out on the page. For example, you might want a big headline to be centered in the middle of the page. You might want a photograph to appear off to the side of the page, with a caption aligned underneath it. HTML gives you the tools to control many aspects of page layout.

The Importance of Closing Tags An HTML tag for alignment must be accompanied by a closing tag. If you do not use a closing tag, this is what will happen: the alignment tag will apply not only to the text you want aligned, but also to everything that appears after it.

✓ Concept Check

What kind of text would you center on a Web page? How could you use left align and right align? Give an example of each.

DIGITAL DIALOG

Web Pages and Teamwork

Teamwork Creating a Web site is often a team effort, involving designers, writers, and even financial experts. Your school has asked you to revise and update its Web site. Work in a small group. Begin by discussing what information you feel should be included in the Web site. Assign one person in your group to take notes as you brainstorm ideas.

Writing Create a graphic organizer displaying the layout of the Web pages, with the title of each page printed at the top of the page. Below the title, describe the information you feel should be on each page. An example has been started for you.

```
                Home Page
                    |
    ----------------|----------------
    |               |               |
Sports Page    Student         Faculty
               Organizations
```

Chapter 4 Create a Web Page with HTML   Lesson 4.4 **104**

Activity 2

If you need additional help on this topic, open the Help menu; or refer to the documentation for your hardware.

DIGI*Byte*

Fastest-Growing Occupations

Did you know that of the ten fastest-growing occupations, the top five are computer-related? They are computer engineer, computer specialist, systems analyst, database administrator, and desktop publishing specialist. Taking this class is great preparation for a future career.

Source: Bureau of Labor Statistics

Step-by-Step

Plan Your Education and Training

1. Open a file in a word processing or spreadsheet program. Save the file as **Education Plan.** Create a table similar to the one below.

2. Enter one of the careers that interests you on your newly created education and training plan.

3. Research online or in the library to determine the degrees or certifications you will need to qualify for the career you chose. The job descriptions you read in the last lesson may also include required education and training. Record the required degrees or certifications on your education and training plan.

4. Research online or in the library to find three educational or training organizations that offer the training you will need. Record information about each organization on your education and training plan. Include the name of the organization, location, programs offered, cost, and time schedule for completion.

5. For future reference, enter the address and contact information for each organization you research in Address Book in your PDA. Remember to synchronize your PDA and your computer whenever you add data.

Education and Training Plan

	Name	Location	Programs Offered Related to Career Choice	Cost to Attend	Time to Complete
Career					
Required Degrees or Certification					
Possible Education or Training Organizations					

Chapter 14 Plan Your Career — Lesson 14.2

Activity 16

If you need additional help on this topic, open the Help menu; or refer to the documentation for your hardware.

Step-by-Step

Align Text

To left or right align text, you will use the codes <align="LEFT"> and <align="RIGHT">.

1. Open the FBLA Notepad document you have been working on.
2. Change the font color to "#004400". Place the insertion point after the <I> tag in the title of your document.
3. Press ENTER and key the tag: <CENTER>
4. Click **File, Save**.

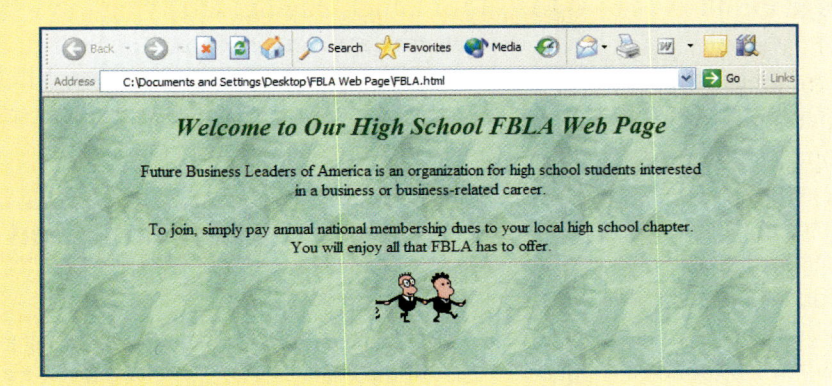

5. Open the FBLA Web page and click the **Refresh** button.

   Note that all the text and the graphics are now in the center of the page. This is because you have not yet inserted a closing center tag.

6. Go back to the FBLA Notepad document. Place the insertion point after the word *Page* in the title. Then press ENTER.
7. On the new line, key the closing tag: </CENTER>
8. Click **File, Save**. Open your browser and click the **Refresh** button. Now only the heading should be centered. Remove both center tags.
9. Apply the <align="LEFT"> attribute to the header, <H2>, tag. Enter:

```
<FONT color="#004400">
<H2 align="LEFT">
<B>
<I>
```

DIGI Byte

View Source Code

Remember that you can see the HTML code of any Web site by clicking View and then Source in the browser's Menu bar.

Continued

Chapter 4

105

Lesson 14.2 Plan Your Education and Training

You will learn to:
- **Identify** degrees and certifications required for your career interests.
- **Develop** an education and training plan.

Plan Your Education How will you gain the skills and knowledge necessary to succeed in the career that interests you? The best opportunities for a high-paying, satisfying job will require you to attain the highest level of education needed to succeed in that job. Since education can be time-consuming, planning ahead will help you to achieve your goals. Plan the courses you need to take in high school. Find out if you need to apply to a college or technical school.

Education and Training Options

Option	Description
Colleges and Universities	Offer a bachelor's degree for four years of study. Some offer master's degrees and doctorates.
Community and Technical Colleges	Offer two-year associate degrees and certification programs. Can transfer credits to a college or university.
Continuing Education	Programs for adult students to help them complete education, review existing skills, pursue new interests, or receive required training to maintain certification in a particular field.
Distance Education and E-Learning	Online classes that students can take to earn credits toward degrees. Offer flexibility in location and schedule.
Vocational-Technical Centers	Offer a variety of skills-training programs.
Trade Schools	Private schools that train students for a certain profession.
On-the-Job Training	Instruction received at a place of employment while working for pay.
Apprenticeships and Internships	Hands-on learning under the guidance of a skilled worker.
Military Service	Education and training offered by the government in exchange for serving in a branch of the military.

✓ Concept Check

Why is it important to plan your education and training?

DIGIByte

College for All
There is always a way to achieve your educational goals. Anyone can attain a college education, according to the American Council on Education and its College Is Possible program. To learn more about this program, search online for "college is possible."

Suite Smarts
Keep a list in your PDA of colleges and schools you are interested in. You may want to visit one or more of them to see what they are like. Check each school's Web site on how to arrange for a visit and for directions to their location.

Step-by-Step continued

```
Welcome to Our High School FBLA Web Page
</I>
</B>
</H2 align="left">
</FONT>
```

Align Graphics

The way you add alignment codes to graphics is similar to the method you used with text. The code needs to be placed inside the image <IMG src=> tag.

1. Open the FBLA Notepad document. Find the image <IMG src=> tag for the clip art.

2. Place the insertion point in the space between IMG and src and key this text: `align="RIGHT"`

    The tag should now look like this, but it will include the filename of your clip art:

    `<IMG align="RIGHT" src="people2.gif">`

3. Click **File, Save**.

4. Open the Web browser and click the **Refresh** button. The graphic is now on the right margin below the horizontal rule.

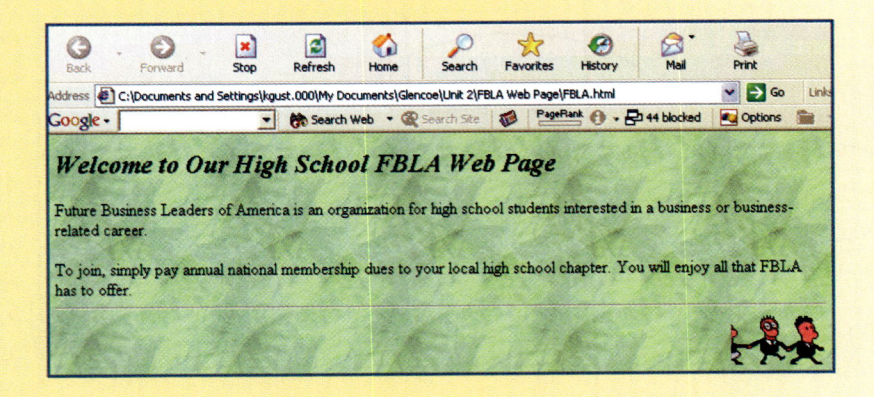

The HTML reads as follows:

`<P>`
To join, simply pay annual national membership dues to your local high school chapter.

Activity 17

DIGIByte

Working with Images

Different HTML instructions can be used to control the way an image looks on a Web page. For example, you can use HTML to put a border around an image or to change the image's height and width.

Continued

Application 1

Scenario You are a recruiter for a company that has an opening for the job you researched in Activity 1, page 450. The position you need to fill is a very important position in the company, so you want to do everything you can to get the most qualified people to apply.

Directions Write a one-page job description that will attract highly qualified people to apply for the job.

Include:
- The job title.
- A sentence describing the job itself.
- A brief description of the educational requirements for the job.
- A brief description of the experience requirements for the job.

Emphasize the tasks and responsibilities that are the most important and attractive. Create a graphic organizer similar to the one below to help you brainstorm ideas. Draw the graphic organizer on a sheet of paper, a whiteboard, or a Tablet PC.

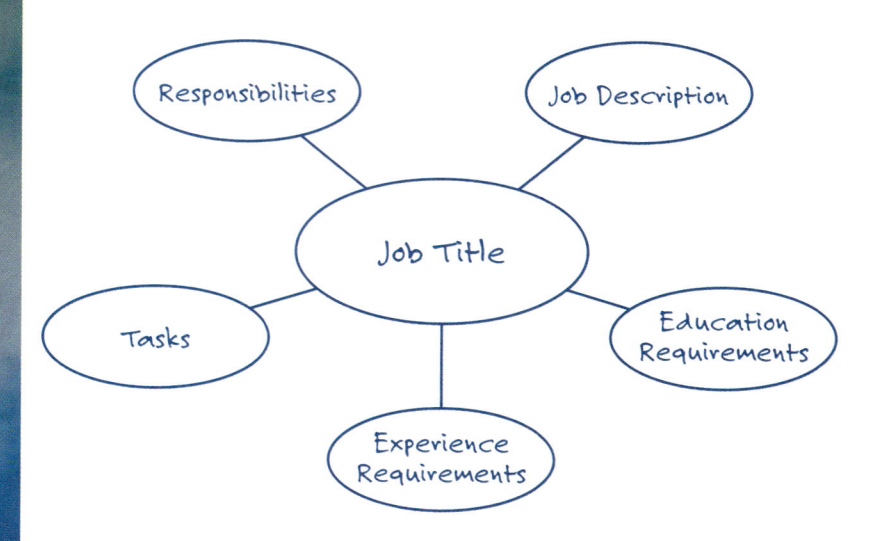

RUBRIC LINK
Go to the Online Learning Center at **digicom.glencoe.com** and click Lesson 14.1, Rubric, to find the application assessment guide(s).

Look Who's DIGITAL

Technology and the Troops

Keeping soldiers safe and in good spirits is a challenge that technology meets from the ground up. Robots equipped with wireless cameras roam the streets like remote-control cars in a search for hidden explosive devices. Soldiers survey the terrain from the safety of their computer monitors.

North Carolina, Governor Mike Easley sent laptop computers to help soldiers carry out their duties and to "allow for more contact between soldiers and their families at home." Soldiers can use e-mail, satellite phones, and videoconferencing. Capt. Jason Hughes in Baghdad exchanged virtual hugs with his sons in California. Like 1,500 other soldiers a day, Hughes was cheered by the chance to visit with his family and show them he was safe.

Activity How could these technologies be used in your area?

DIGI Byte

Trial and Error

When aligning text and graphics, it helps to experiment. You might need to place tags in several different locations in your Notepad document before the tags work the way you want them to.

RUBRIC LINK

Go to the Online Learning Center at **digicom.glencoe.com** and click Lesson 4.4, Rubric, to find the application assessment guide(s).

Step-by-Step continued

```
<BR> You will enjoy all that FBLA has to offer.
<HR>
<IMG align="RIGHT" src="people2.gif">
</P>
```

5. To move the graphic next to the paragraph above the horizontal rule, the code should be rewritten as follows:

```
for high school students interested <BR>in a business
or business-related career.
<IMG align="LEFT" src="people2.gif">
</P>
<P>
To join, simply pay annual national
membership dues to your local high school chapter.
```

6. Click **File, Save**. Go to the FBLA Web page and click the **Refresh** button. By moving the <IMG SRC=> tag to the end of the first paragraph, the graphic now appears below that paragraph.

Application 7

Directions Open the FBLA Notepad document and make changes to the text to change the alignment. In the previous lesson, you added additional graphics to your Web page. Try aligning these in different locations. Use the <align="right"> and <align="left"> attributes. Print your Web page after you have made alignment changes.

DIGITAL DIALOG

Web Site Ethics

Teamwork Create a small group of students. Discuss the topic of ethics as it relates to Web sites.

- What information do you think should be displayed on a Web site?
- If you were creating a Web site for an organization or club, what kind of information would you list about your members?
- Would you include personal information like phone numbers and home addresses? Explain why or why not.
- What other types of information about your members should you be careful about?

Writing Each member of your group should write a few paragraphs to summarize the discussion points. As a class, compare each member's points.

Activity 1

If you need additional help on this topic, open the Help menu; or refer to the documentation for your hardware.

DIGI Byte

Occupational Outlook Handbook

The *Occupational Outlook Handbook* contains detailed job descriptions for thousands of jobs. It also includes earnings, job outlooks, and related occupations. The *Occupational Outlook Handbook* is available at the library. You can also find it online by entering "Occupational Outlook Handbook" in a search engine.

Step-by-Step

Evaluate Job Descriptions

1. Open a file in a word processing or spreadsheet program.
2. Save the file as **Job Descriptions**. Create a table similar to the one below.
3. Find one of the careers in which you are interested in the *Occupational Outlook Handbook*. Obtain a copy of the book from your library, or find it online at **stats.bls.gov/oco**.
4. Enter the job title in the top row of the table.
5. List the tasks included in the job description in the left-hand column in the table. In the right-hand column beside each task, rate the task as follows:

    + = Interests me

    ? = Not sure if it interests me

    − = Does not interest me

   Add more rows to your table, if necessary.
6. Count how many pluses and minuses you marked. If you marked a lot of minuses, are you still interested in the job?

Would I Like the Job?

Job Title:

Job Tasks	Task Rating + = Interests me ? = Not sure if it interests me − = Does not interest me
1.	
2.	
3.	
4.	
5.	
Number of pluses	**Number of question marks**	**Number of minuses**

21st Century Connection

Media Literacy

One hundred years ago, any person who could read was classified as "literate." Now, literacy is a much broader term. Today's culture demands that we know how to "read" information in print, on television, and on Web sites. How can we make sense of all the information around us?

Purposeful Messages

Communication media carry purposeful messages. For example, advertisements try to convince audiences to buy a product. Often, they create a scene based on an emotion to "hook" the audience.

Imagine an ad for home security showing a house with a broad green lawn on which neighborhood children play in the sunshine. Now imagine the same house shown in black and white at night during a storm with ominous music playing in the background. The threatening scene would be more effective in making people believe they need extra security. Many ad campaigns use fear to sell products.

Ad campaigns also use positive emotions like love and adventure. Would you want to buy a car if the commercial showed it sitting in a traffic jam? Of course not! Car commercials feature young, happy drivers on adventures, out on a glamorous date, or loading up the trunk with vacation gear. Those ads want you to think, "If I buy that car, I will have great adventures. I will be happy."

Consumer Be Aware A media literate consumer is aware of the ad's agenda and instead thinks, "That looks like a fun car. Nice ad." See the difference? Being media literate means being aware of the purpose behind the message.

Online Awareness

You need to exercise the same kind of awareness when working online. A generation ago, publishers and editors did most of the work of validating the truthfulness of anything in print. Now it is possible for anyone to publish anything on the Internet, so the burden is on the reader to figure out if the information is believable.

One way to check up on a site's author or owner is to click the About Us link or visit the home page. If the owner is authoritative, like Discovery.com or the Library of Congress, you can trust what you read on the site.

By honing your media literacy skills, you will be able to make good sense of all the forms information takes.

Activity Think of a time when you bought a product because of an advertisement. Did the product do what the advertisement promised? Would you buy the product again? In a paragraph, describe why or why not.

Lesson 14.1 Evaluate Job Descriptions

You will learn to:
- **Analyze** job descriptions for jobs that interest you.
- **Rate** your interest in each job-related task.

job description An explanation of the major areas of a job necessary for an employee to perform the job successfully.

Job Descriptions You are starting to think about some careers you might like. How do you know if you will really like a job? A **job description** is an explanation of the major areas of a job necessary for an employee to perform the job successfully.

Reading a job description can give you a better idea of how much you might like the job. A job description also includes the training required for the job. This can help you to plan your career path.

Parts of a Job Description A job description may include the following items:

- **Job title**—For example, a commercial airline pilot.
- **Job objective**—A brief statement summarizing the main objective or purpose of the job. For example, a commercial airline pilot navigates aircraft, primarily for the transport of cargo and passengers.
- **Job tasks**—An item-by-item list of the duties, responsibilities, and activities the employee needs to carry out in order to produce results. For example, a commercial airline pilot plans flight routes, checks weather conditions, monitors instruments, and contacts air traffic control when necessary.
- **Required knowledge and skills**—What the employee needs to know and be able to do. A commercial airline pilot must know and apply principles and methods for moving people or goods by air. A pilot also needs to know about equipment, policies, procedures, and strategies to promote effective local, state, or national security operations for the protection of people, data, property, and institutions.
- **Required education and experience**—The minimal acceptable level of education, certifications, licenses, or years of experience necessary for employment. For instance, a commercial airline pilot needs to have a commercial pilot's license.

DIGIByte

Job Descriptions Are Flexible

Jobs can change due to personal growth, growth of the company, and new technologies. Flexible job descriptions encourage employees to try new things and to figure out better ways to do the job.

Suite Smarts

A job description also may be called a job profile or position description.

✓ Concept Check

How are job descriptions helpful when you are choosing a career?

Lesson 4.5 ➜ Bulleted and Numbered Lists

You will learn to:
- Create a bulleted list of items.
- Create an ordered list of items.

Organizing Information Creating a list can be a great way to organize information, especially when that information consists of many related items. In addition to organizing information, lists draw attention to the information in a visual way, making the items more distinct from one another and easier to remember. When you need to create a list of several items on a Web page, HTML gives you two main ways to accomplish this:

- **Ordered List** An ordered list is a list that has numbers before each item, and the items are usually in priority order. This is used when the order of the items matters.

    For example, for items on an agenda:

    1. Call to order and welcome
    2. Report of last month's results
    3. New business

- **Unordered List** An unordered list is a list that has bullet points before each item and is used when the items can be in any order.

    For example:
    - Boston Red Sox
    - Seattle Mariners
    - Chicago Cubs

Often, you might find it makes more sense to choose one style over the other. For example, you could use an unordered list for all the ingredients in a recipe. You could then use an ordered list for the step-by-step cooking directions.

Here are the tags that you will use for lists:
- The <UL> tag is used to start an unordered list.
- The <OL> tag is used to start an ordered list.
- The <LI> tag is used for an individual item within a list.

✓ Concept Check

When might you want to include an unordered list or ordered list in your Web page? What type of information would you put in each list?

---

ordered list A list that has numbers before each item, and the items are usually in priority order.

unordered list A list that has bullet points before each item and is used when the items can be in any order.

Read to Succeed

Key Terms
- job description
- chronological order
- prioritize
- backup plan

Focus

Enhance your learning by being actively involved in your reading. Think of ways to create your own interaction with the material. When you write, discuss, and organize the material, you stay focused.

One way to focus your attention while reading is to maintain an interactive notebook. If you are using a paper notebook, use the right side for taking notes during class and while you read. On the left side, write questions you have, express your opinions, or add drawings that illustrate ideas in the text. If you are using a Tablet PC, draw a two-column chart. On a PDA, create a memo for class notes, and write your opinions in the Journal.

Example:

Ideas/Questions	Notes
Ask my family about their jobs.	Setting goals will help me reach my ultimate career goals in realistic stages.

DIGITAL CONNECTION

Take the Reality Test

It is ten years from now. You have a "reality camera," and you are taking digital photos of yourself as you go through your regular workday. What do the shots show? Are you enthusiastic about your work—doing something you really love to do?

Most people spend much of their time at work. When you plan your career path, you are more likely to enjoy your life and feel a sense of accomplishment. Develop your career plan, and take charge of your life.

Writing Activity

Write a paragraph describing how you looked and felt in your "reality camera" photos.

Activity 18

? If you need additional help on this topic, open the Help menu; or refer to the documentation for your hardware.

Step-by-Step

Create an Unordered List

In this activity, you will organize text into an unordered list using the unordered list, or <UL>, tag.

Remember that the <LI> tag is used to specify that each item is in the list.

1. Open the FBLA Notepad document.
2. Place the insertion point on the blank line below the <HR> tag.
3. Enter the following:

```
The FBLA prepares students for job opportunities in:
<UL>
<LI>Accounting and finance
<LI>Information systems
<LI>Business management and applications
<LI>Small business entrepreneurship
<LI>Business administration
<LI>Office systems technology
</UL>
```

4. Click **File, Save**. Open the FBLA Web page and click the **Refresh** button. Your Web page should look similar to the one shown.

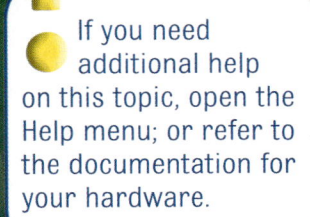

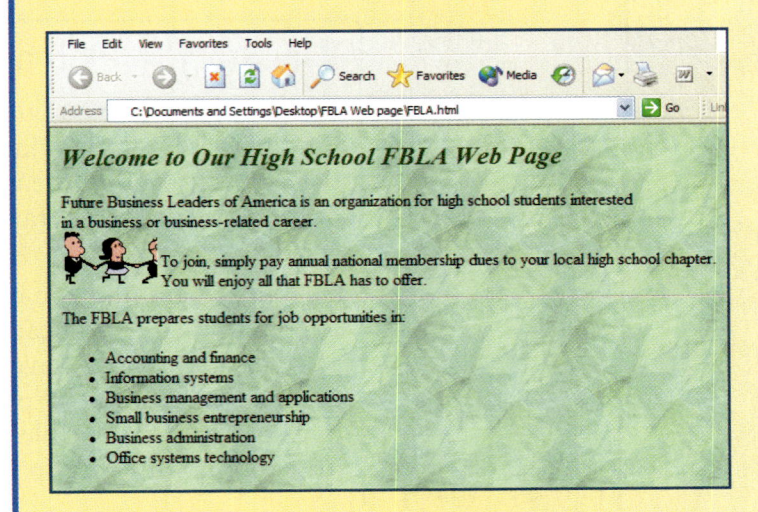

DIGI Byte

Source Code

If you see an interesting design feature on another Web page, you can view the HTML that was used to create it. Click **View, Source**, and then search for the HTML code. Try copying the code to your document to see the effect it has.

Continued

Chapter 4

110

CHAPTER 14
Plan Your Career

Lesson 14.1   Evaluate Job Descriptions

Lesson 14.2   Plan Your Education and Training

Lesson 14.3   Develop Your Career Plan

Lesson 14.4   The Career Decision-Making Process

You Will Learn To:

- **Describe** the parts of a job description.
- **Evaluate** job descriptions.
- **Identify** institutions and organizations that offer the training you need for a job.
- **Develop** an education and training plan.
- **Develop** a career plan.
- **List** short-, medium-, and long-term career goals.
- **Evaluate** career decisions.
- **Develop** backup plans.

Why Start to Plan a Career Now?

It is unlikely that you will fall into your dream job. Finding a job that you know you will enjoy with the right pay and in the right location will take a lot of planning and preparation. It does not happen by accident or luck.

The more planning you do now, the more prepared you will be to make important career decisions. Although it might seem like you have plenty of time to make these decisions, the time will pass quickly. Now that you have researched possible careers that interest you, it is time to put your plan in writing.

Activity 19

Step-by-Step continued

Create an Ordered List

In this activity, you will use the ordered list, or <OL>, tag.

1. Press ENTER and add another horizontal rule tag: <HR>
2. Enter the following text:

```
Some of the chapter events students can compete in include:
<OL>
<LI>American Enterprise Project
<LI>Community Service Project
<LI>Gold Seal Chapter Award of Merit
<LI>Helen Ragan Chapter of the Year
<LI>Local Chapter Annual Business Report
<LI>Local Recruitment of Chapters
<LI>Partnership with Business Project
</OL>
```

3. Click **File, Save**. Go to the FBLA Web page and click the **Refresh** button. Your Web page should resemble the one shown.

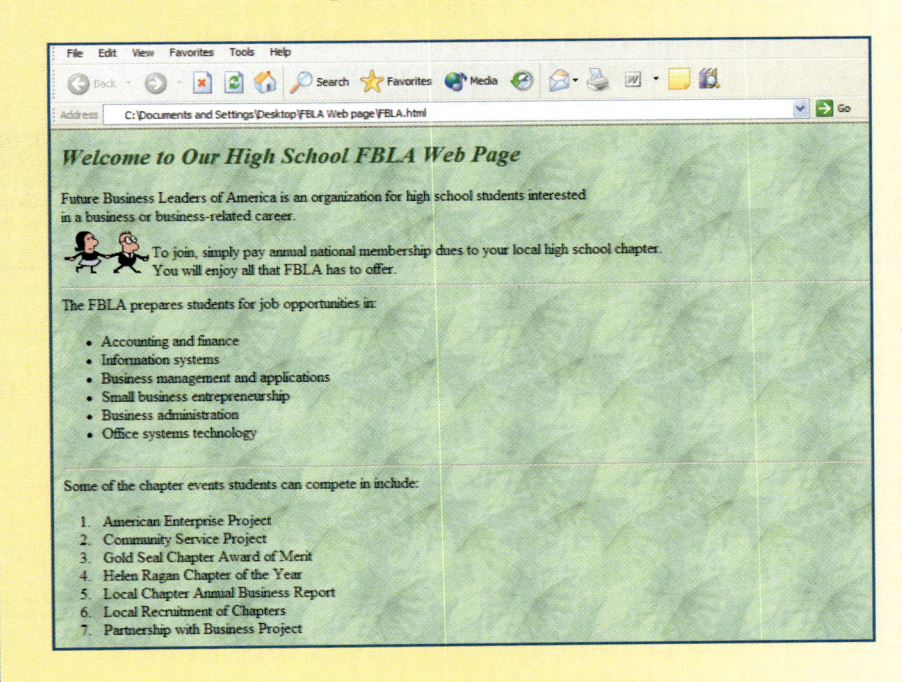

Suite Smarts

Ordered and unordered lists on Web pages are similar to lists you can create in Microsoft Word and PowerPoint.

CHAPTER 13

Self-Assessment

Take a moment to review what you have learned in this chapter. Rank your understanding of the topics below.

4 means, "I understand all of this."
3 means, "I understand some of this."
2 means, "I understand very little of this."
1 means, "I don't remember this."

> **To use a printout of this chart, go to digicom.glencoe.com and click on Chapter 13, Self-Assessment.**
> **Or:**
> **Ask your teacher for a personal copy.**

Rank Your Understanding

Lesson	Topic	4	3	2	1
13.1	• How to find the skills required in jobs that interest you				
	• Define interests, values, skills, and aptitudes				
	• List your characteristics: your interests, values, skills, and aptitudes				
	• Evaluate your abilities in some important skills.				
	• How to compare your characteristics to job requirements				
13.2	• Define career clusters				
	• Name at least five of the Department of Education's 16 career clusters				
	• Give examples of jobs of interest to you within five career clusters				
	• Explain how values can relate to job choices				
13.3	• Define networking				
	• Give reasons why networking is important				
	• List three ways to learn more about the jobs that interest you				

If you ranked all topics 4, congratulations! Consider doing a quick review.
If you ranked yourself 3 or lower in any topic, consider reviewing these topics first.

Application 8

Scenario You are to review the information below about appropriate dress for future business leaders. Using the formatting features you have learned about in previous activities in this chapter, you will format the lists on your Web page.

Directions Decide which text in the information below should be an <H2> headline, which should be in an ordered list, and which should be in an unordered list. Add the new, formatted text to the FBLA Notepad document. Print your Web page when you are finished.

Appropriate business dress for young women:
Business suit with blouse
Business pantsuit with blouse
Business dress
Skirt or dress slacks with blouse or sweater
Dress shoes and nylons

Appropriate business dress for young men:
Business suit with collared dress shirt and necktie
Sport coat, dress slacks, collared shirt, and necktie
Dress slacks, collared shirt, and necktie
Banded collar shirt (may be worn only if a sport coat or business suit is worn)
Dress shoes and socks

Inappropriate business attire, for both men and women:
Jewelry in visible body piercing, other than ears
Denim clothing of any kind
Revealing blouses, tops, dresses, or skirts
T-shirts
Spandex, midriff tops, tank tops
Sandals, athletic shoes, hiking boots
Hats

RUBRIC LINK
Go to the Online Learning Center at **digicom.glencoe.com** and click Lesson 4.5, Rubric, to find the application assessment guide(s).

Look Who's DIGITAL

Digital Olympics

What has 70,000 people, 4,000 cars, 1,000 cameras, 12 boats, 9 helicopters, and a blimp? These elements made up the security system at the 2004 Athens Olympics.

Camera microphones collected conversations that were converted to text by speech recognition software. Security guards sent and received alerts of suspicious activity on their PDAs. By using GPS coordinates, it took only a second for help to be dispatched.

In 2002 at the Salt Lake City Olympics, security forces used Tablet PCs to respond to threats. Officers communicated and sent reports over a wireless network.

Tablet PCs also work in the dark! In stealth mode, officers use night-vision goggles with Tablet PCs to avoid revealing their position.

Activity How could police officers use stealth mode on Tablet PCs? Write a short paragraph.

CHAPTER 13 Review

Read to Succeed PRACTICE

After You Read

Discover Your Strengths in Learning Work with a partner, and write a list of ten strengths that you believe the other person has in learning skills and study skills. For example, you ask questions in class, you focus on assignments, you take notes, and you are good with graphic organizers. Then make a list of ten strengths in learning and study skills you believe that you have.

When you have completed your lists, exchange them and discover what the other person has pointed out about you. Do these strengths match up with what you think about yourself? Explain to your partner why you chose the skills you did, and ask them why they chose theirs.

Example:

Strengths
1. Asks lots of questions
2. Always does homework

Using Key Terms

The following terms will help you understand and communicate effectively as you decide on a career path. Match each term to its definition.

- interests (432)
- values (432)
- skills (432)
- aptitudes (432)
- career clusters (436)
- networking (440)
- internship (440)
- job shadow (440)

1 Beliefs and ideas you live by and think are important.

2 Communicating with people to share information or advice.

3 Groups of similar occupations and industries developed by the U.S. Department of Education.

4 A temporary position, paid or unpaid, where you can gain practical work experience in a career field.

5 Your abilities to perform a task as a result of your talents, training, and experience.

6 Your favorite activities, the subjects you like best in school, and your hobbies.

7 To follow someone for a day or two on the job.

8 Your potential for learning a skill.

Lesson 4.6 — Link to Other Pages and Web Sites

You will learn to:
- **Create** a hyperlink to another Web page.
- **Create** a hyperlink to another page within your Web site.
- **Add** a link to an e-mail address on your Web page.

DIGI Byte

Web Design
Web sites such as Webmonkey.com offer free advice for beginning Web designers.

Anonymity
If you would rather that your name not appear on your Web site, you can use *Webmaster* instead of your name.

Planning a Web Site A Web site consists of a home page and any number of other pages that link from the home page. Creating a storyboard or diagram can help you lay out pages and organize the information you want on your Web site. A well-planned Web site includes links that make it easy for users to navigate from one page to another within the site.

Creating Hyperlinks When text or a graphic on a Web page is linked to another Web page, it is called a hyperlink. These links are created with an anchor <A>, and hypertext reference, href. The <A href> links to the URL of the other Web page. The URL, or uniform resource locator, is the address of the Web page. You learned about URLs in Lesson 3.1, page 60. The <A href> code can be used to connect pages within your own Web site. The A is the anchor portion of the tag and is followed by href, which stands for hyperlink reference.

The <A href> code gives Web sites the ability to connect to any other public Web site on the Internet. Hyperlinks connect millions of pages all across the World Wide Web. Web surfers can go from one page to another in what seems to be an endless galaxy of Web pages. A typical hyperlink reference tag would look like this:

<A href="http://www.glencoe.com">Glencoe</A>

where *http://www.glencoe.com/* is the URL and *Glencoe* is the name of the hyperlink.

In the browser, the word *Glencoe* will be blue and underlined. The user will be able to click the word *Glencoe* and navigate to the correct Web page.

✓ Concept Check

What is the advantage of linking pages together on the Internet? Explain how this makes moving around within a site easier. How does this make navigating the World Wide Web easier?

CHAPTER 13

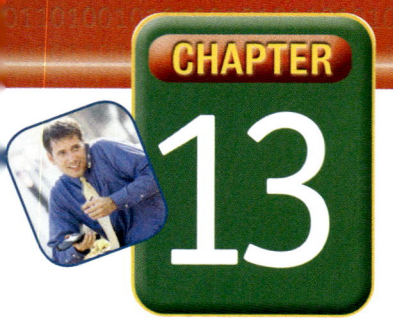

DIGITAL Dimension → Math
INTERDISCIPLINARY PROJECT

Managing Your Money

Using your spreadsheet skills, you will create a worksheet to calculate credit card interest.

Credit Card Interest Credit cards are convenient alternatives to carrying cash. However, many people tend to use credit cards to buy more than they can afford. Then they need to pay high interest charges. Will you choose to be one of those people, or will you use credit cards wisely?

Each month, the credit company sends you a bill showing the balance you owe for the charges you made to the card. If you do not pay the full balance by the due date, then you will be charged a high interest rate.

The formula for interest is:

   Interest = Balance × Annual Percentage Rate (APR)

Calculate Interest Charges Over the past month, you charged the following to your credit card:

Clothes	$150.76	Sports equipment	$83.29
Gasoline	47.56	Gifts for family	121.07
Restaurant meal	17.45	Shoes	163.17

1. Create a worksheet to calculate the total balance. What is the total balance?
2. Calculate your approximate yearly interest charges. Assume that you make no more charges. Include the formula above in your worksheet to calculate the approximate interest on the balance. Use the APR of 21 percent in your calculation (multiply the balance by .21). What is the interest? Note: This is the *approximate* annual interest charge.
3. Assuming you earn $10 per hour, how many hours would you need to work to pay the interest? Use your spreadsheet to make the calculations.

Advisors say, "You should always pay more than the minimum payment and pay off your balance as soon as possible." Do you agree with this statement?

E-Portfolio Activity

Spreadsheet Skills

To showcase your ability to use spreadsheet software, format the worksheet with appropriate headings. Save the document to your e-portfolio.

Self-Evaluation Criteria

Did you set up a spreadsheet to calculate:

- The total amount charged to the credit card?
- The interest charged in a year?
- How many hours you need to work to cover the payment of interest charges?

Activity 20

Step-by-Step

Create a Hyperlink

In this activity, you will turn ordinary text on your Web page into a hyperlink using the <A> beginning tag with the href attribute around the URL, followed by the name of the link and the </A> closing tag. Note that the URL is in quotes.

1. Open the FBLA Notepad document.

2. Near the top of the document, place the insertion point on the line below the closing formatting tags for the message beginning "Welcome."

3. To create a hyperlink with the name National FBLA Web site, enter the following tag and text:

```
<A href="www.fbla-pbl.org">National FBLA Web Site</A>
```

4. Click **File, Save**, then open the FBLA Web page and click the **Refresh** button.

```
<HTML>
<HEAD><TITLE>FBLA Home Page</TITLE></HEAD>
<BODY> <BODY background="green.gif">


<FONT color="#004400">
<H2 align="LEFT">
<B>
<I>

Welcome to Our High School FBLA Web Page

</I>
</B>
</H2 align="LEFT">
</FONT>
<A href="www.fbla-pbl.org">National FBLA Web Site</A>
```

5. You will see the words *National FBLA Web Site* on your Web page. If the text is blue and underlined, you have successfully created a hyperlink!

DIGI*Byte*

Spacing on the Web Page

If you have more than one <A href> code and you want them to appear on separate lines, you will need to insert a <P> or <BR> tag between the <A href> codes.

Suite Smarts

Hyperlinks can be created in Microsoft Word, Excel, Access, and PowerPoint documents by simply entering the URL.

Continued

PERSONAL Career Perspective

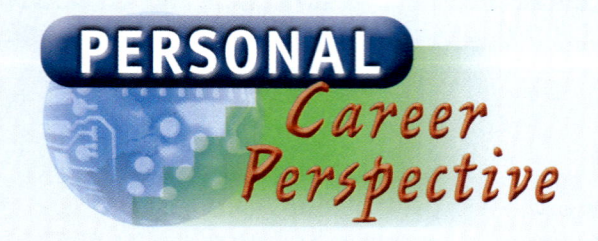

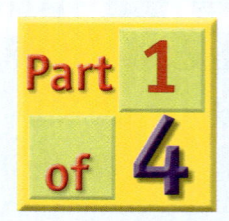

Part 1 of 4

Your Career Path

This four-part project gives you the chance to apply all the skills you have learned in this course in an investigation of your own career choice. Communicate, investigate, decide, plan, interview, write, and present your own career story.

Tapping Your Network

Choose one of the five jobs you listed in Lesson 13.2, Activity 6, page 437. You are determined to learn as much as you can about this job. Choose one of the networking contacts you identified in Lesson 13.3, Activity 7, page 441, who knows about the type of job you chose. Arrange to interview your networking contact. To prepare for the informal interview:

- Write 10 to 15 career questions you want to ask. Use a Tablet PC or Memo Pad in a PDA. You may want to base some of these questions on job-specific skills and some on soft skills. Another question might be about the importance of lifelong learning to the career. For example, is a certification or license necessary to practice in the profession?

- Write an e-mail to your contact asking for permission to interview him or her and listing the career questions. Edit and proofread your e-mail carefully, and make any necessary revisions before you send it.

Finding Good Internships and Volunteer Opportunities

Look at the Web sites on internships and volunteering that you bookmarked in Lesson 13.3, Activity 8 and Activity 9, page 441. Prepare a presentation for a class discussion on internships and volunteer opportunities. In your presentation, include the following:

- A brief explanation of your career interests.
- Which two Web sites on both internships and volunteering did you like the best? Explain why you liked them.
- Which volunteer or internship position included on these Web sites was the most interesting to you and why?

Look Ahead…

Continue your project in the following activities:

Part 2, Chapter 14..........p. 464
Part 3, Chapter 15..........p. 491
Part 4, Chapter 15..........p. 492

E-Portfolio Activity

Your Outstanding Work

To showcase your ability to send a well-composed, proofread e-mail, save the e-mail as an HTML file to your e-portfolio.

Self-Evaluation Criteria

For your e-mail did you:

- Include a minimum of 10 questions—maximum 15 questions?
- Proofread and edit it before sending?

For your presentation have you:

- Briefly explained your career interests?
- Included two Web sites on both internships and volunteering and explained why you liked them?
- Included which volunteer or internship position was most interesting to you?

Step-by-Step continued

Activity 21

DIGI Byte

All Caps Reminder
Remember to key tags in all caps. This will help you locate tags if you need to edit your Notepad document.

Link to a Page Within Your Site

Now you will link this Web page to another one. In Application 6, Lesson 4.3, page 103, you created a Web page called **conferences.htm**. You saved it in the FBLA folder on your desktop.

1. Open the FBLA Notepad document you are currently working on.
2. Place the insertion point below the <A href=> code you just entered. Key a paragraph tag: <P>
3. Enter the tag and text:

```
<A href="conferences.htm">Conferences</A>
```

4. Click **File, Save**. Now open your Web browser and click the **Refresh** button.

Activity 22

Add a Link to an E-Mail Address on Your Web Page

1. Open the FBLA Notepad document. Place the insertion point at the bottom of the document, on the line just above the </BODY> tag.
2. Enter the tag and text below. Instead of *Your Name*, key your actual name.

```
<A href="mailto:yourname@yahoo.com">Your Name</A>
```

3. Click **File, Save**. Now open the FBLA Web page and click the **Refresh** button to make sure your name shows as a link to your e-mail address.

RUBRIC LINK

Go to the Online Learning Center at **digicom.glencoe.com** and click Lesson 4.6, Rubric, to find the application assessment guide(s).

Application 9

Directions Find three Web sites on the Internet that you would like to link to from the FBLA Web page. You can choose any site you like, as long as it is related to the topic of your Web page.

Start Notepad and create another Web page document. For now, you do not need to add anything more than a title to this Web page, but you do need to save it as an HTML document. Give the page a name. You may want to visit another FBLA chapter Web site to get some ideas. After you save this page, create a link to it on the FBLA document you have been working on.

Finally, create an e-mail address link and identify the name as *Webmaster*. You can use your actual e-mail address or make one up, but *Webmaster* should display on the Web page.

Application 4

Directions Find tips on how you can get the most out of volunteering by searching the Internet. Key words for your search could include "volunteering tips" and "volunteering skills." Record the tips in your Tablet PC or in Memo Pad on your PDA.

Application 5

Directions Set up a system to keep track of internship, job shadowing, and volunteer appointments. For example, you could use the Task List, or To Do, application on your PDA. Include reminders to write thank-you letters to anyone who helped you find these opportunities.

To find event reminder software that works best for you, search the Internet for date book software applications that can be downloaded to your computer. These could be date books, appointment calendars, and personal information managers. Key words for your search could include "date book application" and "appointment calendar application."

Prepare a brief description of the application you would choose to use, and explain the benefits to the user. Include how you intend to set up your appointment tracking system.

RUBRIC LINK
Go to the Online Learning Center at **digicom.glencoe.com** and click Lesson 13.3, Rubric, to find the application assessment guide(s).

Digital Decisions

Thank-You E-Mails

Critical Thinking You have just completed your volunteer work for the area food bank. Robin Young, the person you reported to, has always communicated with you by e-mail. You are wondering if an e-mail thank-you note is acceptable or if you should mail a handwritten note.

Communication Survey ten people you know and ask them their opinion on this question: Is it appropriate to send an e-mail thank-you note, rather than a handwritten note?

Ask them to rank an e-mail note as:

- Very appropriate.
- Somewhat appropriate.
- Somewhat inappropriate.
- Very inappropriate.

Combine your results with those of your classmates, and calculate the percentage of each rank. Divide the number of responses for a rank by the total number of responses, and then multiply by 100 to determine the percentage.

For example, if there were 50 responses and 20 respondents ranked an e-mail thank you as very appropriate, divide 20 by 50, multiply by 100, and the percentage response is 40 percent.

Create a pie chart in your spreadsheet software to illustrate the results.

DIGITAL Career Perspective

Manufacturing

Levonda Stewart
Workshop Coordinator

Sharing Information About New Products

"It is essential to learn as much as possible while in school. Technology is changing daily and is very important for jobs," says Levonda Stewart. Stewart coordinates workshops for employees of the manufacturing firm she works for. She also coordinates regional trade shows and the company's apparel program.

Microsoft Office Suite software makes her job run smoothly. The Internet is also important in her work as she schedules hotels and checks the status of product shipments. Stewart is responsible for updating the company's Web site with information about trade shows and workshops. Participants can register online, which helps Stewart in her planning.

Stewart's job keeps her busy. "I coordinate approximately 60 to 80 workshops per year and about 10 trade shows per year. I enjoy working with different types of people from all over the United States," says Stewart. "There is a lot of planning that takes place when a new product line is being developed," Stewart comments. "Everyone in the company has a part, and we have to work as a team to make the product a success."

Stewart oversees the creation of brochures in PageMaker® for new products and workshops. Then the brochures have to be mailed to the customers on the company's mailing list.

"When information about a new product has to be given to the sales reps, I create a PowerPoint presentation and e-mail it to them so they can share it with their customers," adds Stewart.

What does Stewart like best about her job? "Occasionally I am on site at the workshops and shows and get an opportunity to work with folks I don't normally see every day. I enjoy setting up the shows. I get to be creative. I really enjoy working with a project when it begins as an idea and then develops into a successful workshop or trade show."

Training
Many manufacturing jobs require a high school diploma. Most training is received on the job, but basic math, technology, and communication skills are important.

Salary Range
Typical earnings are in the range of $500 to $600 a week.

Skills and Talents
Manufacturing workers need to be:

- Good written and verbal communicators
- Able to apply computer and technology skills
- Willing to adjust their products to changing markets
- Able to learn new processes quickly
- Able to familiarize themselves with new product lines quickly

Career Activity

There are many different products that are manufactured.
- What products would you be most interested in working with?
- What would you teach someone about a new product?

Write a paragraph about the areas you would cover.

116

Activity 7

If you need additional help on this topic, open the Help menu; or refer to the documentation for your hardware.

Step-by-Step

Identify People Who Work in a Career That Interests You

1. Create a category in Memo Pad in your PDA named Career Contacts List.
2. Generate a list of 10 to 15 names of everyone you can think of who works in the field you are interested in or who might know someone who works in that field. The names may include teachers, counselors, relatives, friends, neighbors, and others.
3. Once you have generated a list of 10 to 15 contacts, locate their contact information, such as name, address, phone number, and e-mail address.
4. Enter the contact information in the Address Book application in your PDA as you did in Chapter 5, Lesson 5.6, page 140.
5. To back up the data, be sure to synchronize your PDA with your personal computer. You will use this information later in this chapter.

Activity 8

Find Internship Positions

1. Search on the Internet for internships in your local area. Use the key words internship and the name of your city or an organization in your area.
2. Look at the listings and find three to five Web sites that interest you. Copy, paste, and save details into a Word document.
3. Save the document as **Internships**.
4. Bookmark the Web pages that list internship positions that you might want to learn more about. You will use this information later in this chapter.

Activity 9

Find Volunteer Positions

1. Search on the Internet for after-school community service and volunteer opportunities in your area.
2. Look at the listings and find three to five Web sites that interest you. Copy, paste, and save details into a Word document.
3. Save the document as **Volunteer Positions**.
4. Bookmark the Web pages that list volunteer positions that you might want to learn more about, as you did earlier in Activity 8, Step 3. You will refer to this information again later in Chapter 13.

DIGIByte

Thank-You Notes

Remember to write thank-you notes to your networking contacts to let them know you appreciate their help.

CHAPTER 4

DIGITAL Dimension → Language Arts
INTERDISCIPLINARY PROJECT

Be a Savvy Consumer

Conduct a survey of online purchasing habits. Then, using the digital input skills you have learned in this chapter, key a summary report in WordPad.

Buying Online Find out how many of your friends, family, neighbors, and teachers have bought products online. Ask the participants questions such as:

- Have you bought products online?
- How often do you shop online?
- What types of products do you buy?
- Do you sell items online?
- How has the Internet influenced your buying habits? Do you buy more or less?
- How do you learn about the sites you buy from?

Record the Responses Create a table on a sheet of paper to record the responses that you get. Follow the format used below and add columns for other questions. Survey eight people and fill in your table.

Summarize the Responses Review the information you have gathered. Write a summary report about what you have found from your survey. Use numbers and percentages to quantify the results. For example, you could report that 75 percent of the survey participants have bought products online.

Key your summary in WordPad. Save your document as **Buying Online**.

E-Portfolio Activity

Collecting Data

Proofread and edit the document you saved as **Buying Online**. To showcase the skills you have learned in this chapter, as well as your ability to think critically and be a savvy online consumer, save the document to your e-portfolio.

Self-Evaluation Criteria

In your document, did you:

- Include a summary of the online buying habits of your friends or family?
- Proofread and edit the document?

What is your first name?	Do you shop online? How often?	What type of products do you buy?	Do you buy with a credit card or mail a check?	Do you sell items online?	How did you learn about the sites that you buy from?
Mary	Yes, 1 or 2 times a month	Books	Credit card	No	Read about it in the newspaper

Lesson 13.3 Find Out What People Do on the Job

You will learn to:
- **State** why networking is important.
- **Describe** ways to learn more about the jobs that interest you.

How Do You Learn More About a Career? You have determined your interests, skills, values, and aptitudes and looked at various jobs in the 16 career clusters. It is now time to find out what people do every day on the job. This helps you in two important ways:

- Learning more about what people do on the job will help you find the right career path.
- Getting to know people who are in jobs that interest you gives you leads to possible job opportunities when you need them, which may be soon.

There are four effective ways to learn about careers:

- **Networking** When you are **networking**, you are communicating with people to share information or advice. These people can be family, friends, neighbors, teachers, counselors, coaches, businesspeople, and others.
- **Internship** An **internship** is a temporary position, paid or unpaid, where you can gain practical work experience in a career field. For example, if you want to pursue a career in athletic training, you may want to work as an intern in a sports physical therapy office or for an athletic team during your summer vacation.
- **Volunteer** Volunteering, or working without pay, helps you explore a career, gain practical hands-on work experience, and make invaluable contacts with people in a career field.
- **Job Shadow** When you **job shadow**, or follow someone for a day or two on the job, you can learn about a career by observing, asking questions, and listening.

Other ways to learn about careers are attending career fairs, reading books on careers, talking to your school guidance counselor, or researching the career on the Internet.

✓ Concept *Check*

Why is networking important? How can you learn more about what people do on the job?

---

networking Communicating with people to share information or advice.

internship A temporary position, paid or unpaid, where you can gain practical work experience in a career field.

job shadow To follow someone for a day or two on the job.

Chapter 4 Review

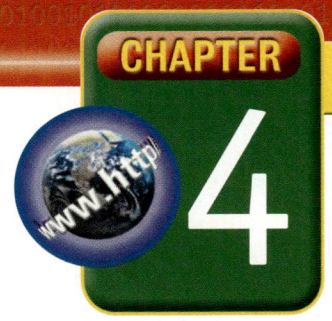

Read to Succeed PRACTICE

After You Read

Visualize—Draw Graphic Organizers Almost everyone is a visual learner, so drawing diagrams helps you to learn, remember, and study. Diagrams, or graphic organizers, show the relationships you find in information.

Create a two-column chart of all the HTML codes in this chapter. On the left side write the heading "HTML Code." On the right side, write the heading "Meaning and Purpose." Fill in the HTML codes and what they do.

Example:

HTML Code	Meaning and Purpose
<P>	Paragraph

Using Key Terms

The odds are, in the future you will need to communicate about Web pages. See how many of these terms you know by writing each statement on a sheet of paper and filling in the blanks.

- Hypertext Markup Language (HTML) (85)
- tag (85)
- attribute (88)
- font (91)
- horizontal rule (91)
- clip art (100)
- ordered list (109)
- unordered list (109)

1 Electronic illustrations that can be inserted into a document are called _____.

2 A(n) _____ is a list that has bullet points before each item and is used when the items can be in any order.

3 _____ is a set of codes used to create documents for the Web.

4 A(n) _____ is a combination of specific visual characteristics of text, including size, typeface style, bold, and italics.

5 A line that runs across the width of a page is called a(n) _____.

6 A list that has numbers before each item, and the items are usually in priority order, is called a(n) _____.

7 A(n) _____ is a specific instruction that tells the browser how to display the text or graphics enclosed by the HTML tags.

8 A(n) _____ is a piece of HTML code that tells a Web browser how to display a particular section of text, a graphic, or other Web page element.

21ST CENTURY CONNECTION

Good Judgment in Today's Workplace

Daily Decisions One of the most useful skills you have been learning since childhood is your ability to exercise good judgment daily. To practice good judgment is to make decisions based on what you know about situations that benefit rather than harm you, your peers, family, and community.

Use good judgment to evaluate your facts and beliefs and move forward in the workplace. Good judgment empowers you in your personal life as well as on the job.

At Work There are many opportunities every day at work to use good judgment when using digital tools. Computers have fundamentally changed the way people learn, communicate, and have fun. You should use technology tools with sound judgment. Many of the choices you make now will affect the quality of your future experiences in the workplace and in your life.

Gossip in the Workplace Poor judgment at work is often found when employees use office equipment for the purpose of spreading gossip or venting personal matters. Do people put each other up, or do they run each other down? The way we talk to ourselves and to others has a powerful effect on what we can accomplish. Using work hours and e-mail, instant messaging, or phone systems to discuss non-work-related issues with coworkers can interfere with employees' ability to perform. It costs organizations time and money, weakens business progress, and frequently leads to employee dismissals.

Agreement to Practice Good Judgment
In most work situations, new employees will sign an agreement to follow the business's practices and contribute to its success. Employers expect that you will think through your actions carefully and exercise good judgment daily.

Steps to Good Judgment
Follow this four-step process to good judgment:
1. **Use Background Knowledge** Consider what you know already and what you know about your organization.
2. **Find Out More** Investigate and find out more, if you have questions.
3. **Consider Your Choices** Think of possible choices and solutions for the information you have gathered.
4. **Act on Your Best Choice** Choose to take the course of action that will be best for you, your coworkers, organization, family, and community.

Activity Think about a time recently when you feel you exercised good judgment. Did you follow the four-step process to good judgment? List the steps and describe what you did in each part of the process.

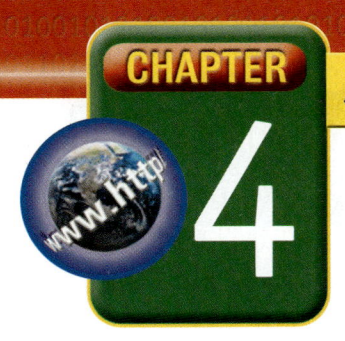

Chapter 4: Self-Assessment

Take a moment to review what you have learned in this chapter. Rank your understanding of the topics below.

4 means, "I understand all of this."
3 means, "I understand some of this."
2 means, "I understand very little of this."
1 means, "I don't remember this."

> To use a printout of this chart, go to **digicom.glencoe.com** and click on **Chapter 4, Self-Assessment.**
> Or:
> Ask your teacher for a personal copy.

Rank Your Understanding

Lesson	Topic	4	3	2	1
4.1	• Create a new HTML document				
	• Apply tags to an HTML document				
	• View an HTML document as a Web page				
	• Add background color to an HTML document				
4.2	• Apply color codes to fonts				
	• Change font size, bold, and italics				
	• Use paragraph and break tags in an HTML document				
	• Add a horizontal rule				
4.3	• Add clip art to an HTML document				
	• Add a digital photograph to an HTML document				
	• Apply a background image to an HTML document				
4.4	• Explain the importance of closing tags in an HTML document				
	• Align text and graphics in an HTML document				
	• Experiment with aligning elements in different locations				
4.5	• Explain the difference between ordered lists and unordered lists				
	• Create an ordered list in an HTML document				
	• Create an unordered list in an HTML document				
4.6	• Create a hyperlink in an HTML document				
	• Link to a Web page within your own site				
	• Add e-mail addresses to an HTML document				

If you ranked all topics 4, congratulations! Consider doing a quick review.
If you ranked yourself 3 or lower in any topic, consider reviewing these topics first.

digicom.glencoe.com

Application 2

Directions Open the *Your Name* **Skills**, *Your Name* **Aptitudes**, and *Your Name* **Values** files.

Choose one of the five jobs from Activity 6, Step 2, page 437. Write one or two sentences for each of the following:

- How could your skills, aptitudes, and values be used in this job?
- Describe how you would use one skill, one aptitude, and one value in the job. Give detailed examples.

Application 3

Directions Survey the other students in your class to find out about the jobs they determined were most interesting among those listed for each career cluster.

Find out how each student's values affected his or her choices.

You can use some of these questions, or you can make up your own questions.

- How did you choose your jobs for each career cluster?
- What career clusters are your top three job choices in?
- How do you think your values are reflected in your job choices?
- How do your skills and aptitudes relate to your job choices?

Prepare a chart to record the responses to your survey, using a Tablet PC and Journal, word processing, or spreadsheet applications, if possible.

Summarize the information, and present the summary in a one-page report.

RUBRIC LINK
Go to the Online Learning Center at **digicom.glencoe.com** and click Lesson 13.2, Rubric, to find the application assessment guide(s).

TECH ETHICS

Keeping Commitments

Scenario You are scheduled to volunteer at Hope Hospital on Friday from 6:00 P.M. to 7:30 P.M. Unfortunately, for the first time in the history of your school, your high school football team is playing the same night in the state finals. You need to leave at 5:30 P.M. to get to the game on time.

Teamwork In a small group, discuss how this situation should be handled. Evaluate each option below and decide why it would or would not be a good choice. Your final choice may be a combination of some of the options.

1. Go to the game even though it means you cannot keep your commitment at the hospital.
2. You have a rule that an appointment that is listed in the To Do List application of your PDA takes priority. You keep your commitment at the hospital and miss the game.
3. Contact the hospital to let them know that you cannot volunteer that evening.
4. Tell the hospital you are sick.
5. Find someone who would be qualified to substitute for you at the hospital.
6. Is there a better option?

As a group, decide on your final suggestion of how best to handle this situation. Report your final choice to the class.

UNIT 1 LAB

DIGITAL Dimension → Language Arts
INTERDISCIPLINARY PROJECT

Be a Savvy Consumer

Background In Unit 1 you have been gathering information about online advertising and consumer awareness:

- In Chapter 1 Digital Dimension, page 32, you compared print and online ads.
- In Chapter 2 Digital Dimension, page 55, you filmed your own commercial.
- In Chapter 3 Digital Dimension, page 80, you analyzed online ads.
- In Chapter 4 Digital Dimension, page 117, you surveyed the online buying habits of friends and family.

Writing Your Own Web Ad Create an advertisement for a product and present it to the class. The goal of the advertisement is to sell the product to your classmates. As you create the advertisement, consider what you have learned about communicating with digital technology and digital communication tools, using the Internet, and applying HTML and Web page design.

Some common advertising techniques to consider using are:

- Humor
- Promises of an improved appearance
- Celebrity endorsements

You will create your advertisement as an HTML document.

E-Portfolio Activity

Web Ad

To showcase your HTML skills, include your **Web Ad** document in your e-portfolio.

- Be sure to view the advertisement in a browser to check that the ad looks the way you want it to.
- Make sure any hyperlinks to other Web sites work.
- Include a link to your e-mail address on your Web page.

Continued

Activity 6

? If you need additional help on this topic, open the Help menu; or refer to the documentation for your hardware.

Step-by-Step

Use Your Interests to Focus on Jobs in Different Career Clusters

1. Open the file *Your Name* **Interests** from Activity 3, page 434. Select one of your main interests.

2. Set up a table similar to the one below in your word processing or spreadsheet software. The example below is for the main interest area of sports. Refer to the career clusters and determine jobs that would take advantage of the main interest you chose. Each job should be in a different career cluster. Fill in jobs in at least five clusters.

Interest Area: Sports

Career Cluster	Examples of Jobs in This Career Cluster
Agriculture, and Natural Resources	Camp counselor
Architecture and Construction	Architect specializing in sports arenas and stadiums
Arts, Audio/Video Technology, and Communication	Sports broadcaster, sports writer
Business and Administration	Sports team manager, athletic association manager
Education and Training	College athletic director, coach, fitness trainer, health educator
Finance	Sports team financial manager, sports arena financial specialist
Government and Public Administration	Gaming administrator, convention/tourism bureau manager
Health Science	Physical therapist, sports trainer, massage therapist, sports medicine assistant
Hospitality and Tourism	Sports events promoter, destination manager, convention or tourism bureau assistant
Human Services	Youth center sports coach
Information Technology	Timing equipment operator, statistics technician for sports broadcasts, computer simulation designer
Law and Public Safety	Equipment supervisor, safety/security specialist for sporting events
Manufacturing	Sports equipment and safety gear designer/manufacturer
Retail/Wholesale Sales and Service	Sports marketing specialist, athletes' agent, sports equipment repair technician
Scientific Research and Engineering	Sports arena design and construction, ceramics/plastics engineer/designer, computer games designer, simulation designer
Transportation, Distribution, and Logistics	Major events coordinator, team transportation coordinator

UNIT 1 LAB *continued*

Follow these steps to complete your HTML advertisement:

Step 1 Decide on the product you want to sell. It does not have to be a real product. You could create a new product or think of a new name for an existing product.

Step 2 Plan the content and layout of your ad. You may want to sketch the ad on a piece of paper.

1. Find or create a graphic of your product to use in your advertisement. You could modify an existing item and take a digital picture. You could also look for clip art.

2. Write several lines of text about your product.
   - Name and describe the product.
   - Convince your audience to buy your product. Give them reasons they should buy your product.
   - Consider the age group you are trying to sell to. What would be appealing to them?

3. Include information about how to purchase the product:
   - Where can it be purchased?
   - How much does it cost?
   - How can someone pay for it?

4. Think about how you want to lay out your information in your HTML document.
   - What colors will you use?
   - What size should the text be?
   - Where will the image or images go?

Step 3 Create the HTML document for your ad. Review Chapter 4, pages 83–119, if you need help. Save your HTML document as **Web Ad**.

Step 4 Present your ad to the class. Be sure to proofread your ad, and make sure that it represents your HTML skills effectively.

Self-Evaluation Criteria

In your HTML document, did you include:

- A graphic of your product?
- The name and description of your product?
- Reasons why someone should buy your product?
- Details on how to buy the product?
- Correct HTML code?

Lesson 13.2 ➤ Explore Career Clusters—Focus on a Career Path

You will learn to:
- **Explore** the Department of Education's 16 Career Clusters.
- **Focus** on your career path based on the career clusters and your interests, values, skills, and aptitudes.

career clusters Groups of similar occupations and industries developed by the U.S. Department of Education.

One Way to Discover Careers: Career Clusters

Career clusters are groups of similar occupations and industries developed by the U.S. Department of Education. Examples are: Architecture and Construction, Education and Training, and Law and Public Safety. See the figure below.

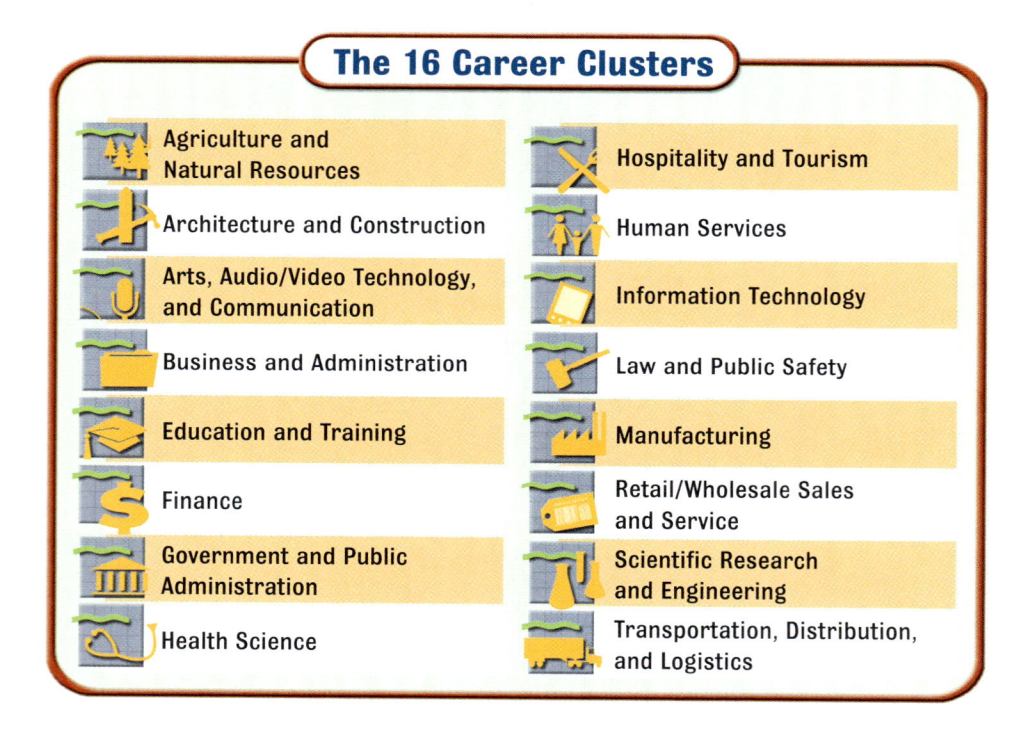

The 16 Career Clusters

- Agriculture and Natural Resources
- Architecture and Construction
- Arts, Audio/Video Technology, and Communication
- Business and Administration
- Education and Training
- Finance
- Government and Public Administration
- Health Science
- Hospitality and Tourism
- Human Services
- Information Technology
- Law and Public Safety
- Manufacturing
- Retail/Wholesale Sales and Service
- Scientific Research and Engineering
- Transportation, Distribution, and Logistics

Looking at the 16 career clusters can help you discover career possibilities in areas that you value. For example, if you value helping people, then the Health Services career cluster would be the first to consider, but you might also consider Human Services or Hospitality and Tourism.

✓ Concept Check

What is helpful about looking at the 16 career clusters? Give an example.

UNIT 2
Digital Communication Tools and Skills

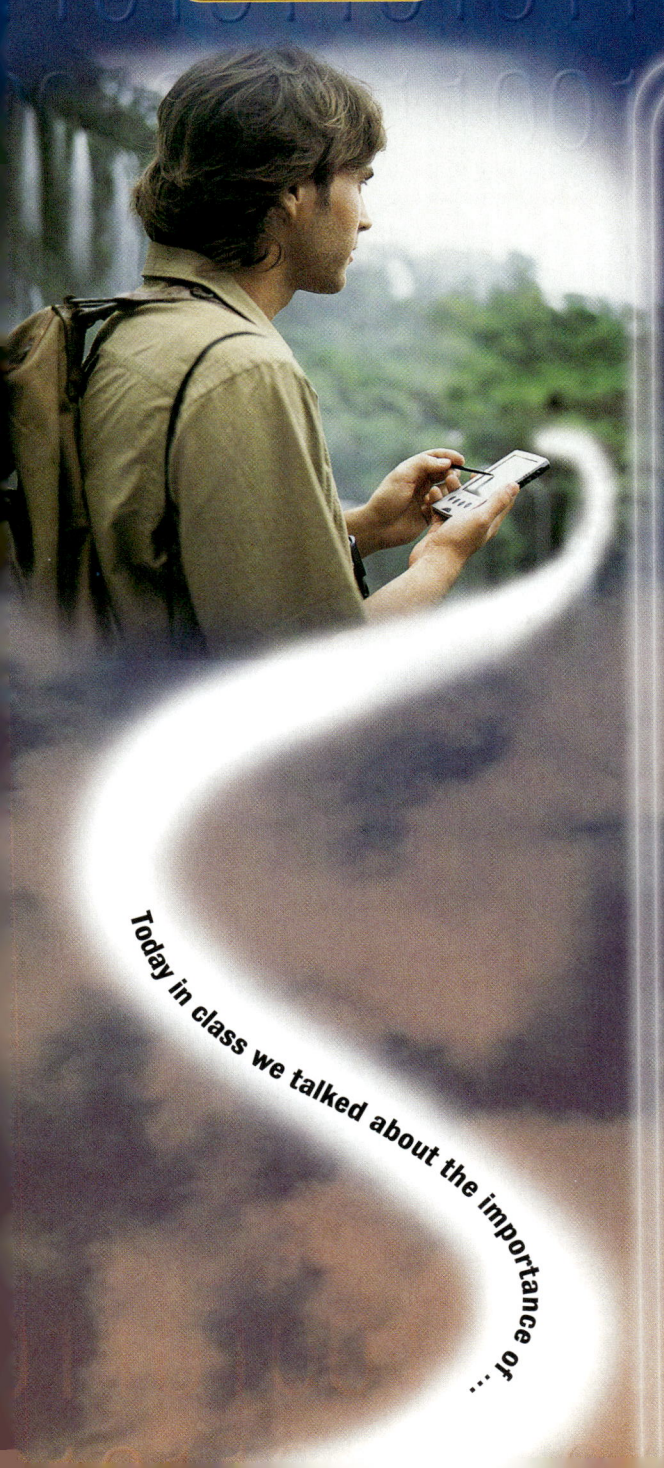

CHAPTER 5 — Handheld Computers—PDAs

CHAPTER 6 — Speech Recognition Tools

CHAPTER 7 — On-Screen Writing—Tablet PCs

CHAPTER 8 — Additional Features of Digital Tools

Curriculum Connections
- Language Arts—Reading
- Language Arts—Writing
- Social Studies
- Technology

Today in class we talked about the importance of...

21ST CENTURY CONNECTION

Leadership

These statements show some of the perceptions people have about leadership. True or false?

- You are born a leader.
- In a company, a leader is a manager.
- A leader needs to order everyone around.

All are false.

Leadership Skills Can Be Learned You do not need to be "born a leader." Leadership skills can be learned. The main characteristics commonly attributed to leaders include the following.

A leader develops:

- A strategic vision for the future.
- Commitment and dedication to the vision.
- Integrity, showing honesty, fairness, loyalty, and character.
- Self-confidence and self-esteem based on job knowledge and abilities.
- The ability to communicate well and persuade others.
- The ability to look for solutions to problems in a positive way.
- The ability to form effective work teams.

Everyone can learn to be a leader. All these leadership skills can be learned as a member of a club or team, by taking courses, reading books on leadership, emulating a teacher or coach, on the job, and so on.

Leadership Is Not Management According to management expert Peter Drucker, "Management is doing things right; leadership is doing the right things." Leadership means providing direction and vision for a company—choosing the right course to take with the big picture in mind. Managers make sure that specific tasks get done.

Leadership in Teams The style of leadership has changed. About 20 years ago, leaders would have been expected to order people around. The trend now is for companies to use teams and shared leadership to effectively lead projects and get results. Companies encourage employees to see themselves as leaders, that is, to act with integrity in alignment with the vision, be a role model, and communicate effectively and persuasively.

Activity Write a list of five experiences you have had when you were a leader. If possible, use a Tablet PC or Memo Pad on your PDA. For each experience, answer the questions:
- What are you proud of achieving as a leader?
- What leadership characteristics did you use?

DIGITAL Dimension → Social Studies
INTERDISCIPLINARY PROJECT

Link the new digital input skills you will learn in Unit 2 with social studies while learning about elected officials. Your project finishes with a class presentation.

Elected Officials

Did You Know? In social studies class you learn about the structure of the United States government and how our leaders are elected to their positions. Did you know you can contact the elected officials who represent you? Citizens should contact their elected officials when they have a concern about proposed laws.

How Do You Contact an Elected Official? Elected officials fill many government positions, such as mayor, councilperson, commissioner, state senator, state representative, governor, U.S. Senator, and U.S. Representative. To voice your opinion on laws that affect citizens of your city, state, and nation, you need to know how to contact the correct elected official. The addresses, office phone numbers, and e-mail addresses of elected officials are *public information*. That means you have a right to know this information. Key words to search under to find this information include "state representation" and "election districts."

Digital Dimension Activities

Start-Up
Your state is divided into political districts from which state senators and representatives are elected. Search online for a map of political districts in your state. Print the map and place an X in the district in which you live.

Look Ahead...
Continue building your project in the following activities:

Chapter 5 .... p. 150
Chapter 6 .... p. 187
Chapter 7 .... p. 224
Chapter 8 .... p. 255
Unit 2 Digital Dimension Lab .... pp. 258–259

Today in class we talked about the importance of working together.

Step-by-Step *continued*

Activity 3

What Are Your Interests?

1. What are your major interests? For example, do you like to play sports, perform in theater productions, tutor and mentor children, or write?

2. Create an electronic document for your answers. Save the document you create as *Your Name* Interests.

Activity 4

What Are Your Aptitudes?

1. Have you ever been told that you have a knack, or talent, for something, such as art, music, building things, fixing things, writing, or something else? These are your aptitudes. What are your aptitudes, or talents?

2. Create an electronic document for your answers. Save the document you create as *Your Name* Aptitudes.

Activity 5

What Are Your Values?

1. What are your values? What work, life, and school-related beliefs do you consider important? These might include being reliable, meeting deadlines, and maintaining close relationships, among others.

2. Create an electronic document for your answers. Save the document you create as *Your Name* Values.

RUBRIC LINK

Go to the Online Learning Center at **digicom.glencoe.com** and click Lesson 13.1, Rubric, to find the application assessment guide(s).

Application 1

Directions Open the *Your Name* Skills and *Your Name* Aptitudes documents you developed in your self-assessment, pages 433–434. Refer to the Web pages describing the six jobs you bookmarked in Activity 1, Step 3, page 433.

Check the requirements for these jobs against your skills and aptitudes.

- Do you see any patterns?
- What in the lists of job requirements matches your current skills and aptitudes?
- What skills will you need to learn to meet the requirements for the jobs you selected?

Make a new chart in your word processing or spreadsheet application that shows the requirements for each job, the skills you have now, and those you would need to learn. Save the document.

Handheld Computers —PDAs

CHAPTER 5

Lesson 5.1	Enter Text in Memo Pad
Lesson 5.2	Explore Memo Pad
Lesson 5.3	Beam a Memo
Lesson 5.4	Use, Create, and Beam Categories
Lesson 5.5	Create and Beam a Note with Note Pad
Lesson 5.6	Create an Address Book and Business Card
Lesson 5.7	Use Date Book
Lesson 5.8	Create a To Do List

You Will Learn To:

- **Identify** the parts of a handheld.
- **Use** Memo Pad, Note Pad, Address Book, Date Book, and To Do List applications on a PDA.
- **Create** and edit text in Memo Pad and Note Pad.
- **Explain** synchronization.
- **Beam** a memo.
- **Create** addresses and business cards in Address Book.
- **Organize** with categories.

When Will You Ever Use Handhelds?

Wouldn't it be helpful if there were an electronic way to organize your notes so you could find them whenever you needed them? A device called a personal digital assistant, or PDA, can help you stay organized in many different ways.

Activity 1

> If you need additional help on this topic, open the Help menu; or refer to the documentation for your hardware.

Activity 2

Step-by-Step

What Types of Jobs Are Out There?

1. Search the Internet for the Web sites of some employers in your area.
2. Most employer sites will have a section called Employment, Job Opportunities, Positions Available, or something similar. Look at the job descriptions posted in that section to find the skills or qualifications required for the jobs.
3. Bookmark the Web pages describing six jobs of particular interest to you. You will use the bookmarked pages again later in this lesson, after you have determined your interests, values, aptitudes, and skills.

What Are Your Skills?

1. Think about your main skills, both job-specific and soft skills. What tools—software, hardware, shop, business, and so on—do you already know how to use? How well do you use them? What subjects do you do well in?
2. Create an electronic document for your answers, and honestly assess your capabilities. Set up a table in your word processing or spreadsheet software similar to the table below. Save the document you create as *Your Name* **Skills**.

Skills	What is my level of performance?		
	Excellent	Good	Needs Work
Computer Skills			
Writing Skills			
Speaking Skills			
Academic Skills			
Sports Skills			
Artistic Skills			
Other Skills			

Continued

Read to Succeed

Key Terms

- stylus
- Memo Pad
- beam
- synchronization
- categories
- Note Pad
- To Do List

Reading Strategy

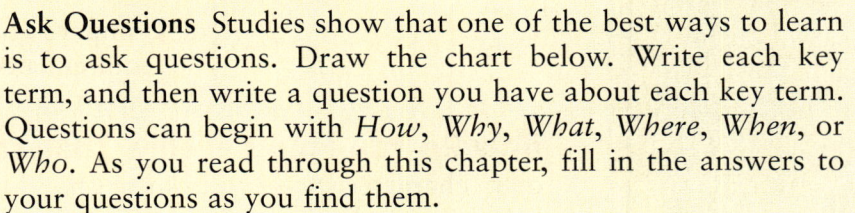

Ask Questions Studies show that one of the best ways to learn is to ask questions. Draw the chart below. Write each key term, and then write a question you have about each key term. Questions can begin with *How, Why, What, Where, When,* or *Who.* As you read through this chapter, fill in the answers to your questions as you find them.

Use the example below as a guide. Fill in for each key term.

Example:

Key Term	Question	Answer
stylus	What is a stylus?	A stylus is a penlike device used with a PDA.

DIGITAL CONNECTION

Wanted: Work in a Paperless Environment

Two days ago, the phone rang and your friend asked if you would like to interview next week for a cool summer job. You made a note of the place, time, phone number, and company Web site, but now you have lost the note. If only you had entered the company information in your computer. What if there were a computer you could hold in your hand or carry in your pocket? There is—a personal digital assistant, or PDA. A handheld computer has many different applications that can keep you organized. You can work more efficiently in a paperless environment.

Writing Activity

Write a paragraph describing how you keep track of important events, appointments, and assignments. Remember to use the correct punctuation and spelling.

Lesson 13.1 Match Careers to You

You will learn to:
- **Find** skills required in jobs of interest to you.
- **Determine** your characteristics: your interests, values, skills, and aptitudes.
- **Compare** your characteristics to job requirements.

interests Your favorite activities, the subjects you like best in school, and your hobbies.

values Beliefs and ideas you live by and think are important.

skills Your abilities to perform a task as a result of your talents, training, and experience.

aptitudes Your potential for learning a skill.

How Do You Find Careers That You Will Enjoy? Knowing your own characteristics—your interests, skills, aptitudes, and values—will help you find careers that are right for you. You can compare your characteristics to the job requirements in the types of careers that you are interested in. If there is a good match, you are likely to enjoy that career.

What Are Your Interests, Values, Aptitudes, and Skills? Your interests include your favorite activities, the subjects you like best in school, and your hobbies. Your values are the beliefs and ideas you live by and think are important. For example, you may value honesty, success, compassion, courage, family, education, or recognition, among others.

Your skills are your abilities to perform a task as a result of your talents, training, and experience. Finally, your aptitudes indicate your potential for learning a skill. Most people have an easier time learning some skills than others. Your aptitudes indicate the skills you will probably learn more easily.

Two Types of Skills Often employers look for employees with the following skills:

- **Job-Specific Skills** These are skills needed to do a certain job. For example, the skills of knowing how to use Microsoft Office, write reports, or analyze a company's financial statements may be specific to a certain job.
- **Soft Skills** These are more general skills that you have already learned at school, in team activities, or at home and can use on the job. Examples include your ability to:
  ▶ Listen and communicate.
  ▶ Work in a group.
  ▶ Have a positive and professional attitude.
  ▶ Make decisions.
  ▶ Adapt to change.

✓ Concept Check

Why is it helpful to you to figure out your interests, values, skills, and aptitudes before looking at possible careers?

Lesson 5.1 Enter Text in Memo Pad

You will learn to:
- **List** reasons why a PDA is useful.
- **Identify** two methods of entering data into a handheld computer.
- **Enter** text in Memo Pad.

Why Use a PDA? When you use a PDA, you are working with a computer that is easy to carry around. Also, you are working in a paperless environment. With a PDA you can easily send text documents to other PDAs or computers. You can quickly create and edit documents on a PDA. The PDA also makes accessing documents simple because it is a small device that fits comfortably in a pocket or briefcase. Instead of carrying around a bulky laptop, you can accomplish many of the same tasks with a PDA.

What Are Two Ways to Enter Text? One of the most common ways to enter text on your PDA is with the stylus. A **stylus** is a penlike device used to select icons on a PDA screen or to write on a PDA screen. The PDA also has an on-screen, or soft, keyboard to enter text. In this lesson you will have the opportunity to practice both of these methods of entering text.

Entering Text To practice entering text on a PDA, you will use the Memo Pad application. **Memo Pad** is an application on a PDA that allows you to create simple documents.

stylus A penlike device used to select icons on a PDA screen or to write on a PDA screen.

Memo Pad An application on a PDA that allows you to create simple documents.

Figure 5.1
A handheld computer, or PDA

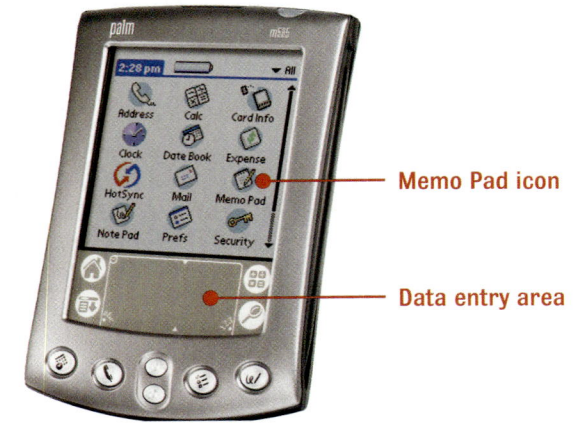

- Memo Pad icon
- Data entry area

✓ Concept *Check*

What are some advantages of storing data electronically in a device that is small enough to carry in your pocket?

Read to Succeed

Key Terms
- interests
- values
- skills
- aptitudes
- career clusters
- networking
- internship
- job shadow

Take Responsibility

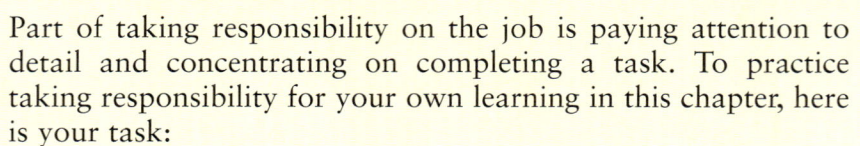

Part of taking responsibility on the job is paying attention to detail and concentrating on completing a task. To practice taking responsibility for your own learning in this chapter, here is your task:

1. For Lesson 13.1, write each key term and its definition on a separate page. If possible, use Memo Pad on your PDA and write separate memos for each key term, or use Journal on your Tablet PC.
2. Write each remaining chapter key term and its definition on a separate page.
3. Create a category or folder named **Chapter 13 Key Terms** so that you can easily find and review them later.

Example:

Key Term	Definition
aptitudes	Potential for learning a skill

DIGITAL CONNECTION

Connect What Makes You Happy to a Career

You are going to one of your favorite events tomorrow—a ball game or a concert. How will you feel when you wake up in the morning? Would you like to feel like that on days when you get up to go to work? Many people do feel that way, whether they are entrepreneurs or employees in large corporations. It is possible to be passionate about your work, especially when you plan a career goal and go and get it. Now is a great time to start.

Writing Activity

Write two paragraphs about jobs that might make you happy. Include at least three reasons why you think these jobs are the type you would enjoy.

Activity 1

? If you need additional help on this topic, open the Help menu; or refer to the documentation for your hardware.

DIGIByte

Graffiti®

If your PDA has an application called Graffiti, you can practice making letters and numbers with your stylus in this shorthand writing.

RUBRIC LINK

Go to the Online Learning Center at **digicom.glencoe.com** and click Lesson 5.1, Rubric, to find the application assessment guide(s).

Step-by-Step

Enter Text

1. With the stylus, tap the **Memo Pad** icon to open the application.
2. Tap **New** to create a new memo.
3. The dark rectangular box, or data entry area, just below the screen of the PDA is the area where you enter text and numbers. To use the soft keyboard, tap the **abc** button in the lower left-hand corner.
4. Tap the letters on the keyboard to enter your first name.
5. Tap the space bar.
6. Tap the shift key once, and enter your last name. Tap **Done**.
7. To use Graffiti, write in the dark rectangle. Write letters on the left side of the rectangle. Write numbers on the right side of the rectangle.

   Write a letter or number by starting at the dot shown in the figure. Do not lift your stylus until the letter or number is completed.

   Use Graffiti to write your first name. Enter a space.
8. To capitalize the first letter of your last name, draw an upward line in the Graffiti area. Continue writing your last name.
9. Enter a return to go to the next line.

Application 1

Scenario Your social studies teacher is describing a new assignment. You decide to enter the details in Memo Pad.

Directions Open Memo Pad and start a new memo. Enter the following notes about the assignment.

```
1. History paper on Bill of Rights
2. About 20 pages
3. Due May 25
```

CHAPTER 13

Explore 21ˢᵗ Century Careers

Lesson 13.1 Match Careers to You

Lesson 13.2 Explore Career Clusters—Focus on a Career Path

Lesson 13.3 Find Out What People Do on the Job

You Will Learn To:

- **Determine** your characteristics: your skills, interests, values, and aptitudes.
- **Compare** your characteristics to job requirements.
- **Explore** the Department of Education's 16 Career Clusters.
- **Identify** career choices by comparing your characteristics with jobs in various career clusters.
- **Explain** ways to learn more about jobs that interest you, such as networking, internships, volunteering, and job shadowing.
- **Create** questions to gather important career information.

What Do You Want to Accomplish in Your Life?

The time when you will need to make your own living and support yourself is not far away. You will need to make choices about careers and learn how to find, and succeed in, a job you will enjoy.

Knowing how to use digital communication tools effectively will help you in your career planning and job search—and on the job. Having a solid knowledge of technology is one key to a great career because most jobs today require at least basic computer knowledge.

Lesson 5.2 ➔ Explore Memo Pad

You will learn to:
- **Describe** the functions and features of Memo Pad.
- **Create** and edit a memo in Memo Pad.

Suite Smarts
The layout of the on-screen, or soft, keyboard on the PDA is the same as the keyboard on a desktop computer.

Memo Pad Functions and Features Memo Pad is a standard application that comes with all PDAs. It can be used to take notes or messages. Memo Pad is similar to a word processor. The blinking insertion point on the screen indicates the placement of the text you will enter. Blank lines help to place the text where you want it. Cut, copy, and paste commands can also be used with Memo Pad. Each memo is stored on the PDA.

When You Open Memo Pad A blinking insertion point will appear on the first line when you open Memo Pad. You cannot move the insertion point to another line unless you open the keyboard and tap the Enter key ⏎ or create a return with the stylus by drawing a forward slash (/) in the writing area on the PDA.

✓ Concept Check

Describe a situation at school where it may be helpful to use the Memo Pad application on a handheld computer.

- How could you use Memo Pad on a job?
- What are some advantages of using the Memo Pad application?

DIGITAL DIALOG

Memo Pad and PDA Applications

Communication Choose a partner to discuss the following questions:
- What computer application does Memo Pad remind you of?
- How is Memo Pad different?
- How is Memo Pad similar?
- What makes Memo Pad easy and convenient to use?
- How would you use Memo Pad at school? At work?

Teamwork Research on the Internet some of the types of applications that can be downloaded to a PDA for use either in the classroom or on the job. Some of the applications may be free; some may need to be purchased. Discuss your findings. Summarize the points you discussed on Memo Pad and PDA applications to share with the class.

DIGITAL Dimension → Math

INTERDISCIPLINARY PROJECT

Link what you will learn about choosing and preparing for a career in Unit 4 with math, by planning your personal finances. Your project will help you learn to manage your money, and you will finish with a presentation to the class.

Managing Your Money

It Takes Math What does it take to support yourself when you are working and out on your own? It is important to learn how to budget effectively to live the life you want. It is critical to calculate what it takes to meet all of your expenses, such as the cost of owning a car, managing a credit card, and the cost of living.

Planning a Personal Budget In Unit 4 you will be creating a personal budget. A budget includes your planned spending on housing, transportation, utilities, entertainment, and so on. You will check your budget against your planned earnings.

Digital Dimension Activities

Start-Up
Go over your monthly expenses, and determine what you spend your money on.

- How much do you earn every month? Does that amount change? For example, you might receive gifts of money.
- What are your monthly expenses? Consider all months of the year, including holiday months and months when a family member or friend has a birthday.
- What is the amount of your monthly phone bill?
- What amount of money do you save each month? If the amount per month varies, average the amount over 12 months.

Look Ahead...
Continue building your project in the following activities:

Chapter 13 .... p. 444
Chapter 14 .... p. 465
Chapter 15 .... p. 493
Unit 4 Digital Dimension Lab ....pp. 496–497

Activity 2

? If you need additional help on this topic, open the Help menu; or refer to the documentation for your hardware.

Suite Smarts

The text you enter on the first line of Memo Pad becomes the name of the document. The file name of a Microsoft Word document is created the same way unless you rename the document.

RUBRIC LINK

Go to the Online Learning Center at **digicom.glencoe.com** and click Lesson 5.2, Rubric, to find the application assessment guide(s).

Step-by-Step

Create and Edit a Memo

On-screen, or soft, keyboard

1. Tap the **Memo Pad** icon or the **Memo Pad** button on the PDA to open the application.

2. Enter text using any one of the following procedures:
   - Open the soft keyboard and tap the keys.
   - Use the stylus to write the letter or number you want in the writing area on the PDA.
   - Attach a portable keyboard for PDAs and key the text.

3. Enter your name on the first line of the Memo Pad application. Memos are automatically saved. The text entered on the first line is the name of the memo.

4. To move down to the next line, or create a return, use one of these two procedures:
   - Tap **Enter** ⏎ on the keyboard.
   - Draw a forward slash (/) in the writing area of the screen.

5. Use the stylus to enter your class schedule on the next few lines. Proofread what you have entered.

6. To edit the text, tap the stylus next to the letter or number you need to change, and then open the keyboard.

Application 2

Scenario As the end of the school year approaches, you decide you want to get a summer job. To help you fill out a job application, your teacher suggests that you make some notes about your school activities.

Directions On the first line of the document, enter the title *Your Name* **School Activities**. Enter your extracurricular activities. If you have not participated in any extracurricular activities, enter the ones below:

```
Student Council representative—freshman year
SADD Club member—freshman and sophomore year
```

Proofread your memo, and edit as needed.

UNIT 4
Developing 21st Century Employability Skills

CHAPTER 13 Explore 21st Century Careers

CHAPTER 14 Plan Your Career

CHAPTER 15 Get the Job You Want

Curriculum Connections
- Language Arts—Reading
- Language Arts—Writing
- Math
- Technology

Lesson 5.3 — Beam a Memo

You will learn to:
- **Explain** beaming.
- **Explain** synchronization.
- **Beam** and receive a memo.

beam To send data from one PDA to another via an infrared beam.

synchronization The process of transferring data between a PDA and a computer.

Why Beam? With a PDA you can beam, or send, data from one PDA to another via an infrared beam. Beaming allows users to quickly and easily share information. Applications can also be beamed from one PDA to another. Beaming is a feature unique to PDAs.

How Do You Beam? Each PDA has an infrared beam that transmits the data or application. The infrared ports on each PDA need to be two to four feet apart and facing each other for beaming to be successful. If additional printing software is purchased for the PDA, documents can also be beamed to a printer.

What Is Synchronization? Synchronization is the process of transferring data between a PDA and a computer. The PDA communicates with the computer during synchronization through a cradle or cable that connects the PDA to the computer.

Information created on a PDA can be synchronized to a computer to ensure that the data is the same on both devices. It is important to synchronize often to protect the data on your PDA. If the data on your PDA were ever lost, you would be able to copy the data on your computer back to the PDA through synchronization. Word documents, Excel spreadsheets, PowerPoint presentations, and e-mail can also be synchronized from a computer to a PDA. The term HotSync® is often used to mean synchronization.

✓ Concept Check

List one major difference between beaming and synchronization. In what situations is it important to synchronize your PDA?

DIGI Byte
Tips on Beaming
1. Make sure both handhelds are on a level plane.
2. Position devices so the infrared beams face each other.
3. Make sure the devices are two to four feet apart.

DIGITAL DIALOG

Beaming

Teamwork With other students, discuss these questions: What advantages can beaming offer? What are possible disadvantages? Explain your answers, giving examples to support your positions. Share your group's responses in a class discussion.

UNIT 3 LAB continued

Follow these steps to track, promote, and launch an e-cycling program:

Step 1 Prepare a spreadsheet to track products that are turned in to be recycled and their estimated recycle value. This will give you a visual representation of the products being recycled. Save your file.

Step 2 Design a database table to track and store information about the products you collect. Add a second table of companies that participate in e-cycling and their addresses. Search the Internet for these companies. Save your file.

Step 3 To promote your e-cycling program, create dynamic flyers in your word processing software. The flyers are to inform your school or community about the importance of recycling electronics and the hazardous elements found within the products. Include information about the location of local or school recycling centers.

Step 4 Prepare an eye-catching presentation. Include the following in your presentation:
- A dynamic title slide.
- Dangerous substances in electronic products.
- A list of local companies that participate in recycling programs.
- Types of products that will be part of the e-cycling program.
- Your plan for the location of recycling centers and collection bins in your school or neighborhood.
- Motivating statements to encourage people to participate in the program.

Step 5 Share your presentation with the class. Arrange with your teacher to distribute the flyers and make scheduled presentations in your school or community to promote and launch your e-cycling program.

Self-Evaluation Criteria

In your Unit 3 Lab have you created:
- A spreadsheet identifying the products you plan to recycle?
- A database with the correct fields to store data about the products you have collected?
- Word documents to use as advertising flyers about your project?
- A motivating and informative presentation?

Activity 3

If you need additional help on this topic, open the Help menu; or refer to the documentation for your hardware.

Suite Smarts

Tapping Memo at the top of the screen on your PDA is the same as using the menu bar in a word processor.

RUBRIC LINK

Go to the Online Learning Center at **digicom.glencoe.com** and click Lesson 5.3, Rubric, to find the application assessment guide(s).

Step-by-Step

Beam and Receive a Memo

In this exercise you will beam a memo to a friend or your teacher.

1. Open **Memo Pad**.
2. Tap the title of one of the memos you have created.
3. Tap the Memo title bar at the top of the PDA screen.
4. Align your PDA with another PDA so the infrared beams are pointing to each other, or aligned.
5. Tap the **Beam Memo** selection on the drop-down menu.
6. The PDA sending the memo will display a message on the screen that says, "Searching ...." The PDA receiving the memo will ask if the person wants to receive the memo. The receiver can tap **Yes**.
7. To delete a memo, tap **Delete Memo** on the Record drop-down menu.

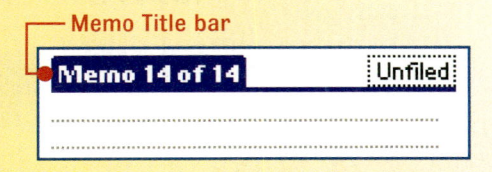

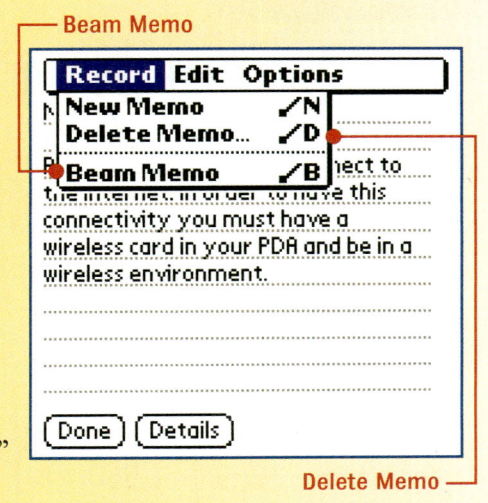

— Memo Title bar
— Beam Memo
— Delete Memo

Application 3

Directions Practice entering text with your stylus by writing letters and numbers in the writing area. Do not use the soft keyboard for this activity. For help on how to enter text with the stylus, see Lesson 5.1, page 127. Enter your name on the first line.

Enter the following text, and then beam this document to another student. Check that your classmate has received the document.

```
We are learning how to use personal digital
assistants in class. A PDA is like a mini-
computer. Many people in business use a PDA to
keep organized. Many students use PDAs to keep
track of appointments and assignments. A PDA
is also a good way to keep track of important
documents.
```

UNIT 3 LAB

DIGITAL Dimension → Science
INTERDISCIPLINARY PROJECT

E-Cycling

Background In Unit 3, you have been gathering information about e-cycling:

- In Chapter 9 Digital Dimension, page 310, you gathered information about companies that participate in e-cycling programs.

- In Chapter 10 Digital Dimension, page 356, you collected information through a survey and entered it in a spreadsheet about the types of electronic products people you know have in their homes.

- In Chapter 11 Digital Dimension, page 390, you collected information about e-cycling in different states and created a database with this data.

- In Chapter 12 Digital Dimension, page 423, you created a presentation which pulled together all of the information you gathered in the previous chapters.

Promote an E-Cycling Program

Consider the skills you have learned in Unit 3 to create documents, spreadsheets, databases, and presentations. Integrate these skills to track, promote, and launch an e-cycling program in your school or neighborhood. The goal of this project is to motivate your audience to recycle electronic products. Include a chart in your presentation. Create a database to store information about products you collect during the recycling effort. Print some of the presentation slides as posters to display.

Continued

E-Portfolio Activity

Showcase Presentation

Include your Unit 3 showcase presentation in your e-portfolio. If you created an electronic presentation, a flyer, a spreadsheet, or a database, save it to your e-portfolio. If you made a digital video or sound recording, save it to your e-portfolio.

Application 4

Directions Enter the following text in a new memo in Memo Pad. Use the keyboard on the PDA, or enter the text with Graffiti. Remember to enter your name on the first line. Add the number 2 after your name to distinguish this memo from one you entered previously.

Beam the memo to your teacher.

```
PDAs have the ability to connect
to the Internet. For this
connectivity, the PDA needs to
have a wireless card and be in a
wireless environment.

Many airports now have wireless
capability for travelers. Some
businesses use wireless access
points to provide connectivity
for employees.

Wireless connectivity makes it
possible to send and receive
e-mail by PDA.
```

Application 5

Scenario You want to add more information to a memo you have already created and saved on a PDA.

Directions Open Memo Pad, and tap the name of the memo you want to open.

Move the insertion point to a new line, and enter the following information.

```
A PDA can print directly to some
printers. To do this, software
must be purchased and installed on
the PDA.

The PDA connects to the printer by
infrared beam, Bluetooth, Wi-Fi,
modem, or cell phone connection.
```

Digital Decisions

Handheld Models

Critical Thinking The purchasing department you work for has decided to buy handhelds for each employee. You have been assigned the task of researching the various models and recommending three of them at the next staff meeting in two weeks.

Internet Investigation Search the Internet for available models and current pricing. Check for software bonus packages that are included with each model. Key words for your search include "PDA prices" and "computer bargains."

Prepare a list of additional hardware that can be purchased for each model, including the price. Prepare a presentation for the next staff meeting to explain the features of the three models you recommend.

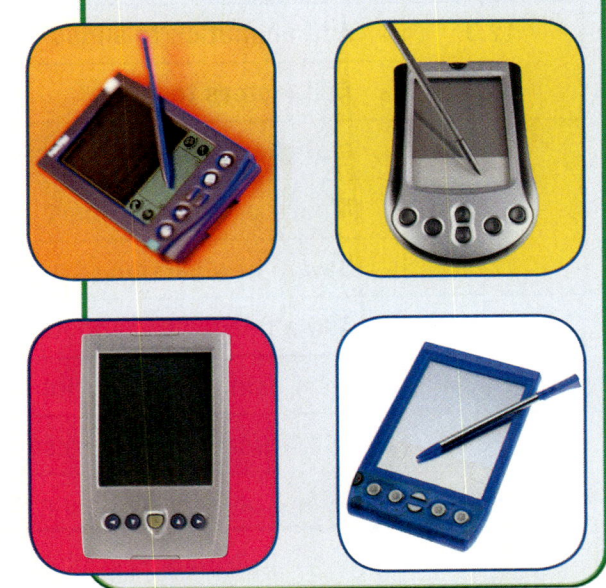

CHAPTER 12

Self-Assessment

Take a moment to review what you have learned in this chapter. Rank your understanding of the topics below.

4 means, "I understand all of this."
3 means, "I understand some of this."
2 means, "I understand very little of this."
1 means, "I don't remember this."

To use a printout of this chart, go to **digicom.glencoe.com** and click on **Chapter 12, Self-Assessment.**
Or:
Ask your teacher for a personal copy.

Rank Your Understanding

Lesson	Topic	4	3	2	1
12.1	• Create and save a new presentation using a design template				
	• Enter text on a slide				
	• Insert a new slide				
12.2	• Add clip art to a slide				
	• Create a graphic object using the Drawing toolbar				
	• Resize and move a graphic object or graphic				
	• Add WordArt				
	• Add an AutoShape				
12.3	• Add animation to slides				
	• Add footers to slides				
12.4	• Add charts to a presentation				
	• Modify or resize a chart				
	• Save a new version of a presentation as a new file				
12.5	• Add a header to handouts				
	• Print using various print layouts				
12.6	• Add hyperlinks to a presentation				
	• Save a presentation as a Web page				

If you ranked all topics 4, congratulations! Consider doing a quick review.
If you ranked yourself 3 or lower in any topic, consider reviewing these topics first.

21ST CENTURY CONNECTION

Manage Your Time Wisely

While you are in school, teachers help you to manage your schedule. When you graduate from school, you must learn to manage your work day on your own. For some people this is a difficult adjustment.

In school, teachers and professors gave you a syllabus or timeline for all the work you would be expected to do and the books you must read. Now there is no guide to follow!

Prioritizing on the Job You may find that working on a job is like doing one project after another. Some of the projects can be done in a day, while others may take weeks to finish.

You must learn to prioritize your work. When you complete your first priority, simply move on to the next most important task.

At first you may be uncertain about what is important and must be done first. You should always ask your supervisor if you are not sure. He or she will respect you for taking pride in your job and wanting to do it correctly.

Avoid the Domino Effect Prioritizing in a team situation is especially important because the next person's task may depend on the timeliness of you completing your task. This domino effect can greatly affect the success of any project. If one piece is not completed on time, the entire project may be jeopardized!

Some delays cannot be prevented. If there is a delay, look for a way to complete your task more quickly so that the project can stay on schedule.

Here is an example of the domino effect in a manufacturing setting. A product must be completed by a particular date. The product must be worked on in different departments as it progresses through production.

However, in one of the departments two employees are out sick for almost a week. All of the work in that department has almost stopped!

If the product is not be completed by the due date, future orders may be affected by the delay. The supervisor moves two other employees to the department for a week to get production back to the appropriate level.

Software for Your Schedule There are many software applications you can use to track your responsibilities.

Microsoft Outlook includes a calendar that is interactive with OneNote and your PDA. Scheduled appointments can easily be synchronized between these programs. You can print a schedule from any of these calendars if you need a hard copy.

Activity Prioritize your homework assignments for the day. Enter them in your PDA's To Do List.

133

CHAPTER 12 Review

Read to Succeed PRACTICE

After You Read

Turn Your Notes into a PowerPoint Presentation Gather the notes you have been taking throughout this chapter. Using the data you have collected, create a PowerPoint presentation. Refer to the lessons in this chapter if you need additional information. Recall the guidelines for preparing effective presentations, such as limiting the text on a slide to about six lines. Also, be sure to keep the focus on each slide to the main point.

To prepare to create the presentation, create a graphic organizer similar to the illustration below to organize your thoughts. Use the Tablet PC Journal feature, if possible.

Example:

Notes	PowerPoint Slide
PowerPoint is software used to create presentations.	PowerPoint presentations contain text and graphics.

Using Key Terms

Communication and presentation skills are essential 21st century learning skills. Check how well you can communicate on the topic of electronic presentations by doing the following. On a Tablet PC, PDA, or piece of paper, rewrite the sentences below. Then use the correct key term to complete each sentence.

- slide (395)
- animation (408)
- legend (412)
- hard copy (412)

1. A sound or visual effect that can be added to text and graphics in a presentation is called _____.

2. A _____ is a table that lists and explains the symbols used in a chart.

3. A _____ is a paper printout of what is shown on a computer screen.

4. An image created in PowerPoint that is displayed on a screen as part of a presentation and can contain text, graphics, sound, video, and animation is a _____.

Lesson 5.4 — Use, Create, and Beam Categories

You will learn to:
- **Use** categories.
- **Create** and beam a category.
- **Copy** a memo to a different category.

Organizing Information Organization is an important skill for any job. Keeping your work organized helps you quickly find information when you need it.

PDAs have features to help you organize the information you store electronically. If you have entered a lot of information into your PDA, how do you find it later?

Organize with Categories On a PDA, one way to organize the information in your different applications is with categories. Categories are used to organize data and are similar to folders on a desktop computer or to files in a file drawer.

categories Used to organize data and are similar to folders on a desktop computer or to files in a file drawer.

Each PDA has several categories that have already been created for you. You can add more categories and give them names that will be useful to you.

The purpose of the categories is to help you organize the information on your PDA so that you can find it later. Categories will help you to manage your time more efficiently.

Using Categories After you place memos, notes, or addresses in categories, you can beam the entire category to another PDA. This saves time when you have more than one document you want to beam.

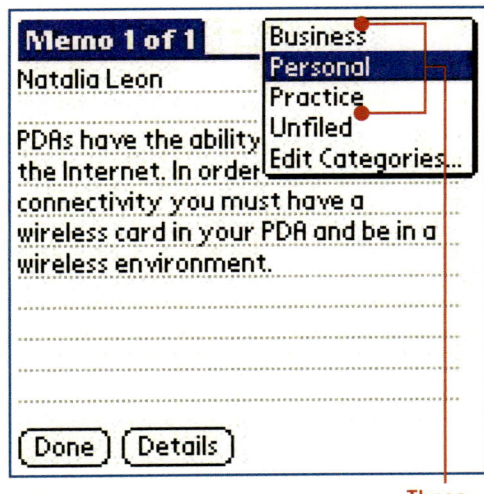

Three categories

Suite Smarts

Categories on a handheld are similar to folders on a PC.

✓ Concept Check

What categories might be helpful in organizing memos on your PDA? Think of three categories.

CHAPTER 12

DIGITAL Dimension → Science
INTERDISCIPLINARY PROJECT

E-Cycling

Using the presentation skills you have learned in this chapter, create a PowerPoint presentation that includes the information you have gathered on e-cycling.

Share and Inform Create a PowerPoint presentation on what you have learned in the Digital Dimension projects in Chapters 9, 10, and 11, pages 310, 356, and 390. Prepare the presentation to educate others about the importance of e-cycling.

As the first step in creating your PowerPoint presentation, take stock of your information. Organize it into categories on your PDA. In the category, summarize the information you have on that topic. Then put your categories in order for your presentation. You can organize them by priority, in the order of the e-cycling process, or in whatever way you think will be most effective in convincing your audience.

The categories become the foundation of an outline. Look over the data in each category again. If there is information that does not relate to the objective, delete the memo in the category. Sort the remaining data around the categories. What photos, graphics, and charts do you have for each category? Estimate how many slides you will need for each topic.

Use the PDA categories as a reference when creating your PowerPoint outline. Make sure you include the following in your PowerPoint presentation:

- Pictures that illustrate your points.
- Charts to visually represent data.
- Graphics of electronic items that can be recycled.
- List of local companies that collect e-cyclable items.
- Hours of operation, if available.
- Dangers of disposing of electronics in landfills.

Make a Lasting Impression Proofread the presentation carefully. Enlist a friend or family member to also proofread your presentation. Before the presentation, prepare by printing out notes for yourself and audience handouts.

E-Portfolio Activity

Presentation Skills

To showcase the PowerPoint skills you have learned in this chapter, create a new presentation describing these skills. List them on the slides. Include graphics with your slides as well as animations.

Give a title to your presentation.

Include a photo of yourself taken with a digital camera.

Save the presentation as an HTML file for your e-portfolio.

Self-Evaluation Criteria

Have you included:

- Graphics and animations for slides?
- A title?
- A picture of yourself?
- A thorough listing of your skills?

Activity 4

Step-by-Step

Use and Create a Category

1. Open **Memo Pad**.
2. Tap the category box. You will see the categories that have already been created.
3. To add a new category, tap **Edit Categories**. Tap **New**.
4. Enter the name of this new category, and tap **OK**.

Activity 5

Move a Memo to a Different Category

When you create a new memo, the memo appears in the current category, but you can move the memo to any other category.

1. Look at the list of memos. Open the memo you want to move by tapping the stylus on the name in the list.
2. Tap the category name in the upper right-hand corner to display a list of available categories.
3. Tap the name of the category where you want the memo placed.

Activity 6

Beam a Category

1. Categories can be beamed to another PDA. View the category you want to beam. To change to another category, tap the category name in the upper right-hand corner.
2. Tap **Memo** in the upper left-hand corner of the screen. You will see the Beam Category menu choice.
3. Align your handheld with another one. Tap **Beam Category**, and the category will be sent.

If you need additional help on this topic, open the Help menu; or refer to the documentation for your hardware.

DIGI*Byte*

Category Box
The category box may be in a different location on the screen, depending on the model of PDA.

Chapter 5 Handheld Computers—PDAs   Lesson 5.4   **135**

Suite Smarts

Any Microsoft Office document can be saved as a Web page.

RUBRIC LINK

Go to the Online Learning Center at **digicom.glencoe.com** and click Lesson 12.6, Rubric, to find the application assessment guide(s).

Step-by-Step *continued*

The title of each slide appears in the Outline pane on the left side of the screen. Notice that slides 4 and 5 do not have a title. If you find a mistake in your Web page, go back to the PowerPoint presentation, make the correction, and then save again as a Web page. When you are asked if you want to replace the current file, click **Yes**. Your new file with corrections will replace the old, incorrect file.

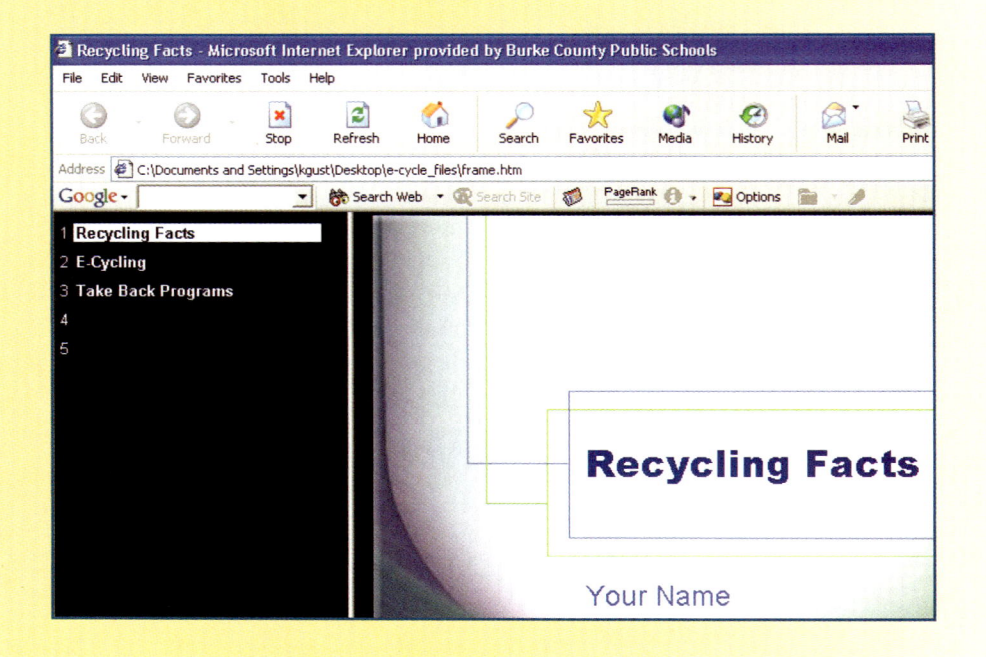

Application 8

Directions Open the **E-Cycle** PowerPoint presentation. Add a title to the slides without titles. Save the presentation again as a Web page, but be sure to use the same name and replace the old file. Double-click the HTML file to open it. Do the new titles display in the Outline pane?

Application 9

Directions Write two paragraphs about creating Web pages with PowerPoint presentations. Consider the Web page you created in Chapter 4. Are there advantages to knowing HTML? Open the **E-Cycle** Web page you just created. Click the View menu and choose Source. What do you see in the source code? How is it similar to or different from the HTML Web page you created in Chapter 4?

Create and fill in a chart that contains advantages and disadvantages of PowerPoint Web pages. Share your ideas with the rest of the class.

RUBRIC LINK

Go to the Online Learning Center at **digicom.glencoe.com** and click Lesson 5.4, Rubric, to find the application assessment guide(s).

Suite Smarts

Moving memos to another category in a PDA is like moving files from one folder to another folder on a PC.

Application 6

Directions Create three additional categories in the Memo Pad application.

- What are the names of the categories?
- What types of documents will you store in the categories?

Application 7

Directions Open a memo that you previously created. Copy this memo to one of the categories you created in Application 6 above.

Application 8

Directions Create a new category, and give it the name of one of your other classes.

- Open the memo you have been working with, and copy it to this new category.
- Find another student in your class who does not have a category with the same name, and beam your category to him or her.

DIGITAL DIALOG

PDA Applications to Help You Organize

Internet Investigation With a partner, search the Internet for applications that can be downloaded to your PDA to help you organize the information you are storing.

Some examples of applications to consider might be:

- Date books
- Diaries
- Databases
- Work logs
- Financial planners

Prepare a brief summary of at least five applications that provide some type of organizational benefit to the user. Include the price of each application and the Web site where you found each application.

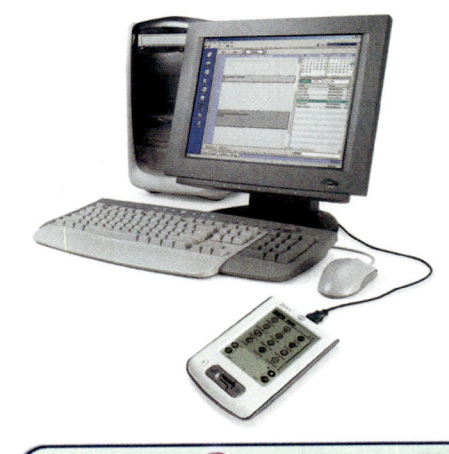

ONLINE Resources

To find Web sites on this topic, visit the Digital Communication Tools Online Learning Center at **digicom.glencoe.com**. Click on Chapter 5, Resources.

Step-by-Step

Activity 19

Add a Hyperlink to a Presentation

1. Open your **E-Cycle** presentation.
2. In the first slide, click after your name and press [ENTER].
3. Enter your e-mail address. If you do not have an e-mail address, enter *yourname@yourschool.net*.
4. Press [ENTER]. The text will turn into a hyperlink.

Activity 20

Save a Presentation As a Web Page

1. Click the **File** menu and choose **Save as Web Page**. You can change the file name if you want or leave it the same as the presentation.
2. Click the **Save as type** drop-down list and choose **Single File Web Page**.

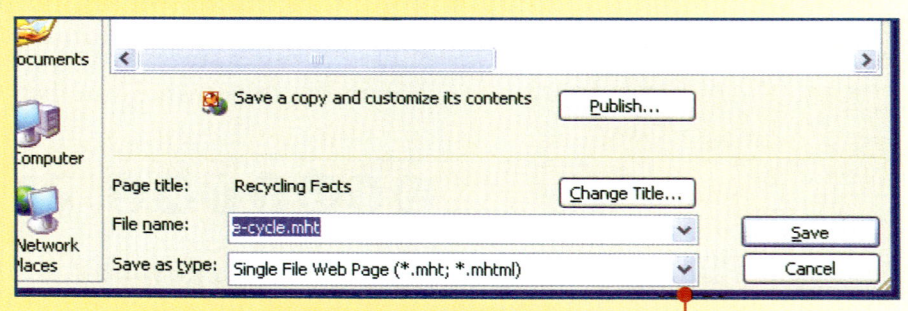

Save as drop-down arrow

3. Look at the folder name listed in the Save In box at the top of the Save window. If necessary, click the drop-down arrow to select a different folder.

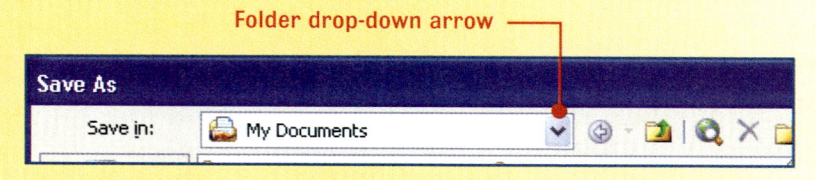

Folder drop-down arrow

4. Click the **Save** button and then close the presentation.
5. Open the folder where you saved the file. You will see the HTML file.
6. Double-click the file you just saved to open it in your Web browser.

DIGI Byte

Background

Use a template with a dark background if you will be showing your presentation in a light room. Use a template with a light background if you will be showing your presentation in a large, dark room.

Continued

Lesson 5.5 → Create and Beam a Note with Note Pad

You will learn to:
- **Describe** the features and functions of Note Pad.
- **Create** a new note in Note Pad.
- **Beam** a note to another PDA.

Note Pad An application on a PDA that allows you to create short notes or simple diagrams.

DIGI*Byte*

Features of Note Pad
- Notes can be stored in categories just as in Memo Pad.
- New categories can be created.
- Notes can be beamed to another PDA.
- Categories cannot be beamed.
- Notes can be made private so that a password needs be entered to open the note.
- An alarm can be set on a note as a reminder.

Functions of Note Pad Note Pad is similar to Memo Pad. **Note Pad** is an application on a PDA that allows you to create short notes or simple diagrams. To use Note Pad, you do not need to use the keyboard or the writing area as in Memo Pad. Instead, you can write or draw directly on the screen.

Why Use Note Pad? You can use Note Pad to jot down a quick message or reminder to yourself or to draw a diagram. For example, in social studies class, you may want to draw a rough sketch of a map of an area your teacher is describing.

On the job, for example, you may be discussing a design element with a client if you are an architect or contractor. Using Note Pad, you can draw the design element and then write notes below it to mark off dimensions.

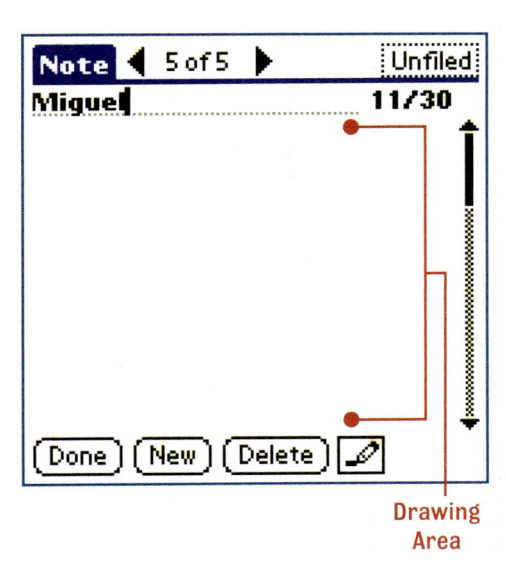

✓ Concept *Check*

Compare the characteristics of Memo Pad and Note Pad. On a sheet of paper, draw the diagram shown below. Write the unique characteristics of each application in the outer areas of the ovals. Fill in the similarities in the overlapped area.

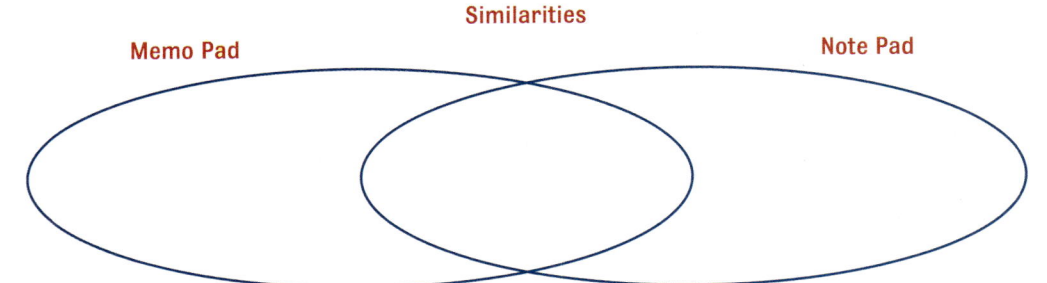

Chapter 5 Handheld Computers—PDAs

Lesson 12.6 Convert a Presentation Into a Web Page

You will learn to:
- **Add** hyperlinks to a presentation.
- **Save** a presentation as a Web page.

Making a Web Page In Chapter 4 you learned about creating Web sites and using HTML, the special language that converts text to Web pages. An alternative to using HTML is to create a Web page from a PowerPoint presentation. The process is fairly simple, especially since you already know how to create a presentation. You can add hyperlinks to slides so that you can include your e-mail address and links to other Web sites on a Web page created in PowerPoint.

Are Changes Needed? When you convert a PowerPoint presentation into a Web page, there are several considerations to keep in mind. As usual, the highest priority is to think about your audience. Who will want to see the presentation online? Will it be people who have already seen your presentation in person and might like a review? If so, then you will not need to add much additional information.

If you are putting the presentation online for people who have never seen it before, however, reread each slide to make sure it makes sense and conveys a complete thought. If necessary, add additional information to make the presentation clear, even if it means you must add more slides.

Presentation Size Consider the size of the presentation. Many people have slow Internet connections. When you put your presentation online, you need to make sure that it is reasonably accessible by anyone. This means using small image files whenever possible and limiting the use of animation, video, and sound.

✓ Concept *Check*

When would it be appropriate to convert a PowerPoint presentation to a Web page? What considerations do you need to keep in mind about the audience, the presentation's purpose, and the size of the presentation?

DIGITAL DIALOG

Exploring PowerPoint

Teamwork With a small group, discuss ways you could use PowerPoint to complete school assignments. Make a list of ways you could use PowerPoint in projects. Share your choices with the rest of the class.

Activity 7

If you need additional help on this topic, open the Help menu; or refer to the documentation for your hardware.

Step-by-Step

Create and Beam a Note

1. Tap the **Note Pad** icon to open Note Pad.
2. Tap after the time displayed. Open the keyboard or use the writing area on your PDA to backspace and delete the time.
3. Use the keyboard or writing area to name this new note.
4. Enter your name.
5. Tap the pencil icon to change the width of the line you will be drawing or writing with.
6. Use the stylus to draw an outline of your state.
7. Beam this note to your teacher by tapping the **Note** title bar and then tapping **Beam Note** on the drop-down menu.

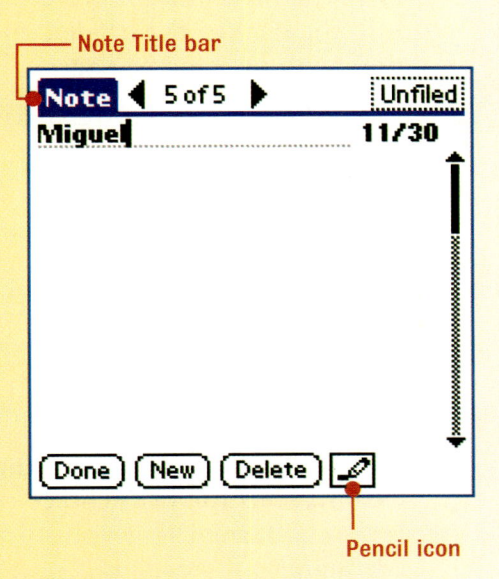

— Note Title bar

— Pencil icon

DIGIByte

Delete a Note
To delete a note, tap the note name and then tap Delete.

Application 9

Scenario You are an architect and need to redesign the room you are in. To do this, you need to create a rough sketch of the room as it is at this time for reference.

Directions Open the Note Pad application. Look around the room and draw the walls, doors, windows, and any other features. Be sure to include the approximate dimensions of the room and to draw everything to scale.

Name this drawing *Your Name* **Room Sketch**. Beam the drawing to your teacher.

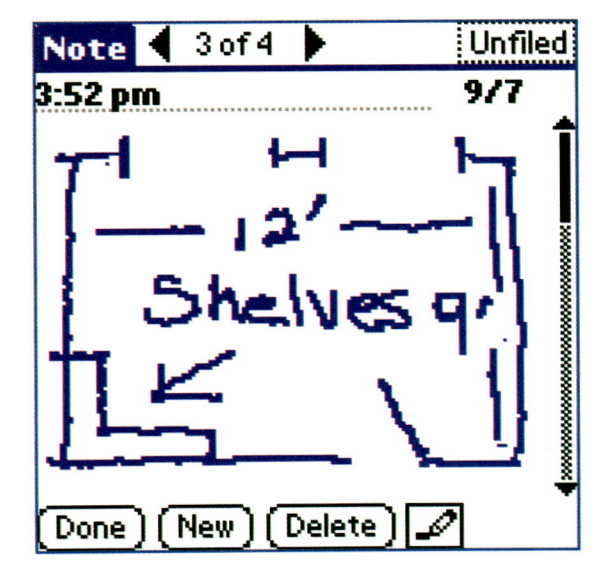

RUBRIC LINK

Go to the Online Learning Center at **digicom.glencoe.com** and click Lesson 5.5, Rubric, to find the application assessment guide(s).

Application 6

Directions Open your **E-Cycle** presentation and follow the instructions below.

- Print the presentation as a handout with two slides per page.
- Print the presentation in outline view.
- Add the date to the header, and preview the presentation as notes pages.

What is the difference between handouts and notes pages?

Application 7

Directions In a small group, discuss the copies of your presentation that you printed in Activity 18 and Application 6, above.

- Which format do you feel is best?
- Give reasons for your choice of format.
- Is there a format that you do not feel is adequate?
- Explain why.

Reproduce the following chart and discuss as a group how to complete it. Fill in your responses as you discuss the printout.

Best Uses for Different Printing Options	
Notes Pages	
Handouts	
Slides	
Outline View	

Summarize your group's findings and report to the class.

RUBRIC LINK
Go to the Online Learning Center at **digicom.glencoe.com** and click Lesson 12.5, Rubric, to find the application assessment guide(s).

Look Who's DIGITAL

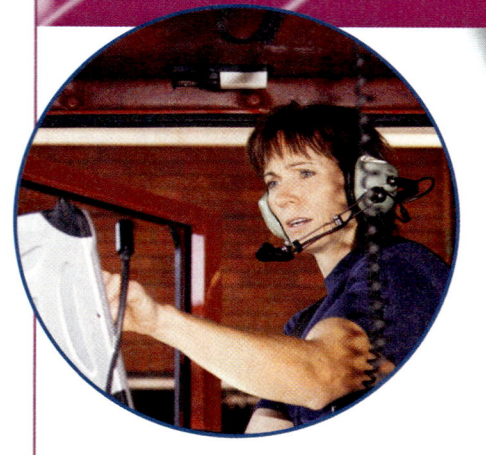

New York Fire Department

Dealing with the aftermath of the September 11, 2001, attack on New York City's World Trade Center motivated the city's fire departments to make their behind-the-scenes work more efficient. They acquired PDAs to document the items recovered from Ground Zero. Special software allowed them to tag each item with a bar code and scan it to record the time, date, and GPS coordinate location.

In addition, each NYFD division installed a digital command board to monitor crew deployment. Each table-like touch screen board can zoom in to an individual building's floor plan and is part of a wireless communication network. Brooklyn Deputy Assistant Chief Joseph W. Pfeifer says a high-tech command board "will make the job safer."

Activity Write a paragraph about how other industries might use a bar code-scanning PDA.

Lesson 5.6 — Create an Address Book and Business Card

You will learn to:
- **Add** and edit addresses in Address Book.
- **Create** and beam a business card.

Use of Business Cards Business cards are a trade symbol for businesspeople. A business card contains information about a person and the company he or she works for. The information listed is usually the person's name, title, company name, business address, phone number, e-mail address, and fax number.

When businesspeople meet new associates, they often exchange business cards. The business card is used in place of an address book. Many people organize their business cards in a wallet-style organizer with dividers that separate the business cards. One problem with this system is that the wallet organizer can become bulky and awkward.

Address Book in PDAs A common application in most PDAs is Address Book. Since the address information is stored electronically, a PDA is capable of storing a large number of addresses. Once a person's address and information is entered into the PDA, an electronic business card can be created. This business card can be beamed to another PDA to share the information.

✓ Concept Check

Create a list of the type of information you would like to save about people you know. What kind of information do you think would be helpful for businesspeople to know about someone they are doing business with?

DIGITAL DIALOG

Wi-Fi in Your Area

Teamwork With a partner, search the Internet to find out about connectivity for handheld computers. Keywords for this search might include "PDA" and "connectivity."

- What can you find out about the speed of the connection?
- What costs are involved in wireless connectivity for a handheld?

Activity 18

If you need additional help on this topic, open the Help menu; or refer to the documentation for your hardware.

Suite Smarts

Documents with a white background use much less ink when printed than documents with colored backgrounds.

You should always use Preview to make sure everything looks correct before you print.

Step-by-Step

Print a Presentation

The following steps will guide you through printing options with PowerPoint. Follow the directions closely so you do not waste paper and ink.

1. Open the **E-Cycle** presentation.
2. Click the **File** menu and choose **Print**.
3. Select **Handouts** from the **Print What** drop-down list in the lower left-hand corner. You can choose how many slides will print on each page in the Handouts box.

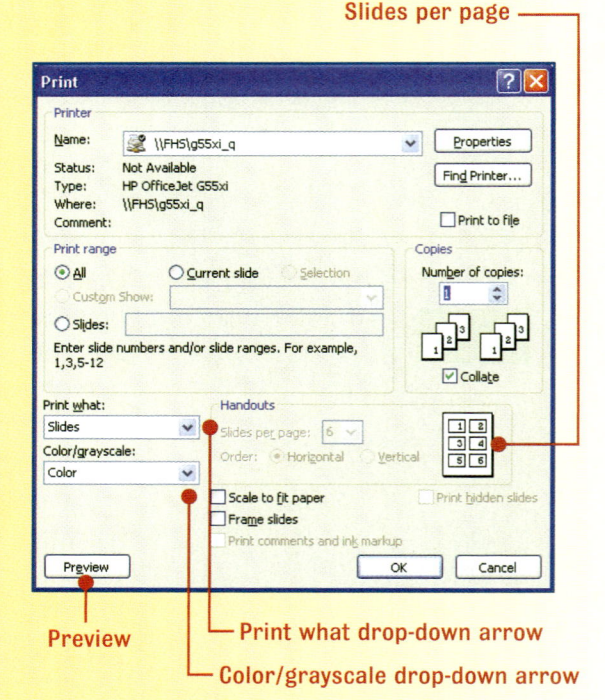

Be sure to check the **Print What** drop-down list and change it, if necessary, every time you get ready to print.

4. Click **Preview** in the lower left-hand corner of the window to see how the slides will look on a printed page. Click **Close**.
5. Click the **View** menu and select **Header and Footer**.
6. Click the **Notes and Handouts** tab. Click the check box next to **Date and Time** to deselect this option. Select **Header** and enter the time of day or period that you have this class in the Header text box.
7. Click the **Apply to All** button.
8. Choose the **File** menu and select **Print**.
9. Make sure **Handouts** is still selected.
10. Change the number of slides to **3** per page.
11. Change the color to **Grayscale**.
12. Click **Preview**.
13. Click the **Print** button at the top of the window.
14. Save your presentation and close it.

Slides per page
Preview
Print what drop-down arrow
Color/grayscale drop-down arrow

Activity 8

If you need additional help on this topic, open the Help menu; or refer to the documentation for your hardware.

DIGIByte

Edit Text
Edit entries in any PDA application as you would in word processing. Highlight text by dragging the stylus across the text. Use a backward stroke to delete the highlighted text or continue to add text, and it will be inserted where the insertion point is located. Using the keyboard may be helpful when editing text.

Step-by-Step

Add and Edit Addresses

1. Tap the **Address Book** icon to open the Address Book application. Tap the **New** button in the lower right-hand corner of the screen.

2. Enter the following information on the line indicated. Tap on the line where you want to enter text.

   Last name: Smith
   First name: Janet
   Title: Administrative Asst.
   Company: ABC Computers
   Work: 888.555.4444
   Fax: 888.555.2442

3. Tap the down arrow in the lower right-hand corner of the address screen to see more lines on the next screen.

4. After you enter the information for the first address, tap **Done**.

5. To enter another address, tap **New** and repeat Steps 2 through 4. If you make a mistake and want to edit the text, tap the stylus next to the letter or number you need to change, and then open the keyboard to make editing changes.

6. You can search for entries in Address Book using the Look Up feature on the main Address page. After you have entered many addresses, this is an effective feature to use instead of scrolling through several pages of addresses.

   Tap the stylus on the blank line after Look Up. Enter the last name you want to find with the keyboard or using Graffiti. When the name is found, it will be selected. Tap the name to view the data.

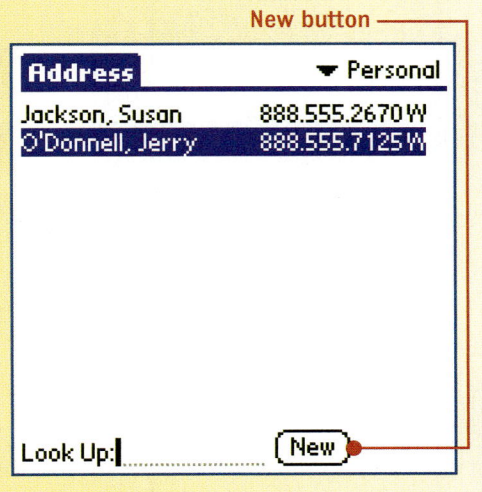

Continued

Chapter 5

140

Lesson 12.5 — Investigate Print Options for Presentations

You will learn to:
- **Evaluate** the printing options in PowerPoint.
- **Add** a header to handouts.
- **Print** handouts for a presentation.

hard copy A paper printout of what is shown on a computer screen.

Making a Hard Copy As you have seen, a PowerPoint presentation is a great tool to share information with an audience. It can also be helpful to give audience members a hard copy of a presentation. A **hard copy** is a paper printout of what is shown on a computer screen.

You have several choices of how you can print a presentation. One way is to print each slide on a separate sheet of paper. This is a good option if you are creating a display or bulletin board. However, if a presentation is long, this option can require a lot of paper.

Handouts Another printing option is to print handouts. Handouts let you print up to nine slides on each page. You can also include a header on handouts. Handouts allow your audience to pay attention to you during the presentation, instead of worrying about taking notes. Later, they can refer to the handout to review the information you provided.

Notes Pages The notes pages printing option prints one slide on each sheet of paper. Each slide is reduced to fit on half of the page, leaving room for notes. This option can be very useful when you are presenting. As you create your slides, you can enter additional information in the notes section. This information could contain the material you will be saying aloud at that screen or could include reminders of additional topics to cover. You can then use a notes pages printout to help you rehearse your presentation.

Outline View In the outline view, only the text on the slides is printed. WordArt, graphics, and charts are not printed, so the text from many slides can fit on each page. If your presentation is very long, or if you want to save paper and printer ink, the Outline View may be a good handout format to use.

✓ Concept Check

Do you think it is helpful to print a hard copy of a presentation to give to your audience? Why or why not?

DIGIByte

Printing Individual Slides
You can print one slide or a group of slides.

Grayscale
If you are going to make black-and-white copies of your slides, print in grayscale instead of color. Your copies will look better.

Print from the File Menu
Select the File menu and choose Print to get options for printing a presentation. If you click the Print button in the toolbar, each slide will print on a separate page.

Activity 9

Step-by-Step continued

Create and Beam a Business Card

1. Open Address Book.
2. Tap the address you want to create a business card for. The complete address of your selection should appear on the screen.
3. Tap **Address Edit** at the top of the screen. A drop-down menu will appear.
4. Tap **Select Business Card**.
5. You will be asked if you want to make this name your business card. Tap **Yes**.

   You will return to the Address Edit window, and an icon of a small card will be visible to the right of the Address Edit title.

6. The business card can be beamed to another PDA from any one of three different views:
   - The main Address screen, when the name of the business card is selected
   - The Address View screen
   - The Address Edit screen

7. Tap **Address**, **Address View**, or **Address Edit** to open the drop-down menu, and then tap **Beam Business Card**. You will see a message that says, "Searching…" on your screen. The receiving PDA will display a message to "Accept" the business card. The person receiving the business card can respond **Yes** or **No** to accept or reject the beamed card.

Application 10

Directions Add a new category to the ones already created on your PDA, and name the category. For help on how to add categories, see Lesson 5.4, page 135. After you add the category, create three new addresses and business cards to be placed in the new category.

RUBRIC LINK

Go to the Online Learning Center at **digicom.glencoe.com** and click Lesson 5.6, Rubric, to find the application assessment guide(s).

DIGITAL Career Perspective

Information Technology

Bridget Burleson
Information Systems Engineer

Keeping People Connected

"Sometimes my job is stressful just because of the nature of it, but I enjoy getting to make technology work for people. I love it when it works and it makes them happy," says information systems engineer Bridget Burleson. "Recently I set up an Intellikeys™ keyboard for a student with disabilities at one of our schools. I don't think she had ever used one, but she learned very quickly how to surf the Internet and play games." Burleson works in a public school system and is responsible for keeping technology running at seven different schools.

Burleson and her colleagues train each other and share ideas on how to make the computers and networks run smoothly. "In our department, everyone is so mobile that we have to use e-mail because we do not always see each other every day. We also have phones and beepers to keep in touch," says Burleson.

"A tool called a Fluke helps me troubleshoot wiring problems and network issues," says Burleson. When a computer is not working or does not connect to the network, it is important to correct the problem as quickly as possible. The Fluke makes the work easier.

Why does Burleson like being an information systems engineer? "I like working with others. I have a lot of interaction with administrators, teachers, and students. I like that I have the ability to make the technology work and to make it do the things they want it to do."

Career Growth Trend

Employment of information systems managers is expected to grow much faster than the average through the year 2012 due to increasing demand by companies conducting business over the Internet and a simultaneous need for greater network security.

Growth Trend: 36% Growth (Year 1 to Year 10)

Training
Most positions require a bachelor's degree and familiarity with the newest technologies and tools. Relevant work experience is also helpful.

Salary Range
Typical earnings are in the range of $30,000 to $50,000 a year.

Skills and Talents
Information systems engineers need to be:

- Able to work as part of a team with other engineers and technicians
- Good communicators, in order to give effective advice to users
- Comfortable with the newest technologies
- Willing to work overtime when unexpected technical problems occur
- Able to learn new applications quickly
- Strong critical thinkers and problem solvers, in order to troubleshoot the problems of users and systems
- Detail-oriented, in order to track the progress of issues with users or within a network

Career Activity
What are the names of some companies in your area that offer careers in information technology?

Application 11

Directions Create a new category named **Colleges** for your address book. Go to the Internet, and look up the addresses of two colleges you would like to apply to or visit. You can choose community colleges, technical colleges, or four-year colleges.

For each college, find the address for either of the following:

- The freshman admission office.
- Campus tours.

If you do not find a contact name, leave the Last Name and First Name lines blank. Fill in as much information as you can find.

Find a classmate who has a college address other than the ones you chose and ask him or her to beam it to you. Then find another classmate who does not have one of your colleges and beam that college address to him or her.

Application 12

Scenario On the weekend, you plan to apply for a job at the local newspaper. Many companies request at least two references on job applications. Also, colleges often require personal references when you fill out their applications.

Directions Create a new category in your address book and name it **References**.

Add two new addresses to your address book, and store them in the new References category. You may use the information below or add your own references.

Amelia Hartso
ahartso@stateschool.edu
1017 Cambridge Court
Berkeley, CA 94703
510.555.9750
510.555.2490 fax
School Guidance Counselor

Chu Lee
clee@leebusiness.com
526 Olive Street
Tampa, FL 33607
813.555.4320
President

TECH ETHICS: Software Licenses

Scenario You are the owner of Wheeler Properties, a real estate company. You purchase PDAs and a special software program that will help employees better organize client addresses and phone numbers. The software even dials clients' phone numbers if they have a compatible cell phone. The PDAs and the software were purchased by Wheeler Properties. The software license, a legal agreement between the software company and anyone who uses the software, allows each user to install and use one copy of the software.

One day you find out that one of the employees has installed the program on his or her personal PDA. What action should you take?

Teamwork In a small group, discuss how this situation should be handled. Discuss the questions below and write a recommended plan of action for this incident. Defend your response with good reasons.

Critical Thinking Can the employee legally install the software on his or her own PDA? Why or why not?

ONLINE Resources

To find Web sites on software licenses, visit the Digital Communication Tools Online Learning Center at **digicom.glencoe.com**.
Click on Chapter 5, Resources.

Activity 16

Step-by-Step continued

Add a Border to a Chart

1. Right-click the chart, and select **Format Object** from the drop-down menu.
2. Select the **Colors and Lines** tab.
3. Choose a dark shade of blue from the Color drop-down list.
4. Choose **3 pt** from the Weight drop-down list and click OK.

Activity 17

Save as a New Presentation

1. In the File menu, click **Save As**.
2. The current file name will be selected in the File name box. Click after the file name to deselect it, and add the number 2 after the file name.
3. Click **Save**.

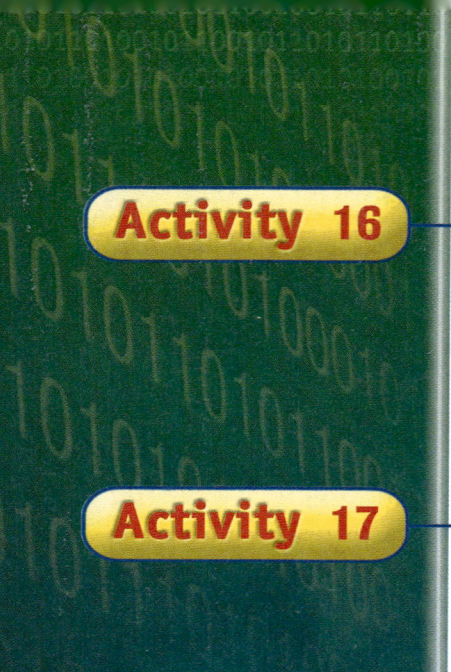

RUBRIC LINK

Go to the Online Learning Center at **digicom.glencoe.com** and click Lesson 12.4, Rubric, to find the application assessment guide(s).

Application 5

Directions Select the chart again. Double-click each of the bars displayed in the chart. In the Format Data Series window, change the color of the bars. Double-click the text in the chart, and change the color and size of the font. Change the color and size of the x- and y-axis fonts. Change the border of the chart. Check your presentation for grammatical and spelling errors, and correct any that you find. Save your presentation as a new file.

DIGITAL DIALOG

Critique Slides

Teamwork Form a small group. Create a set of four 3" by 5" note cards for each group member. Enter your name and one of the following phrases on each card:

Color, Chart Style, Chart Size, Font Size.

Give a set of your cards to each group member.

Writing Take turns viewing each others' slides of the chart you edited in Application 5. How do the colors look? What would you change? Does the size and style of the font look appropriate? Why or why not? Write any comments on the card. Be constructive and respectful, if you criticize someone's work. Add compliments if you feel they are appropriate. Correct any grammatical or spelling errors. Return each card to the person whose name is on it when you are finished.

Lesson 5.7 — Use Date Book

You will learn to:
- **Explain** the features and advantages of Date Book.
- **Create** an event in Date Book.
- **Copy**, paste, and delete events.
- **View** Date Book in daily, weekly, or monthly format.

DIGIByte

Features of Date Book
- Users can view the calendar by the day, week, or month.
- Events can be cut, copied, and pasted from one date to another.
- Events can be beamed to another PDA.
- Notes can be added to events.
- An alarm can be set on an event as a reminder.

Calendars Do you have a calendar hanging on your wall at home to remind you of family events, school events, and holidays? Calendars, as you know, can be wall-sized or small enough to fit in a pocket.

What Is Date Book? In a PDA, Date Book is an application that allows you to keep track of appointments. Although the times are automatically displayed in one-hour segments, different time periods can also be created.

Advantages of Date Book One of the advantages of having the Date Book application on a handheld computer is that the calendar can go several years into the future. Another advantage is that you can record many different events. It is easy to copy and paste appointments between dates and times. If an appointment takes more than one hour, you can block the amount of time needed for the appointment. This helps to make sure you do not create appointments that overlap. Appointments can also be beamed from one PDA to another.

✓ Concept Check

List ways you might use Date Book. Would Date Book help you keep organized? Why or why not?

DIGITAL DIALOG

Applications to Download

Internet Investigation Work with a partner to search the Internet for various applications to download to handhelds that will help you organize appointments.

- Do any of these applications update a date book on your desktop computer?
- What are some of the features of these applications?
- How much do they cost?

Summarize your findings in a one-page report.

Step-by-Step continued

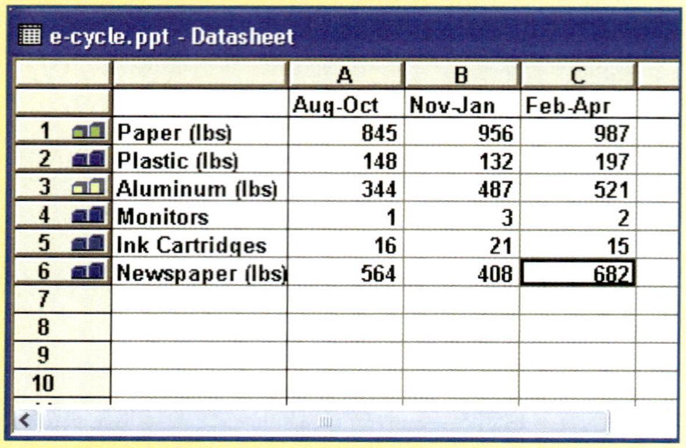

		A	B	C
		Aug-Oct	Nov-Jan	Feb-Apr
1	Paper (lbs)	845	956	987
2	Plastic (lbs)	148	132	197
3	Aluminum (lbs)	344	487	521
4	Monitors	1	3	2
5	Ink Cartridges	16	21	15
6	Newspaper (lbs)	564	408	682

e-cycle.ppt - Datasheet

6. Click the **Close** button in the top-right corner of the datasheet.

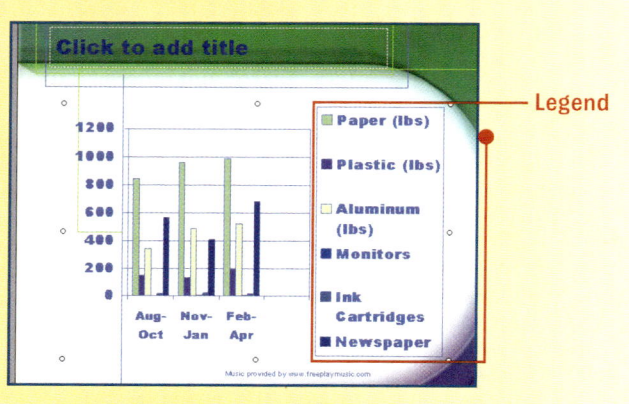

Suite Smarts

When creating a chart in PowerPoint, the Chart menu will appear on the menu bar. This is the same menu used in Excel to edit a chart.

Activity 15

Edit a Chart in a Presentation

1. Double-click the chart to edit it.

2. Resize the chart by dragging one of its corner selection handles.

3. Double-click the legend on the right side of the chart to select it.

4. In the Format Legend box, change the font size to 20 and change the font color. Click **OK**.

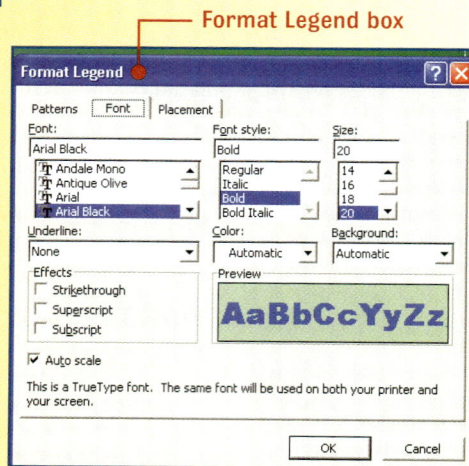

You should be able to see all of the text in the legend. If not, select the chart and make it a little larger.

Continued

Chapter 12

414

Activity 10

If you need additional help on this topic, open the Help menu; or refer to the documentation for your hardware.

Activity 11

Activity 12

DIGI Byte

To Set an Alarm
Tap the box next to Alarm on the Event Details window to set an alarm for an event.

Step-by-Step

Use Date Book in Daily View

When you open Date Book, the entry screen for the current date appears. The appointment times are automatically displayed in one-hour increments. Tap Details to create different time increments.

1. Tap [icon] to open **Date Book** so today's date is displayed.
2. Tap the stylus on the line next to the **3:00** mark.
3. Enter **Student Council Meeting**.

Create a Future Event

1. Press the down button seven times to go to the date one week from today. Tap on the line next to 2:00.
2. Tap **Details**. Tap the time display box.
3. Select the start time **2:15**. Tap **OK** twice.
4. Enter **Class pictures**.

Copy and Paste an Event

To copy an event to more than one date or time, follow the steps below.

1. Drag the stylus across the Class pictures event you just entered. The text will be highlighted.
2. Tap the date at the top of the screen to open the drop-down menu.
3. Tap **Edit**, and then tap **Copy**.
4. Tap **10:00** on the same date to place the insertion point.
5. Repeat Step 2. Tap **Edit**, and then tap **Paste**.

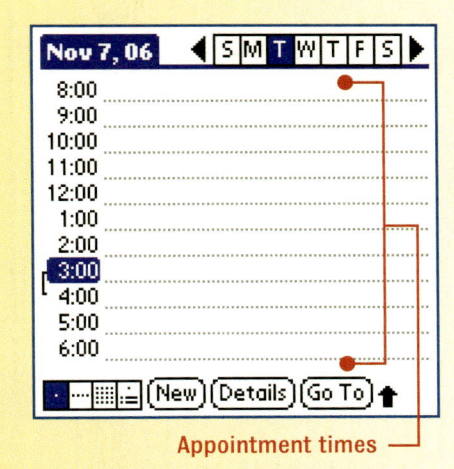

Appointment times

Time Display box

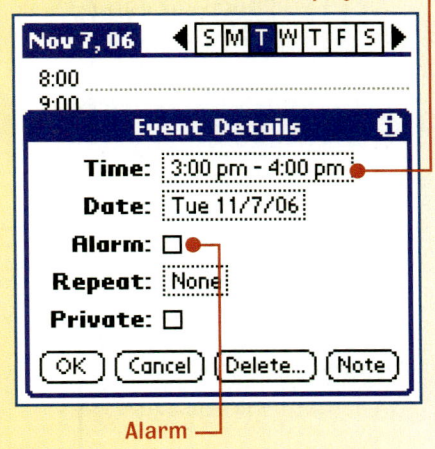

Alarm

Highlighted text

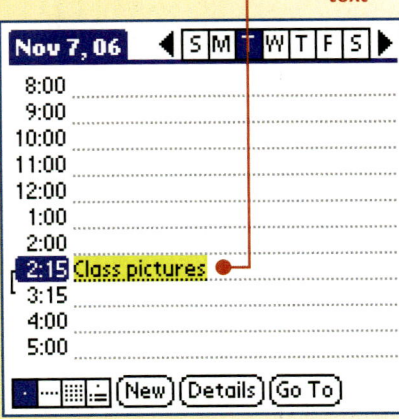

Continued

Chapter 5

Activity 14

If you need additional help on this topic, open the Help menu; or refer to the documentation for your hardware.

Step-by-Step

Add a Chart to a Presentation

1. Open the **E-Cycle** presentation and go to the last slide.
2. Click the **New Slide** button. In the Other Layouts section in the Slide Layout task pane, click on the **Title and Chart layout** to insert a slide with the Title and Chart layout.
3. Double-click the picture of the chart.

   A small sample datasheet will appear on the screen in a new window. The data you enter in the datasheet will automatically be displayed as a chart.

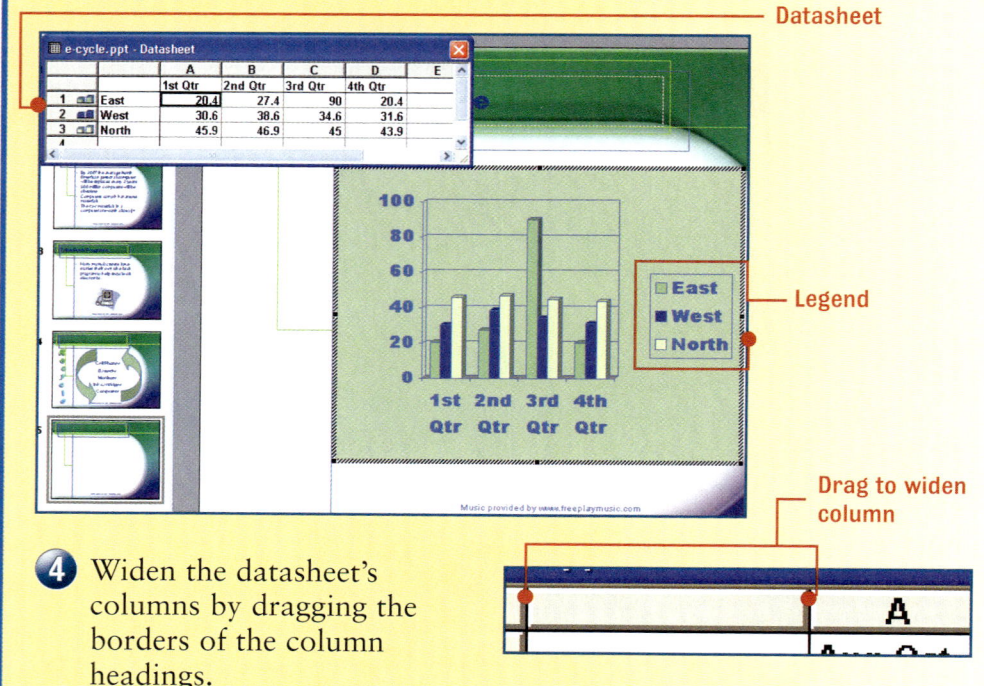

   — Datasheet
   — Legend
   — Drag to widen column

4. Widen the datasheet's columns by dragging the borders of the column headings.
5. Enter the following data into the datasheet.

	Aug-Oct	Nov-Jan	Feb-Apr
Paper (lbs)	845	956	987
Plastic (lbs)	148	132	197
Aluminum (lbs)	344	487	521
Monitors	1	3	2
Ink Cartridges	16	21	15
Newspaper (lbs)	564	408	682

Suite Smarts

A report that has been created in Word can be shortened and designed as a PowerPoint presentation.

Continued

Activity 13

Step-by-Step continued

Delete an Event

1. Tap the time **2:15**, where Class pictures was entered.
2. Tap **Details** at the bottom of the screen.
3. Tap **Delete**. Tap **OK**.

Activity 14

Choose Weekly and Monthly View Format

You can view Date Book in Weekly View or Monthly View if you want to look for dates when an event is scheduled. The shaded squares indicate an event has been scheduled on that date.

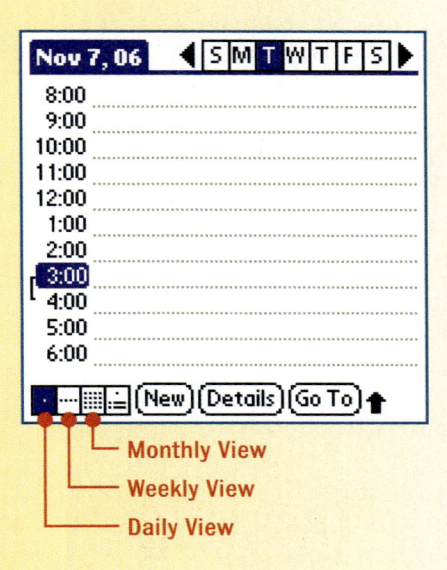

- Monthly View
- Weekly View
- Daily View

1. Look at the four square icons in the lower left-hand corner of the screen. The first one is Date View, ■. It is dark because it is the view you are currently accessing.
2. Tap the second square icon, ▭, to view Date Book in the Weekly format.
3. Tap ▦ to view the Monthly format.

RUBRIC LINK

Go to the Online Learning Center at **digicom.glencoe.com** and click Lesson 5.7, Rubric, to find the application assessment guide(s).

Application 13

Directions In Date Book, add the birthdays of five of your friends and family. Also add any important events that you know about, such as meetings you need to attend for various school activities, school holidays, or deadlines for school projects.

Application 14

Directions Use Date Book to enter the dates for home football games. Get a copy of the football schedule from your school's Web site or your teacher.

Enter the first home game on your PDA. Copy this event and paste it in the correct date and time for each home football game. View in weekly and monthly format. Choose one of the events you just created and delete it.

Lesson 12.4 ▶ Add Charts to Presentations

You will learn to:
- **Insert** a chart into a presentation.
- **Modify** a chart in a presentation.
- **Save** a new version of a presentation as a new file.

legend A table that lists and explains the symbols used in a chart.

Displaying Charts In Chapter 10, you learned how charts can make data easier to understand. Adding charts to PowerPoint presentations is simple to do. A chart can be created in PowerPoint, or it can be created in Excel and pasted into a PowerPoint slide. Clicking on the Chart box on a PowerPoint slide automatically opens Microsoft Graph, a supporting application that you can use to create charts. With the Graph application, you can make the same kinds of charts that you can in Excel.

Often, you need to use symbols to represent the types of data in a chart, as there may not be room for a complete description. A **legend** is a table that lists and explains the symbols used in a chart.

Keep Your Audience's Attention The purpose of a presentation is to grab your audience's attention. The text in a presentation should be short and to the point. Consider who your audience will be when you show your presentation. Use vocabulary words they will understand. Define words when necessary. Insert pictures taken with a digital camera instead of clip art, if possible. Make sure that charts and graphs are clearly labeled so that your audience can interpret what they are seeing.

Saving a Presentation as a New File When you make a big change to a presentation, you might want to save the presentation as a new file. This makes it easy if you change your mind about the change. You could simply use the version that did not have the change instead of having to remove the change manually.

Remember, though, that too many versions of a presentation can be confusing and will take up a lot of disc space. Save a presentation as a new file only when necessary.

✓ Concept Check

Would it be better to display the results of a survey in a chart or as a table in a presentation? Explain your answer. If you used a chart in a presentation, which type of chart, such as pie, bar, or line, would work best? Why?

DIGI*Byte*

Layout Designs
You can choose from many different layout designs when adding a chart to a presentation.

Lesson 5.8 → Create a To Do List

You will learn to:
- **Explain** the use of the To Do List application.
- **Add** items to the To Do List application.
- **Use** the Details feature.
- **Select** preferences from the Show feature.
- **Add** a note to a To Do item.

To Do List An application on a PDA that allows you to create, keep track of, and prioritize important tasks and events.

To Do Lists How do you remind yourself about an appointment or something you need to take to class with you? Some people create a to do list to help remind them of these items. **To Do List** is an application on a PDA that allows you create, keep track of, and prioritize important tasks and events.

Organizing with To Do List The To Do List application helps you organize activities in a number of ways. You can prioritize the items on your list and set preferences for different features to display for each item. Items in To Do List can have due dates and completion dates. To Do List has categories in which to store items.

✓ Concept Check

How is the To Do List application similar to the other applications in a PDA? How is it different? List additional features that can be helpful to you in organizing items.

DIGITAL DIALOG

Compare and Contrast

Teamwork Work in a group to compare these three applications: Memo Pad, Note Pad, and To Do List.

- What features do they have in common?
- What is different about each application?
- Give an example of how each application could be used in school or on a job.

Summarize your findings and report them to the class.

21ST CENTURY CONNECTION

Employee Rights and Responsibilities

Federal and state laws protect employees and their rights. Employees of all companies have rights that can be enforced by the Equal Employment Oppor-tunity Commission (EEOC). States can expand their labor laws based on the initial laws set forth by the federal government. The government makes laws about the age at which you can begin to work, the amount of minimum wage, hiring people with disabilities, and many other issues.

Status Affects Rights

Different types of workers have different rights. Your rights can vary depending on if you work as a contractor, a full-time employee, or a part-time employee. If you have been fired, laid off, or asked to resign, your rights as an employee are also different in each case. These rights may affect whether or not you can receive unemployment benefits. Unemployment benefits are paid to you by the government if you lose your job. If you quit your job, you usually are not entitled to unemployment benefits.

Laws That Protect You

There are also laws that protect you against discrimination. Some forms of discrimination are racial, religious, sex, age, disability, or national origin. Harassment on the job is in the news often. Someone using inappropriate language can be considered to be harassing you or other employees if you are offended by the language. Other areas that have laws protecting employees involve sick leave or disability leave, pay and the hours you work, unemployment insurance, injuries and illnesses, health and safety, and privacy.

Know Your Rights

It is important to be aware of your rights as an employee. Employers are required to furnish employees with information about their rights and how to contact the EEOC or a company official if employees feel their rights have been violated. Visit the EEOC's Web site at www.eeoc.gov to learn more about employee rights.

Report Hazards and Injuries

Employees have a responsibility to report unsafe or hazardous working areas to their employers. If an injury occurs, an employee should report it to a supervisor immediately. If you work in a dangerous environment, it is your responsibility to be aware of emergency plans and make sure you receive information on the company's policy for handling dangerous situations.

When everyone stays informed, a safe work environment can be maintained. Injuries and accidents will be at a minimum.

Activity Why is it important to know your rights as an employee? List two reasons.

Step-by-Step

Activity 15

If you need additional help on this topic, open the Help menu; or refer to the documentation for your hardware.

Add Items to the To Do List Application

1. Open the **To Do List** application by tapping [icon].

2. Tap **New** to start a new To Do item. A blank line will appear in the window with a blinking insertion point.

3. Enter **Meeting**.

To edit a To Do item, tap the item listed and then tap the insertion point where the change needs to be made.

Activity 16

Use the Details Feature

1. Tap **Details…**. Choose the priority level for this item. Choose a category, if desired. Select a due date. If you tap Private, the title will not appear in the opening To Do List menu, and you will be prompted to enter a password.

2. Tap **OK** when you are finished.

Priority levels

Activity 17

Select Preferences

1. To set Preferences for each item, tap **Show…**. You can choose how you want your To Do items sorted and displayed.

2. Tap **OK** when you are finished.

Activity 18

Add a Note to a To Do Item

1. Tap **Meeting** in your To Do List. Tap **Details**.

2. Tap **Note**. Enter a note, such as the place and room number for the meeting, on the Note screen. Tap **Done**.

Activity 19

Mark and Delete an Item

When you have completed an item on your To Do List, you can mark it as complete or delete it.

1. Tap the check box to the left of **Meeting** to mark it as complete.

2. Tap on **Meeting** to select it. Tap the **Details…** button, and then tap **Delete**. Tap **OK** to delete the item.

Application 4

Directions Click on the Slide Sorter View button at the lower left-hand edge of your PowerPoint file to view thumbnails of all the slides you have created. Identify which slides contain graphics that can be animated and which contain text that you can animate. You can create a graphic organizer similar to the one below to help you track the places to add animation:

Animation to Add

Slide Number	Item to Animate
1	
2	
3	

Click on one of the slides with a graphic, and click the Normal View button or double-click on the slide itself to view that slide at full size.

Add animation to the rest of your slides. Add the name of this class to the footer. When you are finished, click the Slide Show button at the bottom of the Animation menu to view your slide show. Save your file.

RUBRIC LINK
Go to the Online Learning Center at **digicom.glencoe.com** and click Lesson 12.3, Rubric, to find the application assessment guide(s).

DIGITAL Decisions

PowerPoint Presentation for Instruction

Teamwork Form a group of four students. Choose one of the following topics and create a PowerPoint presentation: digital cameras, laptop computers, Tablet PCs, PDAs, scanners, cell phones, or MP3 players. Include the following information about your topic in your presentation: what the item does, cost of the item, where to buy the item, and examples of how you can use the item.

Internet Investigation Look for information about your topic on the Internet, in magazines, in the newspaper, and in books. Divide up the following tasks: collecting data, writing text for the slides, entering and arranging the text on the slides, and adding graphics and animation to the slides.

DIGITAL DIALOG

Animation Decisions

Communication In a small group, discuss the animation you just added to your PowerPoint presentations. Take turns viewing each other's presentations. When it is your turn, explain why you chose the animation styles on each slide. The other students in the group can then comment on whether they feel the animation you chose complements your presentation. Your classmates should give reasons why it does or does not enhance the slides. If the group does not agree on an animation, decide on another animation or whether there should even be animation for the text or object.

Application 15

Scenario You have an important meeting with the vice president of your marketing department in one week. You do not want to forget anything, so you decide to use To Do List to note the items you want to share.

Directions Enter each item below in To Do List as a separate item. Choose the priority settings for each item.

- Report of survey results from telephone survey
- Market analysis results of similar products
- Cost of packaging
- Advertising target audience
- Time needed to complete production

Application 16

Scenario You are in charge of arranging the annual company holiday party. As you begin a list of things to do, you realize the best place to save it is on your PDA. That way you can send the list to your office manager.

Directions Open To Do List and name this file **Holiday Party.** Enter the following items. If you can think of anything else that is needed, add it to the list. Prioritize the items after you enter them. Add a note to two of the items.

- Send invitations
- Talk to a caterer about menu choices
- Select and buy invitations
- Check availability of locations for party
- Plan decorations
- Buy decorations
- Choose a date and get approval from management

Digital Decisions

Training Presentation

You work for Stephen Bank International, a large bank that has its own training staff. Each of the department managers and vice presidents has been given a new handheld computer. You have been chosen to prepare and present a workshop on the three most important applications on a handheld computer.

Critical Thinking

Choose from the following applications:

- Memo Pad
- Note Pad
- Address Book
- Date Book
- To Do List

Summarize which three applications you will explain, and why. You may create a table for the information, or you may enter the information in your PDA.

Be prepared to explain why you feel the applications you are going to talk about are important for the managers and VPs. Give at least two examples of how each application could be used.

RUBRIC LINK

Go to the Online Learning Center at **digicom.glencoe.com** and click Lesson 5.8, Rubric, to find the application assessment guide(s).

Activity 12

? If you need additional help on this topic, open the Help menu; or refer to the documentation for your hardware.

Step-by-Step

Add Animation to Slides

1. Open the **E-Cycle** presentation that you have been working on.
2. Select the **Slide Show** menu and click **Animation Schemes**.
3. Click the slide's title so that it is selected.
4. Use the scroll bar on the right of the task pane to view the animation names.
5. Click one of the animation names. A preview of the animation will show on your slide.
6. Animate the rest of slides 1 and 2.
7. Select a graphic or WordArt. In the **Slide Show** menu, click **Custom Animation.** In the **Add Effect** box, click an effect.
8. Click the **Slide Show** button. Each animation will begin when you click the item that you animated.

Activity 13

DIGI Byte

Headers and Footers
Only footers can be added to slides. Headers can be added to notes and handouts.

Add a Footer to a Presentation

1. In your presentation, click the **View** menu and select **Header and Footer**. Click the check box next to **Date and Time** to deselect this option. Select **Footer** and enter **E-Cycle** in the Footer text box.
2. Choose **Don't show on title slide** and click the **Apply to All** button. Scroll through your slides to make sure the footer information is displayed.

Chapter 12 Presentations   digicom.glencoe.com   Lesson 12.3   **409**

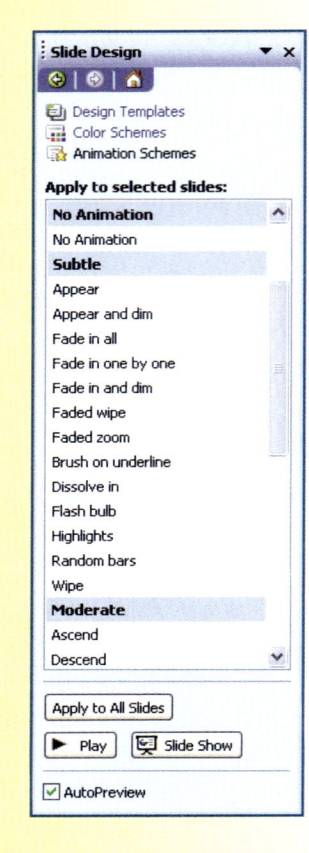

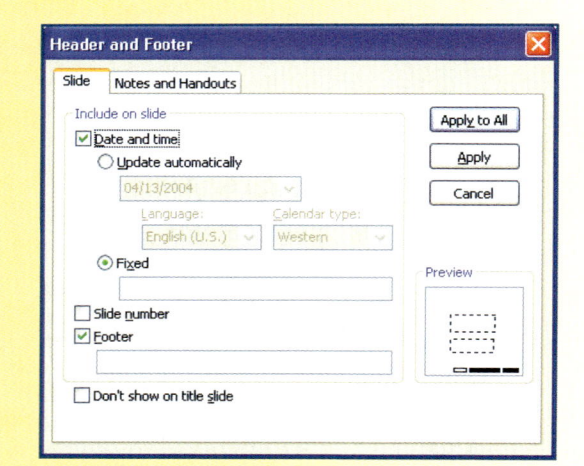

DIGITAL Career Perspective

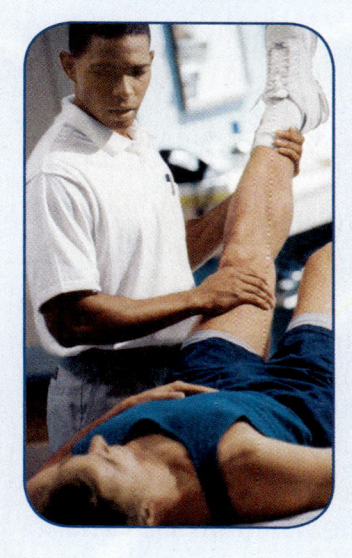

Health Science

**Jared Harris
Physical Therapist**

Helping People Improve Their Lives

"I love to help my patients get better," says physical therapist Jared Harris. "But with the pace of the health system today I need to work fast, so my PDA really helps." Harris, who works in more than one clinic to treat patients, has patients' treatment records easily accessible through his PDA. "I can even look up new treatments for a patient," says Harris.

"I help people who are injured in sports or an accident and people with disabling conditions, such as a stroke, cerebral palsy, or arthritis. I first need to examine patients' histories and then measure the patients' strength and capabilities," says Harris. Next, he creates treatment plans that may include exercises for patients to increase flexibility and strength and ultrasound or massage to relieve pain. Harris says, "I have the newest treatments right here on my PDA. It's quick and easy to use."

"Every week, I see a lot of patients," says Harris, "and I need to track their progress. I document patients' progress on my PDA. Then when a doctor needs a patient's report, it's easy to send."

Why does Harris like being a physical therapist? "Helping people improve their lives is the most rewarding part of my work. Also, I'm never bored. I'm always learning new treatments. My PDA helps with that—I wouldn't go anywhere without it."

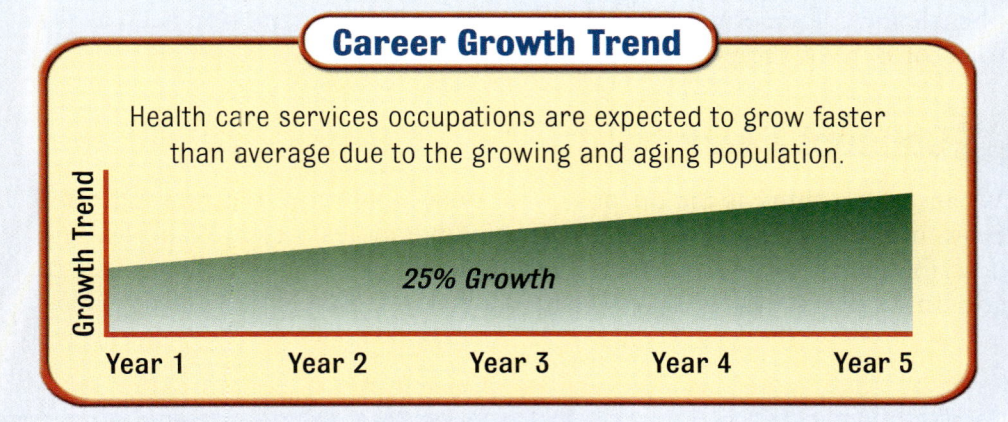

Career Growth Trend

Health care services occupations are expected to grow faster than average due to the growing and aging population.

25% Growth

Growth Trend: Year 1 — Year 2 — Year 3 — Year 4 — Year 5

Training
All physical therapists need to graduate from an accredited physical therapist educational program and then pass a licensure exam in their state. Useful courses to take before applying to a physical therapist educational program include biology, anatomy, chemistry, and mathematics.

Salary Range
Typical earnings are in the range of $50,000 to $70,000 a year.

Skills and Talents
Physical therapists need to be:

- Passionate about improving patients' lives
- Compassionate—to encourage patients
- Able to work in a physically demanding job
- Able to make decisions in developing patients' treatment plans
- Willing to study to keep up with advances in treatments

Career Activity
Find out about a different career in health science, such as being a radiologist or an optometric assistant. What training, skills, and talents are needed?

Lesson 12.3 — Add Animation to Presentations

You will learn to:
- **Add** animation effects to text and graphics in a presentation.
- **Play** the presentation so that you can view the animation effects.
- **Add** a footer to a presentation.

animation A sound or visual effect that can be added to text and graphics in a presentation.

DIGIByte
Animation for Graphics
Clip art and drawing objects can be animated.

What Is Animation? Animation is a sound or visual effect that can be added to text and graphics in a presentation. Animation can be added to text or graphics in a PowerPoint presentation. When adding animation, be careful not to add too much to each slide. Animation can make your presentation more interesting, but too much can ruin a presentation by distracting the audience.

For example, animation could be very effective in the following scenario: You work in the marketing department of a company that advertises its products on national television. You and your team have conducted research about your customers' viewing habits, and it is your job to present the findings to the rest of the department. Your goal is to convince the department that you have identified the best time slot for your commercials.

As you develop your presentation, you consider ways to present the data about television viewing habits. You could use a line graph or a bar graph, but you want something that is guaranteed to catch the attention of the audience. You decide to present the data in the form of a graph built up of small television images, with portions of the data appearing as you talk about the morning, midday, afternoon, evening, prime-time, and late-night time slots. To make this happen on a single slide, you can use animation.

Footers A footer shows a piece of important information at the bottom of each slide. The footer is a good place to include information such as the topic or date of the presentation.

✓ Concept Check

What do you think is the purpose of adding animation to a presentation? How might it enhance a presentation? Give at least two examples and describe the general guidelines to follow when deciding whether or not to add animation.

CHAPTER 5

DIGITAL Dimension → Social Studies
INTERDISCIPLINARY PROJECT

Elected Officials

Using the digital input skills you have learned in this chapter, record information about four elected officials.

Investigate Collect the names and addresses of four representatives in your local area or at the state level. You can find this information on the Internet, in newspapers, and in magazines.

Add the names to the Address Book application of your PDA. Include the following data about each representative:

- Political party
- Length of term
- Re-election date
- Any bills or legislation he or she participated in—that is, voted on, promoted, lobbied for, or wrote

Use the PDA When you begin to add the names to Address Book, look for a place to add extra information about a representative.

Does Address Book have a special place where you can add narrative information about an individual? What kind of narrative information would be most important to add? If so, where is it?

List the steps to add narrative information in Address Book. Make sure the steps are clear and easy to follow. Beam the steps to another student to check that you have included all the necessary information.

E-Portfolio Activity

PDA Skills

To showcase the new digital input skills you've learned in this chapter, use Note Pad to create a list of your skills with a PDA. Include skills such as: use Address Book, use Notes feature, and so on.

- Give a title to your document.
- Include a graphic or photo.

Save the document as an HTML file for your e-portfolio.

Self-Evaluation Criteria

Have you included:

- Name and address?
- Length of term?
- Re-election date?
- Any bills or legislation he or she participated in?
- The name of a place in Address Book to add narrative information?
- The steps to add narrative information in Address Book?

150

DIGITAL Career Perspective

Education and Training

Thelma King
Professor

Teaching Students with Technology

"What did we do before e-mail?" asks associate professor Thelma King. "It is economical and convenient. I use e-mail every day." King teaches college students. She even teaches an online class.

"I use the e-mail system to communicate with students, peers, and associates, as well as send file attachments," says King. In addition to her teaching responsibilities, King advises students and visits schools to observe her students that are interning or student teaching.

As part of her professional development requirements with the university, King attends and leads workshops at conferences. She finds that taking a notebook computer and handheld when she does presentations at conferences or conducts workshops is convenient. King is also required to do research on the Internet and publish articles related to teaching. "I also use the computer's Internet connection to conduct research using library resources," says King.

"Online courses are increasing in popularity and are similar to classroom-based instruction but do provide a few challenges. Because you do not see your students regularly, you have to compensate by finding other ways to communicate. With e-mail, blogs, text messaging, and discussion groups, this is not a big problem," says King.

King uses many kinds of digital communication tools. She uses a desktop computer, handheld, Tablet PC, and notebook computer, along with scanners, copiers, printers, digital cameras, Web cameras, and video cameras. These devices "make my job better because I can enter things quickly, copy things, and make changes effortlessly. Also the fact that Microsoft Office applications features are similar means I don't have to learn so many commands," states King. "Windows makes it easy to click on icons and not have to remember commands."

Why does King enjoy being a professor? "I like being on the cutting edge and learning about and using new technology," states King. "Knowing that I make a difference in people's lives and sharing my knowledge and experience is rewarding."

Training
Requirements for positions vary widely. At a four-year college, a doctoral degree is usually required. At career or technical institutes, relevant work experience is the most important factor. Experience as a graduate teaching assistant is also helpful. Teaching certifications are often required and must be updated periodically.

Salary Range
Typical earnings are in the range of $35,000 to $85,000 a year.

Skills and Talents
Professors need to be:
- Good communicators, in order to work with and motivate students effectively
- Up to date on advancements in their field
- Willing to work on a flexible schedule
- Able to balance teaching and publishing responsibilities
- Able to work without direct supervision

Career Activity
What qualifications are required for professors who teach your favorite subject?

Chapter 5 Review

Read to Succeed PRACTICE

After You Read

Create a Personal Study Aid Create your own study aid by drawing a line down the middle of a sheet of paper. Write **Questions** on the left side and **Answers** on the right side. Find the "You will learn to:" section of each lesson. Write the bullet points as questions on the left side of your paper. For example, for Lesson 5.1, you could write "What are two methods of entering data into a handheld?" Then answer the questions on the right side of your paper. Do this for each lesson.

Check your answers by reading the lessons or asking your teacher. To study, cover the answers with another piece of paper. Look at the questions, practice answering them, and you have created your own study aid!

Example:

Questions	Answers
What are two methods of entering data into a handheld?	With a stylus and with the soft keyboard

Using Key Terms

Understanding the terms to use with PDAs is important. See how many of these terms you know by matching each one with the correct statement.

- stylus
- Memo Pad
- beam
- synchronization
- categories
- Note Pad
- To Do List

1. To send data from one PDA to another via an infrared beam.
2. An application on a PDA that allows you to create short notes or simple diagrams.
3. An application on a PDA that allows you to create simple documents.
4. A penlike device used to select icons on a PDA screen or to write on a PDA screen.
5. An application on a PDA that allows you create, keep track of, and prioritize important tasks and events.
6. Used to organize data and are similar to folders on a desktop computer or to files in a file drawer.
7. The process of transferring data between a PDA and a computer.

Application 2

Directions Add a new slide, and enter the text shown below.

```
Over 20 million personal
computers became obsolete in
1998 and only 13% were reused or
recycled, according to the U.S.
Environmental Protection Agency.
```

Application 3

Directions In the slide you created in Application 2, use an AutoShape such as an oval or rectangle to create a background for a slide title. Insert a WordArt title, and place it on top of the AutoShape. Add an appropriate clip art image to the slide. Save the presentation.

> **RUBRIC LINK**
> Go to the Online Learning Center at **digicom.glencoe.com** and click Lesson 12.2, Rubric, to find the application assessment guide(s).

Digital Decisions

Six-Slide Presentation

Teamwork Work with two other students to create a PowerPoint presentation that is at least six slides long about one of your favorite activities, such as a hobby or sport.

Internet Investigation Search the Internet for information to include in your presentation.

Critical Thinking As a group, decide how you will divide the information among the slides. You should use all of the features you have learned about in Lessons 12.1 and 12.2, pages 395–406. Review these lessons, if needed. Each person in your group will create at least two slides. Follow the presentation tips that you have learned.

DIGITAL DIALOG

PowerPoint Online!

Teamwork Find a partner to work with. Start a new presentation. If the Getting Started task pane does not appear when you open your new presentation, click the Home icon to open the Getting Started screen.

Click the link that says *Get the latest news about using PowerPoint*, and you will be connected to the Internet. Navigate to the section on Working With Graphics and Charts.

Click Shapes, Pictures, and Drawing Objects. Choose one of the topics to read about and then discuss with the rest of the class.

Writing Create a PowerPoint presentation about what you have learned to share with your classmates.

Chapter 5 Self-Assessment

Take a moment to review what you have learned in this chapter. Rank your understanding of the topics below.

4 means, "I understand all of this."
3 means, "I understand some of this."
2 means, "I understand very little of this."
1 means, "I don't remember this."

To use a printout of this chart, go to **digicom.glencoe.com** and click on **Chapter 5, Self-Assessment.**
Or:
Ask your teacher for a personal copy.

Rank Your Understanding

Lesson	Topic	4	3	2	1
5.1	• Four reasons to use a PDA				
	• Two methods to enter data on a PDA				
5.2	• Uses of Memo Pad				
	• How to create a new memo				
5.3	• Two uses of beaming				
	• How to beam a memo				
5.4	• Uses of categories in PDAs				
	• How to create new categories				
5.5	• Uses of Note Pad				
	• Six features of Note Pad				
	• How to create a new note in Note Pad				
	• How to beam a note to another PDA				
5.6	• How to create an entry in Address Book				
	• How to create a business card from an address				
5.7	• Five features of Date Book				
	• Two advantages of Date Book				
	• How to create an event in Date Book				
	• How to copy, paste, and delete events				
	• How to view Date Book formats				
5.8	• Uses of To Do List				
	• How to add items to To Do List				
	• How to use the Details feature				

If you ranked all topics 4, congratulations! Consider doing a quick review.
If you ranked yourself 3 or lower in any topic, consider reviewing these topics first.

digicom.glencoe.com

DIGI Byte

Proportional Resizing
When resizing a graphic, drag one of the corner selection handles to keep the picture in proportion.

Suite Smarts

Clip art that you use in PowerPoint can also be used in Excel, Word, and Access.

Step-by-Step *continued*

3. Insert a Curved Left Arrow to the right of the Curved Right Arrow. Make sure that the arrows are approximately the same size.

4. Move the arrows apart to leave space in the center of the slide. Try to keep the tops and bottoms of the arrows aligned.

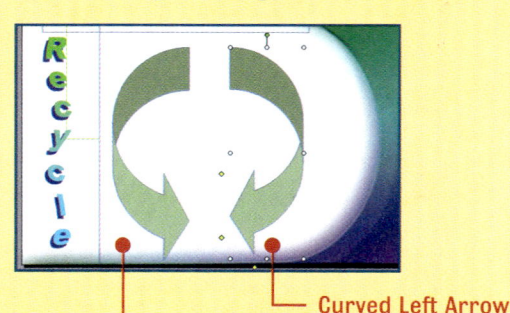

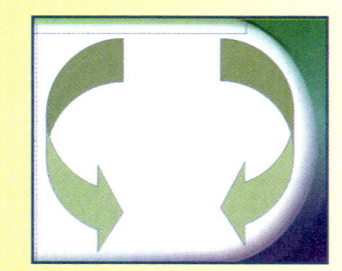

— Curved Left Arrow
— Curved Right Arrow

5. Select the Curved Left Arrow, and click the **Draw** menu in the Drawing toolbar. Choose **Rotate or Flip** and click **Flip Vertical**.

6. Click the **Text Box** button in the Drawing toolbar and create a text box between the two arrows.

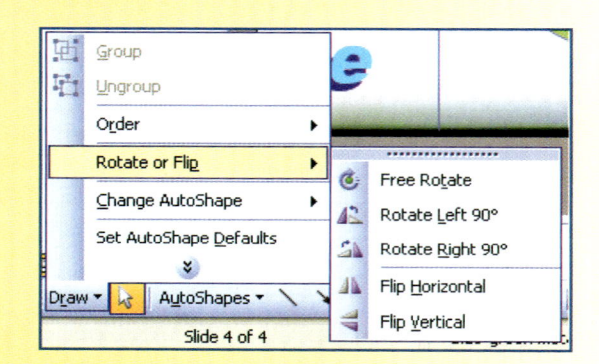

7. Enter the text shown below in the text box. Key each word on a separate line in the text box.

```
Cell phones
Batteries
Monitors
Ink cartridges
Computers
```

8. Format the text in the text box so that it is centered, bold, and a 28-point font size.

9. Save your file.

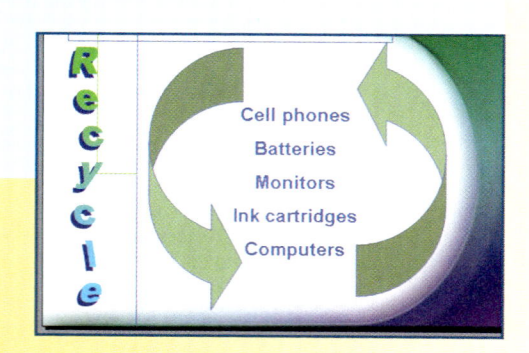

Chapter 12 Presentations

digicom.glencoe.com

Lesson 12.2

405

CHAPTER 6
Speech Recognition Tools

Lesson 6.1	Prepare to Use Speech Recognition
Lesson 6.2	Use Voice Command Mode
Lesson 6.3	Navigate and Edit Text
Lesson 6.4	Add Punctuation and Make Corrections
Lesson 6.5	Use the Scratch That and Tab Commands
Lesson 6.6	Dictate Numbers
Lesson 6.7	Dictate Special Symbols and Emoticons

You Will Learn To:

- **Configure** the microphone and set up and train a user profile.
- **Dictate** text.
- **Use** Voice Command mode to perform computer operations, such as accessing menus, saving files, and printing.
- **Edit** and format text.
- **Dictate** numbers, symbols, and emoticons.

When Will You Ever Use Speech Recognition?

Can you imagine operating a computer without a keyboard? With speech recognition software, you can enter text three or four times faster than with a keyboard! Using speech recognition can increase your productivity at school, at home, and on the job.

153

Step-by-Step *continued*

7. Resize the text by dragging the center selection handle at the bottom of the WordArt until the text stretches down nearly the entire length of the slide.

8. Close the WordArt toolbar by clicking the **Close** button in the top-right corner of the toolbar.

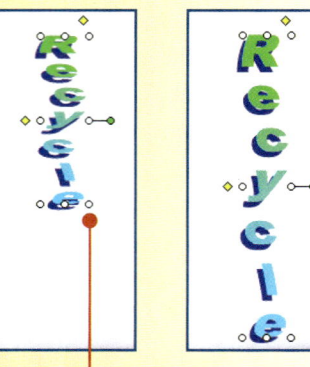

Corner selection handle

Close button

9. Drag the WordArt to the left side of the slide. Click in an empty portion of the slide to deselect the WordArt.

Insert and Format AutoShapes

1. Click **AutoShapes** and select **Block Arrows** from the Drawing toolbar. Select the **Curved Right Arrow**.

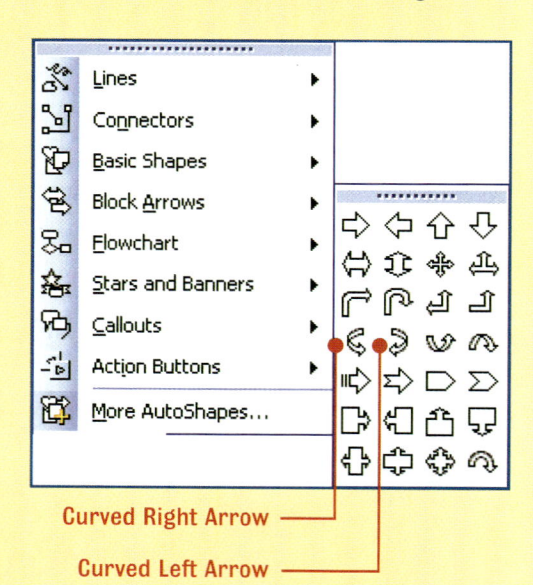

Curved Right Arrow

Curved Left Arrow

2. Notice that the pointer takes the shape of a plus sign. Position the pointer to the right of the word *Recycle* and drag to create a large arrow next to the WordArt.

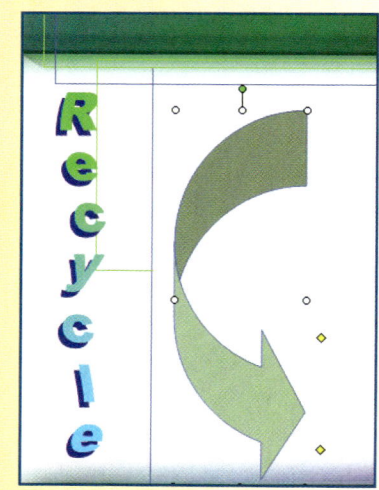

Activity 11

DIGI Byte

Resizing

You can resize AutoShapes in the same way you resize WordArt.

Continued

Chapter 12

404

Read to Succeed

Key Terms

- dictate
- user profile
- enunciate
- Dictation mode
- Voice Command mode
- Correction feature

Check Your Understanding

As you read, stop at the end of a sentence or paragraph and ask yourself questions to check your understanding, such as:

- Does this make sense?
- How would I say this in my own words?
- What part, or words, do I not understand?
- Have I forgotten what I just read?

Create a graphic organizer similar to the illustration to help you check your understanding of each lesson as you read this chapter. Fill in the chart as you read the lessons, and apply any "fix-it" strategy needed, or make your own.

Example:

Page Number	Topic	In My Own Words
155	Prepare to use speech recognition	To use speech recognition software, I must dictate clearly so the software will understand my words.

DIGITAL CONNECTION

Saved by Speech Recognition

You break your arm skateboarding and find out that you will be in a cast for three weeks. Unfortunately, you broke the arm you write with. You have midterm reports due in both your history and computer classes. Keyboarding with one hand is difficult and takes a lot of time. If you cannot write or use a keyboard, how are you going to complete the reports? Using speech recognition software, you can operate your computer by speaking commands and text and complete your reports in time!

Communication Activity

Time yourself while you key the paragraph above. Now time yourself while you read the paragraph out loud. Share your times with the class. In your class, what is the average time to key the paragraph? What is the average time to read the paragraph out loud?

Step-by-Step *continued*

Title Only layout

② Resize the graphic by dragging a corner selection handle to enlarge the graphic.

③ Save your presentation.

④ Move to the second slide, and add at least one clip art graphic. Choose a graphic that is appropriate for the slide text.

Use the Drawing Toolbar and Insert WordArt

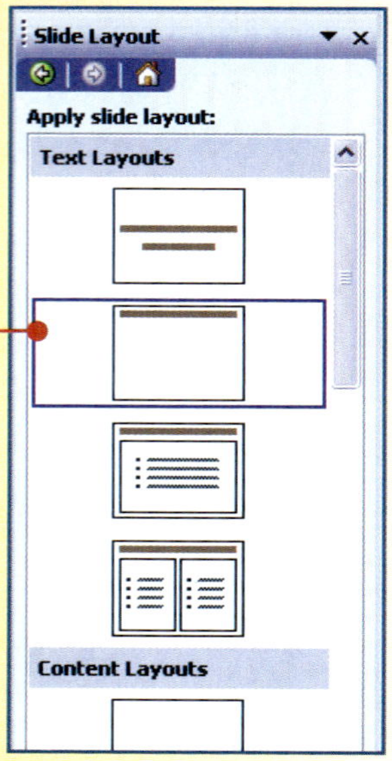

① Create a new slide and choose the **Title Only** layout.

② Examine the Drawing toolbar at the bottom of your screen. Slowly move the pointer over each button to see the names of the buttons.

③ Click the border of the title box at the top of your slide to select it, and then press DELETE.

④ Click the **Insert WordArt** button on the Drawing toolbar.

⑤ From the WordArt Gallery window, choose the green vertical text style and click **OK**.

⑥ In the Edit WordArt Text window, enter the word *Recycle* and click **OK**.

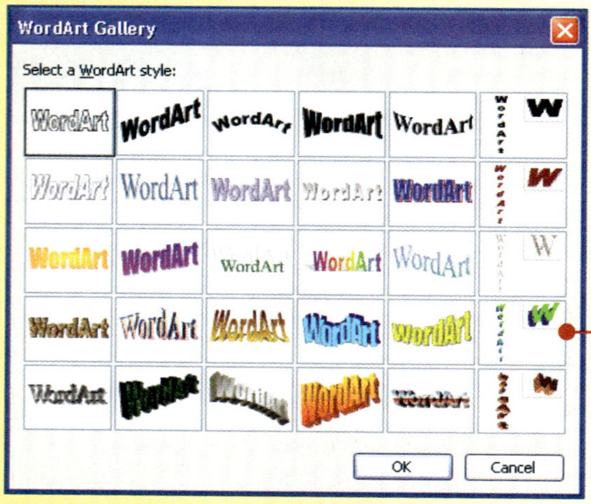

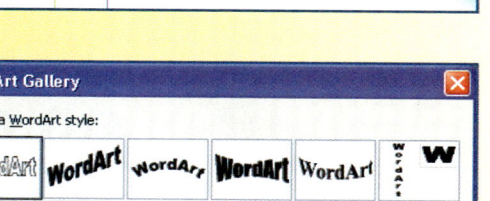

Green vertical text style

Activity 10

DIGI*Byte*

Drawing Toolbar
If the Drawing toolbar is not visible, click the View menu, choose Toolbars, and then select the Drawing toolbar.

Text Boxes
The text boxes in a slide can be deleted or resized as needed.

Suite Smarts
After an object or text is selected, it can be deleted by pressing the Delete or Backspace keys.

Continued

Lesson 6.1

Prepare to Use Speech Recognition

You will learn to:
- **Configure** the microphone.
- **Create** a user profile.
- **Deactivate** the microphone.
- **Retrain** a user profile.
- **Dictate** text.
- **Edit** text with voice commands.

dictate To speak words for a computer to enter as text in a document.

user profile A record kept on a computer of the way a user pronounces words and the tone and speed with which the user speaks.

enunciate To pronounce words clearly, including the beginning and ending sounds of the words.

DIGI Byte

User Profile Location
Your user profile can be accessed later by clicking Tools and then Current User from the Language bar.

Who Can Use Speech Recognition? Everyone can benefit by using speech recognition software. This type of software allows you to **dictate**, or to speak words for a computer to enter as text in a document. It takes less time to learn how to dictate text than it does to learn how to key a document. Speech recognition software also allows you to speak computer commands. People with physical handicaps can use speech recognition to operate a computer and create a variety of documents efficiently.

User Profile The key to successful speech recognition is to train the computer to recognize your individual speech patterns. You do this by creating a user profile. A **user profile** is a record kept on a computer of the way a user pronounces words and the tone and speed with which the user speaks. When you create a user profile, it is important to **enunciate**, or to pronounce words clearly, including the beginning and ending sounds of the words.

Each time you use speech recognition, you will need to choose your user profile. As you become more proficient in your dictation, the efficiency of your speech recognition software will also improve. The software will learn and adjust to match your speech patterns.

✓ Concept Check

What are some of the advantages of using speech recognition? Why is it important to always speak clearly?

Step-by-Step

Activity 8

If you need additional help on this topic, open the Help menu; or refer to the documentation for your hardware.

Insert Clip Art into a Slide

1. Open the **E-Cycle** presentation you created in the last lesson, and move to the last slide you added using the slide thumbnails in the Outline pane.
2. Select the **Insert** menu, choose **Picture**, and then click **Clip Art**.
3. In the Clip Art task pane, enter *computer* in the Search For box and click **Search**.
4. Use the scroll bar to review the graphics, and click one to select it.
5. Click the **Close** button in the top-right corner of the Picture toolbar, which automatically appears when you insert a graphic.

Format a Graphic in a Slide

1. Position the pointer over the picture. When the pointer becomes a four-sided arrow, drag the graphic to the bottom of the slide so that it is not covering any text.

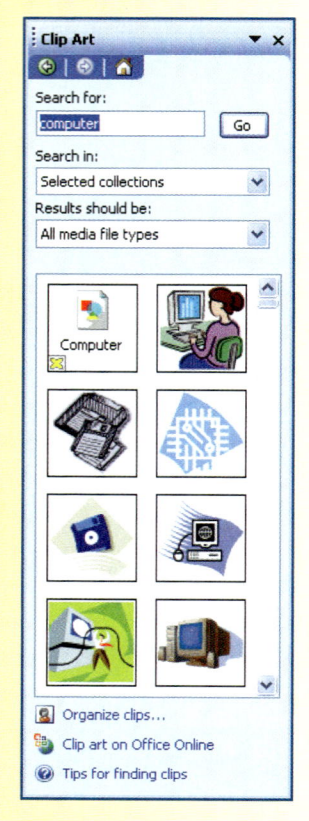

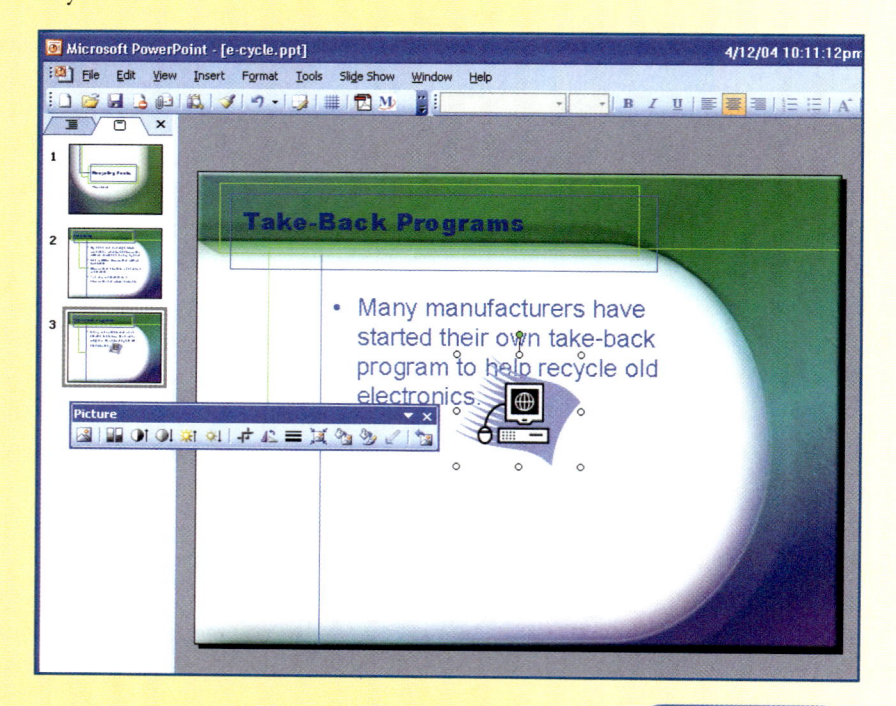

Activity 9

DIGI *Byte*

Simplicity

Do not add too many graphics to a presentation. They will distract your viewer from the content.

Continued

Chapter 12

402

Activity 1

Step-by-Step

Position the Microphone

1. Position the headset with the microphone to the side of your mouth, not directly in front of your mouth. The microphone should be about one inch away from your mouth to avoid interference when you breathe.

2. If there is a mute button on your headset, make sure it is off. If there is a volume control, set it midway between low and high.

Activity 2

Create Your User Profile

Before you begin using speech recognition software, you must complete a training tutorial so the computer can learn to translate your voice into text.

The first time you use the Speech feature in Microsoft Word, you will be prompted to begin the tutorial. It is important to complete the tutorial so that you can train the computer to understand the way you speak.

When you take the tutorial, speak clearly and talk at a normal speed as you dictate. Do not whisper or shout. Say each sentence without stopping. Speech recognition software recognizes words best within the context of a sentence.

? If you need additional help on this topic, open the Help menu; or refer to the documentation for your hardware.

1. Start Microsoft Word.

2. Select the **Tools** menu and click **Speech** if it is not already selected. A checkmark next to Speech indicates that it is selected.

3. Notice the Language bar in the Word title bar.

Language bar

[Microphone | Dictation | Voice Command | Tools | Handwriting]

4. On the Language bar, click the **Tools** button and select **Options**.

5. Click **Advanced Speech**, and then click **New…**.

6. Enter your name in the Profile text box and click **Next**.

7. Follow the instructions on the screen to complete the Microphone wizard. You will be prompted to read several passages to create a custom speech profile and adjust the volume of your microphone.

8. Choose **Apply** and then **OK**.

You can also create a user profile by clicking the Start menu, selecting Control Panel, and then double-clicking the Speech icon. From the Speech Recognition tab of the Speech Properties window, click New. Follow the instructions to complete the Profile wizard.

DIGIByte

Integrating Skills

Use the keyboard and mouse as needed. To be truly proficient, you will need to utilize all three forms of input: speech recognition, keyboard, and mouse.

Continued

Lesson 12.2 — Add Graphics to Presentations

You will learn to:
- **Add** graphics to a slide.
- **Move** and resize graphics.
- **Insert** WordArt, AutoShapes, and text boxes.

Graphics Make a Presentation More Interesting Adding graphics or digital pictures to a PowerPoint presentation is easy to do. Pictures can be downloaded from the Internet, scanned with a scanner, or inserted from a clip art collection or a digital camera. Graphics can be added when a slide is created or at any time later on.

Graphics can have a positive or negative effect on a presentation. Graphics enhance a presentation when they match the content or convey a related idea.

To make sure you are using graphics effectively, follow these guidelines:

- Use graphics that are related to the presentation topic.
- Use graphics that convey a presentation idea visually.
- Be sure that graphics and text are balanced on the slide.
- Be sure that graphic colors match or complement the slide colors in a way that makes the presentation appealing.
- Do not use graphics that are not related to the presentation topic simply because they are appealing or funny.
- Do not use graphics that are distracting or make the presentation confusing.
- Do not let graphics crowd the slide or get in the way of text.
- Do not use too many graphics.
- If graphics are detailed, be sure that they are large enough to be clear to the audience.

Create Shapes The Drawing toolbar allows you to create symbols and shapes and add them to a slide. The shapes can be sized and shaded. Color can be added to the outline of the shape as well as the interior area of a shape. The shapes allow you to add your own artistic design to a slide.

✓ Concept Check

How have you used clip art? Explain why you used the clip art. Why do you think it is important to add clip art to slides?

DIGI Byte

Clip Art Online

Many clip art pictures can be found on the Internet and downloaded to the Microsoft Clip Organizer or to a folder you create. Microsoft's Web site provides thousands of free images that can be downloaded and used in Microsoft Office applications.

Step-by-Step *continued*

Turn the Microphone Off and On

1. If you need to pause during speech recognition, turn the microphone off by clicking the **Microphone** button on the Language bar.
2. Click the **Microphone** button again to turn the microphone back on.

Continue Training Your Profile

1. On the Language bar, click the **Tools** button and choose **Training**.
2. Choose another selection to read.
3. Follow the instructions to complete the training exercise. Click **Finish** when you are done.

Use the Period and New Line Commands

In creating and training your user profile, you have practiced dictating text for the computer to enter. When you dictate text in a document, you need to tell the computer when to punctuate sentences with periods and when to create new lines. To do this, you can use the Period and New Line commands. See below for some other common commands.

Common Commands

Command	Action
New Line Enter Return	Moves the insertion point to the next line of a document.
New Paragraph	Moves the insertion point down two lines.
Select Paragraph	Selects the paragraph where the insertion point is located.
Select All	Selects everything in a document.
Select (word)	Highlights any word that is said after the Select command.
Backspace	Moves the insertion point back one space and deletes a character; this command will also delete any highlighted text.
Delete	Moves the insertion point forward one space and deletes the next; this command will also delete any highlighted text.
Delete Next Word	Deletes the word to the right of the insertion point.
Delete Previous Word	Deletes the word to the left of the insertion point.
End of Line	Moves the insertion point to the end of a line.
Beginning of Line	Moves the insertion point to the beginning of a line.

Activity 3

Activity 4

Activity 5

DIGI *Byte*

Headset Controls

Check your headset for an on/off or mute button that you can use to temporarily turn off the microphone.

Suite Smarts

The Language bar will appear in other Microsoft Office programs. You can add buttons to or delete them from the Language bar. These buttons provide quick access to speech recognition features.

Continued

21ST CENTURY CONNECTION

Company Benefits for Families

Maintaining a full-time job and raising a family can be a challenging lifestyle. Many large companies provide resources for workers with families. As children begin school, after-school activities demand more time from parents. Some larger companies provide day care facilities for employees to use when children are young and not yet in school. These day care facilities are usually located in the same building or one nearby, making it convenient for parents to drop off and pick up their children. This also gives parents the opportunity to visit with their children during the day or to have lunch with them.

School Programs Schools often have before-school and after-school programs where children can stay so parents can drop them off and pick them up before and after work instead of having the children spend time alone at home. These programs are very popular with parents and often have a waiting list of children who want to join. The schools provide fun and educational activities for students, as well as occasional field trips.

Flexible Hours Companies often offer employees flexible hours so employees can adjust the beginning and end of their day. For example, instead of working from 8:00 A.M. to 5:00 P.M., an employee could choose to work from 8:30 A.M. to 5:30 P.M. so he or she could drop his or her children off at school without being late for work. Still other companies allow employees to work at home one or two days a week, if their job responsibilities permit this type of work. This is often favored by employees who have young children or older relatives at home to take care of.

Family Leave The Family and Medical Leave Act was passed into law in 1993. According to the U.S. Department of Labor, businesses with more than 50 employees are subject to the law. The law allows employees to take unpaid leave when they adopt or give birth to a child. Research on businesses that provide this and other benefits for families indicates that providing benefits is good for the company as well as for the individual and his or her family. Employees tend to take fewer sick days, to be more productive, and to stay on the job longer. Investing in family benefits means spending less on new employee training, resulting in cost savings for the company.

Activity Write a short paragraph about how you or someone you know has benefited from one of these programs.

DIGI*Byte*

Additional Training

Additional training will improve the accuracy of speech recognition software. Although the training exercises offer the best way to improve the functioning of the system, the software will also become more accurate as you continue to use it.

Activity 6

RUBRIC LINK

Go to the Online Learning Center at **digicom.glencoe.com** and click Lesson 6.1, Rubric, to find the application assessment guide(s).

Step-by-Step continued

1. Position your microphone and headset. Make sure the microphone is turned on.
2. Start Word if it is not already running.
3. Click the **Dictation** button on the Language bar.
4. Read the sentence below aloud.

> When you use speech recognition, you need to speak in a normal tone and at a normal speed.

5. Say *period*.
6. Say *new line* to move the text insertion point to the next line.
7. Read the sentence below aloud.

> Do not run your words together when you speak.

8. Say *period* and then *new line*.
9. Read the sentence below aloud.

> Try to say each word clearly as you dictate.

Use the Backspace Command

1. Select the word *dictate* with the mouse.
2. Say *backspace*. The selected word will disappear.
3. Say *speak* and then *period*.

Application 1

Directions Start Word and make sure the Language bar is displayed. In the Language bar, click Tools, Current User to check that your profile is active. Adjust your headset and microphone, and dictate the paragraph below. Remember to say *period* to end each sentence. Click the Dictation button on the Language bar and begin reading. Use the backspace command to delete the last sentence after you dictate. Then read one additional selection to train your profile.

> Speech recognition is fun and easy to learn. I can enter text quickly. The more I practice, the better the system recognizes what I say.

Step-by-Step continued

3. Edit the last bulleted item to read:
   - Raw materials in one computer are worth about $4.00.
4. Click the **Slides** tab in the Outline and Slides pane to return to Slide View, from which you can make changes to the text, graphics, or background on all slides at once.
5. On the View Slide toolbar, click the **Slide Sorter View** button. In Slide Sorter View, it is easy to reorder slides or delete slides that are not needed.
6. Click the **View** menu, choose **Master**, and click **Slide Master**. Notice the boxes designating different areas on the slide. Slide Master View allows you to make changes to the entire presentation at one time.
7. On the Slide Master View toolbar, click the **Close Master View** button.
8. Save this file as **E-Cycle** and close the file.

Use Two Methods to Delete a Slide

1. Add two new slides to your presentation.
2. Click on the first new slide in the Outline pane to display it on the screen.
3. Select the **Edit** menu, and then choose **Delete Slide**.
4. In the Outline pane, click the remaining new slide to select it, and press DELETE.

Activity 7

RUBRIC LINK

Go to the Online Learning Center at digicom.glencoe.com and click Lesson 12.1, Rubric, to find the application assessment guide(s).

Application 1

Directions Open your E-Cycle presentation. Add another new slide to the presentation, and insert the text shown below in the slide. Use the Title and Text Slide layout. Delete the bullet on the slide pane before you enter the text. Enter an appropriate title for the new slide, and save the presentation.

Add a new slide, and then delete it.

```
Many manufacturers have started their own take-
back programs to help recycle old electronics.
```

Application 2

Directions Read the passage from Application 1 again. Before you begin, say *new line* to move the insertion point down one line.

Application 3

Directions Dictate the passage below. Use the commands you have learned so far to correct any errors as you dictate or when you have finished reading. Turn the microphone off if you need to pause.

> The computer does not always understand what I say. Sometimes the words that appear on the computer screen do not make sense. As I practice, the computer will learn more about my speech patterns.
>
> I have to be patient and keep practicing to improve. Speech recognition can save me a lot of time. I want to become good at speech recognition so I will not have to depend on the keyboard all of the time.

Digital Decisions

Speech Recognition Programs

You and another employee at your company have been assigned the task of researching speech recognition software.

Internet Investigation Search the Internet for speech recognition programs that will work on your computer.

Critical Thinking Choose a partner to work with and create a chart similar to the one below to display your findings.

Speech Recognition Software

Name of Software	Cost	Operating System Requirements

Digital Dialog

When Is It Good to Use Speech Recognition?

Teamwork With one or two classmates, brainstorm a list of situations in which you think speech recognition would be helpful. Be prepared to explain your ideas. Discuss situations in which speech recognition may not be appropriate or useful.

Writing Start a Positive list and a Negative list, and list the ideas your group generates. Use speech recognition or a PDA to record your answers, if possible. If you use a PDA to record your lists, you can beam your responses to your teacher or other classmates.

Step-by-Step *continued*

Add a New Slide to Your Presentation

1. Click the **New Slide** button [New Slide] in the upper right-hand corner of the screen.

   When you click the New Slide button, the Slide Layout task pane will appear.

2. A new slide will be added to your presentation with the default slide layout setting, Title and Text.

   To choose a different layout when you insert a new slide, click on the layout you want in the Slide Layout task pane. The layouts include areas to insert such items as text, bulleted text, graphics, tables, and charts. You can also insert a blank slide.

Slide Layout task pane

Enter Text into the New Slide

1. In the title text box, enter **E-Cycle**.
2. Click the text box containing bulleted text. Enter the text shown below. Press ENTER at the end of each item.

```
By 2007, the average North American personal
computer will be replaced every two years.

500 million computers will be obsolete.

Computers contain hazardous materials.

The raw materials in a computer are worth
about $4.
```

You do not need to add bullets in this text box. Bullets will appear automatically when you press the Enter key.

Explore PowerPoint Views

1. Click the **Outline** tab in the Outline pane on the top left side of your screen to see an outline of your presentation.
2. Move the insertion point before the last bullet in the slide in Outline view.

Continued

DIGITAL Career Perspective

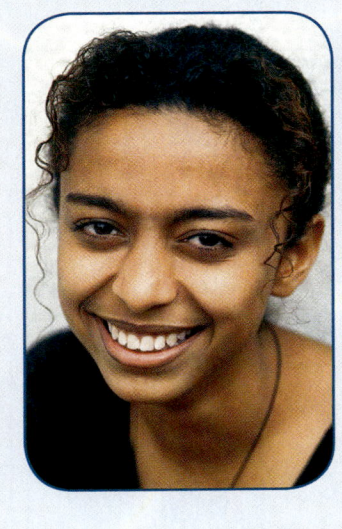

Human Services

Anita Guzman
Fundraising Professional

Helping Nonprofit Organizations to Thrive

When individuals or communities are caught in emergencies or other difficult times, nonprofit organizations, or human services agencies, provide valuable support to help them through. These organizations may arrange for shelter, counseling, job training, healthcare, transportation, home care, or other vital services for people in need. There are also other kinds of nonprofit organizations that support the arts or educational programs. While nonprofit organizations employ licensed social workers or other highly trained staff, administrative employees carry out important tasks that make the services provided by these organizations possible. A fundraising professional is someone employed by an organization to ensure it has the funding necessary to fulfill its purpose.

"I've had a lifelong passion for helping people in tough situations," says Anita Guzman, a fundraising professional for a cancer patient and family support center. "This is what attracted me to a career in community service, where my primary job is to raise money and other resources for the organization. I depend on the Internet and Excel a lot for my job. I use the Internet to research new funding sources and grant application information. Then I organize the information I find into Excel spreadsheets that allow me to record contacts, track deadlines, and generate financial reports."

"People who need specialized services in difficult times appreciate personal communication from caring human services professionals," Guzman says. "An agency's access to digital tools allows everyone in the organization to maintain a high level of efficient, quality service to clients."

Career Growth Trend

Fundraising occupations are expected to grow faster than average. This is because nonprofit organizations are providing more services once offered by government agencies.

Training
A bachelor's degree in any field is required. Volunteering and entry-level positions with nonprofit organizations are a good way to get experience. Graduate degrees may be helpful for career advancement but are not necessary.

Salary Range
Typical earnings are in the range of $60,000 to $80,000 a year.

Skills and Talents
Fundraising professionals need to be:

- Passionate about helping others
- Excellent written and oral communicators
- Highly organized
- Able to work with and supervise others
- Familiar with basic accounting principles
- Persuasive
- Possessed of a high degree of integrity
- Able to act quickly to benefit from possible opportunities
- Familiar with relevant fundraising legislation

Career Activity
Research some volunteer opportunities in your community. In a short paragraph, explain how one volunteer opportunity would help you in a career.

Suite Smarts

The toolbars in PowerPoint are the same ones you have been using in other Microsoft Office programs.

Activity 2

DIGI Byte

Rearranging

The order in which slides appear can be rearranged at any time. In the Outline pane, simply drag and drop the slides into the order you want.

Activity 3

Step-by-Step *continued*

4. Drag the Picture toolbar to the bottom of your screen and place the toolbar next to the Drawing toolbar. Release the mouse button once the toolbar is in position.

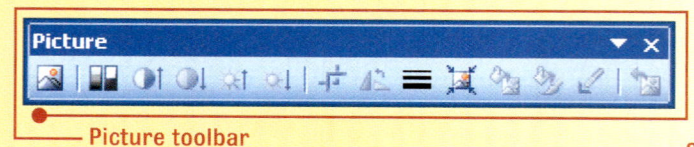

— Picture toolbar

Scroll bar —

Create a New Presentation

1. In the Getting Started task pane, click the **Create a new presentation** link.

2. Click **From design template** to create the presentation based on a design template.

3. In the Slide Design task pane, scroll to find a design you like. Drag the scroll bar down to see other designs.

4. Click the design you choose to insert a new slide with the selected design. Let your mouse hover over a design, and the name of the design will be dispayed.

   By default, the first slide in a presentation is always inserted with the Title Slide layout.

Title Your Presentation

1. Click the text *Click to add title*.
2. Enter the title **Recycling Facts.**
3. Click the text *Click to add subtitle* and key your name.

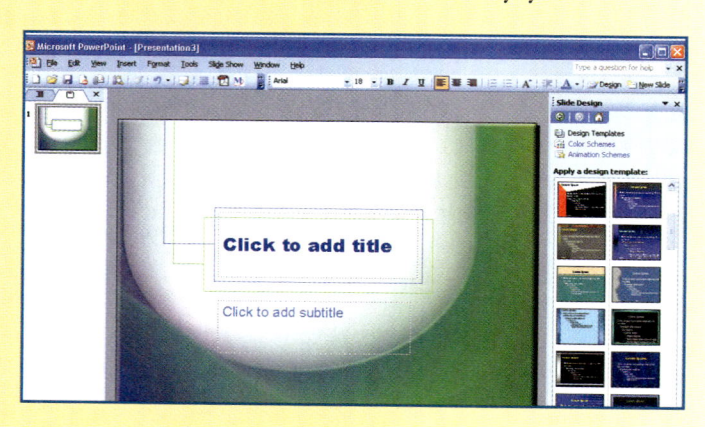

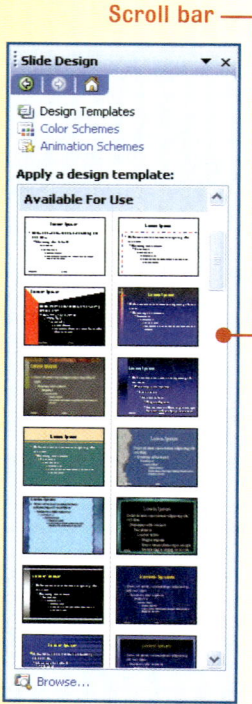

Continued

Chapter 12

397

Lesson 6.2 — Use Voice Command Mode

You will learn to:
- Use Voice Command mode to access menu choices.
- Use Voice Command mode to underline, bold, italicize, and align text.
- **Save** and print in Voice Command mode.

Dictation mode The mode in which the computer enters words and characters on the screen exactly as they are spoken.

Voice Command mode The mode in which the computer uses spoken words as instructions rather than as literal text.

Why Do We Use Speech Recognition? Speech recognition eliminates complete dependency on the keyboard. Programmers have been working with speech recognition for more than 30 years. It is now appearing as part of many everyday technologies, such as automobiles, home security systems, cell phones, computers, and even children's toys. People can learn to dictate at 120 wpm with 95 percent accuracy or greater in a short time.

Programs There are many programs that use speech recognition, including Microsoft Office XP, Dragon Naturally Speaking, IBM Via Voice, and Microsoft Windows XP Tablet. You will learn about Windows XP Tablet in Chapter 7.

Dictation and Voice Command Modes When you use speech recognition, you alternate between two modes of speaking. **Dictation mode** is the mode in which the computer enters words and characters on the screen exactly as they are spoken. **Voice Command mode** is the mode in which the computer uses spoken words as instructions rather than as literal text.

Voice Commands See **below** for some commands you can use in Voice Command mode. Pause briefly before and after each command.

Voice Commands

Voice Command	Action	Voice Command	Action
Cut	Deletes selected text	File, Save	Saves a document
Copy	Copies selected text	File, Print	Sends a document to a printer
Paste	Inserts text that has been cut or copied	File, Exit	Closes a document

✓ Concept Check

When would you use Dictation mode? When would you use Voice Command mode?

Activity 1

If you need additional help on this topic, open the Help menu; or refer to the documentation for your hardware.

DIGI Byte

PowerPoint and Handhelds

Special software can be purchased so PowerPoint presentations can be synchronized with handhelds. In some instances, the handheld can be connected to an LCD projector so that the presentation can be shown directly from the handheld.

Step-by-Step

Display and Hide Toolbars

1. Open PowerPoint.

   The screen is divided into three sections. The Outline and Slides panes are on the left. The Slides pane shows a miniature of each slide in a presentation, and the Outline pane shows the slide text in outline format. The slide pane is in the middle. The task pane is on the right. Notice that the Getting Started task pane is open. It may be helpful to have additional toolbars open while working in PowerPoint.

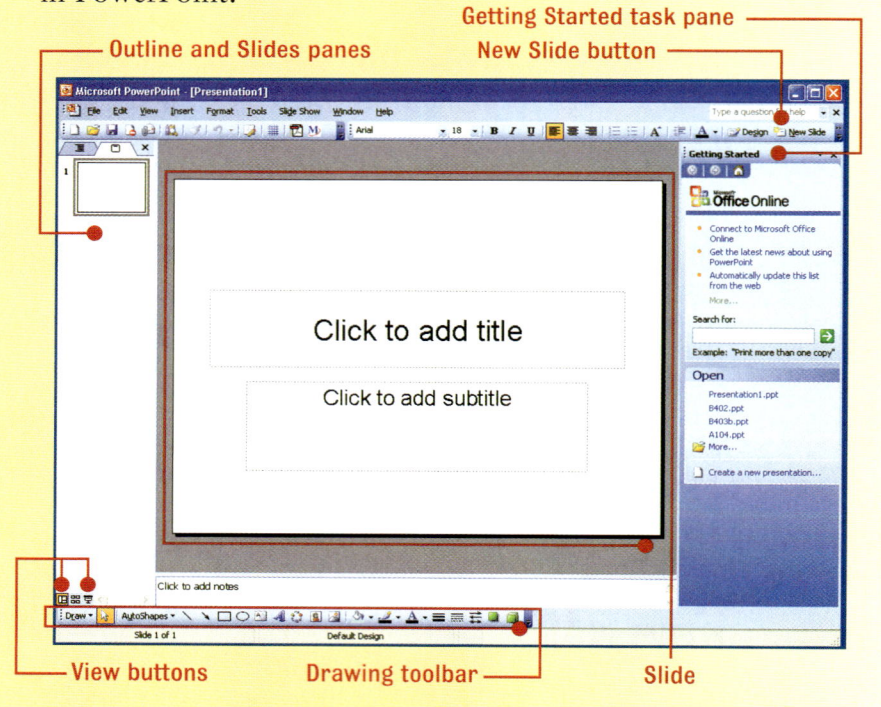

2. Click the **View** menu and select **Toolbars** to see which toolbars are currently showing.

   You will notice that some of the toolbar names in the menu have a check mark next to them. These are the toolbars that are currently being displayed. If you do not want to display a toolbar, click the selected toolbar name again and it will not be displayed.

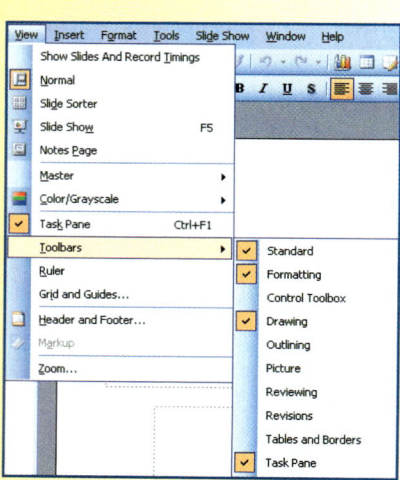

3. Click **Picture** to display the Picture toolbar.

Continued

Chapter 12

396

Activity 7

? If you need additional help on this topic, open the Help menu; or refer to the documentation for your hardware.

Step-by-Step

Access Menus

1. Start Word if it is not already running.
2. Click the **Voice Command** button on the Language bar.
3. Say the following commands to access menus.
   File
   Edit
   View
   Insert
   Format
   Tools
   Table
   Window
   Help
4. Say *escape*, pause, and then say *escape* again.
5. Say *Help*, then *cancel*, then pause, and then say *cancel* again.

You can also click outside a drop-down menu to close it.

Use the Underline Command

Activity 8

1. Switch to Dictation mode if you are in Voice Command mode. To switch to Dictation mode, say *dictation* or click the **Dictation** button on the Language bar.
2. Say the following:
   school, new paragraph
   classroom, new paragraph
   holidays, new paragraph
3. Say *select school*.
4. Say *voice command*.
5. Say *underline*.
6. Repeat Steps 3–5 to underline *classroom* and *holidays*.

To remove underline formatting, say *underline* again.

7. Switch to Dictation made, and read the following sentence aloud. Then follow Steps 3–5 to underline the word *also*.

```
Speech recognition can also be used with a
Tablet PC.
```

DIGI Byte

Select Command
You can use the Select command in both Dictation and Voice Command modes.

Suite Smarts
Speech recognition can be used in other Microsoft applications.

Continued

162

Chapter 6

Lesson 12.1 → Begin a Presentation

You will learn to:
- **Create** a new presentation using a design template.
- **Enter** text in placeholder text boxes.
- **Add** a new slide and choose the slide layout.

slide An image created in PowerPoint that is displayed on a screen as part of a presentation and can contain text, graphics, sound, video, and animation.

DIGIByte

Text Boxes
All text in PowerPoint presentations is entered into text boxes.

Formatting Text
Text in PowerPoint can be formatted using the Formatting toolbar or the Format menu. Select your text before you format it.

Getting Started A PowerPoint presentation is a collection of slides containing text and graphics. A **slide** is an image created in PowerPoint that is displayed on a screen as part of a presentation and can contain text, graphics, sound, video, and animation. When creating a presentation, you need to know how to use the PowerPoint application itself and know the basic principles for creating effective presentations. The most important issue to consider is your audience. You want your presentation to be interesting to them.

Organizing Ideas Each slide should concisely deliver information on a single topic. Limit the amount of text on a slide to about six lines. To help you to organize and condense the information in your presentation, outline the presentation on paper. Think of each sheet of paper as a slide as you make notes about what you will say in your presentation. Find the main point you want to get across in each slide and stay focused on it.

Understand Your Audience As you plan a presentation, ask yourself some questions: How can I make my presentation interesting to my audience? What do they already know, and what do they need to know? These questions will help you empathize with your audience. Empathizing is putting yourself in someone else's place. Empathizing with your audience will help you to create a presentation that gives your audience the information they need in a way that is interesting to them.

Entering Text To enter text on a slide, you enter the text in a text box. A slide may have more than one text box. You do not have to use complete sentences. In fact, sometimes bullet points are the most effective way to get your point across. You can add graphics to help explain the text. You can also format text using different font sizes and colors. These features will make your presentation easy to read and interesting.

✓ Concept Check

What are some advantages of using graphics when you are trying to explain something? How could a PowerPoint presentation be used to explain and define a science experiment? Give an example.

Step-by-Step continued

Activity 9

DIGI Byte

Commands
If you cannot get a command to work in Dictation mode, try it in Voice Command mode.

Use the Bold Command

You can use bold formatting to draw attention to text.

1. Switch to Dictation mode.
2. Say the following:
   vacation, new paragraph
   camp sites, new paragraph
   summer, new paragraph
3. Say *select vacation*.
4. Say *voice command*.
5. Say *bold*.
6. Repeat Steps 3–5 to bold *camp sites* and *summer*.

To remove bold formatting, say *unbold*.

7. Read the following sentence aloud, and follow Steps 3–5 to bold the words *Tablet PC*.

```
We will be learning about the Tablet PC.
```

Activity 10

Use the Italics Command

Italic formatting is another way to make text stand out.

1. Switch to Dictation mode.
2. Say the following:
   swimming, new paragraph
   hiking, new paragraph
   biking, new paragraph
3. Say *select swimming*.
4. Say *voice command*.
5. Say *italics*.
6. Repeat Steps 3–5 to italicize *hiking* and *biking*.

To remove italic formatting, say *italics* again or *unitalicize*.

7. Read the following sentence aloud, and follow Steps 3–5 to italicize the words *speech recognition*.

```
I wish I could use speech recognition in all of
my computer applications.
```

Continued

Read to Succeed

Key Terms
- slide
- animation
- legend
- hard copy

Connect to Prior Knowledge

There may be topics within a chapter that you are already familiar with. A KWHL (kwell) chart can track prior knowledge and tell you which new information to focus on. KWHL stands for the questions to ask as you complete the chart: What do I already **know**? What do I **want** to know? **How** can I find out? What have I **learned**?

Draw a chart similar to the one shown below on a Tablet PC, if possible. Before reading each lesson, ask yourself the first three questions above and fill in the K, W, and H columns. When you complete each lesson, summarize what you learned in the L column.

Example:

Chapter 12, Presentations	K	W	H	L
Lesson 12.1	PowerPoint is a presentation software program	How to create a presentation with text and graphics	Complete all activities and applications	How to add slides and enter text into slides using Outline View

DIGITAL CONNECTION

Presenting Information

Your science club is going to be recognized at the next school board meeting for the outstanding recycling project it has organized in your school. As president of the science club, you have been asked to prepare a five-minute presentation about the recycling program. The science club historian has many digital photos of students working on this project.

Writing Activity

Write a paragraph on your plan for your presentation. How could you display the digital photos of fellow club members in an interesting and eye-catching manner? Think of how you have used graphics in previous lessons.

Step-by-Step *continued*

Activity 11

Use the Center Command

1. Switch to Dictation mode.
2. Read the following aloud.
3. Say *new paragraph*.

The students going on the field trip are

4. Say *voice command*.
5. Say *center*.
6. Say *dictation*.
7. Say the following.
   Mindy, new paragraph
   Tina, new paragraph
   Javier, new paragraph
   Michael, new paragraph
   Yuki, new paragraph
   Yolanda, new paragraph

To resume left alignment, say *align left*.

Activity 12

Use the Align Right Command

1. Switch to Dictation mode.
2. Read the following aloud.

The students going on the field trip are

3. Say *new paragraph*.
4. Say *voice command*.
5. Say *new paragraph, align right*.
6. Say *dictation*.
7. Say the following.
   Mindy, new paragraph
   Tina, new paragraph
   Javier, new paragraph
   Michael, new paragraph
   Yuki, new paragraph
   Yolanda, new paragraph

To resume left alignment, say *align left*.

DIGI Byte

Mode Keys
To quickly switch between Dictation and Voice Command modes, assign mode keys. From the Language bar, select Tools and then Options. Click Mode Key.

Suite Smarts
To access a menu, say the name of the menu when in Voice Command mode.

Continued

CHAPTER 12

Presentations

Lesson 12.1   Begin a Presentation

Lesson 12.2   Add Graphics to Presentations

Lesson 12.3   Add Animation to Presentations

Lesson 12.4   Add Charts to Presentations

Lesson 12.5   Investigate Print Options for Presentations

Lesson 12.6   Convert a Presentation Into a Web Page

You Will Learn To:

- **Create** presentations with design templates.
- **Choose** slide layouts that will work with your text.
- **Add** graphics to a presentation.
- **Add** animation to a presentation.
- **Insert** and **modify** a chart in a presentation.
- **Print** a presentation in various formats.
- **Add** hyperlinks to a presentation.
- **Save** a presentation as a Web page.

When Will You Ever Create a Presentation?

When you have important information to share, you can use Microsoft PowerPoint to create presentations that help capture an audience's attention. A presentation generally consists of a concise outline of information that is broken up onto slides. Slides can contain text, graphics, sound, video, and animation. You use these slides as a guide for your audience as you give a speech.

Activity 13

DIGIByte
Tips for Improving Speech Recognition
- A headset with both earphones and a microphone can help minimize external noise that may interfere with dictation.
- Do not pause between words within a voice command.

Activity 14

Activity 15

RUBRIC LINK
Go to the Online Learning Center at digicom.glencoe.com and click Lesson 6.2, Rubric, to find the application assessment guide(s).

Step-by-Step continued

Use the Align Left Command
1. Switch to Dictation mode.
2. Read the following aloud.

> The students going on the field trip are

3. Say *new paragraph*.
4. Say *voice command*.
5. Say *new paragraph, align left*.
6. Say *dictation*.
7. Say the following.
   Mindy, *new paragraph*
   Tina, *new paragraph*
   Javier, *new paragraph*
   Michael, *new paragraph*
   Yuki, *new paragraph*
   Yolanda, *new paragraph*

Save a Document
1. Say *voice command*.
2. Say *file, save as*.
3. Use the keyboard to enter a file name, such as **Field Trip List**.
4. Say *Save*.

Print a Document
1. Switch to Voice Command mode if necessary.
2. Say *file, print*.
3. Use the mouse to select a printer if necessary.
4. Say *OK*.

Application 4

Directions Dictate a short paragraph on the things you like about using speech recognition so far. Include underlined, bold, italic, centered, right-aligned, and left aligned text. Proofread the paragraph, and make any necessary corrections.

CHAPTER 11

Self-Assessment

Take a moment to review what you have learned in this chapter. Rank your understanding of the topics below.

4 means, "I understand all of this."
3 means, "I understand some of this."
2 means, "I understand very little of this."
1 means, "I don't remember this."

> **To use a printout of this chart, go to digicom.glencoe.com and click on Chapter 11, Self-Assessment.**
> **Or:**
> **Ask your teacher for a personal copy.**

Rank Your Understanding

Lesson	Topic	4	3	2	1
11.1	• Identify fields, records, and entries				
	• Enter data into a database				
	• Design a data table				
	• Edit data types				
11.2	• Sort records in ascending or descending order				
	• Create a query				
	• Search for specific data in a query				
	• Create a report				
	• Edit a report in Design view				
11.3	• Locate downloadable online databases				
	• Download database templates				
	• Edit a downloaded database				
11.4	• Add a database report to a Word document				
	• Copy database information into Excel				
11.5	• Create a Web page from a database table				
	• View a data access page in a Web browser				
	• Navigate a database Web page				

If you ranked all topics 4, congratulations! Consider doing a quick review.
If you ranked yourself 3 or lower in any topic, consider reviewing these topics first.

digicom.glencoe.com

Application 5

Directions Dictate the following text. Use the commands you have learned so far to make any necessary corrections. You may also use the mouse to make corrections.

```
The more you practice speaking to
your computer, the better you will
become at dictating text. Do not
forget to switch to Voice Command
mode when necessary. Some commands
work in both modes.

You may still want to use your
keyboard and mouse to help you
move around in your document while
you are getting comfortable with
speech recognition.
```

Say *voice command*.

Say *file*, *print*, *OK*.

Application 6

Scenario Your new job at the local newspaper is to write the Tech Tip column. Your supervisor tells you that your first story will be about Pocket PCs and speech recognition capabilities.

Directions Research using the key words "Pocket PC" on the Internet.

- Do Pocket PCs have speech recognition capability?
- What kinds of commands can you use with these devices?
- What are some of the advantages of using speech recognition with Pocket PCs?
- What are some disadvantages?

Use speech recognition and the commands you have learned to dictate a paragraph or two that answers the questions above. Use speech recognition to save the document as *Your Name* **Pocket PC Speech Recognition** and to print the document.

Digital Decisions

Portable Speech Recognition Devices

You are the administrative assistant to the vice president of estimating in a large commercial construction company. One of the responsibilities of the vice president is to inspect the progress of the company's various building projects while taking notes and pictures.

The vice president has heard about PDAs that have speech recognition and digital photo capabilities and would like you to find out if this type of PDA would make the inspections safer and more efficient.

Internet Investigation Search the Web sites of office supply stores and PDA manufacturers to learn about PDAs with speech recognition and digital photo capabilities. Investigate PDAs that can also be used as cell phones as well.

Writing Prepare a three-page report on the prices, capabilities, and availability of PDAs and cell phones with speech recognition and camera functions. Include pictures that you find on the Internet, if possible. Decide which PDA or cell phone you would recommend, and explain why in your report.

Use speech recognition to dictate your report.

CHAPTER 11 Review

Read to Succeed PRACTICE

After You Read

Make It Easy—Chunk into Small Parts Studying the material after reading a chapter can sometimes seem a huge task. How about making your task easy by chunking each lesson into smaller parts? Write each paragraph heading from Lesson 11.1 and summarize the main points.

Do the chunking in a similar way for Lessons 11.2–11.5. To help you, use word processing or spreadsheet software or Journal on a Tablet PC. You could also use speech recognition to enter the information.

Example:

Paragraph Headings	Summary
What Is a Database?	A database is a software application that stores a collection of related information and allows easy retrieval of the information.

Using Key Terms

On a separate piece of paper or in Journal, match the key term with its definition. Use your word processing software and speech recognition, if you prefer.

- database (361)
- field (361)
- record (361)
- primary key (364)
- ascending sort (370)
- descending sort (370)
- query (370)
- report (370)

1. A sort of data arranged in alphabetical (A–Z) order or numerical (0–9) order.

2. A database object that enables you to locate records that meet a certain criteria.

3. A field that contains a unique identification for each record in a database.

4. A database object used to present the information in a database in a way that is easy to read.

5. A category that holds one piece of information in a database.

6. A collection of fields for one item in a database.

7. A sort of data arranged in high to low alphabetical (Z-A) order or numerical (9-0) order.

8. A software application that stores a collection of related information and allows easy retrieval of the information.

21ST CENTURY CONNECTION

Develop Good Oral Presentation Skills

One of the most valuable skills you learn in school is the ability to give an oral presentation, or speak to an audience on a topic. You will use this skill for the rest of your life. You already have many of the skills needed to be an effective presenter. These include participating in class or other group discussions, planning a written report, and talking about homework or weekend plans with friends or family. Improving your oral presentation skills will build your self-confidence, develop your overall communication skills, and give you solid training for any career you choose.

Research Choose a topic that interests you or that you would like to learn more about. Find interesting articles, books, and Web sites on the topic. Ask a parent or other adult if he or she has given presentations at work. How did he or she research, prepare, and give the presentation? What digital tools and applications did he or she use? You can take notes on a PDA during the interview.

Prepare Decide which points you will cover in your presentation. As you plan a presentation, ask yourself some questions about your audience. Knowing who your audience is will help you make a presentation that gives your audience the information they need in a way that is interesting to them. How can you make this topic exciting and interesting to them? Create an outline, list, or diagram on your PDA to help organize your presentation. Write a brief sentence or two on each point you plan to cover. Determine the order in which you will discuss the points. Begin your presentation with an interesting fact that will grab the attention of your audience. Create visual aids that will keep your audience interested. You could create a PowerPoint presentation that shows your key points as you cover them or shows digital pictures on your topic. You could also create charts, graphs, or maps on a Tablet PC.

Practice Practice several times in front of a mirror. Refer to the outline in your PDA to guide you through each point. Time yourself with a stopwatch application for your PDA, and stay within the time limit. If your presentation runs too long or too short, remove or add material. Ask a parent, adult, sibling, or friend to be your practice audience. Practice with all of your visual aids. Ask your practice audience if he or she was able to follow your presentation, if you seemed to know your topic, and if your speech was clear and at the right pace.

Present Get plenty of rest the night before. Try to relax, and take a deep breath before you begin the presentation. Stand up straight, face the audience, and make eye contact. Be yourself and be comfortable, while focusing on your presentation. Speak in a conversational tone, as if you are talking to friends or family, and speak clearly, varying your tone.

Activity Think of a topic that you would like to use for an oral presentation. Dictate a short paragraph that explains why you chose the topic.

CHAPTER 11

DIGITAL Dimension → Science
INTERDISCIPLINARY PROJECT

E-Cycling

Using the database skills you have learned in this chapter, create a database of information about e-cycling programs in the United States.

Investigate Collect information for 10 to 20 different companies in at least 5 different states that provide recycling assistance for products that can be e-cycled. Find this information on the Internet and in newspapers, telephone books, and magazines.

Enter the information into a new database that you create and name the database e_cycling.

Include the following information in your database:

- Company name.
- Company address.
- Company Web site.
- Company e-mail.
- Company phone number.
- Type of products collected.

Create a Report After you enter your data, create a report for each type of product collected. Create a report for each state.

- How is this information helpful to you?
- How could you share this information effectively with the people you surveyed in Chapter 10 Digital Dimension, page 356?
- Where could you display this information so more people can find out about e-cycling centers and products that can be recycled?
- What would be the best way to display the information?

Write several paragraphs that include your answers to the questions above.

E-Portfolio Activity

Database Skills

To showcase the new database skills you have learned in this chapter, save the e_cycling database to your e-portfolio. When describing your database to prospective employers, or others, be prepared to tell them how you created the reports.

Self-Evaluation Criteria

Did you include:

- Information for 10 to 20 companies?
- A report for each type of product?
- Answers to the questions to explain how this information could be useful and how it could be shared and displayed?

Lesson 6.3 — Navigate and Edit Text

You will learn to:
- **Navigate** within a document.
- **Open** files and move, copy, paste, and clear text.
- **Use** the Help menu to learn more about speech recognition.
- **Add** words to the speech recognition dictionary.

Improving Dictation As you begin using speech recognition, do not be overly concerned about correcting mistakes. The more you use speech recognition, the better your computer will understand you. You can improve your computer's accuracy by speaking words clearly and distinctly. Talk at a normal speed. Do not pause between syllables in a word.

Do not blend words together. For example, say *go to the school*, not *gototheschool*. As you speak, try to complete each sentence without pausing. You do not have to speak loudly into the microphone. Speak at a normal volume.

When you are not using speech recognition, you should turn off your microphone. Minimizing the Language bar will turn off the microphone.

Navigating a Document You can easily move the insertion point through a document by speaking navigation commands while in Voice Command mode. See the chart below.

DIGI Byte
Use the Mouse
Use the mouse to move the insertion point if you cannot remember the correct command.

Navigation Commands
Move to Beginning of Line
Move to End of Line
Move to Next Line
Move to Previous Line
Move to Beginning of Document
Move to End of Document

✓ Concept Check

List ways to improve the accuracy of your dictation. Can you recall the navigation commands you have learned? Close your book and list as many commands as you can remember.

DIGITAL Career Perspective

Agriculture and Natural Resources

David Burman
Research Biologist

Monitoring and Minding the Environment

David Burman gets paid to do what he loves best—interact with the environment. "It's great!" says Burman. "I love the people I work with. We all share the same passion for the environment." Burman works as a research biologist monitoring the fish population of the Bush River in Maryland. Warm weather keeps Burman outside actively monitoring fish and collecting sedimentation samples.

Burman's favorite piece of digital communication technology to work with is his handheld GPS unit. "A lot of time when you are doing field research, markers or stakes let you know where your site is located, but curious people love to come through and mess with them. Now since we have GPS units, we can use them to find our site."

Communication is important in Burman's department. Colleagues are spread out all over the state. They keep each other informed about new findings by e-mail.

Burman and his colleagues also need to communicate with landowners about research. According to Burman, "If a proposed site is close to someone's property, we write them a letter to let them know what we are doing and ask their permission. The letter is then kept on file as documentation."

Why does Burman like being a research biologist? "I like not having to wear a tie every day and being able to go out into the field and get dirty. I actually do not look at my job as work. I get paid to do what I love to do."

Career Growth Trend

Growth for scientists in the Agriculture and Natural Resources career cluster will be above average and will be driven by environmental regulations. This is because federal, state, and local governments employ about half of all environmental scientists.

Training

The majority of research biologists have a master's degree in biology, and high-level positions require a doctoral degree in biology. Useful courses to take include biology, environmental science, mathematics, and computing. An internship with field experience is also helpful.

Salary Range

Typical earnings are in the range of $40,000 to $60,000 a year.

Skills and Talents

Research biologists need to be:

- Able to work as part of a team with other scientists, engineers, and technicians
- Good communicators and writers, in order to create effective reports and proposals
- Inquisitive, able to think logically, and open-minded, in order to get the best results for their research
- Physically fit and willing to work long hours when doing fieldwork

Career Activity

Find out about a different career in agriculture and natural resources, such as land conservation or farming. What training, skills, and talents are needed?

Step-by-Step

Activity 16

If you need additional help on this topic, open the Help menu; or refer to the documentation for your hardware.

Dictate and Proofread

1. Start Word, if it is not already running.
2. Select the **Tools** menu, and click **Speech** if the Language bar does not load automatically.
3. Dictate the following.

> I can navigate a document using voice commands.
>
> Punctuation marks can also be added to my document.
>
> It takes time to practice dictation and voice commands when you begin using speech recognition. After a short period of training, a user can become very proficient at entering text.
>
> Now that I am getting better at dictation, I am going to use speech recognition all the time.

4. Proofread the text you dictated, and use the keyboard to correct errors.
5. Do not clear your screen. You will use the text you just entered in your next practice.

Activity 17

Navigate a Document

1. Switch to Voice Command mode. Say *move to beginning of document*.
2. Say the following, and observe how the commands move the insertion point.
   Move to next line
   Move to end of line
   Move to beginning of line
   Move to previous line
   Move to end of document

Activity 18

Open a File

1. Switch to Voice Command mode, if necessary. Say *file, open*.
2. Use the mouse to select the **Field Trip List** document that you saved in Lesson 6.2, Activity 14, page 165.
3. Say *open*.

Continued

Activity 22

Step-by-Step *continued*

View the Data Access Page in a Web Browser

1. Minimize the Access program and, from the folder on your desktop, double-click the **Careers HTML Document** file.

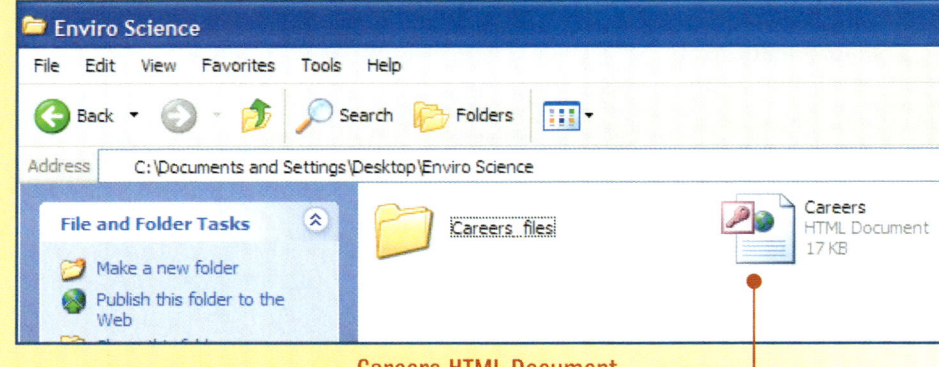

   Careers HTML Document

2. Navigate, or move through, the records in the data access page by clicking the arrows in the toolbar under the data that is displayed.

3. Since this page is an HTML document, you can add it to any Web site.

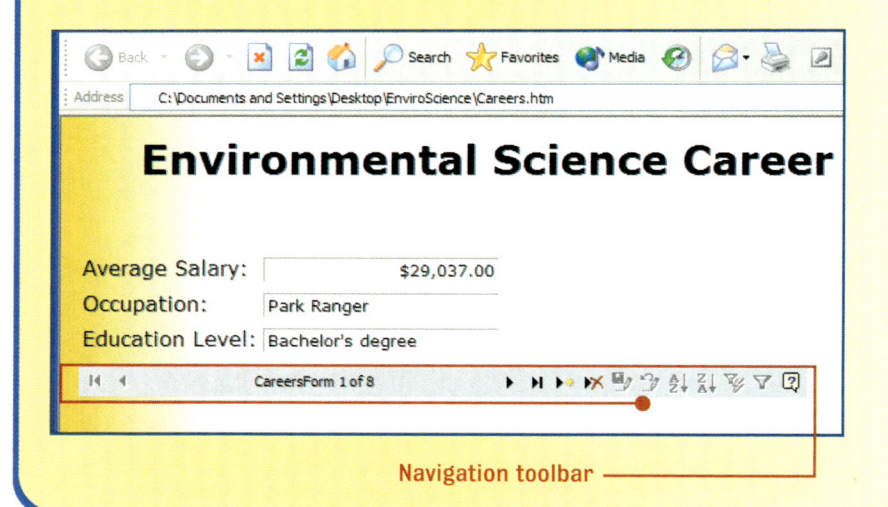

   Navigation toolbar

RUBRIC LINK

Go to the Online Learning Center at **digicom.glencoe.com** and click Lesson 11.5, Rubric, to find the application assessment guide(s).

Application 11

Directions Create a Web page based on the query that lists occupations with average salaries of greater than $50,000 that you created in Lesson 11.2, Activity 10, pages 371–372. Assign a theme to the Web page and then view the page in a Web browser.

Step-by-Step *continued*

Activity 19

Cut and Paste Text

When you cut text, it is deleted from the document and saved temporarily. You paste to insert the text in a new location.

1. Say *select Mindy*.
2. Say *edit, cut*.
3. Say *select Tina*.
4. Say *end, enter, enter*.
5. Say *edit, paste*.

Activity 20

Clear Text

1. Say *select Mindy*.
2. Say *edit, clear contents*.

Activity 21

Copy Text

When you copy text, it is saved temporarily to be pasted in another location.

1. Say *select Michael*.
2. Say *edit, copy*.
3. Say *select Yuki*.
4. Say *end, enter, enter*.
5. Say *edit, paste*.

Activity 22

Use the Help Menu and Add Words to the Dictionary

If you are unsure about some aspect of speech recognition, you can get more information from the Help menu.

1. Switch to Voice Command mode, if necessary. Say *help*.
2. Select **Microsoft Word Help**. Search for *speech recognition*.
3. Select **Format text by using speech recognition**, and read the instructions.

Search for box

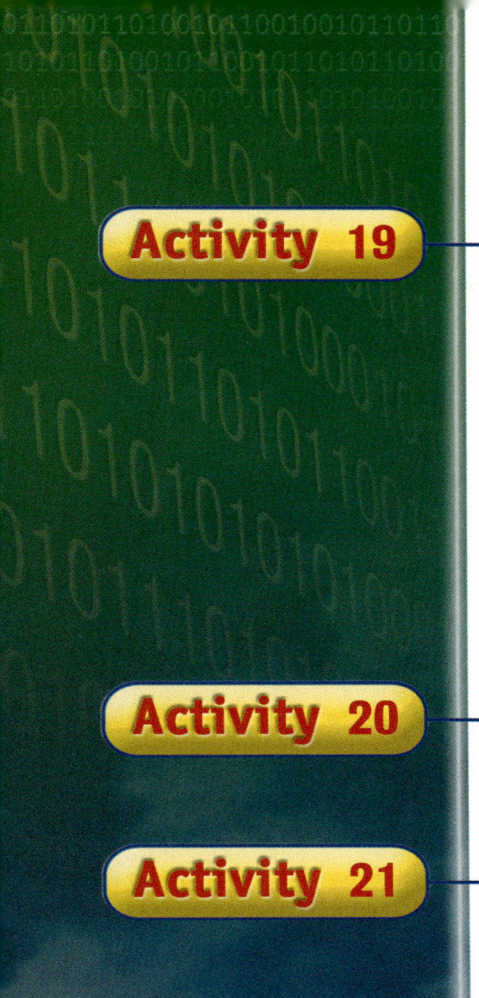

Format text by using speech recognition

Continued

DIGI*Byte*

Clear
Say *select all, edit,* and *clear* to clear the entire document.

Undo
In Voice Command mode, say *edit, undo* to undo the last action.

Chapter 6

170

Activity 21

Step-by-Step continued

Format the Data Access Page

1. Click the text **Click here and type title text** and enter `Environmental Science Career Information`.
2. Close the page. Click **Yes** to save it. 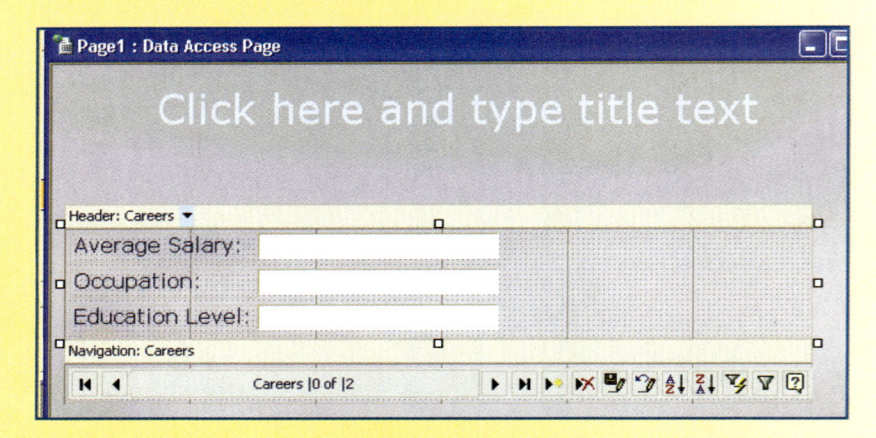
3. Name the file **Careers**.
4. Save the file in the folder you created on your desktop at the beginning of Activity 20, page 385. Click **Save**. 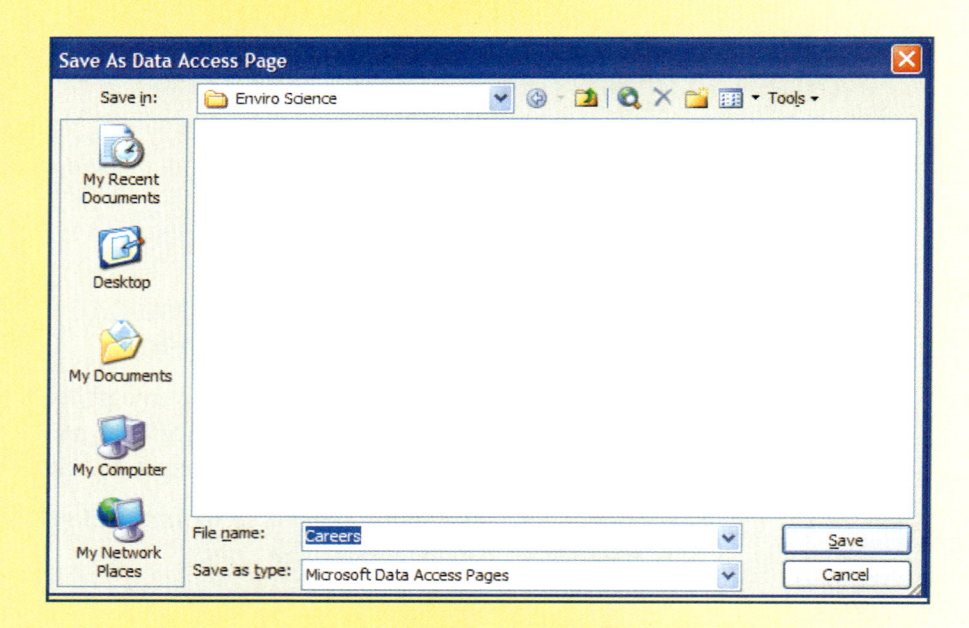
5. Click **Yes** when asked to set this folder as the default location.

Suite Smarts

Word and Excel documents can also be saved as Web pages.

Continued

Suite Smarts
You can use shortcut keys to move through your document.

RUBRIC LINK
Go to the Online Learning Center at **digicom.glencoe.com** and click Lesson 6.3, Rubric, to find the application assessment guide(s).

Step-by-Step *continued*

4. Select **Add to or delete from the speech recognition dictionary**.
5. Click **Add individual words as you dictate**.
6. Read the instructions.
7. Add the following words to the dictionary.
   publicity
   dictionary
   capitalize

   You can also add any other words you have been having a hard time saying.

Application 7

Directions Continue dictating in your document, and do the following:

- Cut and paste one item.
- Clear one item.
- Copy and paste one item.
- Add two more words to the dictionary.

Use navigation commands as you proofread the document, and correct any errors. Switch to Voice Command mode, if necessary, and save the text you have entered. Name the file **Your Name Navigating**. If you have accidentally deleted the text, dictate it again before saving the file. After you have saved the file, use Help to look up two topics of your choice. They can be topics that you have already learned or new topics. Dictate a short paragraph on what you learned from Help. Use speech recognition to open and proofread one other file you have saved.

DIGITAL DIALOG

Speech Recognition Time Line

Internet Investigation In a small group, search the Internet for early versions of speech recognition software. You can use the key words "history of speech recognition" to help you find these early versions.

Writing Create a time line of the information you find. You may also want to search for adaptive technology. Include in your time line the names of all devices and software that you find.

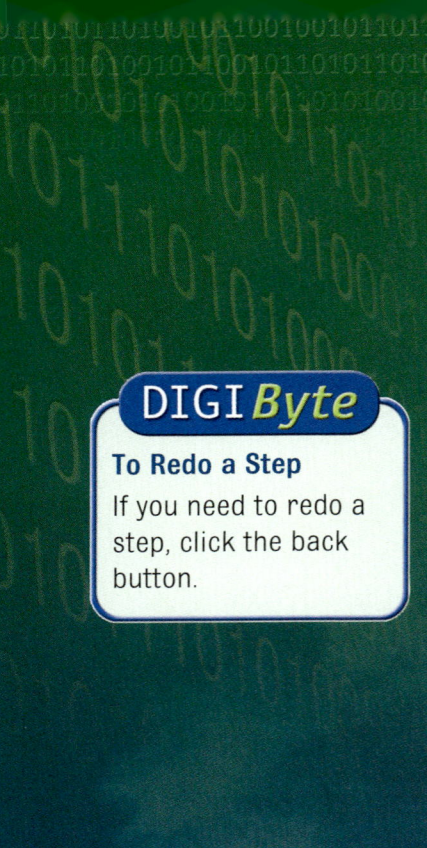

Step-by-Step *continued*

DIGI*Byte*

To Redo a Step
If you need to redo a step, click the back button.

6. Select the **Occupation**, **Average Salary**, and **Education Level** fields in the Available Fields list and click the arrow button to move each field to the Selected Fields list. Choose **Next**. Click **Next** again.

7. In the first drop-down list, choose **Average Salary** and check that the box shows the word Ascending. Click **Next**.

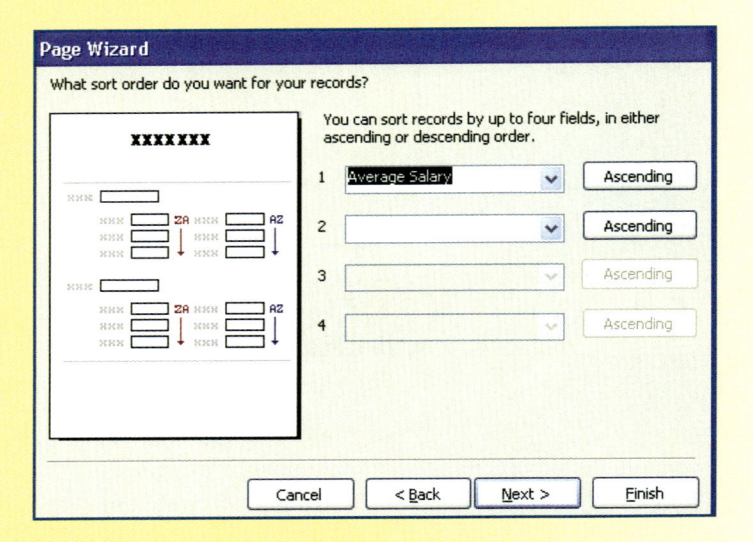

8. Enter Careers as the page title. Select **Do you want to apply a theme to your page?** Click **Finish**.

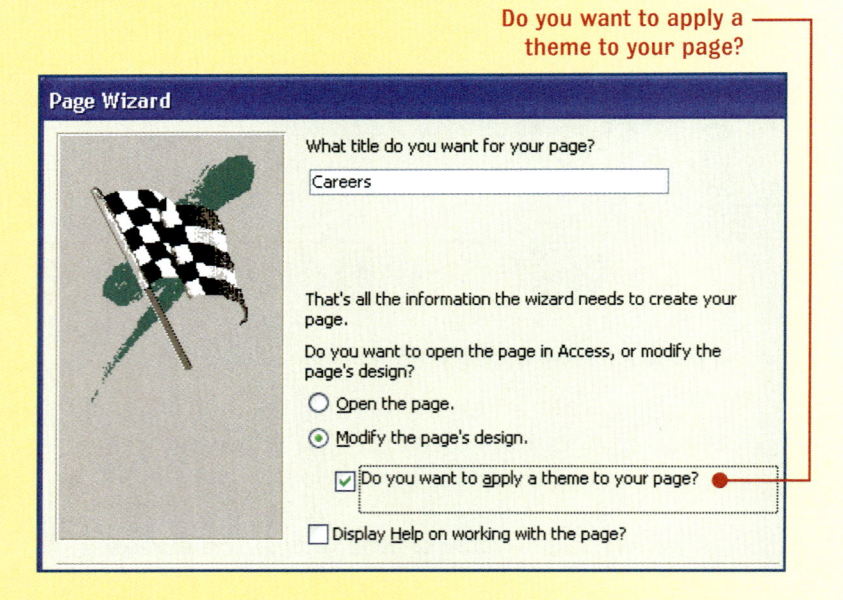

9. Select a theme. Click **OK**.

Continued

Lesson 6.4 — Add Punctuation and Make Corrections

You will learn to:
- **Add** punctuation to your document.
- **Add** bullets.
- Use the Correction feature.
- **Correct** capitalization.
- **Change** case.

Punctuation Commands In this chapter, you have already been using the Period punctuation command. Now you will learn to add other punctuation marks as you dictate text. Pause briefly before saying punctuation commands, and say each command as one complete term. See the figure below. Be careful not to overemphasize individual syllables.

Correcting Text As you work with speech recognition, you will need to make corrections to the text that you dictate. The **Correction feature** allows a user to correct mistakes in dictated text while also recording the corrections as part of the user profile settings. To use the Correction feature, click the Correction button on the Language bar. When you use this feature, the corrections will be stored as part of your user profile settings, improving the accuracy of your speech recognition software.

Correction feature
Allows a user to correct mistakes in dictated text while also recording the corrections as part of the user profile settings.

DIGI Byte

Punctuation
- Pause briefly before you say punctuation you want entered on the screen.
- Punctuation marks can also be entered with the keyboard.

Punctuation Commands

Punctuation Command	Symbol
Backslash	\
Colon	:
Open Quote	"
Close Quote	"
Period	.
Question Mark	?
Comma	,
Dash	-
Semicolon	;
Slash	/
Exclamation Point/Exclamation Mark	!

✓ Concept Check

How can using the Correction feature benefit you in two ways?

Step-by-Step

In this step-by-step, you will create a Web page from an Access database and then view the data access page in a Web browser.

Create a Web Page from an Access Database

1. Create a new folder on your desktop so you will be ready to save the HTML page and its associated resource files in the new folder.
2. Open the **EnviroScienceJobs** database.
3. Click the **Pages** object in the database window.
4. Double-click **Create data access page by using wizard**.
5. Choose **Table: CareersForm** from the Tables/Queries drop-down list.

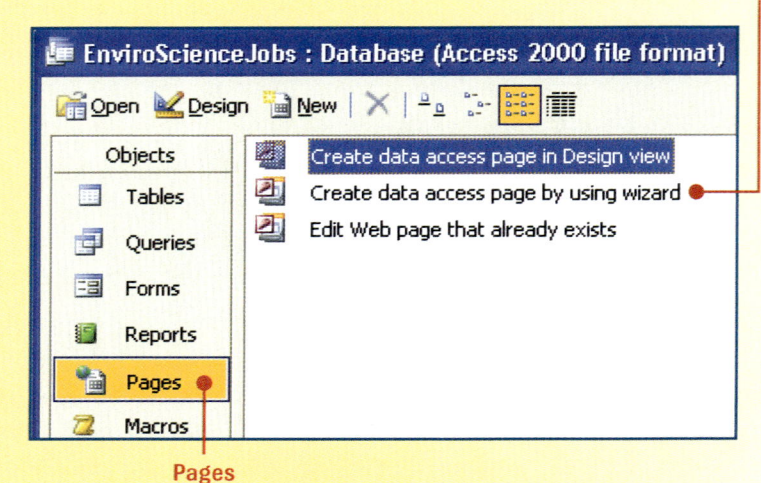

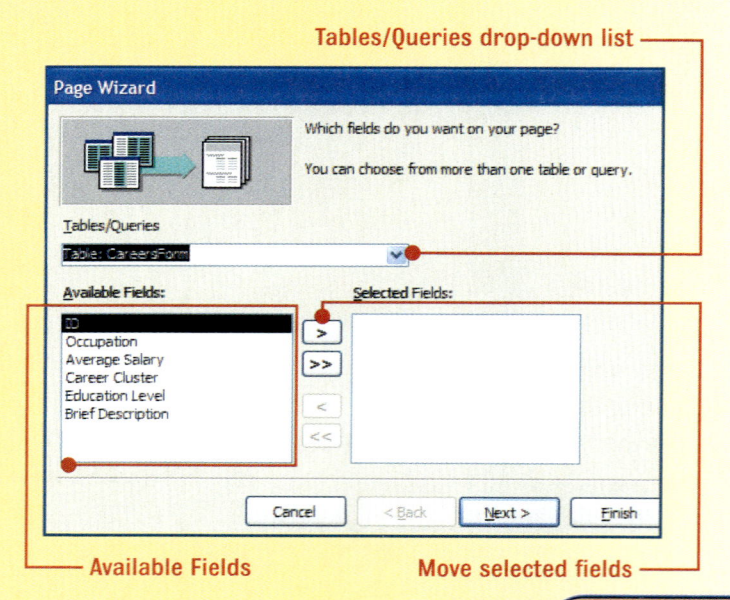

Activity 20

If you need additional help on this topic, open the Help menu; or refer to the documentation for your hardware.

Continued

Chapter 11   385

Activity 23

If you need additional help on this topic, open the Help menu; or refer to the documentation for your hardware.

Step-by-Step

Entering Punctuation

1. Dictate the following two passages, including the punctuation marks.
2. As you dictate, remember to pause briefly before you say the punctuation commands. Remember to also say *new paragraph* at the end of each sentence so the next sentence will start on a new line.

Punctuation commands are displayed as red text in the passages below.

```
There is an art club meeting after school today.
PERIOD

Will you be going on the class field trip?
QUESTION MARK

Dan, COMMA Mark, COMMA Mario, COMMA and Jeff
tried out for the basketball team. PERIOD

All Web addresses start with http:
COLON / SLASH / SLASH
```

```
Our team won first place at the track meet!
EXCLAMATION MARK

OPEN QUOTE "Julie, COMMA what time are you
going to the dance?" QUESTION MARK  CLOSE QUOTE

Are you traveling by plane, COMMA car, COMMA or
train? QUESTION MARK
```

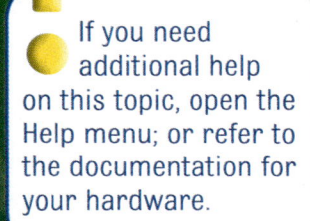

Activity 24

Add Bullets and Numbered Lists

Bullets make lists of items easy to read. Numbered lists are useful for prioritizing the items in a list.

1. Move the insertion point down the page with the New Paragraph command.

```
I have to order the following items for
graduation: COLON
```

2. Dictate the lines that appear above.
3. Say *new paragraph*, *voice command*, *bullets*.

Continued

Lesson 11.5: Create a Web Page with a Database

You will learn to:
- **Create** a data access page to view a database via the Web.
- **Modify** the design of a data access page.

DIGIByte

Web Page
Creating a Web page from the data in your database table is an easy way to display database information on the Internet.

Publishing Access Data on the Web Access data can be easily displayed as a Web page by creating a data access page based on a table in the database. Data access pages are used to publish live data that changes regularly and needs to be available for many people. Once the page is created, the data in the page can be sorted in ascending or descending order. In this lesson, you will create and design your own data access page.

Creating a Data Access Page A data access page is an HTML document, or a Web page, published from Access that is linked to a table. You will recall that HTML is Hypertext Markup Language, which is the standard formatting language for Web pages. The data access page allows you to review and use the data in a table created in a database. You can add to, edit, and manipulate the data stored in the database.

If you create a data access page when a database is open, Microsoft Access creates a shortcut to the page and displays the shortcut in the database window. If you create a data access page without opening a database, Access will create a standalone page.

✓ Concept Check

Why do you think a company might want to display database information on a Web site?

DIGITAL DIALOG

Databases for Handhelds

Internet Investigation In a small group, search the Internet for databases that can be downloaded to a handheld computer. Make a chart like the one shown below of the names of the applications that you find, and briefly describe what each application does.

Software Name	Handheld OS	Brief Description	Web Site
DB tools	Palm OS	Basic databases	www.palmgear.com

Step-by-Step *continued*

4 Switch to Dictation mode and dictate the following.

- Announcements `NEW LINE`
- Cap `NEW LINE`
- Gown `NEW LINE`
- Party invitations `NEW LINE`
- Thank-you notes `NEW LINE`

5 Repeat Steps 1 through 4. In Step 3, say *numbered list* instead of *bullets*.

Use the Correction Feature

1 If the Correction button does not appear on the Language bar, click the down arrow on the right edge of the Language bar and choose **Correction** to add the button to the bar.

Language Bar down arrow

Correction

2 Look over the text you have dictated in this lesson. Are there any errors?

3 If so, select a word that you need to change using the mouse or by saying *select (word)*.

4 Click the **Correction** button or say *correction*.

5 You will hear your pronunciation of the word replayed. In the drop-down list that appears below the Correction button, click the correct choice or say *select (choice number)*. If the correct word is not listed, key it in.

Continued

Activity 25

DIGI Byte

Mouse
Use the mouse if you have trouble getting the computer to recognize your commands.

Application 8

Directions Open the EnviroScienceJobs database. Create a report based on the query that lists occupations that have an average salary of greater than $50,000. Sort the occupations by average salary in descending order. Publish the report in Word and then prepare a new memo in a Word document for Maria Ruiz, the office manager at EnviroScience Recruiting. Tell Maria that this information is in response to a question from her on which positions pay the most. Save the memo as Salaries.

Application 9

Directions Open the EnviroScienceJobs database and open the query that lists occupations that have an average salary of greater than $50,000. Copy the data in the table and paste it into an Excel spreadsheet. Create a bar chart based on the data. Prepare a new memo in a Word document for Maria Ruiz, the office manager at EnviroScience Recruiting. Tell Maria that this information is in response to a question from her on which positions pay the most. Be sure to include the chart in the memo. Save the memo as Salaries 2.

Application 10

Directions Plan a database named Student Jobs that organizes information about the part-time jobs of your classmates. Once you have surveyed at least ten students about their part-time jobs, create five to eight field names and field types that you can use to organize your database, including:

- Student name
- Type of job
- Number of work hours per week

Insert the Access data into Excel and create a chart to show your data. Integrate your data within a Word document of your choice. Save the document as Jobs.

Digital Decisions

Designing Queries

Teamwork Your team is responsible for creating a database with information about all of the people who have registered with EnviroScience Recruiting. The database needs to include the following information:

- Name
- Address
- Education level
- Phone number
- Years of experience
- Salary requirements
- Special licensing or certification

Remember that this information needs to be broken down into the smallest possible units. For example, the name should be included in two fields: one for the first name and one for the last name. Indicate which field will be the primary key of the database. Save the database as Registrants.

Critical Thinking Using your Tablet PC, make a list of the fields that you will need to create for this database and then list at least three queries that could be created from this database.

RUBRIC LINK

Go to the Online Learning Center at **digicom.glencoe.com** and click Lesson 11.4, Rubric, to find the application assessment guide(s).

Activity 26

Step-by-Step continued

Correct Capitalization

1. Create a new Word document, or delete the contents of the document in which you have been working.
2. Say *The art club traveled to the Smithsonian Institute for a field trip*. Your computer may recognize *Smithsonian* as a proper noun and capitalize it, but institute will be lowercase.
3. Say *voice command* to switch to Voice Command mode.
4. Say *select, institute, capitalize*.
5. Say *move to end of line, new paragraph*.
6. Say *I am going to the basketball game on Tuesday. We play the Eagles*.
7. Say *select, Eagles, capitalize*.

Activity 27

Change Case

Another way to correct capitalization is to use the Change Case command.

1. Move the insertion point to a new line and say *The science club wants to go to the Smithfield natural history museum*.
2. Say *voice command*.
3. Say *select, natural history museum*.
4. Say *format, change case, title case, OK*.

RUBRIC LINK

Go to the Online Learning Center at **digicom.glencoe.com** and click Lesson 6.4, Rubric, to find the application assessment guide(s).

Application 8

Directions Dictate the following text. Change capitalization and use the Correction feature when needed. Then dictate the text again as a bulleted list. Save your file using speech recognition when you are finished.

```
Each time you begin using speech recognition,
you need to make sure your user profile has
been selected. You can do additional training
with your profile at any time to improve the
proficiency of your system.

On the Language bar, you can click Tools and then
Training to access the training screens. You can
choose from several different stories to read for
the training exercise.
```

Step-by-Step continued

3. Click the **OfficeLinks** drop-down arrow and select **Publish It with Microsoft Office Word**.

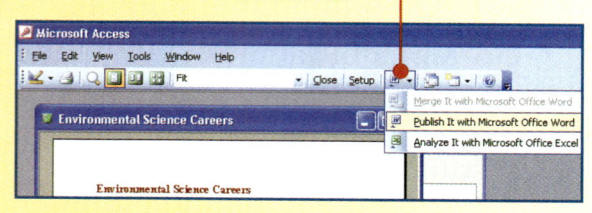

4. In the Word document that is created, select the **Edit** menu and choose **Select All**. Select the **Edit** menu again and choose **Copy**.

5. Switch to the Word memo that you created in Activity 16, page 381. Click to place the insertion point at the end of the text and press ENTER twice.

6. Select the **Edit** menu and click **Paste**. Delete the title Environmental Science Careers. Align the date using your word processing skills.

7. Save this file as **Recruiting Memo**.

Activity 18 — Insert Access Data into Excel

1. Open the **EnviroScienceJobs** database and open the **CareersForm** table.

2. Select the **Edit** menu and choose **Select All Records**.

3. Select the **Edit** menu again and choose **Copy**.

4. Start Excel. In a new Excel spreadsheet, select **Edit, Paste**.

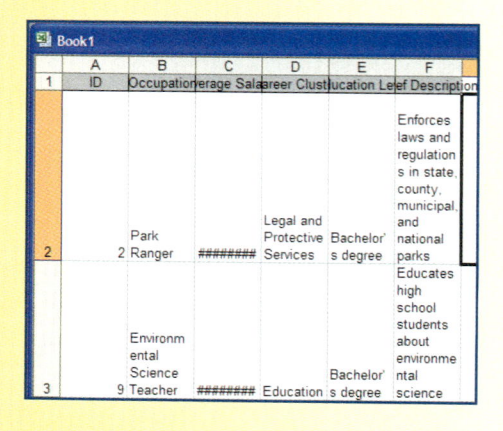

Activity 19 — Format the Spreadsheet and Create a Chart

1. If any columns display number signs (#) rather than recognizable data, widen the columns to display the data.

2. Adjust the format of the cells in the Average Salary column so that the numbers are displayed as whole dollar amounts, with no decimal places.

3. Save the Excel file as **EnviroScience Careers**.

4. Select Cells B2 through C9 and click the **Chart Wizard** button. Proceed through the Chart wizard to create a bar chart of the data. Title the chart **Average Salary by Occupation**.

5. Adjust the width of the chart to show the occupations. Save the file again and print the chart. Close the file.

DIGITAL Career Perspective

Architecture and Construction

Eric Searcy
Architect

Creating Spaces That Inspire

"I enjoy the positive reactions my team gets from clients when they are happy with a finished building," says Eric Searcy, an architect and project manager for a commercial and residential architectural firm. "My objective as an architect is to make buildings and spaces that are functional and uplift the human spirit."

Searcy and his team use digital drawing tools to help make the design process more efficient. Team members use Tablet PCs to sketch and record ideas during project meetings, which they can incorporate into plan drawings to show clients later on.

They also use other digital tools that make communicating with clients and vendors easier. Searcy relies on his PDA. "I often spend about 70 percent of my workday on a job site and never leave the office without my PDA. It stores all of my business contact information, appointments, and project notes so I can access this data quickly, any time I need it." Searcy and his team also value Microsoft Word. "We need to be able to explain ideas, plans, and work in progress to clients in a language they can read easily."

"All the tools available today make our jobs better in many practical ways," he says. "In the end, it's what we do with these tools that gives us the power to improve the quality of life."

Career Growth Trend

Demand for architecture and construction occupations is expected to grow about as fast as average (10 to 20 percent). Opportunities for new architects will be best for those who are familiar with computer-aided drafting and design technology.

10% to 20% Growth

Growth Trend — Year 1, Year 2, Year 3, Year 4, Year 5

Training

A bachelor's or master's degree in architecture is required, as is becoming licensed. Courses to take in high school and beyond might be drafting, life drawing, art history, geometry, geography, cartography, and social studies. Basic business administration classes can also be helpful for anyone interested in forming his or her own architectural firm.

Salary Range

Typical earnings are in the range of $45,000 to $75,000 a year.

Skills and Talents
Architects need to be:

- Strongly interested in beautiful and practical design
- Skilled in visual perception and design
- Able to understand client needs
- Able to work well as part of a team
- Flexible when changes to plans or budgets are necessary
- Excellent written and oral communicators
- Interested in the physical environment

Career Activity

How do you think studies in history and languages might be helpful in a career in architecture? Write a short paragraph.

Step-by-Step

In Activities 16–19, pages 381–382, you will integrate databases with word processing and spreadsheets. You will create a memo in Word and then insert Access data into the memo. You will then insert Access data into an Excel spreadsheet and create a chart to represent the data.

Create a Memo

Activity 16

> If you need additional help on this topic, open the Help menu; or refer to the documentation for your hardware.

1. Start Word.
2. Select the **File** menu and click **New**. In the New Document task pane, select **On my computer**.
3. Click the **Memos** tab. Choose **Professional Memo**, and click **OK**.
4. Select the text **Company Name Here** and enter `EnviroScience Recruiting`.
5. Select the placeholder text in the memo template and enter text to create a memo like the one shown below. You will need to delete the cc:, or copy, line in the greeting section of the memo.

```
To:     Environmental Science Department
        City College

From:   Your Name

Date:   Current date

Re:     Spring Job Fair

Anticipated Openings for Summer Employment

During our company's attendance at the Spring Job
Fair at City College, we will be interviewing
students for the positions listed below. Students
interested in these positions should bring
their resumes to the job fair as well as copies
of their transcripts. Representatives will be
available for the duration of the job fair to
answer any questions.
```

Insert Access Data into the Memo

Activity 17

1. Open the **EnviroScienceJobs** database.
2. Click the **Reports** object in the Database window and open the Environmental Career Education2 report that you created in Lesson 11.2, Application 6, page 375.
3. Format the database so that each field is viewable. Use Print Preview to view the report

Continued

Lesson 6.5 — Use the Scratch That and Tab Commands

You will learn to:
- Use the **Scratch That** command.
- **Indent** with the Tab command.

DIGI Byte

Microphone
You may need to turn off your microphone if you use the keyboard.

Several Ways to Erase Text You have already learned that you can erase text using the Backspace and Clear commands. The Backspace command erases the character to the left of the insertion point. The Clear command erases selected text.

Scratch That Command Another way to erase text is to use the Scratch That command, which removes the last word or combination of words spoken. Repeating the Scratch That command several times will remove multiple words or phrases.

Tab Command You can use the Tab command to indent text. This is useful for indenting paragraphs, creating lists, and creating outlines.

✓ Concept Check

How can you erase text using a keyboard? How is this process similar to using speech recognition to erase text?

DIGITAL DIALOG

Investigate Computer Rules

Communication As a class, discuss rules for using computers and the Internet at your school. You could have the school principal, media coordinator, computer lab teacher, or technology director come in for a short interview with the class. Ask about the privacy of student files saved on school computers. Find out about your school's student e-mail policy.

Here are some questions to get your discussion started:

- Are students allowed to have e-mail accounts from the school district, or are they required to use public accounts, such as Yahoo and Hotmail?
- How is network security maintained?
- What are the reasons for such security or lack of security?

Teamwork As a class, decide if you have any suggestions for additional rules. Your rules should make the school's network more secure, if possible.

Lesson 11.4 Integrate Databases with Applications

You will learn to:
- **Insert** an Access report into a Word document.
- **Convert** an Access report into an Excel document.

Integrating Access, Word, and Excel Access reports can be easily inserted into a Word document. The Access data does not need to be rekeyed in order to use it in Word. Integrating Access and Word in this way can be very useful. For example, you may need to insert Access data into a memo or report that is created in Word.

Access data can also be easily translated into an Excel spreadsheet, making it easy to use Excel functions on data that has been stored in a database. In this lesson, you will insert Access data into both Word and Excel.

DIGIByte
Formatting Reports
Format the text in Access reports in styles and colors that make the report easy to read.

✓ Concept Check
What are some reasons why you might want to use Access data in Word or Excel documents?

DIGITAL DIALOG

Databases Versus Spreadsheets

Teamwork In a small group, discuss the similarities and differences between a database and a spreadsheet application. Open each application on a computer to help you compare the features of each one.

Communication Design a table similar to the one shown below to display your answers. Give examples of situations in which you would use one application in preference to the other. Include examples of when you might use either application. Give reasons for your answers.

Application	Uses	Reason to Use the Application
Spreadsheet	Create a budget	Easy to enter formulas to calculate expenses

Activity 28

If you need additional help on this topic, open the Help menu; or refer to the documentation for your hardware.

Activity 29

Suite Smarts

A quick way to select a single word is to place the insertion point anywhere within the word and double-click.

RUBRIC LINK

Go to the Online Learning Center at **digicom.glencoe.com** and click Lesson 6.5, Rubric, to find the application assessment guide(s).

Step-by-Step

Use the Scratch That Command

1. Dictate the following text.

   When you first begin using speech recognition, the computer may have a difficult time identifying some of the words you speak.

2. Switch to Voice Command mode. Say *scratch that* until all of the text you just entered disappears.
3. Say *The last day of school this year is Thursday.*
4. Switch to Voice Command mode. Say *scratch that* and then say *Friday.*

Use the Tab Command

1. Say *new paragraph.*
2. Say *voice command, tab.*
3. Say *dictation.*
4. Dictate the following text.

   The drama club is planning a weekend trip to New York City to see a play on Broadway. Students will be selling candy to help raise money for the trip. Everyone wants to go!

Application 9

Directions Use speech recognition to dictate the following paragraph. Be sure to use indentation, capitalization, and punctuation as indicated. Use the Scratch That command as necessary. Use speech recognition to save and print.

   If the computer enters an incorrect word, remember to use the Delete, Backspace, Correction, or Scratch That command to correct the mistake. These commands will become easier to use the more you practice them. Learning speech recognition is just like learning how to ride a bike, swim, or play a musical instrument. You need to practice to improve.

Chapter 6 Speech Recognition   Lesson 6.5

21ST CENTURY CONNECTION

Setting and Meeting Your Goals

Creating a strategy for following through and achieving your goals is the key to realizing your dreams. Setting goals includes recording the things you want to accomplish, calculating the timeframe by which you hope to do this, and tracking your progress to see where you have been on the path and where you will go next. Following this plan will help you stay on course. Keep in mind that life circumstances change, and you may find over time that your goals have changed. Have a flexible, positive attitude and be willing to make adjustments for such changes. With hard work and perseverance, you can achieve any goal.

Be Realistic Long-term goals can sometimes seem too big to achieve, if you do not know where to start. When you break goals into smaller parts and follow through step by step, you will more likely get the results you want.

Digital Tools Help Turn Ideas Into Reality You are learning to use a number of digital tools that will help you plan for your future. Make the most of these tools by using them to record and maintain your plans.

- Use the Web as a source for postsecondary educational programs and career information. To find out which colleges and universities offer degrees in your fields of interest, check online for guides to these schools, or ask your academic counselor or librarian if he or she has hard copies. An excellent place to find information about different kinds of jobs is the *Occupational Outlook Handbook*. You will learn more about this resource in the next unit.
- Make a detailed map of your career goals in an Excel workbook, with separate sheets for your occupational interests, educational information, job facts, and relevant target dates. Include a separate page for recording your thoughts about the process as you progress toward your goals. Remember that you can always change the details of your outline as you learn more about new developments in different industries.
- Record your target goal dates and reminders in an Outlook calendar or PDA.

Activity Write a paragraph about a goal that you met. What planning did you do to meet that goal?

> "Whatever you can do or dream you can, begin it. Boldness has genius, power, and magic in it."
>
> — *Johann Wolfgang von Goethe, German writer and philosopher (1749–1832)*

Lesson 6.6 — Dictate Numbers

You will learn to:
- Use speech recognition to enter numerals.
- Use speech recognition to enter decimals and fractions.
- Use speech recognition to enter dates and phone numbers.
- Use speech recognition to enter currency and time.
- Use speech recognition to enter mathematical operators.

Enter Numbers with Speech Recognition Numbers are used in a variety of ways and there are many different ways to say them, depending on the context. Because of this complexity, dictating numbers can be challenging and may require extra training of your user profile. In this lesson, you will learn how to enter numbers not only as numerals, but also as dates, fractions, decimals, currency, phone numbers, and time.

Techniques Remember that you can use the keyboard with speech recognition to be most effective in dictation. Say a date as one term. Do not pause between the month, day, and year.

✓ Concept Check

How do you use numbers in your daily life? What types of numbers do you think you would most likely need to use when dictating or writing?

DIGITAL DIALOG

The Future of Speech Recognition

Communication Use a search engine to find some articles about the future of speech recognition. Discuss them with the class.

Here are some questions to get your discussion started:

- What does the future hold for speech recognition?
- Will computers soon be able to have intelligent conversations with us?
- Would this be a good feature to have on a computer?
- Why or why not?

As a class, decide how you think speech recognition might be used 10 years from now.

Step-by-Step *continued*

6. Click the right arrow button at the bottom of the screen until a blank form is displayed.

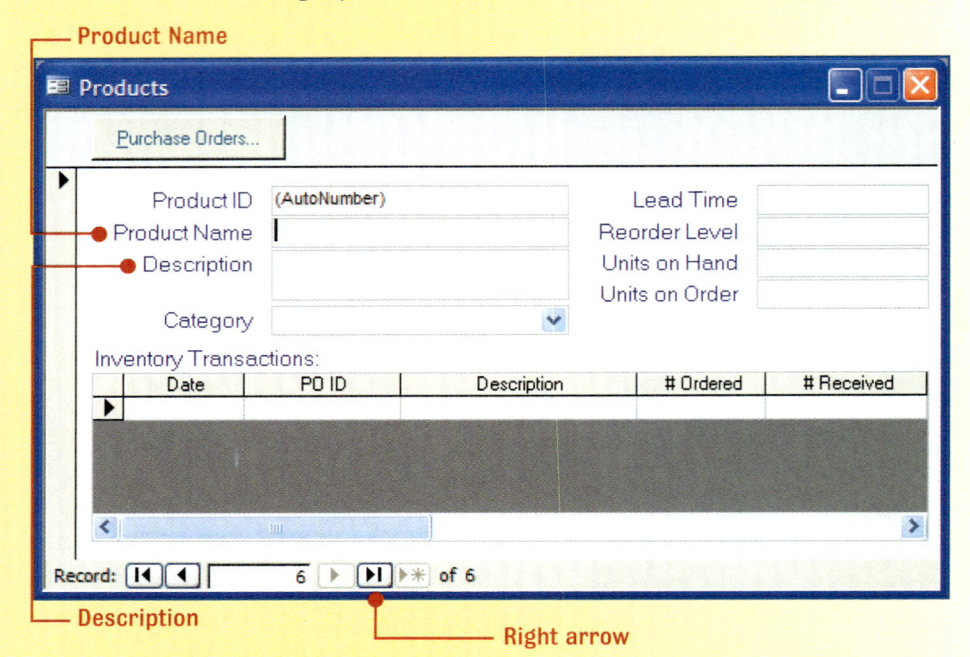

   - Product Name
   - Description
   - Right arrow

7. In Product Name, enter `Yang Lee's Custom Tea Blends`.
8. In Description, enter `Variety of tea blends`.
9. Choose **Beverages** from the Category list.
10. Enter today's date in the Date field.
11. Enter 44 in the # Ordered, 44 in the # Received, and 21 in the # Sold field.
12. Close the window.
13. To preview the report, repeat Steps 5 and 6. Close the file.

Application 7

Directions Search the Microsoft Templates Web site for additional Access templates.

- What types of templates do you find?
- How are these templates used?
- What types of companies could use these templates?

Choose a template and download it. Open it and practice entering data.

Suite Smarts

Often, a particular business form could be created in either Word, Access, or Excel. Large files created in any program can be zipped or compressed to make them easier to save or send via e-mail.

RUBRIC LINK

Go to the Online Learning Center at **digicom.glencoe.com** and click Lesson 11.3, Rubric, to find the application assessment guide(s).

Activity 30

If you need additional help on this topic, open the Help menu; or refer to the documentation for your hardware.

Step-by-Step

Enter Numerals

1. Start Word, if it is not already running.
2. Create a new document.
3. Start Dictation mode.
4. Dictate the following.

```
FORCE NUM 1   (PAUSE)   FORCE NUM 2   (PAUSE)
FORCE NUM 3   (PAUSE)   FORCE NUM 4   (PAUSE)
FORCE NUM 5   (PAUSE)   FORCE NUM 6   (PAUSE)
FORCE NUM 7   (PAUSE)   FORCE NUM 8   (PAUSE)
FORCE NUM 9   (PAUSE)   FORCE NUM 10
```

Saying *force num* before the number tells the computer to display the number itself instead of spelling it out.

5. Now dictate numerals 21 through 30.

You do not have to say *force num* before each of these numerals. They will automatically appear as digits.

Activity 31

Enter Decimals

1. Switch to Dictation mode, if necessary.
2. Dictate the following numbers. Say *point* to enter the decimal point.

```
4 POINT 5    3 POINT 9
6 POINT 2    6 POINT 5
8 POINT 1    4 POINT 7
9 POINT 4    7 POINT 1
```

Activity 32

Suite Smarts

Speech recognition can be used with other Microsoft applications in addition to Word. Try it with Excel and PowerPoint.

Enter Fractions

1. Switch to Dictation mode, if necessary.
2. Dictate the following fractions. Say *slash* to enter fraction lines.

```
1 SLASH 2    3 SLASH 4    1 SLASH 4
5 SLASH 6    7 SLASH 8    1 SLASH 10
2 SLASH 5    1 SLASH 6
```

Continued

Chapter 6

Step-by-Step

In this step-by-step you will start Access and then download a database form, or template. You will then customize the form to your needs.

Activity 14

Locate and Download an Online Access Database Form

1. Start Access. Select the **File** menu and choose **New**.
2. From the New File task pane, click **Templates on Office Online**.
3. From the Business and Legal, Orders and Inventory category, choose **Inventory Control**. Click **Inventory management database**.
4. Click the **Download Now** button and follow the directions for downloading the file. Rename the file when you are asked to save it to your computer. Be sure to choose the correct location on the hard disk to save the file.

Activity 15

Customize the Access Template

1. Click **Open** in the Download complete window and use WinZip to extract the file. Double-click the file to open it.
2. Click **Preview Reports**, and then click **Preview the Product Summary Report**.

? If you need additional help on this topic, open the Help menu; or refer to the documentation for your hardware.

Preview Reports

3. Write down the names of the products that are listed.
4. Close the report and click **Return to Main Switchboard**.
5. Click **Enter/View Products**.

DIGI*Byte*

WinZip

WinZip is a program that opens files that have been compressed, or zipped. You can obtain a free evaluation copy of WinZip on the Internet. Use a search engine to locate a site where you can download a trial version of WinZip.

Continued

Step-by-Step *continued*

Activity 33 — Enter Dates

1. Switch to Dictation mode, if necessary.
2. Dictate the following dates. Do not say the comma in dates.

```
December 25th 2005    12 SLASH 25 SLASH 2005
```

Activity 34 — Enter Phone Numbers

1. Switch to Dictation mode, if necessary.
2. Dictate the following phone numbers. Pause between sets of numbers to indicate the hyphen.

```
five five five  (PAUSE)  two one four one
eight eight eight  (PAUSE)  five five
five  (PAUSE)  two one four one
```

Activity 35 — Enter Currency and Time

1. Switch to Dictation mode, if necessary.
2. Dictate the following.

```
Six dollars and fifty cents
Twenty dollars and sixty-five cents   NEW LINE
Eight PM  Two-thirty AM  Six twenty-two PM
Nine o'clock  five o'clock  three o'clock
```

Activity 36 — Enter Mathematical Operators

1. Switch to Dictation mode, if necessary.
2. Dictate the following. Say *hyphen* for the subtraction operator, *asterisk* for the multiplication operator, and *slash* for the division operator.

```
FORCE NUM 10  HYPHEN  FORCE NUM 6  EQUALS
FORCE NUM 4

FORCE NUM 2  PLUS SIGN  FORCE NUM 6  EQUALS
FORCE NUM 8

FORCE NUM 4  ASTERISK  FORCE NUM 6  EQUALS 24

FORCE NUM 9  SLASH  FORCE NUM 3  EQUALS
FORCE NUM 3
```

Lesson 11.3 — Use Online Database Forms

You will learn to:
- **Locate** Access database forms online.
- **Download** an Access database form.
- **Enter** data in an Access database form.

Find Database Forms Online Just as Word and Excel have downloadable templates online, Access database forms that have already been designed, or templates, can be found online. These template files have been created to be shared. Once you download a file and save it to your hard drive, floppy disk, or USB drive, you can edit the form to fit your specific needs.

Saving Time Downloading and editing these files can save an employee hours of time compared with designing a custom form. These database forms are easily accessed and quickly downloaded from Microsoft's Web site as well as independent Web sites. Some Web sites may charge a fee for using their forms, so be careful to read any disclaimers that may be posted on a Web site.

DIGIByte
Business Forms
Access databases can be used for many types of business forms.

✓ Concept Check
Which of the business forms that you have learned about could have been created as databases?

DIGITAL DIALOG

Benefits of Using Templates

Teamwork In a small group, brainstorm how using templates can make your work easier and more professional. Consider how others' experiences went into creating the database form or other type of template you are using. Give at least four different examples of the benefits of using a template versus creating a form from scratch.

Writing Create a diagram on the benefits your group came up with similar to the one on the right.

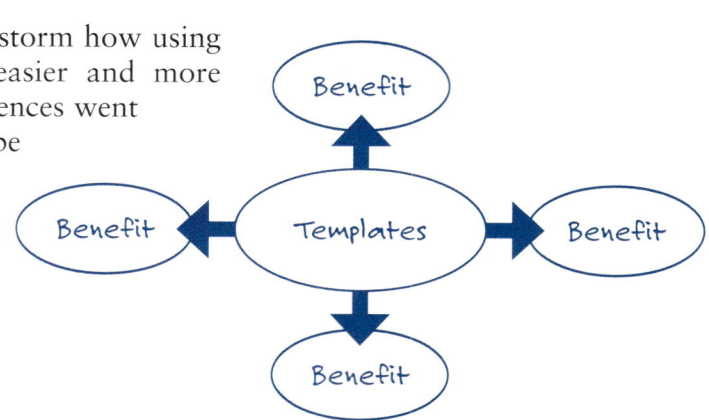

Application 10

Directions Enter the following sentences and numbers with speech recognition. If you have problems with any of the entries, keep practicing. You may even find it helpful to go back and complete another training session so your computer will better understand you. Remember to pause when necessary and to correct mistakes as you go along.

```
Independence Day is celebrated on
July 4.

We are having a dance on
Valentine's Day, February 14.

The new computer I want will cost
$599.95.

I bought a new backpack for $29.95.

Class begins at 7:55 AM.

We get out of school at 2:40 PM.

I am meeting my friends at the
movie theater at 8:00 PM.

My new cell phone number is 888-
555-4121.

The phone number for the
restaurant is 555-2234.

2.5   4.6   9.1   22.54   7.68   31.20
5.7   9.6   41.10

1/2   2/3   4/5   9/10   3/10   5/8   7/8

10,239   345   9,155   29,500   251,678
82,050   598   625

9 + 9 = 18   9 − 9 = 0   9 * 9 = 81   9 / 9 = 1
```

RUBRIC LINK

Go to the Online Learning Center at **digicom.glencoe.com** and click Lesson 6.6, Rubric, to find the application assessment guide(s).

Look Who's DIGITAL

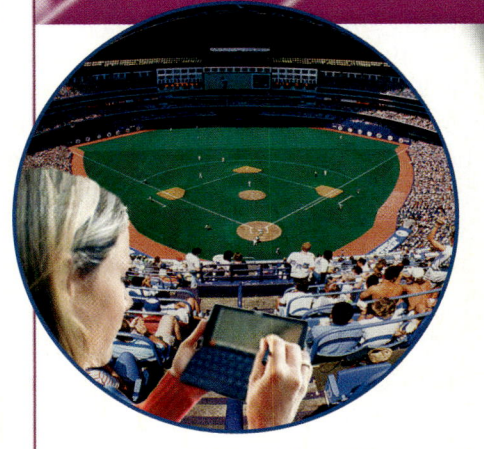

The Giants' Digital Dugout

The San Francisco Giants are playing on a field of digital dreams. Baseball fans can connect to wireless devices through Wi-Fi at the stadium. Fans can bring their PDAs, Tablet PCs, or laptops and access the Internet.

Stats, video clips, interactive games, and team information can all be viewed using the Giants' Digital Dugout network. Players also get digital video training.

The story of the San Francisco Giants is one example of how wireless technology is becoming a part of our world. Across the nation wireless connectivity is being introduced to improve the quality of life and, in this case, the enjoyment of both fans and ballplayers.

Activity With a partner, think of different ways sports teams could use PDAs and laptops for game play. Be ready to discuss your ideas with the class.

Application 4

Directions In the CareersForm table, create a query that finds occupations that have average salaries of less than $40,000. How many occupations fit this criteria? In the CareersForm table, sort the careers by salary. How easy is it to locate the number of occupations that pay less than $40,000 using a sort rather than a query? Would your answer be the same if there were 100 or 1,000 records in the database? Which would be the easier way to locate specific information in a large database, a sort or a query?

Application 5

Scenario You work for a new environmental science recruiting firm. You have been asked to create a database of environmental science careers that includes: salary information, educational level, job skills, job descriptions, and job outlooks.

Directions Search online for information on ten occupations. Use the keywords "environmental career" and gather data to include in the database. Create a query and two reports.

Application 6

Directions In the CareersForm table, create a query that finds occupations that require a bachelor's degree. Save the query with the name **Education**.

Practice creating reports that show what kind of education is needed for environmental careers. Use the CareersForm table. Create one report using the Report Wizard, including the fields Occupation, Average Salary, Career Cluster, and Education Level. Sort the Educational Level field in ascending order. Title the report **Environmental Career Education**.

Create a second report based on the Education query. Include only the Occupation and Education Level fields. Save it as **Environmental Career Education2**.

How would you use the two reports? Which do you prefer and why? Write several paragraphs to answer the questions above. Use a Tablet PC, a PDA, or speech recognition, if possible.

TECH ETHICS: Personal Data Online

Teamwork In a small group, discuss what you have learned in this lesson. Consider what the ethical issues are for a company deciding whether to post information from a company database online. For example, what might happen if a company publishes personal information about employees?

Internet Investigation Search online to find articles and Web sites about the consequences of posting personal information online. Key words to use in your search could be "posting personal information online" and "online privacy violations."

Communication Create a table similar to the one below, and complete it as you discuss these issues and the articles and Web pages that you found. Are the effects in your table generally positive or negative? Do you think the positive effects outweigh the negative, or does the negative outweigh the positive?

Action	Effect
Personal phone numbers posted on Web site	Company could be held liable for violating the privacy of employees or customers

21ST CENTURY CONNECTION

Protect Your Voice

Ergonomics experts and speech pathologists treat some patients with strained vocal cords resulting from improper use of speech recognition products. This type of injury is known as laryngeal stress injury.

What Causes Damage? Some older speech recognition programs require you to pause between speaking words. The pausing action can damage your vocal cords. Many current products use a newer technology called continuous speech, where you dictate at a normal pace without interruption. While continuous speech technology is better for your vocal cords, you still need to be careful.

Symptoms Symptoms of damage to your vocal cords include chronic hoarseness, fatigue when speaking, and even a temporary loss of your voice. These potential problems are very real. If you begin to experience any of the following warning signs, talk to your doctor:

- A dry scratchy throat or a tickle along the back of your windpipe.
- A cough that turns into violent spasms.
- Difficulty varying your pitch or projecting your voice.
- Constant throat clearing.
- Achy neck muscles.
- A lower voice than normal.
- Intermittent loss of voice.

Speak Safely There are methods you can use to lessen damage to your voice. Keep the tips below in mind when you use speech recognition.

Activity See for yourself how using your voice improperly can be tiring. Read out loud for one minute while doing one of the following: whispering, speaking in a high squeaky voice, pausing after each word, and hunched with your knees to your chest. After resting for at least five minutes, try the same passage again in your normal voice.

Protect Your Voice

Avoid Speaking in a Monotone	Change your pitch as you dictate.
Take Breaks	Stop frequently to give your voice a rest.
Drink Fluids	Room-temperature water may be helpful.
Speak Naturally	Dictate in a normal tone.
Avoid Clearing Your Throat	This action bangs your vocal cords together just like coughing and can cause irritation.
Keep Good Posture	Keep your chin level and do not hunch over. If your chin is tilted up, it becomes more strenuous to speak.

DIGI Byte

Use View

As you make changes to your report, it is important that you view the changes as you go. Use the View button to switch between Print Preview and Design View so that you can see the changes.

Activity 13

RUBRIC LINK

Go to the Online Learning Center at **digicom.glencoe.com** and click Lesson 11.2, Rubric, to find the application assessment guide(s).

Step-by-Step continued

4. You can resize the field names in the Page Header section. Click **Education Level** in the Page Header section. Position the mouse pointer on one of the handles at the sides of the box. The pointer will change to a double arrow. Drag the pointer to resize the box.

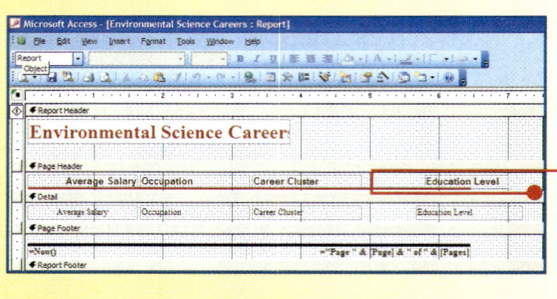

   Field name (selected)

5. Move the Education Level text box in the Detail section of the report so it aligns below Education Level in the Page Header area.

6. Move and resize other text boxes to make sure there is enough room to show all the columns on the page.

7. Click **View** to switch between Design View and Print Preview to observe your changes.

Change Text Formatting

When a Page Header or Detail text box is selected, you can also change the text formatting.

1. Switch to Design view. Select the **Average Salary** text box in the Detail section. Click the **Center** button to center the text. In a similar way, center the text in the Average Salary Page Header section.

2. Click the **View** button to switch between Design View and Print Preview to observe your changes.

3. Close the Report box. Save the changes to the Report.

Application 3

Directions Search the Internet using the key words "environmental science career" to find information on two more environmental science careers. Add this information to the CareersForm table.

Create another report using the Report wizard. In the first step of the wizard, choose to base the report on the Salaries query. Review the report design, and make any changes to it that are necessary for all the columns to show on the page.

Lesson 6.7 — Dictate Special Symbols and Emoticons

You will learn to:
- Use speech recognition to enter special symbols.
- Use speech recognition to enter emoticons.

DIGI Byte

Dictionary
Proper nouns and unusual words may be added to the speech recognition dictionary by selecting the Tools button on the Language bar and choosing Add/Delete Words.

What Are Symbols? Symbols are special characters that are not part of the standard alphabet. Examples of symbols include punctuation marks and mathematical operators, such as !, $, +, and *. On a standard keyboard, many special symbols appear above numerals on the number keys. For example, the dollar sign appears above the numeral 4 on the number 4 key. Special characters can be viewed by selecting the Insert menu and choosing Symbol.

What Are Emoticons? Emoticons are character combinations often used in e-mail and instant messaging to express an emotional message, such as a smile or wink. Emoticons have become so popular that some people even include them in handwritten notes.

✓ Concept Check

How can emoticons be used? Do you ever use them? List the types of documents they are often used in.

DIGITAL DIALOG

Connect Speech Recognition Commands and Keystrokes

Writing Create a table similar to the one shown. In the Command column, list new speech recognition commands as you learn them. In the column next to each command, identify the key or key combination you would use to accomplish the task on the keyboard.

Some other commands you may want to include are: Backspace, Delete, Cancel, Move to Beginning of Document, Move to End of Document, Move to Next Line, Move to Previous Line, Scratch That, Correction, Capitalize, Force Num, and Point.

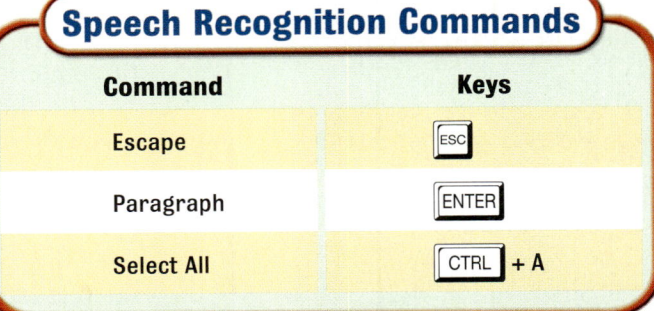

Speech Recognition Commands

Command	Keys
Escape	ESC
Paragraph	ENTER
Select All	CTRL + A

DIGI Byte

Sort Descending

By clicking the Ascending button you can choose to sort descending

Activity 12

Suite Smarts

Wizards provide step-by-step guidance for completing a task.

Step-by-Step *continued*

4. Select **Average Salary** from the first drop-down list. To sort in ascending order check that the word Ascending shows in the box. Click **Next**.

5. Click **Next** again to accept the default page layout settings.

6. Choose the Bold style for the report and click **Next**.

7. Key the report title `Environmental Science Careers` and click **Finish**.

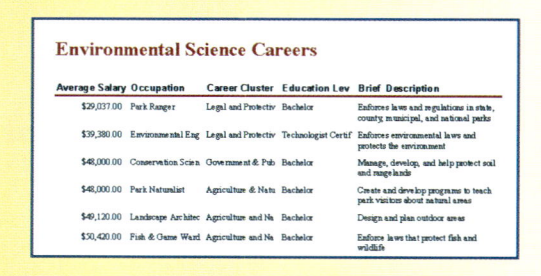

Modify Report Design

In this activity, you will change the width of the report and move the position of field names in the Page Header section. You will move the position of the fields in the Detail section to line up under the header.

1. Switch to Design view by clicking the **View** button.

2. Change the width of the report by dragging the right edge of the page until the edge reaches the 7.5 inch mark on the ruler. When the pointer hovers over the edge of the page, it will appear as a double-sided arrow.

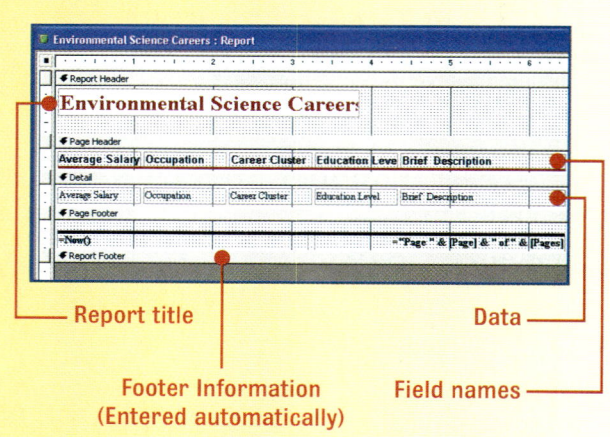

3. To move the Education Level field name in the page header section of your report, click to select the box. Move the mouse over the box. When the mouse pointer changes to a hand, click the left mouse button and drag the box to the right.

Continued

Step-by-Step

Dictate Special Symbols

1. Switch to Dictation mode, if necessary.
2. Dictate the following symbols. You may have to say the text more than once before your computer will recognize it.

```
DOLLAR SIGN   EXCLAMATION MARK   OPEN PAREN
CLOSE PAREN   DOUBLE DASH   ELLIPSIS   PERCENT SIGN
AMPERSAND   CARET   ASTERISK   AT SIGN   OPEN QUOTE
CLOSE QUOTE   TILDE   GREATER THAN   LESS THAN
VERTICAL BAR   UNDERSCORE   EQUAL SIGN
```

Dictate Emoticons

Emoticons are character combinations that are commonly used in e-mail and instant messaging. They allow writers to express emotion.

1. Switch to Dictation mode if necessary.
2. Say *colon*, *hyphen*, *right paren*.
3. Say *semicolon*, *hyphen*, *right paren*.
4. Say *colon*, *caret*, *right paren*.
5. Say *semicolon*, *caret*, *right paren*.

Other symbols that you can combine to create emoticons are shown at the right.

Emoticon Symbols

Symbol	Command Name
<	Less Than or Open Angle
>	Greater Than or Close Angle
_	Underscore
-	Hyphen
|	Vertical Bar
(	Open Parenthesis or Left Parenthesis
)	Close Parenthesis or Right Parenthesis
{	Open Brace, Left Brace, or Curly Brace
}	Close Brace, Right Brace, or End Curly Brace
[	Open Bracket or Left Bracket
]	Close Bracket or Right Bracket
^	Caret
&	Ampersand

Dictate Web Sites and E-Mail Addresses

1. Switch to Dictation mode, if necessary.
2. Say *www dot senate dot gov, Ravi at senate dot gov.*

Activity 37

If you need additional help on this topic, open the Help menu; or refer to the documentation for your hardware.

Activity 38

Suite Smarts

Symbols and special characters can be used in all Microsoft applications, including Excel, PowerPoint, Outlook, and Access.

DIGI Byte

Web Sites and E-Mail Addresses

Web sites and e-mail addresses are dictated just as you would say them naturally.

Activity 39

Step-by-Step continued

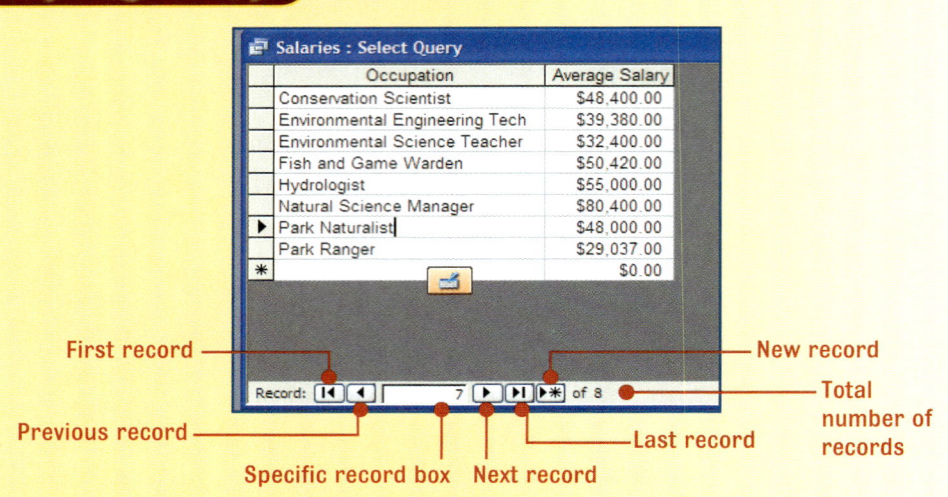

Labels on screenshot: First record, Previous record, Specific record box, Next record, Last record, New record, Total number of records

4. Sort the **Average Salaries** field in descending order. Which record is listed first? Which record is listed last?

5. Switch to Design View by clicking the **View** button.

6. In Design View you can enter selection criteria for your query. To look at occupations in your database with an average salary greater than $50,000, enter **>50000** in the Criteria row of the Average Salary column.

7. Click the **Run** button to see the results of the query. The query will return only occupations that have average salaries greater than $50,000. How many records fit this criteria?

8. Close the query window. Save the changes to the query.

Labels on screenshot: Run button, Criteria >50,000

Create a Report

You select the fields you want displayed in your report.

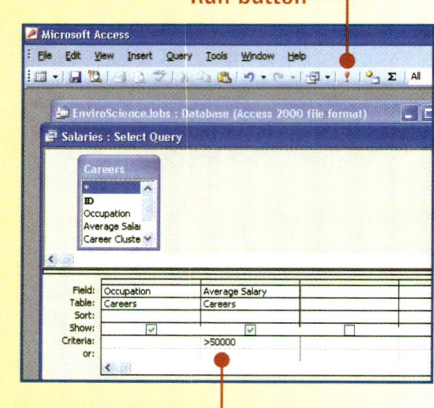

1. Click **Reports** in the Objects pane, and double-click **Create report by using wizard**. Choose **CareersForm** from the **Tables/Queries** list.

2. In the Available Fields list, select each of the fields except the ID field and the Brief Description field, and click the arrow to move each to the Selected Fields list. Click **Next**.

3. Click **Next** again to indicate that none of these fields will be grouped in this report.

Activity 11

Suite Smarts
Database reports can be copied and pasted into Word.

Continued

Application 11

Directions Start Word and create a new document. Dictate the following paragraphs.

```
The student council meeting will
begin promptly at 3:00 P.M. on
Thursday. We will be making plans
for the Valentine's Day dance on
February 14. All representatives
are asked to bring construction
paper in the following colors:
```

- Red
- White
- Pink

```
We will begin to make decorations
for the dance! :) If you have any
ideas for hiring a band, call Mary
(our president) at 555-1414.
You can also e-mail her at
mary@studentgov.org.

We are expecting about 30% of
the students to attend the dance
this year. Tickets for the dance
will cost $5.00 each or $8.00/
couple. Light snacks will be sold,
including soft drinks, pizza,
candy, and popcorn. All proceeds
from the dance will be used to
fund the senior class trip to
Washington, D.C. To learn more
about one of the places they will
visit, go to www.usmint.gov.
```

RUBRIC LINK

Go to the Online Learning Center at **digicom.glencoe.com** and click Lesson 6.7, Rubric, to find the application assessment guide(s).

DIGITAL Decisions

Speech Recognition in the Office

You are the lead word processor for a large law firm. Lately you have been reading articles in technology magazines that claim speech recognition software can be more effective and proficient than traditional keyboarding. You begin to wonder how much more effective your staff of six could be if the stories you have been reading are true about the proficiency of speech recognition.

Internet Investigation Ask another student to work with you to research speech recognition on the Internet for feedback from people who have used it.

Critical Thinking Explore the following questions.

- Is one program better than another?
- Which programs will work best on your computers?
- Is other equipment needed?
- What environment is most adaptable to speech recognition?
- How much could speech recognition improve the efficiency of your department?
- How do people who have used speech recognition feel about it?

Prepare a two-page report on your findings.

Step-by-Step

In this step-by-step, you will sort data in a table, build a query, and create a report.

Sort Data

1 Open the **EnviroScienceJobs** database and open the **CareersForm** table. Click to place the insertion point inside the Average Salary column.

2 Choose the **Records** menu, click **Sort**, and then click **Sort Ascending**. You can also sort records by clicking the Sort Descending button. Close the table and save changes.

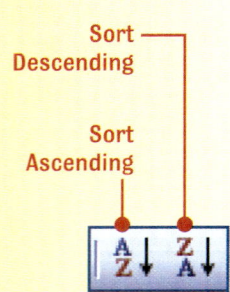

Sort Descending
Sort Ascending

Create a Query

1 From the Database window Objects pane, choose **Queries**. Double-click **Create query by using wizard**. Choose the **CareersForm** table from the Tables/Queries list.

2 Select the **Occupation** field in the Available Fields list. Click the arrow button to move the field to the Selected Fields list.

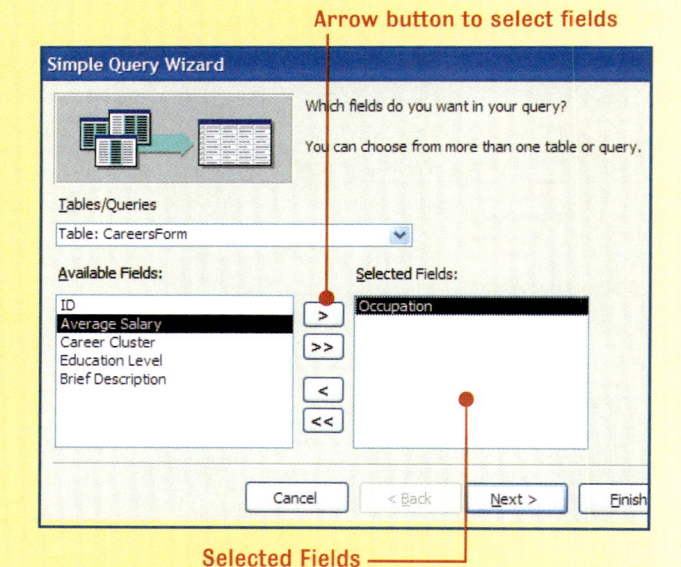

Arrow button to select fields

Selected Fields

3 Select the **Average Salary** field in the Available Fields list and click the arrow button to move it to the Selected Fields list. Click **Next**, click **Next** again. Name the query **Salaries**. Click **Finish**.

The total number of records is displayed in the status bar at the bottom of the window. The arrows allow you to move between records.

Activity 9

Activity 10

If you need additional help on this topic, open the Help menu; or refer to the documentation for your hardware.

DIGI Byte

Queries and Tables
If a database contains more than one table, you must choose which table to base a query on.

Sorting in a Query
Data in a query can be sorted the same way that you sort records in a database table.

Continued

Chapter 11

371

Chapter 6

DIGITAL Dimension → Social Studies
INTERDISCIPLINARY PROJECT

Elected Officials

Using the digital skills you have learned in this chapter, prepare to do a phone interview with one of the elected officials in your area. Plan carefully as to the type of information you would like to gather.

Interview Create a list of questions for the interview using speech recognition to dictate the questions. See how much you can find out about the elected official before the interview by researching on the Internet, in newspapers, or in magazines. Some ideas for questions might be:

- What are some of the responsibilities of your office?
- What is the most challenging aspect of your job?
- What do you enjoy most about your job?
- What committees do you participate in?
- What advice do you have for someone who might want to get into politics?
- What types of technology tools do you use?

Check After you prepare your questions, print a copy and proofread carefully. Exchange questions with a classmate, and discuss the questions each of you has prepared. If your classmate makes good suggestions to improve your questions, incorporate them.

Be sure to leave three or four blank lines between each question before you print the final copy of the questions to use in the interview. You can take notes in the blank spaces during the interview.

Use Speech Recognition After the interview, use speech recognition to dictate the answers in your word processing application. Make corrections as you are dictating.

Proofread the document when you are finished. Save your document as *Your Name* Elected Official Interview.

E-Portfolio Activity

Showcase Your Dictation Skills

To showcase your dictation skills, save your *Your Name* **Elected Official Interview** document in your e-portfolio.

Self-Evaluation Criteria

In your document, did you:

- Include some of the suggested questions and some of your own questions?
- Include some of your classmate's suggestions?
- Dictate the answers to the questions?
- Proofread the document?

Lesson 11.2 — Sort Data and Create Queries and Reports

You will learn to:
- **Sort** data in a database.
- **Create**, run, and modify a query.
- **Create** and modify a report.

ascending sort A sort of data arranged in alphabetical (A–Z) order or numerical (0–9) order.

descending sort A sort of data arranged in high to low alphabetical (Z–A) order or numerical (9–0) order.

query A database object that enables you to locate records that meet a certain criteria.

report A database object used to present the information in a database in a way that is easy to read.

Find Data Quickly After you have created a table of the information in your database, you can sort, or arrange, the data to quickly find information you may need. There are two ways to sort: ascending and descending. An **ascending sort** is a sort of data arranged in alphabetical (A–Z) order or numerical (0–9) order. A **descending sort** is a sort of data arranged in high to low alphabetical (Z–A) order or numerical (9–0) order. Once you have sorted the data, you can quickly identify data that is at the beginning or end of a range of entries in a field. For example, as a sales manager in a company you could see the smallest and largest sales volume of products.

Query—An Important Database Object Perhaps the most important feature of databases is the query. A **query** is a database object that enables you to locate records that meet a certain criteria. Queries are questions about the data stored in a database. For example, a query can be used to quickly answer the question, "What was the company's top-selling product last month?" When you call Information to find a phone number, you are creating a query. Queries will find the exact data you ask for.

Report—A Useful Database Object A **report** is a database object used to present the information in a database in a way that is easy to read. You can build a report and give the report a professional appearance by adding titles, column headings, and so on. There is no limit to the number of queries and reports that a database can contain.

✓ Concept Check

Brainstorm five different queries for a database created at the grocery checkout when purchases are scanned. Give examples of types of reports you might want to create if you were the grocery store marketing manager.

Chapter 11 Databases · Lesson 11.2 · 370

CHAPTER 6 Review

Read to Succeed PRACTICE

After You Read

Summarize and Communicate Information After you read, it is helpful to summarize and communicate the information you have read. When you summarize, you interpret the main idea and then briefly describe the main idea and how the details support it. Communicate the information to others by explaining the main idea in your own words.

For this chapter, prepare a five-minute talk on speech recognition. Decide the main points you want the audience to know for each of the lessons in this chapter. For example, you might want the audience to know who can use speech recognition (Lesson 6.1), why we use speech recognition (Lesson 6.2), and about dictation and voice command modes (Lesson 6.2), and so on.

In your speech:

- Use all of the chapter's key terms.
- Emphasize your likes in using speech recognition.

Dictate notes to give your talk on speech recognition in your speech recognition software.

Using Key Terms

The following terms will help you to remember the different aspects of using speech recognition.

In Memo Pad or on a separate piece of paper, match the terms with their definitions.

- dictate (155)
- user profile (155)
- enunciate (155)
- Dictation mode (161)
- Voice Command mode (161)
- Correction feature (172)

1. The _____ allows a user to correct mistakes in dictated text while also recording the corrections as part of the user profile settings.

2. The mode in which the computer uses spoken words as instructions rather than as literal text is _____ .

3. A record kept on a computer of the way a user pronounces words and the tone and speed with which the user speaks is a(n) _____ .

4. To _____ is to speak words for a computer to enter as text in a document.

5. To pronounce words clearly, including the beginning and ending sounds of the words, is to _____ .

6. _____ is the mode in which the computer enters words and characters on the screen exactly as they are spoken.

21ST CENTURY CONNECTION

Integrity at Work

A person with integrity approaches everything—work, personal relationships, and connection to the community—with honesty, good will, and positive thinking. Honesty is more than telling the truth—it is about building trust. Trust is a very important part of business and personal relationships. When you live with integrity, you are true to yourself, and you and others benefit.

Good General Practices
Life is filled with events that will test your integrity, and it is helpful to have some general guidelines to live by:

- Work to achieve your own personal goals, as well as those of others, so all will benefit.
- Share your knowledge with others and be open to learning from them.
- Put your best effort into every project you do.

Technology and Tools
One important way to show integrity at work is to follow your organization's policies for the appropriate use of equipment and technology. These policies protect the safety and well-being of employees and the interests of the business. Failure to follow the guidelines of the policies can cause a loss of business productivity and even loss of employment.

- Use all employer-owned digital tools and other equipment for purposes of doing your job. Avoid using computers, phones, scanners, pagers, photocopiers, the Internet, the company intranet, or other equipment for personal projects.
- Use your own digital tools on your own time. Avoid placing or receiving nonessential personal calls from your business or cell phone while working. Use breaks for this purpose.
- At work, use e-mail for work-related correspondence only. Avoid answering or forwarding spam, instant messages, or other non-work-related information.

Communicating with Integrity
Communicate in a professional manner at work, both when you are speaking and writing. Your character is reflected in the way that you communicate with others.

- Consider the main point, or "bottom line" of your message. Check facts carefully. Be clear and brief. When writing, take care to spell-check and proofread for correct grammar usage.
- Respect your coworkers. Show consideration for their responsibilities and goals in your speaking and writing.
- Be careful with humorous comments as they may be misunderstood by coworkers or employers. If you are not sure how a story or joke will affect others, it is best to leave it out.

Activity What have school activities taught you about exercising integrity? How can these skills be used in your career? Write a paragraph that includes your answers.

CHAPTER 6 Self-Assessment

Take a moment to review what you have learned in this chapter. Rank your understanding of the topics below.

4 means, "I understand all of this."
3 means, "I understand some of this."
2 means, "I understand very little of this."
1 means, "I don't remember this."

To use a printout of this chart, go to **digicom.glencoe.com** and click on **Chapter 6, Self-Assessment.**
Or:
Ask your teacher for a personal copy.

Rank Your Understanding

Lesson	Topic	4	3	2	1
6.1	• Uses of the Language bar				
	• Creating and training a user profile				
6.2	• Two methods to erase text				
	• How to navigate within a document				
	• How to switch between Dictation and Voice Command modes				
	• Describe the methods to align, underline, italicize, and bold text				
6.3	• Using speech commands to select, copy, delete, and move text				
	• Commands for moving the insertion point in your document				
6.4	• How to add punctuation				
	• Use the Correction feature				
	• How to add bullets				
	• Describe two speech recognition methods for capitalizing letters				
6.5	• How the Scratch That command works				
6.6	• How to enter numbers as fractions, decimals, dates, time, phone numbers, currency, and numerals				
	• How to dictate mathematical operators				
	• Describe what the phrase force num does when used with numbers				
6.7	• How to enter special symbols and emoticons				

If you ranked all topics 4, congratulations! Consider doing a quick review.
If you ranked yourself 3 or lower in any topic, consider reviewing these topics first.

digicom.glencoe.com

Step-by-Step *continued*

10 Close the AutoForm window. You will be asked if you want to save the autoform. Click **Yes**. **CareersForm** will be the default name. Click **OK** to close the CareersForm table.

Add a Record with AutoForm

1 You should be at the Database window.

2 Click **Forms** in the Objects pane. CareersForm should be listed to the right.

3 Double-click and open **CareersForm**.

4 Click the button on the bottom of the window with the ▶✻ to move to a blank input screen.

5 Enter the following information. Press **Tab** TAB to move from one field to another.

```
Record 8:
Occupation:          Environmental Science Teacher
Average Salary:      $32,400
Career Cluster:      Education
Education Level:     Bachelor's degree
Brief Description:   Educates high school students
                     about environmental science
```

6 Close CareersForm. Close the database.

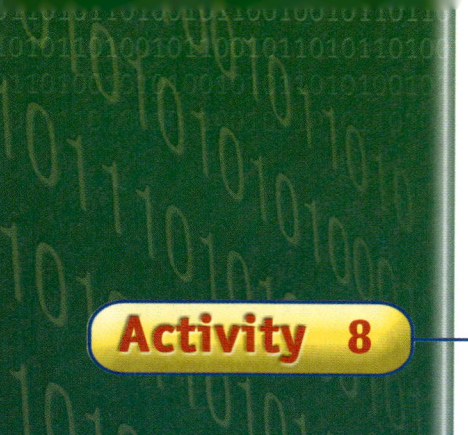

Activity 8

DIGI Byte

Record ID

Remember that you deleted Record 1, Landscape Architect, in Activity 6, page 366. In Step 5, the database interprets the entry as Record 8 because it is the next entry after the existing seven records. The record ID is 9.

RUBRIC LINK

Go to the Online Learning Center at **digicom.glencoe.com** and click Lesson 11.1, Rubric, to find the application assessment guide(s).

Application 1

Directions Create a new database to organize your friends' names, phone numbers, and addresses. Name the database **Contacts**. Choose the Text data type for all the fields in the database. Save the table as **My Friends**, and define a primary key for the table. See Activity 3, Step 8, page 364 for help in assigning a primary key. Enter data for five friends. Save the table.

Application 2

Directions In the **Contacts** database you created in Application 1, above, add an autoform. Enter data for three more friends using the autoform. Save the table and exit.

CHAPTER 7
On-Screen Writing —Tablet PCs

Lesson 7.1	Get Started with the Tablet PC
Lesson 7.2	Enter Text with Input Panel
Lesson 7.3	Write and Edit On-Screen in Journal
Lesson 7.4	Draw On-Screen in Journal and in Word
Lesson 7.5	Edit Text with Gestures

You Will Learn To:
- **Enter** text by writing on-screen.
- **Customize** tablet and pen settings and Input Panel options.
- **Create,** edit, and save Journal notes.
- **Convert** handwriting to typed text.
- **Add** a flag in Journal.
- **Create** and edit Word documents with a Tablet PC.
- **Create,** edit, and save drawings in Word and Journal.
- **Edit** text with gestures.

When Will You Ever Use a Tablet PC?

Would you like to be able to create a piece of artwork, take detailed math or science notes, and search the Internet with just one portable tool? You can easily perform all of these tasks and even save your work in an electronic format when you use a Tablet PC. Tablet PCs combine the convenience of paper with the power of a desktop computer. Because Tablet PCs have a digital pen instead of a keyboard, they are more portable than laptop computers. Most Tablet PCs are about the same size as a paper notebook.

Step-by-Step continued

Activity 7

Create an AutoForm

Additional records can be added to the table in a database at any time. Working in the table to add records may be cumbersome. **AutoForm** makes it much easier to add records.

1. Open the **EnviroScienceJobs** database, and click **Tables** to display the table named Careers.

2. Click one time on the **Careers** table to select it (Do not open the table).

3. Click **Edit, Copy** from the menu bar.

4. Click **Edit, Paste**. A window will open for you to name the new table. Enter: `CareersForm`

5. Be sure **Structure and Data** is selected. Click **OK**. CareersForm will be added as another table in the Tables window.

6. Double-click and open the **CareersForm** table.

7. Go to the Database toolbar, and select the drop down arrow next to the **New Object** button on the menu bar.

8. Select AutoForm. An automatic form, or template, will appear on your screen.

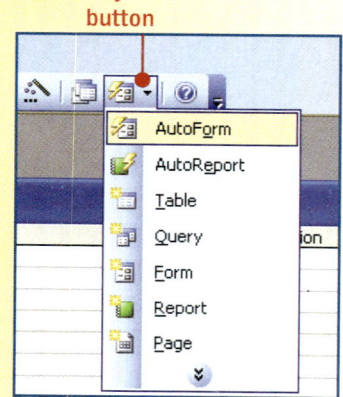

9. AutoForm can be used to enter new records or move through current records. Click the arrow keys at the bottom of the form to move through the records.

   The ID, or identity number, is the primary key. Note that record number 1 has ID number 2 because you deleted record number 1 in Activity 6, page 366. After a record is deleted, the ID numbers of the remaining records will not change.

Continued

Read to Succeed

Key Terms

- Input Panel
- Writing Pad
- Journal
- Quick Keys
- Selection Tool
- gesture

Compare and Contrast

Studies show that comparing and contrasting topics is an effective technique to aid comprehension. Use this technique to help you in all of your reading.

Create a study guide to help compare Tablet PCs with PDAs and speech recognition. Make a chart similar to the one below, and fill it in as you work through this chapter. Refer to previous chapters, if necessary.

Example:

Technology	Ways to Enter Text	How to Edit and Correct
Tablet PC	Write in Journal	Use Scratch Out and gestures
PDAs		
Speech Recognition		

DIGITAL CONNECTION

Wanted: An Easy Way to Manage Class Notes

Not again! You are late for the student council meeting, and you cannot find your notes on the upcoming election. As you rummage through your backpack, you realize you left your student council notebook at home. If only you had one notebook that could hold the notes for all your classes and activities. A Tablet PC can hold all your notes, eliminating the need for multiple paper notebooks.

Writing Activity

Write a paragraph describing how you organize your class notes. Do you use a separate notebook for each class?

Step-by-Step continued

```
Career Cluster:        Agriculture and Natural
                       Resources
Education Level:       Bachelor's degree
Brief Description:     Creates and develops programs
                       to teach park visitors about
                       natural areas

Record 7:
Career Name:           Natural Science Manager
Average Salary:        $80,400
Career Cluster:        Science, Technology,
                       Engineering and Math
Education Level:       Bachelor's degree
Brief Description:     Oversees research for private
                       companies or government
                       agencies

Record 8:
Career Name:           Conservation Scientist
Average Salary:        $48,400
Career Cluster:        Government and Public
                       Administration
Education Level:       Bachelor's degree
Brief Description:     Manages, develops, and helps
                       protect soil and range lands
```

Activity 5

Switch to Design View and Change a Field Title

1. Select the **View** menu and choose **Design View**.
2. Change the title of the Career Name field to Occupation.
3. Close the table and save changes.

Activity 6

Delete a Record

1. Open the **EnviroScienceJobs** database. In the Objects pane, click **Tables**. Open the **Careers** table.
2. Click on any field of the Landscape Architect record. Click **Edit, Delete Record**. A warning window will open saying you are about to delete a record. Click **Yes**.
3. Close the Careers table.

DIGIByte

Deleting a Record
Once a record is deleted, the deletion cannot be undone. The only way to reproduce the record is to reenter the data.

Continued

Lesson 7.1 — Get Started with the Tablet PC

DIGI Byte

Digital Pen Tips
- You can write quickly with the digital pen, but for best results try to form letters carefully.
- To tap accurately, hold the pen upright rather than at an angle in relation to the screen.
- Use only the digital pen to write on the Tablet PC screen; an ordinary pen or pencil might damage the screen.

You will learn to:
- **Use** the Tablet PC Input Panel.
- **Write** text with the digital pen.
- **Open** Journal.
- **Calibrate** the digital pen.
- **Dock** and undock Input Panel.
- **Change** screen orientation and brightness.

Why Use a Tablet PC? You can use a Tablet PC to quickly and easily write notes or draw diagrams directly on the screen. Tablet PCs are small and lightweight, so you can carry one just as you might carry a paper notebook.

Tablet PC Input Panel — Microsoft Word document

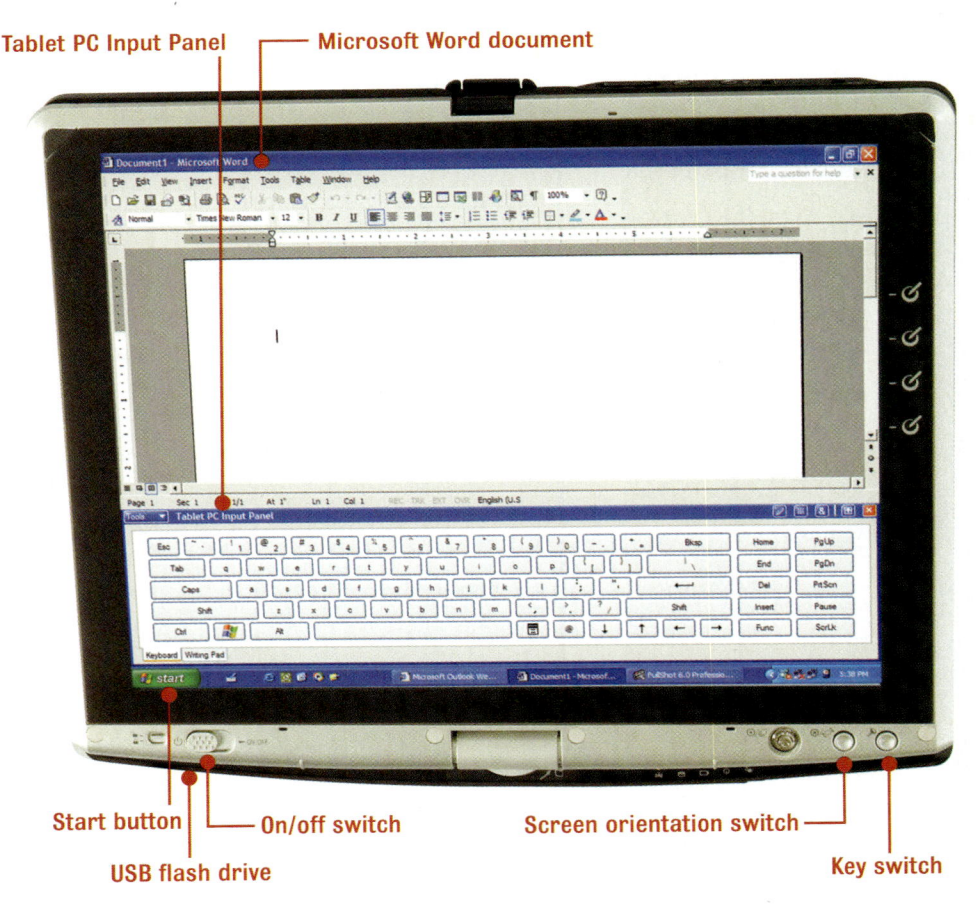

Start button — On/off switch — Screen orientation switch — Key switch
USB flash drive

Figure 7.1 *Tablet PC*

Chapter 7 On-Screen Writing—Tablet PCs

Step-by-Step *continued*

Record 2:
Career Name: Park Ranger
Average Salary: $29,037
Career Cluster: Legal and Protective Services
Education Level: Bachelor's degree
Brief Description: Enforces laws and regulations in state, county, municipal, and national parks

Record 3:
Career Name: Fish and Game Warden
Average Salary: $50,420
Career Cluster: Agriculture and Natural Resources
Education Level: Bachelor's degree
Brief Description: Enforces laws that protect fish and wildlife

Record 4:
Career Name: Environmental Engineering Tech
Average Salary: $39,380
Career Cluster: Legal and Protective Services
Education Level: Technologist certificate
Brief Description: Enforces environmental laws and protects the environment

Record 5:
Career Name: Hydrologist
Average Salary: $55,000
Career Cluster: Engineering and Scientific Research
Education Level: Bachelor's degree
Brief Description: Studies distribution, circulation, and physical properties of underground and surface water

Record 6:
Career Name: Park Naturalist
Average Salary: $48,000

Continued

Input Panel A window where you can enter data using an on-screen keyboard or handwriting and then send the data to a document.

How Is Data Entered in a Tablet PC?
Input Panel is a window where you can enter data using an on-screen keyboard or handwriting and then send the data to a document.

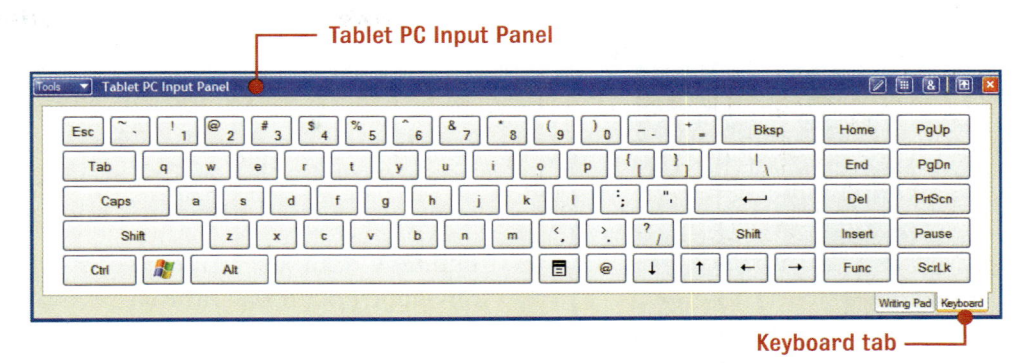

— Tablet PC Input Panel

Keyboard tab —

A digital pen is used for all input. The digital pen can be used for tapping, writing, or drawing on the screen of a Tablet PC. One way to enter data is to tap characters on the on-screen, or soft, keyboard with the digital pen. Another way is to write with the digital pen in the Writing Pad portion of Input Panel. **Writing Pad** is an area of Input Panel that recognizes handwritten input. You can choose to convert handwritten data to typed text or to keep it in the handwritten style.

Writing Pad An area of Input Panel that recognizes handwritten input.

Tablet PC Input Panel

Handwriting area

Write your text here.

— Writing Pad tab

Journal In the Journal application, you can write directly on the screen and then choose if you want to convert the handwriting into typed text. **Journal** is a standard Tablet PC application that saves data as handwriting, drawings, or typed text.

Journal A standard Tablet PC application that saves data as handwriting, drawings, or typed text.

Docking Station Another way data can be entered is by using the Tablet PC docking station, which has a small keyboard. The Tablet PC appears similar to a laptop computer when it is attached to a docking station.

✓ Concept *Check*

Think about how you would use a word processing program. Would the ability to write directly on the computer screen help you to create and edit documents? Give examples.

Step-by-Step continued

8 You will see the message **There is no primary key defined**. Choose **Yes** to add a primary key to each record in the database table.

A **primary key** is a field that contains a unique identification for each record in a database. For example, the Social Security number is commonly used as the primary key when people are part of a database table. You can let Access assign a primary key. In this case, a sequential number is added to each record in the database.

The new database table Careers is now listed in the EnviroScienceJobs Database window.

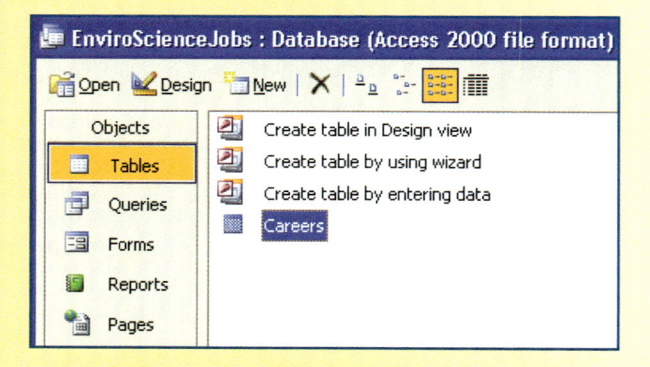

primary key A field that contains a unique identification for each record in a database.

Activity 4

DIGI Byte

Primary Key

The primary key must be a field that contains information that is unique to each record. For example, a Social Security number field could be used as a primary key, because the entries in the field will be unique to each person listed in a database.

Enter Data

To enter data into a table, first you need to open the table.

1 In the Objects pane, make sure that Tables is selected. Double-click on the table named Careers to open it, and add the information below.

2 Adjust the column width as needed by selecting the **Format** menu and choosing **Column Width**. You can also adjust column width by moving the mouse to the top right hand edge of the column. The insertion point will change into a two-headed arrow. Then click and drag the column. Press TAB to move from one field to another.

```
Record 1:
Career Name:          Landscape Architect
Average Salary:       $49,120
Career Cluster:       Agriculture and Natural
                      Resources
Education Level:      Bachelor's degree
Brief Description:    Designs and plans outdoor
                      areas
```

Continued

Step-by-Step

Practice the following steps to become familiar with the Tablet PC.

Write and Send Text

Activity 1

If you need additional help on this topic, open the Help menu; or refer to the documentation for your hardware.

1. With the digital pen, open Word by tapping the **Start** menu, selecting **All Programs**, and then choosing **Microsoft Word**.

2. Open Input Panel by tapping the **Input Panel** button next to the start menu on the taskbar.

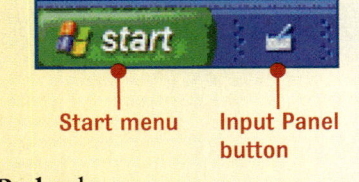

3. Input Panel will appear at the bottom of the screen. If the on-screen keyboard is displayed in Input Panel, tap the **Writing Pad** tab.

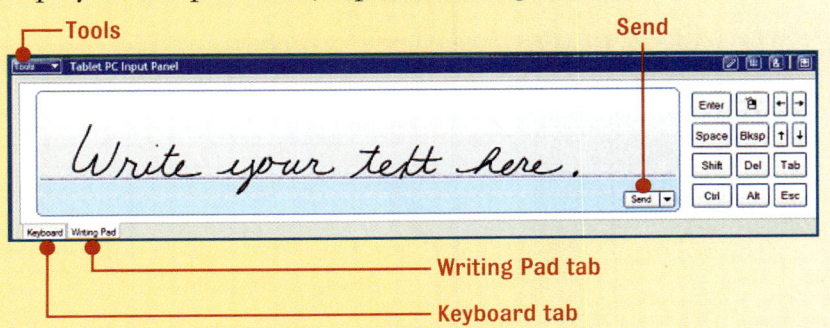

   Start menu — Input Panel button

   Tools — Send — Writing Pad tab — Keyboard tab

4. Use the digital pen to write some text, such as your name, on the entry line in Writing Pad and then tap the **Send** button to send the handwritten text to the open Word document.

   When you tap **Send**, notice that your handwriting is converted to typed text and placed in the open document where the insertion point is located.

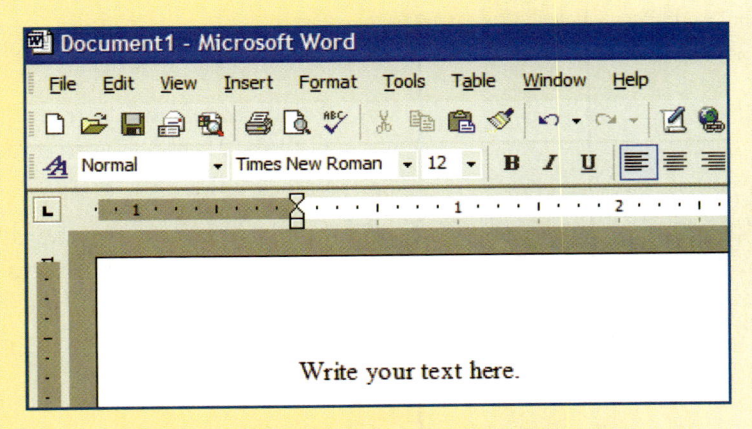

5. Now, instead of using cursive, print the same text on the entry line of Writing Pad and send it to the document. Notice that Input Panel easily converts both cursive and printed entries.

DIGI Byte

Pen-Eye Coordination

Watch the pointer on the screen, not the tip of the digital pen, to point or tap more accurately.

Continued

Activity 3

Step-by-Step continued

Choose the Data Type for a Field

1. Choose **Text** from the Data Type drop-down menu. Examples of data types are: Text, Number, Date/Time, and Currency for dollar values.

   Notice that when a data type is chosen, the Field Properties section appears at the bottom of the screen. This section allows specific changes to each data type.

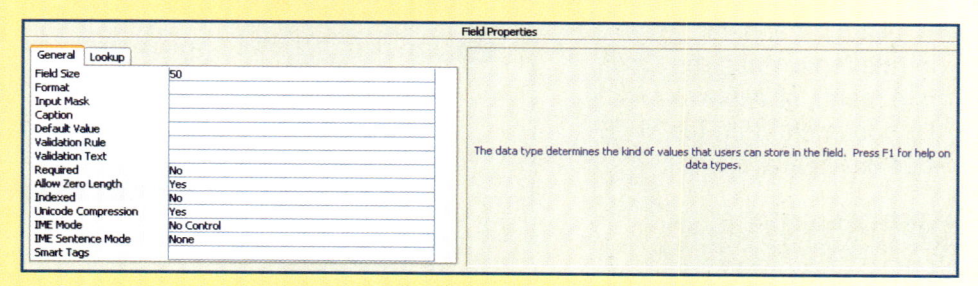

Field Properties	
General Lookup	
Field Size	50
Format	
Input Mask	
Caption	
Default Value	
Validation Rule	
Validation Text	
Required	No
Allow Zero Length	Yes
Indexed	No
Unicode Compression	Yes
IME Mode	No Control
IME Sentence Mode	None
Smart Tags	

   The data type determines the kind of values that users can store in the field. Press F1 for help on data types.

2. Press **TAB** to move to the Description column.

   The Description column can be used to enter a description of the field.

3. Press **TAB** again or press **ENTER** to move to the next row.

4. Add the following fields in your database table:

   ```
   Average Salary
   Career Cluster
   Education Level
   Brief Description
   ```

   Each record in your database table is made up of these fields: Career Name, Average Salary, Career Cluster, Education Level, and Brief Description.

5. Select **Currency** as the data type for the Average Salary field. Select **Memo** as the data type for the Brief Description field. The other fields will be the default Text data type.

6. Select the **File** menu and choose **Close**. Choose **Yes** to save changes to the table.

7. Name your table in the EnviroScienceJobs database. Enter `Careers` as the table name and click **OK**.

Continued

Step-by-Step continued

6 Tap the **Tools** button in the top-left corner of Input Panel. Choose **Options**.

7 At the Writing Pad tab, select **Automatically insert text into the active program after a pause.**

8 Adjust the slider control to choose a short or long delay. Now text will be automatically inserted into the open document as you use Writing Pad. You no longer need to tap Send.

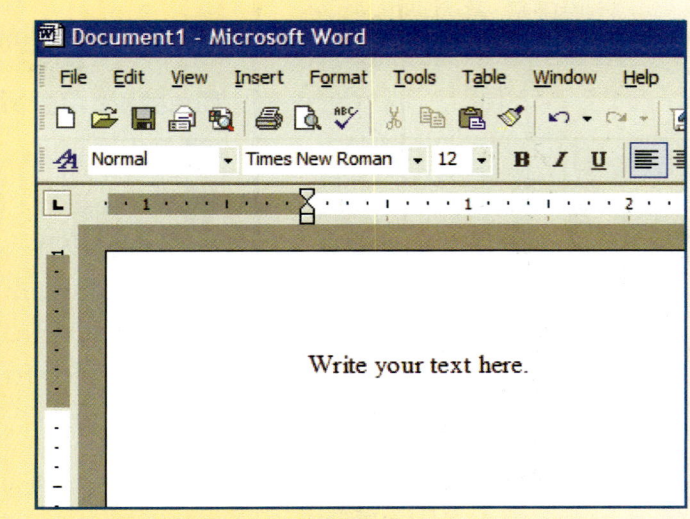

Open Journal

1 Open Journal by tapping **Start, All Programs,** and then **Windows Journal,** or by tapping the **Journal** button on the frame of the Tablet PC. This option may vary depending on the model.

A Journal page looks like a sheet of notepaper. You can write or draw directly on the Journal page with the digital pen to enter data into Journal. You will learn more about this application in Lesson 7.3, page 204.

2 Close Journal by tapping the **Close** button in the top-right corner of the Journal application window.

Continued

Activity 2

Suite Smarts

Hold down the digital pen to produce a right-click and activate shortcut menus, if the shortcut menu feature is activated.

Step-by-Step

In this step-by-step, you will create a database, define its structure, and enter data into the database.

Activity 1

If you need additional help on this topic, open the Help menu; or refer to the documentation for your hardware.

Create a Database File

1. Start Access.
2. Choose the **File** menu and click **New**.
3. Select **Blank Database** from the New File task pane.
4. You need to name your database. Enter `EnviroScienceJobs` as the database name in the File name text box at the bottom of the File New Database window and click **Create**.

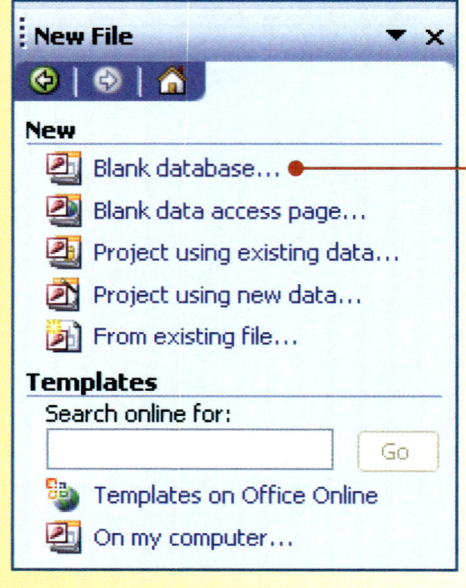

Activity 2

Create a Database Table and Define the Fields

All databases contain at least one table, and often databases contain more than one. In this activity, you will create a database table in Design view.

1. Double-click **Create table in Design view**. A blank table will appear.
2. Enter `Career Name` as the first field name in the first column of this table, titled Field Name.
3. Press TAB to move to the next column, titled Data Type.
4. Click the down arrow in the cell to display data type choices.

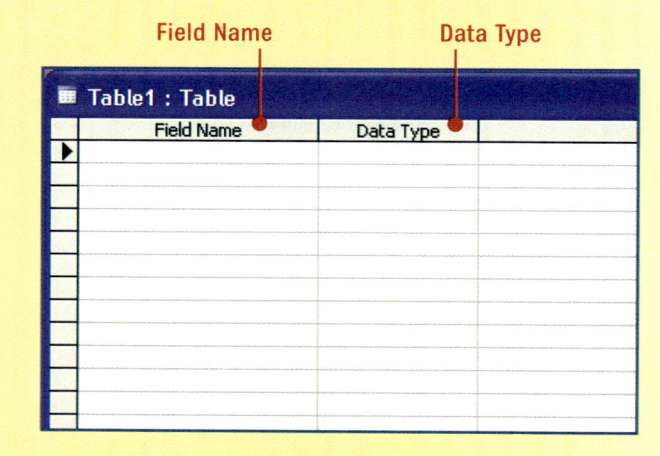

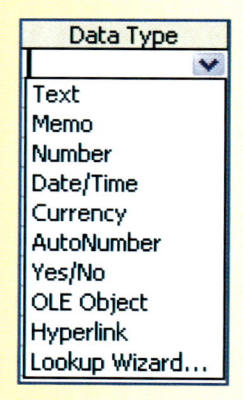

DIGI Byte

Database File Format
Access database files have the file extension .mdb.

Continued

Chapter 11

362

Step-by-Step *continued*

Calibrate the Digital Pen

Activity 3

① To calibrate the digital pen, tap the **Start** menu and select **Control Panel**.

② Double-tap **Tablet and Pen Settings**. To double-tap, tap the digital pen twice on the screen.

③ Tap the **Calibrate** button to improve the accuracy of the digital pen for writing or drawing on the screen. Follow the on-screen prompts. The calibrate function automatically improves the way the pen writes on lines and selects options.

Dock and Undock Input Panel

Activity 4

When Input Panel opens, it is automatically docked, or attached, to the bottom of the Tablet PC screen. You can undock Input Panel and move it to another area on the screen. When Input Panel is undocked, it is said to be floating.

① To undock Input Panel, tap the **Tools** menu in the top left corner of Input Panel and tap **Dock**. The check mark next to the Dock command should disappear, indicating that the command is no longer selected.

② Drag the title bar of Input Panel to move the panel to another location on the screen. To move a window, touch the digital pen to the title bar of the window, keep the pen on the screen, and drag the window to a new location. Lift the pen to drop the window in position.

> **DIGI*Byte***
>
> **Input Panel**
> An easy way to dock or undock Input Panel is to double-tap its title bar.

Change Screen Orientation

Activity 5

① To change the screen orientation of the Tablet PC, tap the **Start** menu, choose **Control Panel**, and double-tap **Tablet and Pen Settings**.

② Choose the **Display** tab.

③ Tap the arrow next to **Orientation** and choose **Portrait—Primary**. Tap the **Sequence Change** button.

④ Set 1 as **Portrait—Primary**, and set 2 as **Landscape—Secondary**. Applications will open in the portrait orientation.

⑤ Tap the hardware **Screen Orientation** button on the side of the Tablet PC. Availability of this button will depend on the Tablet PC model. Landscape is now the primary orientation, so applications will initially open in the landscape orientation.

Continued

Lesson 11.1 → Get Started with a Database

You will learn to:
- **Create** a database and design a database table.
- **Choose** the data type for fields and create a primary key.
- **Enter** data in a database and delete records.
- **Adjust** column width.

database A software application that stores a collection of related information and allows easy retrieval of the information.

field A category that holds one piece of information in a database.

record A collection of fields for one item in a database.

What Is a Database? A **database** is a software application that stores a collection of related information and allows easy retrieval of the information. In Microsoft Office, the database software is Access.

Business Use of Databases Databases often contain millions of records. Businesses use databases to keep track of customer orders, personnel records, inventory, sales, and so on. The databases can be used to provide reports to answer questions such as: What is the company's top-selling product? What were the total sales in April?

How Are Databases Structured? The basic structure of a database is a table consisting of columns and rows of data about one topic, such as a business's inventory. A simple database table is illustrated below. The columns contain fields. A **field** is a category that holds one piece of information in a database. For example, a customer's name or order number. A table's rows contain records. A **record** is a collection of fields for one item in the database. The information that appears in the fields for each record is known as entries.

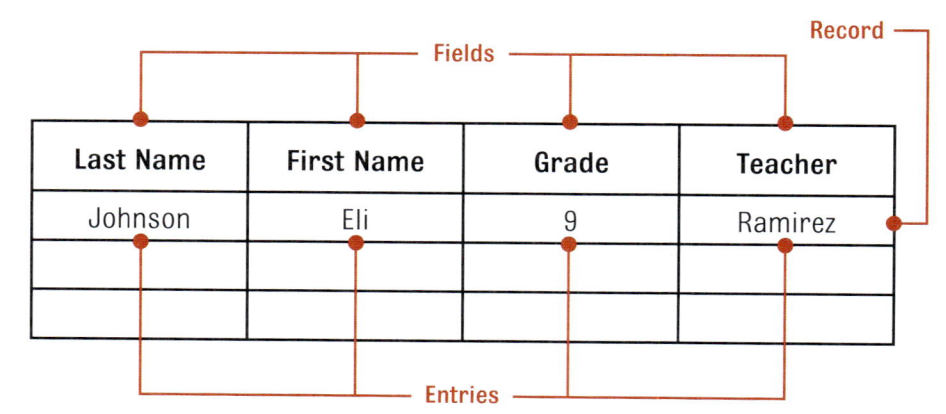

Last Name	First Name	Grade	Teacher
Johnson	Eli	9	Ramirez

Fields — Record — Entries

✓ Concept *Check*

What types of databases, including both paper and electronic, have you used either at home or school?

Activity 6

Step-by-Step *continued*

Change Screen Brightness

1. To change the screen brightness, select **Tablet and Pen Settings** in the **Control Panel** window and choose the **Display** tab.
2. Under **Screen brightness**, drag the **Brightness** control to the right to make the screen brighter. Leaving the brightness setting on low will help conserve the battery.

RUBRIC LINK

Go to the Online Learning Center at **digicom.glencoe.com** and click Lesson 7.1, Rubric, to find the application assessment guide(s).

Application 1

Directions Open Word and open Input Panel. Use Writing Pad to write your name, address, and class schedule.

Now open Journal, and write the same information. Remember that you can write directly in the Journal window. Close Journal.

Experiment with recalibrating the digital pen and changing the screen orientation and brightness to make your working environment more comfortable. Undock and move Input Panel to create more working room on the screen.

DIGITAL DIALOG

Comparison Shop Tablet PCs

Communication Form a small group. Search the Internet for information on the types of Tablet PCs that are available. Create a table similar to the one shown. Fill in the table as you comparison shop.

Add more columns and rows to the table if you need to. Compare what you find with what other groups in your class found.

Tablet PC Model	Memory	Hard Drive	Processor	Cost

Chapter 7 On-Screen Writing—Tablet PCs

Lesson 7.1   **197**

Read to Succeed

Key Terms
- database
- field
- record
- primary key
- ascending sort
- descending sort
- query
- report

Teach Others

Working together in small study groups can help you remember what you have learned. Learning part of the material in a chapter in order to teach others in your group will give you an added incentive to really understand the material.

For this chapter, work in a small group with each group member taking the responsibility to teach one of the lessons to the other group members. Create a graphic map similar to the one below to help organize your thoughts before you begin to teach. Use a Tablet PC, if possible.

Example:

Story Map for Lesson 11.1	
Lesson Section	Get Started with a Database
Key Term	Database—A software application that stores a collection of related information and allows easy retrieval of the information.
How to Do	1. Create a database table. 2. Define the fields in the table.

DIGITAL CONNECTION

Taking Inventory and Getting Organized

You have just been hired to work after school in the science lab. The science teachers share a large workroom where equipment is stored. They are not sure what equipment can be used and what needs to be thrown away. The teachers need to order new equipment, but first they want to find out how much they should order. The teachers ask you to organize the equipment in the cabinets. First, they suggest that you identify each cabinet with a letter and each shelf in the cabinets with a number so a list can be kept of where everything is located.

Writing Activity

List the type of information you think would be important to include to organize the equipment, such as Cabinet Letters. Draw a sketch, if possible using your Tablet PC, showing how the cabinets and shelves will be identified.

Lesson 7.2 — Enter Text with Input Panel

You will learn to:
- **Configure** Input Panel for right- or left-handed use.
- **View** Input Panel options.
- **Choose** two entry lines in Writing Pad.
- **Enter** data using Graffiti strokes.
- **Navigate** in a document with Quick Keys.

Getting Started Input Panel, which can be configured for left- or right-handed use, allows you to enter data by writing in Writing Pad or by tapping the on-screen keyboard. To switch between the keyboard and Writing Pad, tap the Keyboard or Writing Pad tabs. **Quick Keys** are a group of keys that provide an easy way to move the insertion point through a document. The Quick Keys pad is located next to the text entry area of Writing Pad.

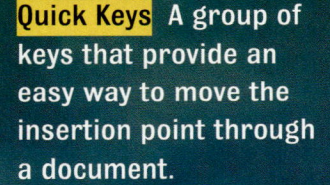

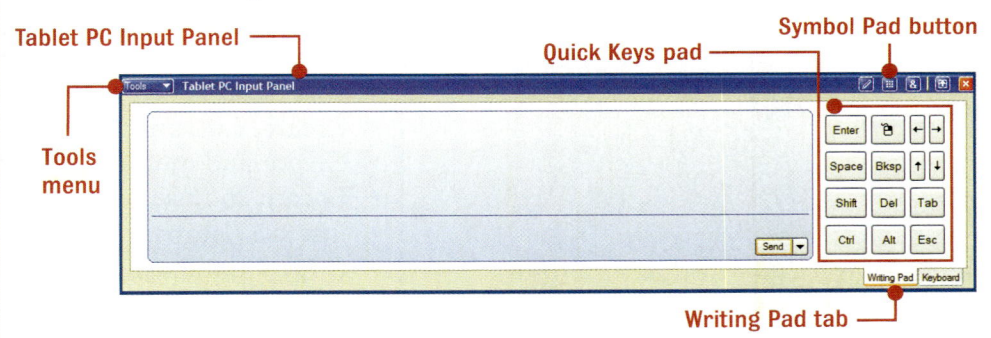

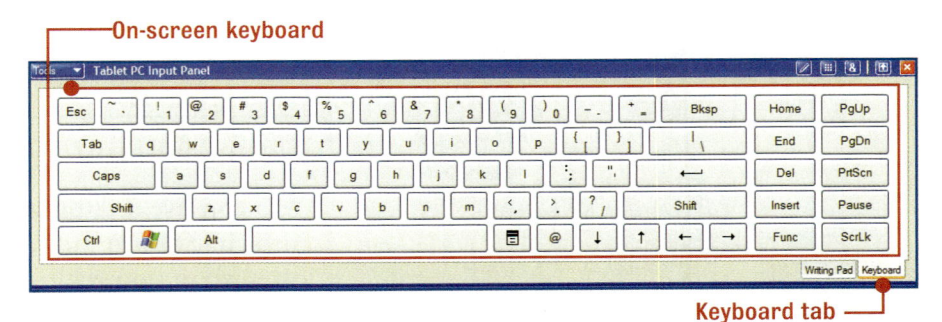

> **Quick Keys** A group of keys that provide an easy way to move the insertion point through a document.

DIGI Byte

Hide Quick Keys

To allow more space for writing in Writing Pad, you can hide the Quick Keys pad. To do this, select the **Tools** menu in Input Panel and choose **Options** and **Writing Tools**. Under **Quick keys** deselect **Show quick keys next to the writing area**. Show the Quick Keys pad again by tapping the **Quick Keys Pad** button that appears in the Input Panel title bar.

✓ Concept Check

In what ways is a Tablet PC similar to and different from a PDA? When would you use a Tablet PC instead of a PDA? What are the advantages of using a Tablet PC rather than a PDA or other portable device?

CHAPTER 11

Databases

Lesson 11.1   Get Started with a Database

Lesson 11.2   Sort Data and Create Queries and Reports

Lesson 11.3   Use Online Database Forms

Lesson 11.4   Integrate Databases with Applications

Lesson 11.5   Create a Web Page with a Database

You Will Learn To:

- **Create** a database and **enter** data.
- **Sort** data.
- **Create, run**, and **modify** a query.
- **Create** and **modify** a report.
- **Locate** and **use** online database templates.
- **Insert** a database report into a Word document.
- **Copy** a database table and **paste** it into a spreadsheet.
- **Create** and **modify** a data access page to view a database via the Web.

When Will You Ever Use a Database?

You can use a database to organize all types of information. In fact, you probably use a database more often than you realize. For example, a card catalog in a library is an example of a database. A merchandise catalog in an auto parts store is a database. A computer that provides automated information over the phone uses a database. If you have shopped on the Internet, you have used a database of the business that you purchased from.

Step-by-Step

Activity 7

Choose Left- or Right-Hand Settings

1. To configure Input Panel for right- or left-handed use, tap the **Start** menu and choose **Control Panel**.
2. Double-tap **Tablet and Pen Settings**. At the **Settings** tab, a black dot next to the right- or left-handed option indicates which option is selected. Tap to select the appropriate option.

Activity 8

View Input Panel Options

1. Tap **Start, All Programs, Microsoft Word** to open Word. Open Input Panel by tapping the **Input Panel** button next to the Start menu.
2. Tap the **Tools** menu in the top-left corner of Input Panel. Select **Options**.
3. From the Options box, you can adjust many input features. Notice the tabs: **Writing Pad, Writing Tools, Write Anywhere, Speech,** and **Advanced**.

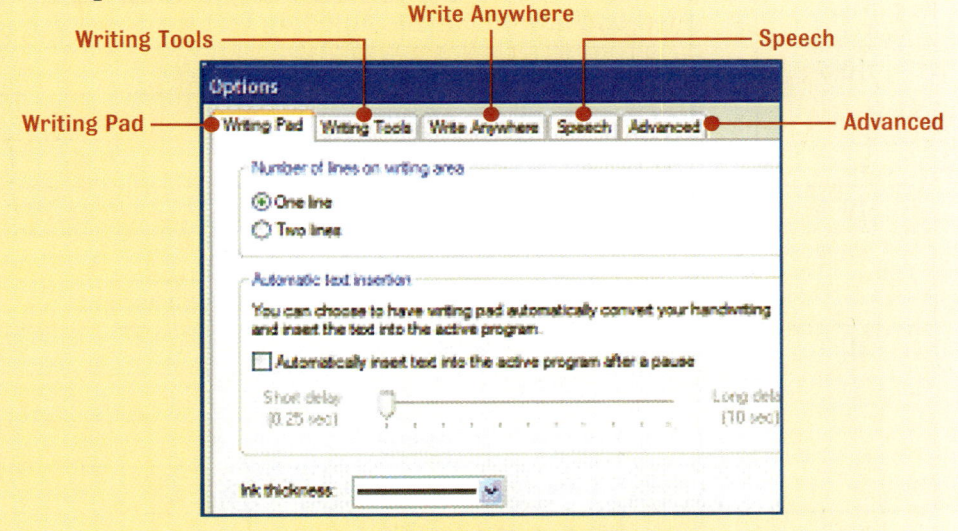

> If you need additional help on this topic, open the Help menu; or refer to the documentation for your hardware.

Activity 9

Choose the Writing Pad Tab

1. Choose two lines to write on instead of one. On the Writing Pad tab of the Options box, select the **Two lines** option and then tap **OK**.
2. When you use two entry lines in Writing Pad, notice that the first line is automatically inserted into the document as soon as you begin writing on the second line. In Writing Pad, write your name on the first entry line and your address on the second line. Tap **Send** to enter your address from the second line.

Continued

Chapter 10 Self-Assessment

Take a moment to review what you have learned in this chapter. Rank your understanding of the topics below.

4 means, "I understand all of this."
3 means, "I understand some of this."
2 means, "I understand very little of this."
1 means, "I don't remember this."

To use a printout of this chart, go to **digicom.glencoe.com** and click on **Chapter 10 Self-Assessment.**
Or:
Ask your teacher for a personal copy.

Rank Your Understanding

Lesson	Topic	4	3	2	1
10.1	Change the formatting of text and numbers in a cell				
	Edit cell contents in the cell or Formula bar				
	Change margins and page orientation				
10.2	Enter formulas using the Formula bar				
	When and how to use the AutoSum feature				
10.3	Add shading and borders to cells				
	How to use cell alignment				
	How to use Autoformat				
10.4	How to use a built-in Excel template				
10.5	Create a chart using the Chart wizard				
	Name a worksheet and move between worksheets				
10.6	Wrap text and merge cells				
	Insert rows and columns				
	Fill in a series of numbers				
	Copy a chart to a Word document				
10.7	Align title text at an angle				
	Copy a spreadsheet to a Word document				
	Add headers and footers to a chart				
	Change chart scale for printing				
	Download and use an Excel template				

If you ranked all topics 4—congratulations! Consider doing a quick review.
If you ranked yourself 3 or lower in any topic, consider reviewing these topics first.

Step-by-Step *continued*

Activity 10

Change Options in Writing Tools

The Writing Tools tab allows you to change the Input Panel screen to look like a Pocket PC or Palm input screen, where you can enter data in Graffiti.

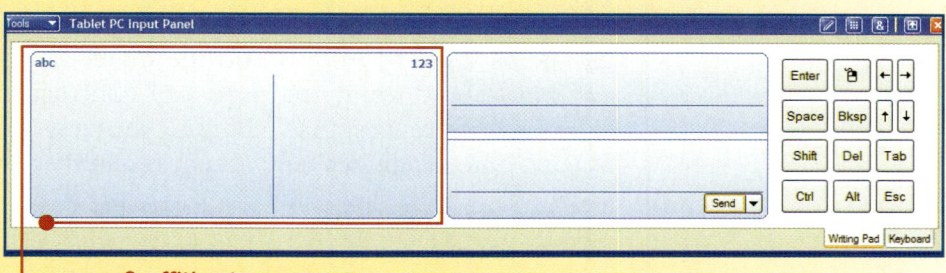

— Graffiti entry area

DIGI*Byte*

Journal Tip
Text and diagrams created in Journal can be copied to other Microsoft applications.

1. To enable Input Panel to recognize Graffiti, tap the **Tools** menu in the upper left-hand corner of Input Panel and choose **Options**.
2. Tap the **Writing Tools** tab and select **Show character recognizer on writing pad**.
3. Choose one of the PDA settings that enables Graffiti and tap **OK**.
4. Enter your name and address in Input Panel using Graffiti strokes. See Chapter 5, page 127, to review Graffiti strokes.

Activity 11

View the Write Anywhere Feature

1. Tap the **Tools** menu in Input Panel, and select **Options**.
2. Tap the **Write Anywhere** tab and view the available options. When the Write Anywhere feature is enabled, you can write anywhere on the Tablet PC screen. You will use Write Anywhere in Lesson 8.6, page 247.
3. Tap **Cancel** to close the window without changing any settings.

Activity 12

View the Speech Recognition Feature

1. Tap the **Tools** menu in Input Panel, and select **Options**.
2. Tap the **Speech** tab, and view the available options. When the Speech feature is enabled, you can use speech recognition with the Tablet PC. You may need a headset with a microphone to use this feature.
3. Tap **Cancel** to close the window without changing any settings.

Continued

CHAPTER 10 Review

Read to Succeed PRACTICE

Compare and Contrast

Many students find that analyzing relationships in material they are reading helps them to understand the material. Examples of relationships include: before and after, compare and contrast, cause and effect, and similarities and differences. As a bonus, showing the relationships in a drawing or graphic organizer helps them remember the information.

Use the experience and knowledge you have gained through this chapter on spreadsheets, and Chapter 9, Word Processing, to compare and contrast key concepts. Create a diagram similar to the illustration to describe and compare characteristics of spreadsheet and word processing applications.

Example:

Word:
Word processing software
Used to enter text documents

Microsoft Office software applications
Format text and data

Excel:
Spreadsheet software
Used to enter data for calculations

Using Key Terms

The following terms will help you to work with spreadsheets more efficiently. In Memo Pad or on a piece of paper, rewrite the sentences below, using the correct key term to complete each one.

- **spreadsheet** (315)
- **formula** (315)
- **label** (316)
- **value** (316)
- **formula bar** (319)
- **function** (324)
- **AutoFormat** (328)

1. A(n) _____ is text that is entered in a cell.

2. A(n) _____ consists of predefined shading and border formats that can be applied to cells in a spreadsheet.

3. A predefined formula in a spreadsheet is called a(n) _____.

4. A mathematical expression, such as adding or averaging, that performs calculations on data in a spreadsheet is a(n) _____.

5. The _____ displays the data within a cell and allows you to edit the contents of the cell.

6. Any number entered in a cell is called a(n) _____.

7. A software application used to list, analyze, and perform calculations on data is a(n) _____.

Step-by-Step *continued*

Activity 13 — Explore Advanced Tab Options

Create and edit documents more efficiently by changing the settings in the Advanced window. For example, to create more room on the screen to work, you can hide the pen input area when you are not using it. Also, you can display buttons for Quick Keys Pad and Symbols Pad on the Input Panel title bar, allowing easy access to these tools while you create and edit documents.

1. Tap the **Tools** menu in Input Panel and select **Options**.
2. Tap the **Advanced** tab.

    Under **Title bar buttons,** make sure that the following are selected: Quick Keys, Symbols, and Hide/Show Pen Input Area. Buttons for these tools now appear on the Input Panel title bar.

3. Tap **OK**.

Activity 14 — Practice Data Entry

1. Restart Word if the application is not still running.
2. Enter the e-mail address **yourname@yahoo.com**. Use the on-screen keyboard to enter letters. Tap the **Symbols Pad** button on the Input Panel title bar to open the Symbols pad and then enter the @ , or at, symbol.

Activity 15 — Select and Edit Text

1. You can double-tap or tap and drag with the digital pen to select any misspelled words in a document. Select part of the e-mail address you entered in Activity 14 above.
2. Rewrite the words in Writing Pad and tap **Send**, or use the soft keyboard, to replace the selected words in the document.

Activity 16 — Use Quick Keys

1. Place the insertion point somewhere in the middle of the text.
2. On Writing Pad, tap one of the arrows in Quick Keys with the digital pen. Watch what happens to the insertion point on the screen.
3. Tap the Enter key with the digital pen.

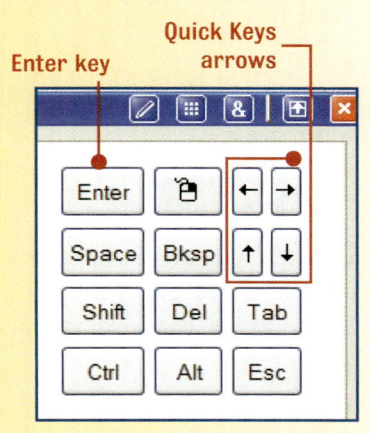

Continued

CHAPTER 10

DIGITAL Dimension → Science
INTERDISCIPLINARY PROJECT

E-Cycling

Using the spreadsheet skills you have learned in this chapter, record information about e-cycling.

Survey Collect information about electronics your classmates have. E-cycling programs focus on collecting discarded electronic devices and reusing them. Survey your friends, family, and teachers to see how many electronic devices they currently have in their homes. Assure the people you survey that the survey is anonymous and that their name will not be linked to the information. Explain that you are learning about e-cycling and are trying to find out how much potential electronic waste exists in your community. Be sure to record the number of each item people say they currently own. Organize this data in a spreadsheet.

Include the following in your survey:

- Radios
- Televisions
- Computers
- Batteries
- Screens and monitors
- Cell phones
- Print cartridges
- Portable CD players
- VCRs
- DVD players
- Printers
- Electronic gaming systems
- Video cameras

Chart Create a chart of the data you collect. Try different charts. Look at a pie chart, bar chart, and line chart. Which chart gives the best representation of the data you have collected?

Format Your Chart Add formatting to your chart. Include a title. Show amounts in your key. Move your key to another location.

E-Portfolio Activity

Spreadsheet Skills

To showcase the new spreadsheet skills you have learned in this chapter, use a spreadsheet to create a list of your skills with a spreadsheet. Include skills such as formatting text, adding borders and shading to cells, and using formulas.

Give your spreadsheet a title.

Save the document as an HTML file for your e-portfolio.

Self-Evaluation Criteria

Have you included:
- Amounts for each item?
- A chart representing your data?
- Formatting for your chart?
- Formulas to total your data?

Suite Smarts

- Tablet PC Input Panel can be configured to resemble a PDA screen.
- Input Panel has a Text Preview pane. To display this pane, tap the Tools menu on Input Panel and select Text Preview.

RUBRIC LINK

Go to the Online Learning Center at **digicom.glencoe.com** and click Lesson 7.2, Rubric, to find the application assessment guide(s).

Step-by-Step *continued*

4. You can choose to make Quick Keys a floating key pad. To do this, double-tap the **Quick Keys** button on the Input Panel title bar and drag the Quick Keys pad to a convenient location on the screen.

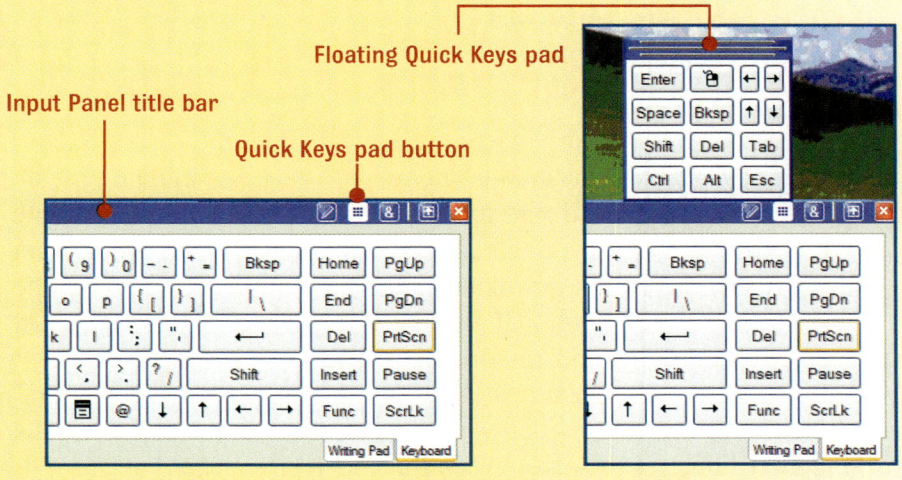

Floating Quick Keys pad
Input Panel title bar
Quick Keys pad button

Application 2

Directions Practice on-screen writing with Input Panel. Open Word and open Input Panel. Select **Options** from the Input Panel **Tools** menu, and change Writing Pad to allow text entry on two lines.

Using both lines, enter the text below. If you make a mistake, use the Quick Keys on the right side of Input Panel to make necessary edits.

After you finish entering the text, proofread carefully, select, and edit as needed.

```
When you use Tablet PC Input Panel, you need to
shape letters carefully; however, you can still
write quickly. I think using a Tablet PC in class
to take notes would be helpful. I wish I had one
to use in all my classes.

A Tablet PC could also be useful in many careers.
Artists, scientists, and writers are three groups
that already use Tablet PCs often. Tablet PCs are
extremely versatile and easy to use.
```

DIGITAL Career Perspective

Law and Public Safety

Viviana Pedroso
Attorney

Technology Aids Justice

The Internet is a valuable tool for attorney Viviana Pedroso. "I can review a pending file on the relevant department's Web site. Documents are uploaded to the site where I can view them or download them, allowing me to confirm the status of a case," says Pedroso. She adds, "Most courts have a Web site that you can access to obtain all kinds of important information, because what occurs in most lawsuits is public information." Pedroso can access this information whether she is home working on a case or working after hours at the office when the courts are closed.

Pedroso works for a law firm that represents insurance companies and employers when employees are injured on their job. This type of legal representation is known as worker's compensation. Pedroso needs to determine if the employee's claims for injuries are valid. "I file pleadings and appear in court on behalf of the insurance company and employer," explains Pedroso. Word processing makes it easy to create pleadings.

"I also use e-mail, a calendar on my computer, and a program to keep track of the time spent on each client's case," says Pedroso. "The Internet is a valuable research tool, and e-mail makes it quicker and easier to communicate with clients. I don't have to scour through a paper file to find the document I need, because everything is right there in one place."

What does Pedroso like best about her job? "The satisfaction of obtaining a just result for someone who really deserves it."

Career Growth Trend

Opportunities for attorneys will continue to increase, due to growth in the population and increase in demand for legal services in elder, antitrust, environmental, and intellectual property law.

Growth Trend: 10%–20% Growth (Year 1 to Year 10)

Training

A bachelor's degree, law degree, and the passing of the bar examination are required. Any college major is acceptable, but a multidisciplinary background is recommended. Some useful courses to take are English, foreign languages, public speaking, government, philosophy, history, economics, mathematics, and computer science.

Salary Range

Typical earnings are in the range of $60,000 to $120,000 a year.

Skills and Talents

Attorneys need to be:

Strong, persuasive, verbal and written communicators

Creative

Able to think logically and quickly

Able to exercise good judgment

Excellent readers and researchers

Highly ethical

Career Activity

Find out about different areas of law. What are the two main areas?

21ST CENTURY ⟷ CONNECTION

Are Phones Getting Smarter?

Smartphones combine the capabilities of your PDA with a cellular phone. You can stay organized and keep in touch at the same time. The smartphone gives you everything you need and is easy to use. Smartphones are available in the Palm OS or Windows Mobile Software. A built-in QWERTY keyboard on some models makes entering text easy. On other models, text is entered with the number/letter phone keypad. You can even customize a smartphone with downloadable ringtones. Expansion slots provide additional storage memory for applications, pictures, and other information you may need. The smartphone is indispensable for people on the go.

Capabilities of a PDA The handheld applications you have become familiar with are also available on a smartphone. Included are a Date Book, Contact List, Memo Pad, Calculator, To Do List, word processor, and electronic book reader. In addition, you have the capability to synchronize your smartphone to your desktop or laptop computer. You can download freeware or shareware applications for PDAs to your smartphone, too.

Some providers have features that will allow you to synchronize your data over the air. Data can be beamed to another infrared device or received from another device.

Wireless Connectivity Wireless connectivity is provided through your cellular provider. With this connectivity, you can do multiple tasks. You have access to browse the Internet from anywhere. If you are traveling, you can check your Outlook e-mail account and send and receive e-mail messages.

A built-in camera allows you to take pictures and send them to others via e-mail or Picture Mail. Pictures can even be attached to phone numbers so you can see who is calling you with a Picture Caller ID.

By using the Internet connection, you can send instant messages. In addition to placing or receiving calls with friends, you can keep friends and contacts in your contact list to send them an SMS (Short Message Service) text message anytime. A chat window allows you to send or read messages.

Listen to Music Some smartphones have Windows Media® Player. You can listen to your favorite music while sending a text message. Music can be downloaded to your desktop computer and then synchronized to your smartphone. You can watch news clips, music videos, film trailers, and other video clips on your smartphone with Windows Media Player.

Activity Are phones getting smarter? List some of the newest features of smartphones available today. Rank the features in order of priority of usefulness to you.

21ST CENTURY CONNECTION

Accountability and Career Success

Being accountable means that you can be counted on to accept responsibilities, handle them competently, and deliver positive results. Whether in a school study group or a department at a company, you are a team player who shares a stake with other members in the outcome of your projects. Each of you is connected to the effort, so you want the end product to be a reflection of everyone's excellent work.

Accountability at Work Here are some ways to be accountable on the job:

- Respond to e-mail and voice mail promptly. Even if you cannot respond in detail immediately, you can let the sender know you have received his or her message and are working to provide the information.
- Communicate on a regular basis with team members and superiors to give project status reports, ask questions, and give feedback.
- If you are unsure of how to solve a problem or if you do not know something, admit it. Be willing to find the answer or to learn how to resolve a situation. Research the problem if necessary, and discuss your questions with superiors or others on your team. You will be respected for your honest questions and willingness to learn.
- When you make a mistake, take responsibility for it. Then take whatever steps are necessary to correct it.

Accountability in Everyday Life Making an effort to be accountable from day to day can help you build good habits for managing your work and your personal affairs.

- Send thank-you notes. When gifts are sent from a distance, a thank-you note lets the sender know it was received. If someone has done you a favor, a note via e-mail or postal mail shows appreciation.
- Return borrowed items promptly.
- Follow through on your promises and obligations. Everyone appreciates a dependable person whose word is as good as his or her deed.

Using Digital Tools Responsibly

- Save your work often. Back up all documents and e-mails on your hard drive and on disks when you finish assignments. Unsaved material could lead to hours or weeks of lost labor and productivity. When you save your work, you are also creating a chronological history of your projects and a record of your progress.
- Back up information you have stored in a PDA or digital notepad on your hard drive and on disks. Print copies for your files occasionally.
- Read and re-read all outgoing e-mails to make sure the tone as well as the words are consistent with what you intended.

Activity The 19th century Norwegian playwright Henrik Ibsen said that being part of a community is like being on a ship where, "Everyone ought to be prepared to take the helm." Write a paragraph about how this applies to being an accountable person.

Lesson 7.3 → Write and Edit On-Screen in Journal

You will learn to:
- **Choose** pen and highlighter settings in Journal.
- **Write** and save a note in Journal.
- **Convert** handwriting to typed text.
- **Add** a flag to a Journal note.
- **Insert** line spaces in Journal.

DIGIByte

Print from a Desktop Computer

Flash Drive You can save a Journal note on a flash drive so you can easily print it from another computer.

Journal Viewer The free Journal Viewer application can be used to view and print Journal notes from a computer that does not have the entire Journal application installed.

Working with Journal The Journal screen looks like a sheet of notebook paper. As you write with the digital pen in Journal, your handwriting appears on the screen just as your handwriting might appear on a sheet of paper. The handwriting is not converted to typed text automatically, although you can convert it to typed text later if you choose to.

Features of Journal You can change the settings of the digital pen to give different colors and thicknesses of ink. If a note is important, you can mark it with a flag. Journal notes can be saved and printed.

Inserting Notes Have you ever used a spiral notebook for notes only to realize that you needed to insert more notes later? How did you add them? Maybe you tried to squeeze them into the margins but found that this made your notes messy and unorganized. With a Tablet PC, you can insert space for additional notes exactly where you need to without rewriting any existing text or making your notes unorganized.

✓ Concept Check

When would it be helpful to convert handwritten input to typed text with the Journal application?

DIGITAL DIALOG

Proofreading and Editing

Teamwork You have been hired by the local newspaper to proofread copy before it goes to press. Newspaper articles will be sent to you via e-mail so that you can proofread them and make corrections before they are printed. Work with a partner to think of some ways to make the proofreading most effective. Discuss why it is important that mistakes be found and corrected. In your opinion, would a Tablet PC help in the proofreading process? Support your answer. Share your ideas with the class.

Application 14

Directions Start Excel. Search the Microsoft Web site to find a template called Event budget, which is in the Business Financial Planning group in the Finance and Accounting category. Download this template, and save it to a floppy disk or to your directory. Enter the following figures in the appropriate cells in the Actual column.

Room and hall fees	$900
Equipment	$250
Tables and chairs	$1500
Lighting	$375
Stationery supplies	$135
Food	$575
Drinks	$250
Staff and gratuities	$1800

Save your spreadsheet, Print Preview, and then print.

Application 15

Scenario The operations manager of the company you work for, Interior Designs by Kay, has asked you to search the Internet for payroll record-keeping and purchasing forms that the company can download and then use.

Directions Start Excel. As you search the Internet, create a spreadsheet that lists the forms you find, their cost, the Web site address where you found the forms, and a brief explanation of how each form can be used.

When you are finished, save and print your spreadsheet. Share your list with a classmate to compare the types of forms you discovered.
- Are they similar?
- How many are the same?
- How many are different?

Discuss your answers with the class.

DIGITAL Decisions

College Computing

As you prepare to go to college, your parents tell you that they will buy you a computer and printer to take with you. Now you will need to choose the right computer from all of the models available. Before shopping at the discount store or large retail store, you decide to do some comparison shopping online.

Internet Investigation Set up an Excel spreadsheet. Along the top row, enter these headings:

- Price.
- Memory.
- Monitor size.
- Hard drive size.
- Internet access.
- USB ports.
- CD or DVD drive.

Navigate online to the Web sites of two computer manufacturers. Select a basic and a full-featured model of each type of computer. Enter the name of the manufacturer and the model name as the left column label.

Fill in the cells on your spreadsheet with data from the descriptions of the computers you are comparing. After you have completed your spreadsheet, compare the data to see which computer model is the best value. Share your analysis with the class.

Activity 17

If you need additional help on this topic, open the Help menu; or refer to the documentation for your hardware.

Activity 18

DIGI Byte

Format Toolbar

Add the Format toolbar to the Journal screen to allow you to change ink colors quickly. To do this, touch the digital pen on the Toolbar area and right-click; then choose Format.

Step-by-Step

Open Journal

1. Open Journal by tapping the **Start** menu, selecting **Programs**, and choosing **Journal** or by tapping the hardware **Journal** button on the frame of the Tablet PC. This button is not available on some models.

2. Notice the Note Title text box at the top of the Journal screen. Saved files are called notes. If you write something in the Note Title text box, that text will become the name of the file. Journal automatically saves notes.

Eraser
Highlighter
Selection Tool
Pen

Change Pen and Highlighter Settings

1. Tap the **Pen** button on the Toolbar of Journal.

2. Select one of the pen settings from the Pen drop-down list or choose **Pen Settings** to select ink color or line thickness and then tap **OK**.

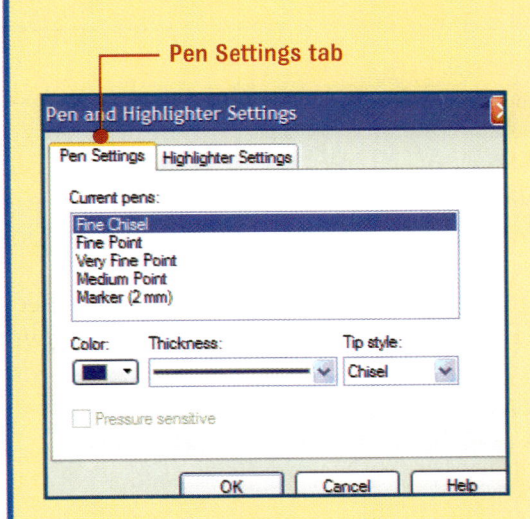

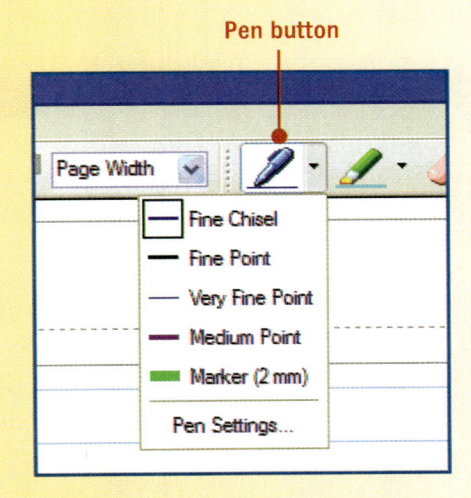

Pen Settings tab
Pen button

Continued

Chapter 7

205

Step-by-Step continued

```
Date              Today's date
Invoice #         2954
Description       10 Print cartridges
Amount            $1,099.50
```

Suite Smarts
This template could also be copied and pasted into a Word document.

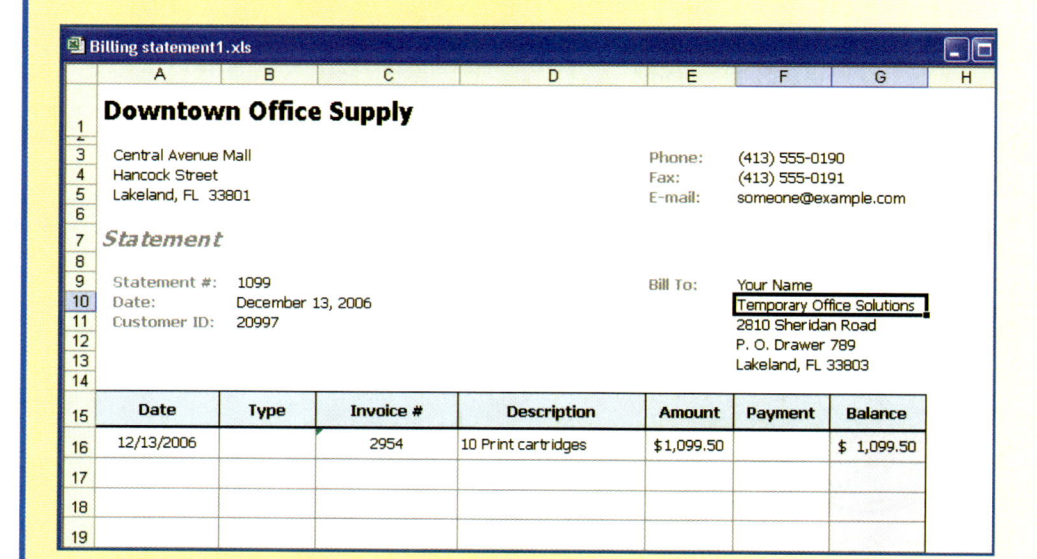

② Select the **File** menu and choose **Save As**.

③ Name the spreadsheet **Temporary Office**.

④ Print preview the spreadsheet and then print it.

RUBRIC LINK
Go to the Online Learning Center at **digicom.glencoe.com** and click Lesson 10.8, Rubric, to find the application assessment guide(s).

Application 13

Directions Think about how you could use a spreadsheet for an assignment that you have right now. Perhaps you need a spreadsheet to create a calendar or to prepare a budget.

Create a list of the kinds of spreadsheets that could be helpful, using a PDA or Tablet PC, if possible. Download and fill in a spreadsheet template from the Microsoft Web site that matches one of your descriptions. Customize the template to make the spreadsheet as useful as possible to meet your needs.

Step-by-Step *continued*

3. Tap the **Highlighter** button to change the color or thickness of the highlighter.

Highlighter
Highlighter colors
Highlighter Settings

Activity 19

Write and Save a Journal Note

1. Using the digital pen, write the information below in Journal.

> Make notes in this area by writing directly on the screen with your digital pen.

2. Tap the **File** menu and select **Save As** to save this note.
3. Name the note **Practice**. Use Writing Pad or the on-screen keyboard to name the note.

Activity 20

Select Handwriting and Convert It to Typed Text

1. **Open** the note named Practice, if you closed it.
2. Tap the **Selection** Tool button on the toolbar that is the shape of a lasso. The **Selection Tool** is a tool that enables you to select text, handwriting, or graphics in Journal. The selected text can then be moved, deleted, resized, cut and pasted, or copied.

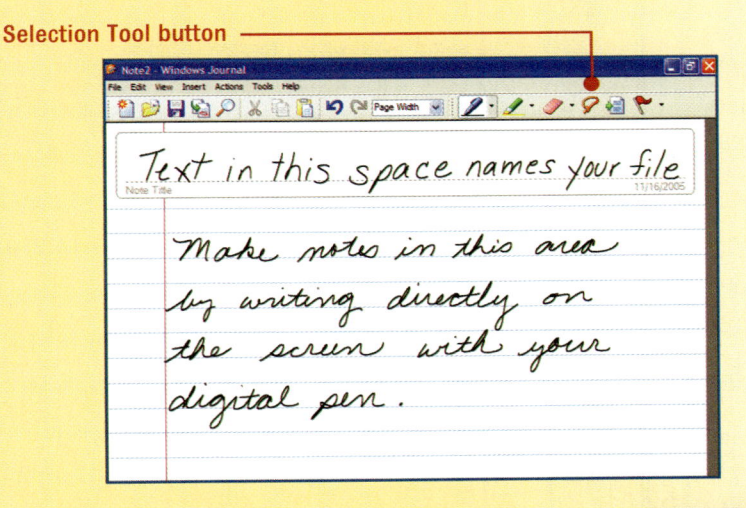

Selection Tool A tool that enables you to select text, handwriting, or graphics in Journal.

Selection Tool button

Continued

Activity 36

If you need additional help on this topic, open the Help menu; or refer to the documentation for your hardware.

Step-by-Step

Download an Excel Template

You can search for business templates on the Internet, or you can access online templates from a link to the Microsoft Web site from within Excel.

1. Start Excel.
2. Select the **File** menu and choose **New**.
3. In the New Workbook task pane, click **Templates on Microsoft.com** or **Templates on Office Online**.
4. In the Finance and Accounting category, click the **Accounting and Reporting** link.
5. Download the **Billing Statement** template, and save it to a floppy disk or the hard drive.

    This file downloads as a zipped, or compressed, file. You will need to use zip software such as WinZip to unzip the file before you complete the next step. You can find a free evaluation version of WinZip on the Internet by conducting a keyword search.

6. Open the file.

Activity 37

Customize a Template

1. Enter the following data into your spreadsheet. Do not enter anything in the Total column.

```
Your Company Name     Downtown Office Supply
Street Address        Central Avenue Mall
Address 2             Hancock Street
City, ST ZIP Code     Lakeland, FL 33801
Statement #           1099
Date                  Today's date
Customer ID           20997
Bill To               Your name
                      Temporary Office Solutions
                      2810 Sheridan Road
                      P.O. Drawer 789
                      Lakeland, FL 33803
```

DIGI*Byte*

WinZip

WinZip is a program that opens files that have been compressed, or zipped. You can obtain a free evaluation copy of WinZip on the Internet. Use a search engine to locate a site where you can download a trial version of WinZip.

Continued

Step-by-Step continued

DIGI Byte

Journal Tip

Word documents can be imported into Journal so that you can mark them up or make notes on them during a meeting.

3 Draw a circle around the handwritten text in the note. Notice how the appearance of the text changes to show that it is selected.

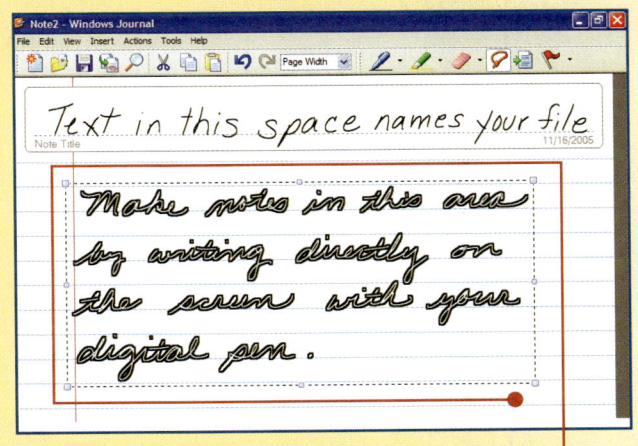

Selected text

4 Tap the **Actions** menu, and choose **Convert Handwriting to Text**.

5 Follow the directions in the Text Correction window that appears. Words that the computer is unable to understand will be highlighted in this window, and you will be able to make any necessary corrections.

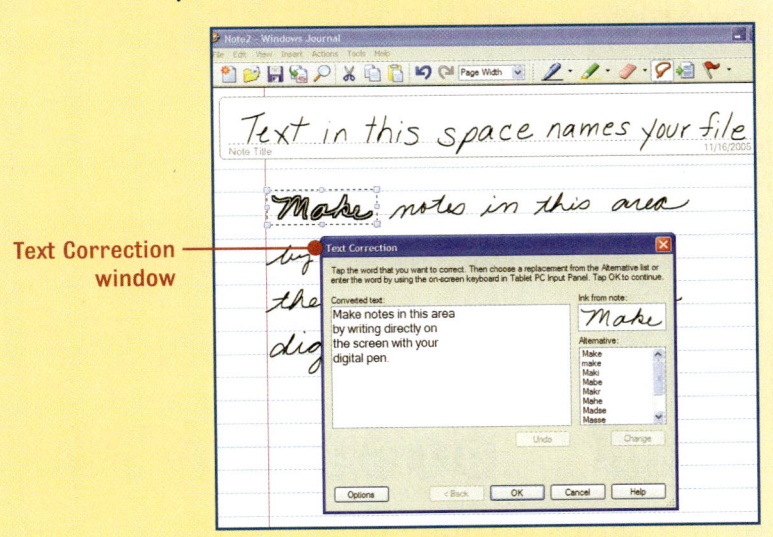

Text Correction window

Double-tap a word you want to replace. Tap to position the insertion point to insert letters or spaces. You can move the Text Correction window to see your handwritten text.

6 After you have finished making corrections, tap **OK**.

Continued

Lesson 10.8 Use Online Spreadsheet Templates

You will learn to:
- **Locate** online Excel templates.
- **Customize** an Excel template to create a spreadsheet.

Why Use Templates? Spreadsheets can be designed to accommodate a variety of business needs that an individual or office may have. Many businesses use a particular set of Excel forms again and again. Instead of creating a new spreadsheet every time one is needed, you can save a spreadsheet as a template and reuse it. There are many ready-made templates available on the Internet that can be downloaded and then customized to fit individual needs. In this lesson, you will go to the Microsoft Web site to download and then make modifications to a template.

Types of Business Templates Some common uses of Excel spreadsheets in business include:

- Accounts receivable.
- Asset depreciation.
- Business calendar.
- Business trip budget.
- Cash flow statement.
- Customer management.
- Expense budget.
- Invoices.
- Profit projection.

✓ Concept Check

Can you think of any forms that you have looked at that could be useful as a spreadsheet template? Describe the forms.

DIGITAL DIALOG

Planning with a Spreadsheet

Teamwork In a small group, discuss the following questions:
- Why would using a spreadsheet to keep track of the budget for an event be helpful to a company?
- Why is using a spreadsheet to plan for an event helpful?

Explain how you can use spreadsheets to estimate how much an event will cost.

Step-by-Step continued

7. In the Text Correction window that appears, under **Choose what you want to do with the converted text**, select **Insert in the same Journal note**. Tap **Finish**.

8. Notice that the handwriting is now converted to text.

9. Resize the text box by dragging one of the corner sizing handles.

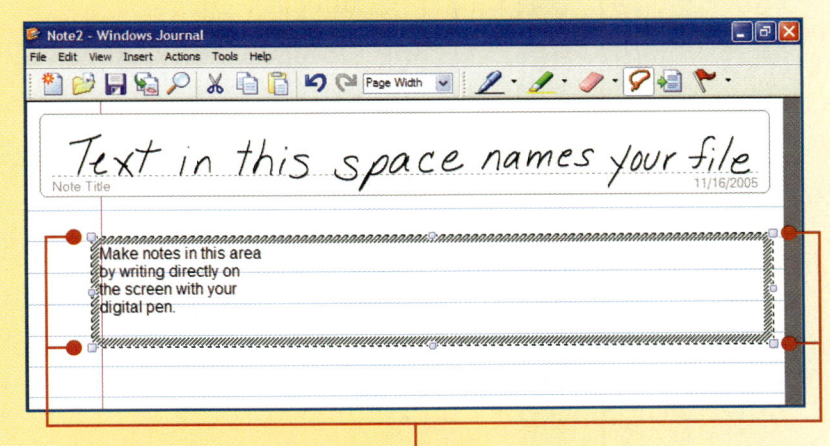

Corner sizing handles

Add a Flag

You can add a flag to an important Journal note to help you easily find it later.

1. Tap the **Flag** button arrow and choose a flag color.

2. Tap to place the flag next to information in the note that you may want to find later. You can drag the flag to any location with the digital pen. You will learn more about flags in Chapter 8, Lesson 8.4, page 239.

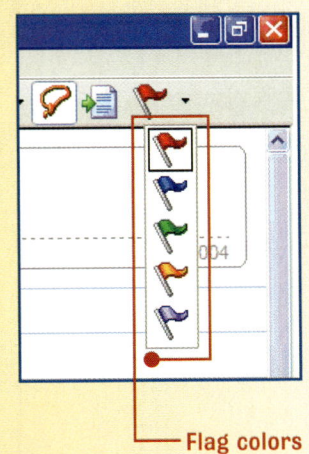

Flag colors

Insert Line Spaces in a Journal Note

1. Open Journal and enter the text shown below.

```
If you create a Journal note and decide to add
extra space between the written lines, you can
do this by using the Insert Space button on the
Toolbar.
```

Activity 21

Activity 22

Continued

208

Chapter 7

Step-by-Step *continued*

2. Click the **Setup** button, and then click the **Header/Footer** tab.
3. From the Header drop-down list, select **Annual Data Page 1** and click **OK**.
4. Open the Page Setup window again, and choose the **Header/Footer** tab. From the Footer drop-down list, choose **Sunny Day** and click **OK**.
5. Open the Page Setup window and click the **Page** tab.
6. Change the page orientation to Portrait.
7. Click the **Chart** tab. Make sure that **Use full page** is selected and click **OK**.
8. Click the **Close** button to return to your spreadsheet.
9. Rename the worksheet tabs. Change *Chart 1* to `Annual Chart` and *Sheet 1* to `Annual Data`.

DIGI*Byte*

Scaling

Scaling ensures that a chart will fit the size of the page on which you choose to print your chart.

RUBRIC LINK

Go to the Online Learning Center at **digicom.glencoe.com** and click Lesson 10.7, Rubric, to find the application assessment guide(s).

Application 11

Directions Open one spreadsheet you created in this chapter. Change the column headings so they are at a 90 degree angle. Does this allow you to fit many more columns on a single printed page? Insert a row above the column headings, and merge the cells in the row. Enter a title for the spreadsheet, and format it to stand out from the rest of the spreadsheet.

Application 12

Directions In your Sunny Day spreadsheet, create a chart for each quarter that shows the total sales for that quarter. You will end up with four charts. Rename each chart tab according to the quarter they represent.

Hint: Select *Product* and the products listed in Column A by holding down `CTRL` and then selecting the data in the *2nd QTR column*. Then click the Chart Wizard button. Repeat this process for the third and fourth quarter data.

Copy and paste each quarter chart into a Word document. Add the title *Sunny Days Annual Sales by Quarters* to the Word document, and save this file as **Sunny Days Sale**.

In Excel, print preview the four charts. Change the page orientation to landscape, and change the chart scale to fit the page. Add a header to each chart that describes which quarter the data is taken from.

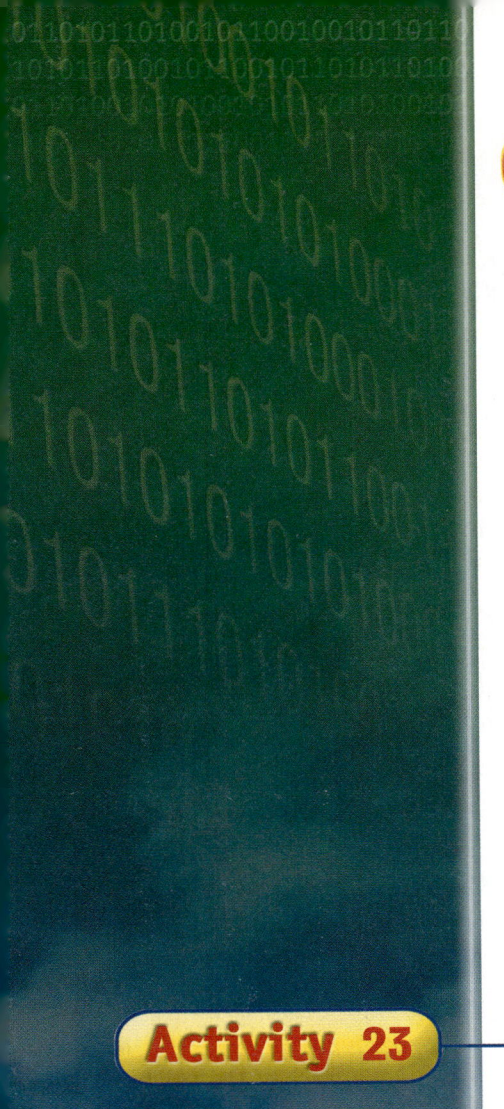

Step-by-Step continued

2. Select the **Insert Space** button.

3. Move the digital pen to the location in the note where you want to add a line space. A dotted line will appear on the screen.

4. Drag the digital pen down the page until enough space appears on the screen.

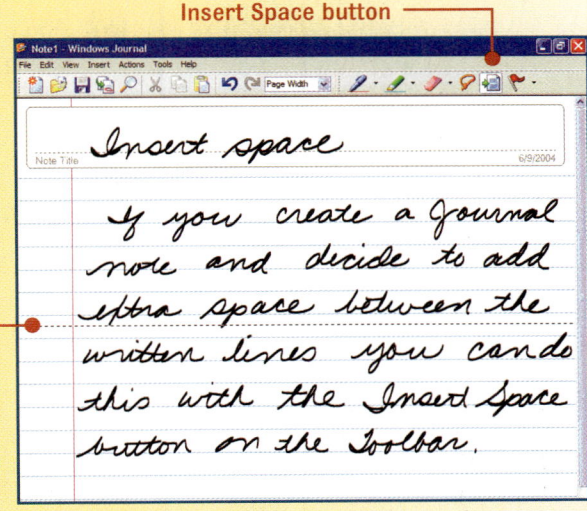

Insert Space button

Inserting line space

5. When you lift the digital pen from the screen, the space will be inserted. You can now add more handwritten notes.

6. To close up the empty space, tap the **Insert Space** button again. Place the digital pen just above the line of text you want to move up. Drag the pen up to close the space.

Erase and Undo

1. To erase handwriting, tap the **Eraser** tool.

2. Notice that the pointer takes the shape of a square eraser. Move the eraser over the lines you want to erase.

3. If you make a mistake while erasing, tap the **Undo** button to bring back any lines or words that you accidentally erased.

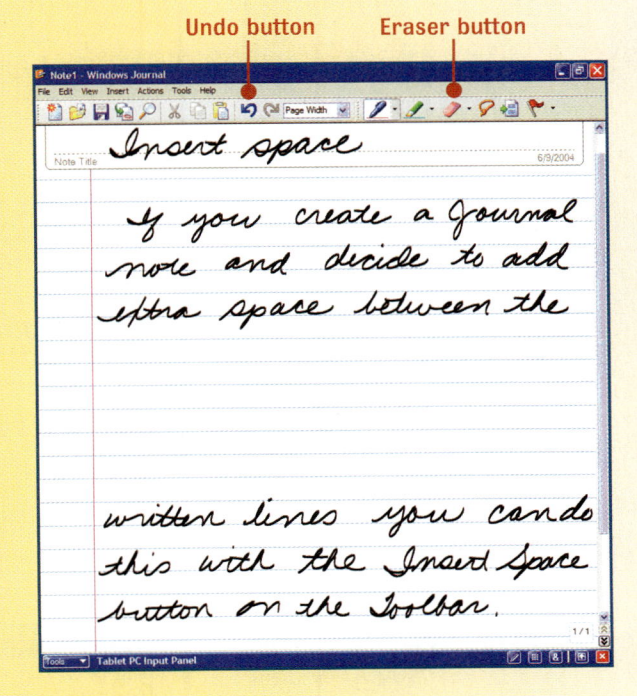

Undo button   *Eraser button*

Activity 23

Continued

209

Chapter 7

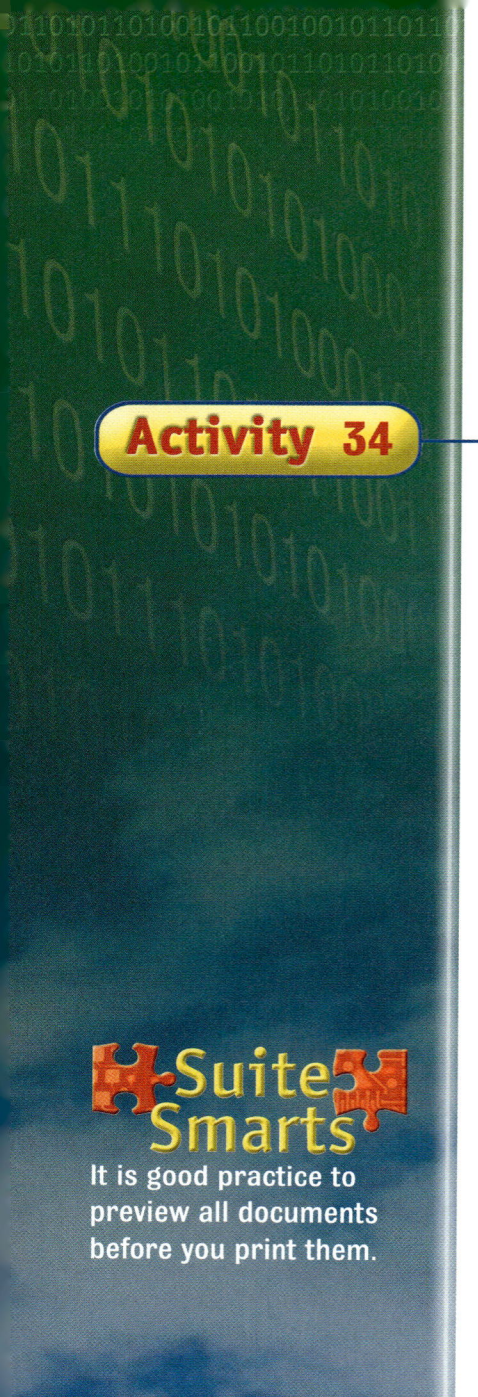

Step-by-Step continued

③ Place the insertion point where you want to insert your spreadsheet data, and then select the **Edit** menu and click **Paste**. The spreadsheet will appear in your document.

④ Save your Word document as **Marketing Memo**.

Apply Formatting to a Chart

In this activity and Activity 35, you will go back to the Sunny Day spreadsheet and apply formatting and use Print Preview features, including adding headers and footers.

① In the **Sunny Day** file, select Cells A3 through E9.

② Click the **Chart Wizard** button.

③ In Step 1 of the wizard, choose **Bar Chart**.

④ Click **Next** until you reach Step 3.

⑤ Click the **Gridlines** tab and select **Minor Grid Lines**. Click **Next**.

⑥ In Step 4 of the Wizard, select the **As new sheet** option and click **Finish**.

Which product sold the best in the 4th Quarter? Which product sold the best in the 2nd Quarter?

⑦ Select the **Chart** menu and click **Source Data**.

⑧ In the Source Data window, choose the **Rows** option and click **OK**.

What does the chart show now?

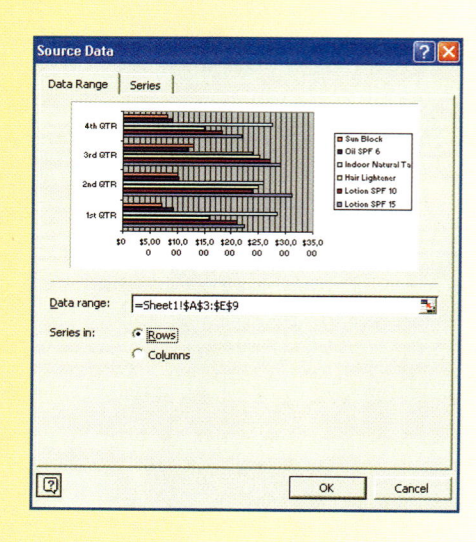

Add a Header and a Footer to a Chart

① Select the **File** menu and choose **Print Preview**.

In the Print Preview screen, you can zoom in or out of your document using the Zoom button.

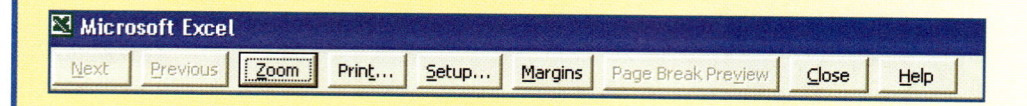

Activity 34

Suite Smarts
It is good practice to preview all documents before you print them.

Activity 35

Continued

Chapter 10

348

Activity 24

RUBRIC LINK

Go to the Online Learning Center at **digicom.glencoe.com** and click Lesson 7.3, Rubric, to find the application assessment guide(s).

Step-by-Step *continued*

View Journal Notes

1. To view Journal notes, navigate to the My Notes folder in the My Documents folder. Journal notes are automatically saved in My Notes.

2. To quickly access recently used Journal notes, tap the **File** menu and select **Recently Used Notes**. Notice that the titles of the notes appear in the Recently Used Notes list as they are in the note itself—either in your handwriting or as typed text.

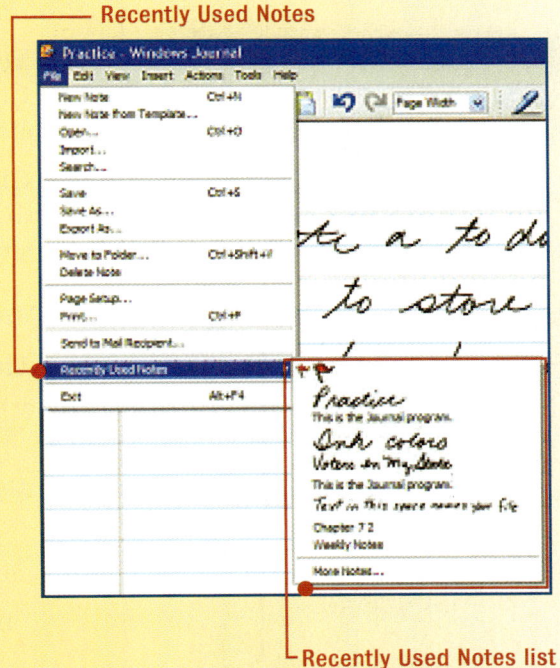

Application 3

Directions Open Journal and write the following paragraph. Use the Selection Tool to select your writing. Convert the handwriting to typed text. Add a flag to the note. Save the note as *Your Name* Journal 1.

```
Have you ever wondered why you cannot access some
Web sites from school computers? Congress passed
the Children's Internet Protection Act in 2000.
According to this law, all schools in the United
States must provide technology that prevents
student access to dangerous or offensive Web
sites.

Most schools have firewall software that helps
block these harmful sites. This software can also
monitor schools' Internet traffic and identify
anyone trying to use the Internet for illegal
activities, such as downloading copyrighted music
files without permission.
```

Step-by-Step continued

Activity 31

Use Page Break View

1. You have been working in the Normal view for your spreadsheet. Click **View**. You will see Normal is selected because the icon next to Normal is shaded.

2. Click **Page Break View** to select it. You will see the extra cells disappear. Only the cells containing the data you entered will show. These are the cells that will print.

Activity 32

Magnify Your Spreadsheet

You can increase the size of your spreadsheet on your screen.

1. Click the drop-down arrow in the Zoom window.

2. Choose the amount of magnification you want.

   Choosing a smaller number decreases the size of your spreadsheet on the screen.
   Adjusting the zoom level on your spreadsheet helps you to see more or fewer cells, depending on the size of your spreadsheet.

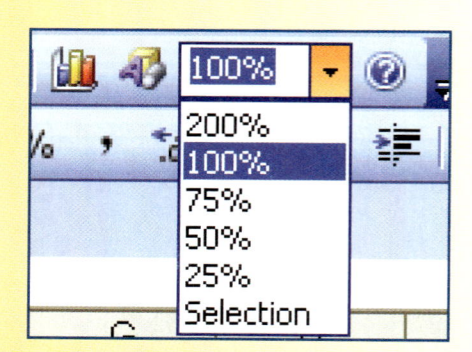

Activity 33

Paste an Excel Spreadsheet into a Word Document

In this activity you will create a chart and then copy and paste it into a Word document, Marketing Memo, which you will create.

1. Open the **Sunny Day** file again, if necessary. Select Cells A1 through E9, click the **Edit** menu, and choose **Copy**.

2. Open a new document in Word, and create a memo that contains the text below.

```
To:     Marketing Dept.
From:   Accounting Dept.
Re:     Annual Sales
Date:   Today's Date

Please review the information below. This table
represents last year's sales data on our top
products. We will discuss these figures at our
weekly sales meeting next Monday.
```

Continued

Application 4

Directions Use Writing Pad to enter the following text in Word.

> Have you ever wondered why you get so much junk e-mail? One reason could be that information about you is being sold by one company to other companies. Many Web sites gather information about you when you visit them. This data is stored and used without you realizing that it has even been collected. The data may contain information about where you shop, what foods you like, and what services you are searching for. This data can be shared with other companies, and they may begin sending you e-mails about their products.

What happens when you tap the control (Ctrl) Quick Key and then tap one of the arrow Quick Keys? Use Quick Keys to move around the paragraph.

Think of some information to add to the selection you just entered. For example, include the types of junk e-mail you have received. Insert space in the note and enter the additional information.

TECH ETHICS

Downloading Music and Games

Scenario Paula Culton is a distribution manager for Interstate Food, a food distribution company. The managers use Tablet PCs to review online order requests.

Culton has a slow Internet connection at home. At work, where the Internet connection is fast, Culton often downloads programs from game and music Web sites.

In a recent staff meeting, another manager complained that the Internet seems to be really slow. The technology engineers investigated and discovered that several employees are downloading games and music during company time. This activity is slowing down the Internet for other employees because the bandwidth is being used for downloading.

? Critical Thinking Should Culton and others be using the company Internet connection to download games and music? Why or why not?

DIGITAL DIALOG

Internet Monitoring

Teamwork Do you think schools should monitor the Web sites students and teachers visit? Why or why not? Find two classmates and brainstorm, as a group, reasons to support or disagree with the use of firewalls that monitor Internet activity.

Writing List the results of your group discussions on a sheet of paper under the headings For and Against. When you are done, tape your results to the board to share with your classmates. Read the lists that your classmates created. Did your classmates tend to agree or disagree with the law?

Step-by-Step continued

Activity 29

Suite Smarts
Spreadsheets can be synchronized to your PDA. Data, including formulas, can be entered into the spreadsheet on a PDA.

Activity 30

DIGIByte

Decimal Places
When formatting numbers in Excel, you can choose how many decimal places will display.

Align Title Text at an Angle

1. Select Cells B3 through E3.
2. Click the **Format** menu and select **Cells**.
3. In the Format Cells window, click the **Alignment** tab.
4. In the Orientation box on the right side of the window, click the red dot and drag it to the 45 degrees position and click **OK**. Drag the row border as needed to see the title text.

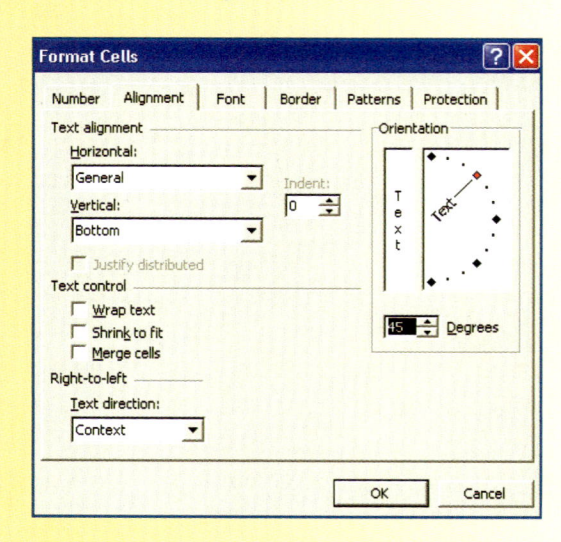

Apply Formatting to the Spreadsheet Data

1. Format the text *1st QTR*, *2nd QTR*, *3rd QTR*, and *4th QTR* so it is dark blue, bold, and 12 point.
2. Select Cells A3 through E3. Click the **Format** menu and select **Cells**.
3. In the Format Cells window, click the **Border** tab.
4. Click the thickest border in the Line Style box, and under Color choose **Dark Blue**.
5. Click the **Outline** and **Inside** buttons and click **OK**.
6. Select Cells B4 through E9.
7. Open the Format Cells window, and select the **Number** tab.
8. Under Category, make sure **Currency** is selected, and set the decimal places to 0. Click **OK**.
9. Click the **Save** button.

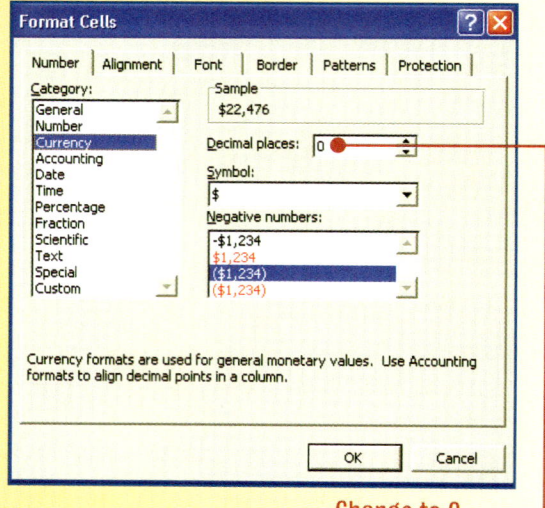

Change to 0

Continued

Chapter 10

Lesson 7.4 → Draw On-Screen in Journal and in Word

You will learn to:
- **Create** drawings in Journal.
- **Copy** drawings in Journal.
- **Paste** drawings into Word.
- **Create** and edit drawings in Word.

Drawing On-Screen Creating drawings or charts in Journal is easy with a Tablet PC. The digital pen makes drawing on the screen feel very similar to drawing on paper. Graphic artists can use this technology to create quick sketches. In the classroom, you can use this technology to quickly copy diagrams or charts that your teacher draws on the board. Your handmade drawing is in electronic form and can easily be copied and pasted to another application. For example, you can copy a chart that was drawn in Journal to a PowerPoint slide.

Drawing in Journal and in Word Original diagrams or drawings can be created in Journal and in Word using the digital pen. The line thickness and color of the pen can be adjusted as you draw.

Internet Diagrams Why might you need to create original drawings rather than copying them from the Internet? One reason is that most pictures and drawings on the Internet are copyrighted, so you cannot use them as though they are your own work without special permission. Another reason is that you may not be able to find existing drawings that exactly meet your needs.

DIGI Byte

Drawings and Diagrams

Handmade drawings and diagrams can easily be created in both Word and Journal.

✓ Concept Check

What skills do you think a person needs to use a Tablet PC? What two applications on a Tablet PC can you use to create electronic copies of original drawings?

DIGITAL DIALOG

Graphing Current Events

Teamwork Work with a classmate. Using the Internet or newspaper, locate a news story that contains a graph. Typical topics that are often graphed include: rainfall, temperatures, unemployment rates, and sports scores. Save graphs that you find. You will use them later in this lesson.

Activity 28

Step-by-Step

In this step-by-step, you will format your spreadsheet and then copy and paste the spreadsheet into Word.

Format a Spreadsheet Title

1. Open a new worksheet in Excel.
2. Enter the data below into your new spreadsheet. Begin with the title **Sunny Day Tanning Products** in Cell A1. Skip down to Row 3 to start entering the rest of the data.

```
Sunny Day Tanning Products

Product           1st QTR      2nd QTR      3rd QTR      4th QTR
Lotion SPF 15     22,476.00    31,078.00    29,056.00    21,774.00
Lotion SPF 10     21,098.00    23,865.00    26,991.00    18,239.00
Hair Lightener    15,834.00    24,987.00    25,125.00    14,882.00
Natural Tan       28,443.00    25,890.00    23,765.00    27,546.00
Oil SPF 6          9,245.00    10,198.00    11,745.00     8,974.00
Sun Block          7,245.00     9,976.00    12,876.00     8,104.00
```

3. Format the product sales as currency. Save your file as **Sunny Day**.
4. Select Cells A1 to E1. Click the **Format** menu and choose **Cells**.
5. In the Format Cells window, select the **Alignment** tab.
6. Select **Merge Cells** and click **OK**.
7. Center the contents of Cell A1.

	A	B	C	D	E
1		Sunny Day Tanning Products			
2					
3	Product	1st QTR	2nd QTR	3rd QTR	4th QTR
4	Lotion SPF 15	$22,476.00	$31,078.00	$29,056.00	$21,774.00
5	Lotion SPF 10	$21,098.00	$23,865.00	$26,991.00	$18,239.00
6	Hair Lightener	$15,834.00	$24,987.00	$25,125.00	$14,882.00
7	Natural Tan	$28,443.00	$25,890.00	$23,765.00	$27,546.00
8	Oil SPF 6	$9,245.00	$10,198.00	$11,745.00	$8,974.00
9	Sun Block	$7,245.00	$9,976.00	$12,876.00	$8,104.00
10					

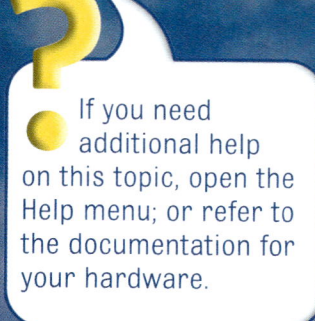

> If you need additional help on this topic, open the Help menu; or refer to the documentation for your hardware.

Continued

Activity 25

If you need additional help on this topic, open the Help menu; or refer to the documentation for your hardware.

DIGI Byte

File Name
The file name of a Journal note will be the same as the text that appears in the Note Title text box.

Step-by-Step

Create and Edit a Drawing in Journal

In the following steps, you will create a diagram about part-time jobs.

1. Open Journal.
2. Write `Student Survey—Part-Time Jobs` in the Note Title text box at the top of the note.

   — Note Title text box

3. Save the Journal note by selecting the File menu and tapping **Save As**.
4. The name of the file should already be filled in. Check it for accuracy and then tap **Save**.
5. Use the **Pen** tool to draw the outline of a pie chart in the middle of the Journal page.

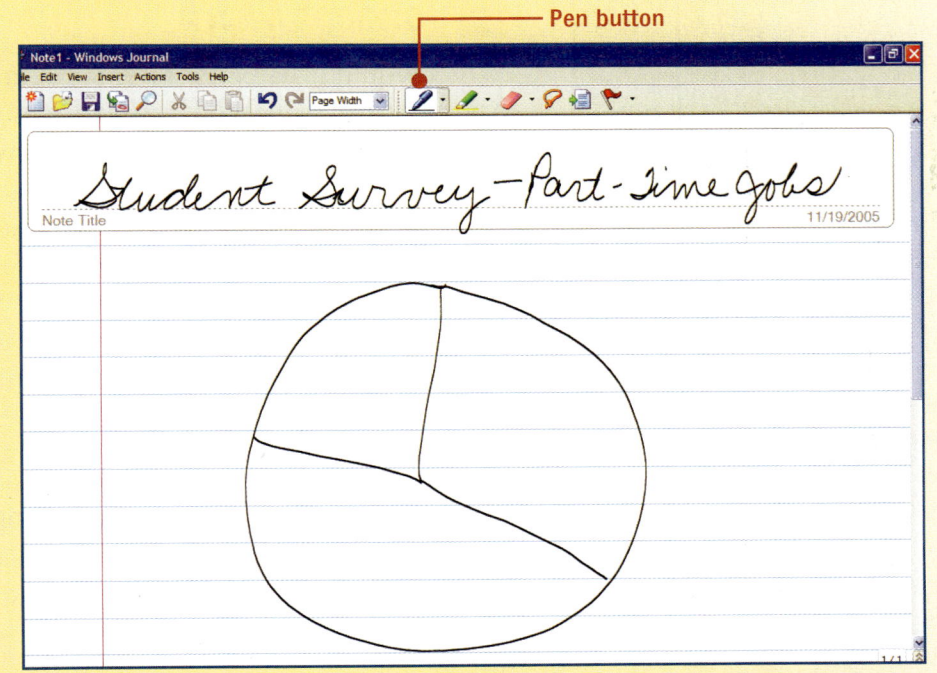

   — Pen button

Continued

213

Lesson 10.7 → Integrate Spreadsheets with Word Processing

You will learn to:
- **Align** title text at an angle.
- **Use** more spreadsheet formatting features.
- **Copy** and paste spreadsheets into Word.
- **Apply** formatting and a header and footer to a chart.

Spreadsheets in Word Processing In Lesson 10.6, pages 339–343, you learned how to integrate charts into Word. Spreadsheets can also be copied into Word. In some instances, it may be easier to create a spreadsheet in Excel and then copy the spreadsheet into Word than it would be to create a table in Word. The formatting applied to cells will copy with the data when you paste the spreadsheet into Word. The original spreadsheet that you create will still remain an independent document in which more data can be added as needed.

In this lesson, you will not only learn how to copy a spreadsheet into Word, but you will also learn more ways to format text and numbers in Excel and how to change the size of a chart so it will fit on the page.

✓ Concept Check

When would it be better to enter data into a spreadsheet rather than into a table created in Word? What advantages are there to using an Excel spreadsheet over using a table in Word?

DIGITAL DIALOG

Charting Data

Teamwork Form a small group of classmates and discuss how you can keep track of data by using a spreadsheet.

Writing On a separate piece of paper, or using a Tablet PC, draw a diagram of a spreadsheet you could use for recording the recycling information.

- Discuss graphing your data, and draw sketches of three different types of charts.
- Which type of chart would display the information best?
- Why would you choose this type of chart?

Share your answers with the class.

Step-by-Step *continued*

6 Color the pieces of the pie using the Highlighter tool. Tap the **Highlighter** button to select thickness and color.

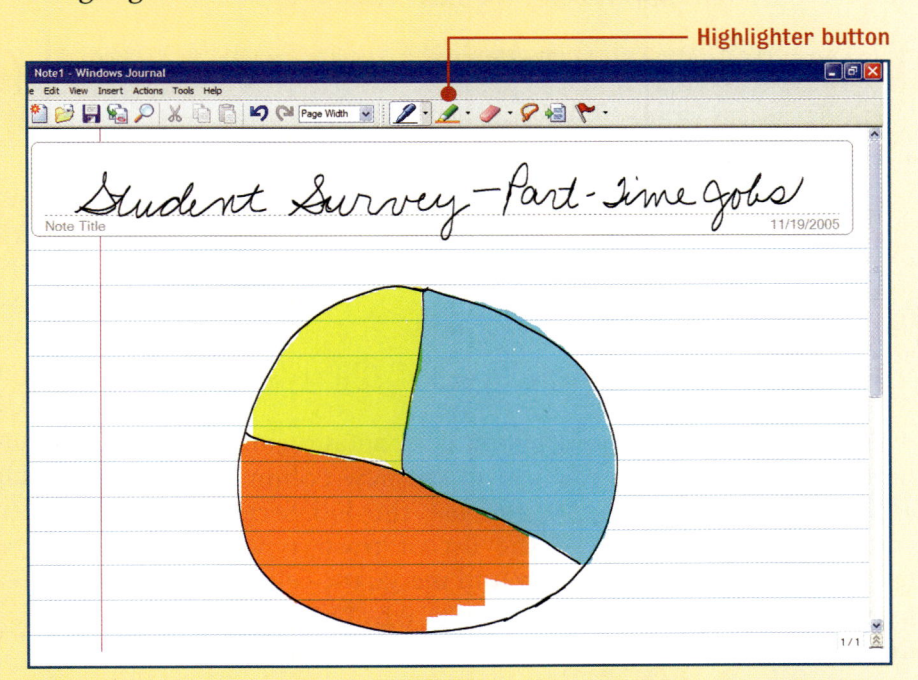

— Highlighter button

7 Add labels to the pie chart using the **Pen** tool.

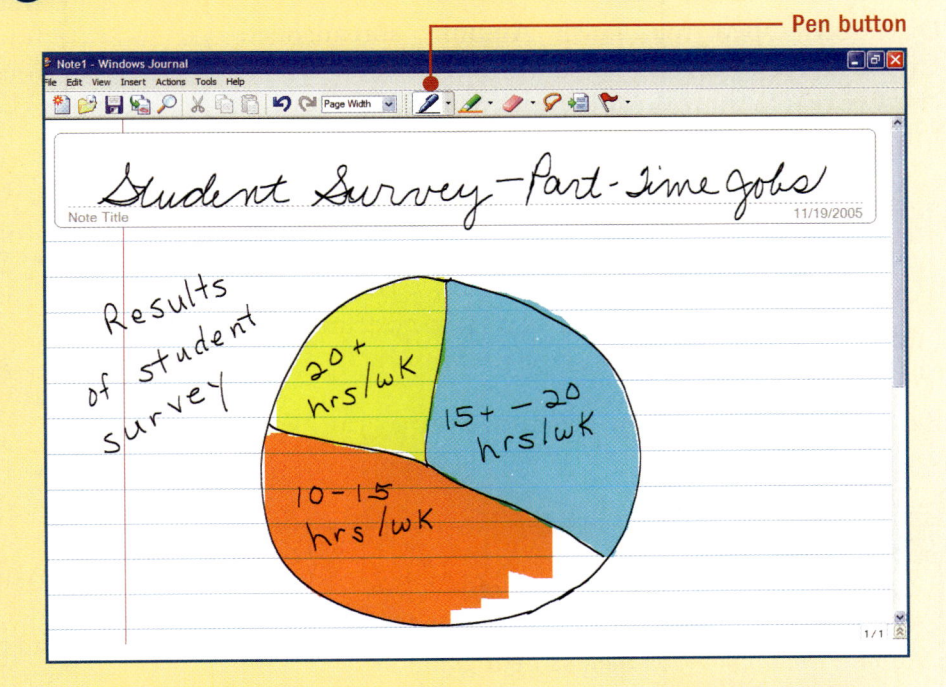

— Pen button

DIGI *Byte*

Highlighter

The Highlighter tool colors are transparent, just like a real highlighter. When two colors overlap, they will combine to create a new color.

Continued

Chapter 7

214

Application 9

Directions Enter the data below into a new spreadsheet. Enter the tonnage numeral in one column and the description *million tons* in the next column.

```
Trash Type         Tonnage
Paper              71.6 million tons
Yard Trimmings     31.6 million tons
Metals             15.3 million tons
Plastics           14.4 million tons
Glass              12.5 million tons
Other              20.8 million tons
```

Insert a column in front of the first column and enter the number 1 in the second cell in the new column. Select Cells A2 through A8 and use the Fill Series feature to add numbers to the list of trash types. Insert a row above the first row and merge the cells in this new row. Enter the title Amount of Trash in this row. Save the file as **Trash Amount** and close it.

Application 10

Directions Open the **Trash Amount** spreadsheet. Create a pie chart of the data and then copy the chart. Create a new document in Word and enter the following text.

```
U.S. citizens produce millions of
tons of trash each year. Some of
these products can be recycled.
The average U.S. citizen produces
4.4 pounds of trash per day.
```

Paste the pie chart below the text. Format the Word document as a poster.

RUBRIC LINK
Go to the Online Learning Center at **digicom.glencoe.com** and click Lesson 10.6, Rubric, to find the application assessment guide(s).

Digital Decisions

Templates

Internet Investigation Work with a small group to search the Internet for templates that have been created with Excel and are used for various business activities. Record the types of templates found and their purpose. Also record whether the template is free and the URL of the Web page on which the template is described. Open Word and create a table to record this information. When you complete your table, save and print it.

Critical Thinking Discuss the templates you found with the rest of the class, or post your table on a class bulletin board. If you participate in a discussion, form small groups and select three or four of the templates you found to focus on. Organize the discussion by considering the following questions:

- What kinds of businesses would be likely to use the templates you located?
- Who in the company would be most likely to use the templates, such as researchers, accountants, salespeople, or executives?
- Who would be the intended audience for the information?
- Do the templates appear to be easy to use? Explain why.

After the discussion, write a paragraph that summarizes your answers.

Activity 26

Step-by-Step continued

Copy the Chart and Paste It into Word

1. Tap the **Selection Tool**, and select the pie chart and its labels.

Selection Tool

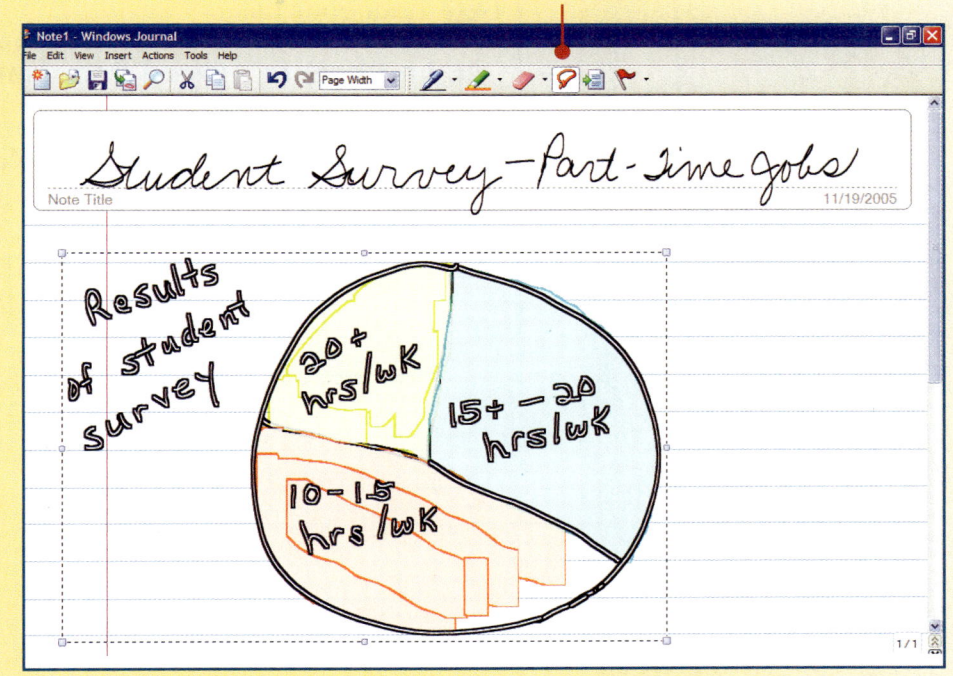

DIGIByte

Selection Tool

When you use the Selection Tool to select a graphic along with text, you cannot convert the handwriting to typed text.

2. Tap the **Copy** button.
3. Open Word.
4. Place the insertion point where you want the pie chart to appear.
5. Tap the **Paste** button.
6. Save the Word document as *Your Name* **Graph**.

Activity 27

Draw in Word

1. Open Word.

Ink Drawing and Writing button

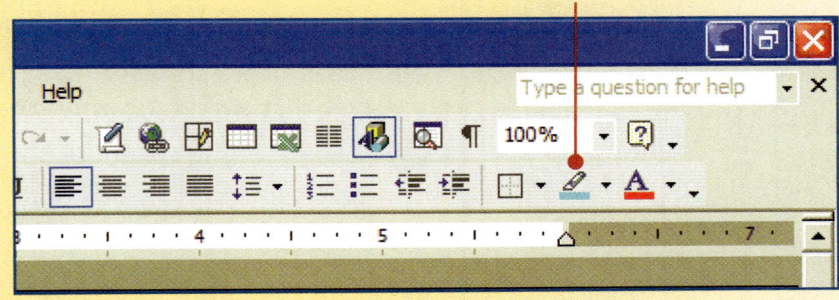

Continued

Chapter 7

Step-by-Step *continued*

6. In Step 4, choose the **As new sheet** option and click **Finish**.
7. Save your file.
8. Print your file.

Add a Chart to a Word Document

Activity 27

In this activity you will create a chart and then copy and paste it into a Word document named **10-6 Recycle** that has been prepared for you.

1. Click the **File** menu and select **New** to open a new worksheet.
2. Enter the following data into the worksheet.

```
Trash Type      Pounds
Newspaper       320
Metals          500
Plastics        189
Glass           224
Office Paper    120
Cardboard       317
```

3. Use the Chart Wizard to create a bar chart of this data. To do this, choose **Bar** from the list of chart types in the Standard Types tab in Step 1.
4. In Step 4 of the Chart Wizard, choose the **As object in Sheet** option. Click **Finish**.
5. Save this file as **Recycle**.
6. Click the chart to select it and click the **Copy** button.
7. Minimize Excel.
8. Open the Word document **10-6 Recycle** and select **Save As** so you can name this Word file *Your Name* Recycle.
9. Scroll down to where the chart should be inserted. Delete the text box that says *Copy and paste chart here*.
10. Select the **Edit** menu and click **Paste**.

    The chart you created in Excel will now appear in your Word document.
11. Print preview and then print the letter. Save and close the files.

Step-by-Step *continued*

2. Tap the **Ink Drawing and Writing** button. The Ink toolbar and a drawing area will appear. The Ink toolbar allows you to choose pen thickness and color.

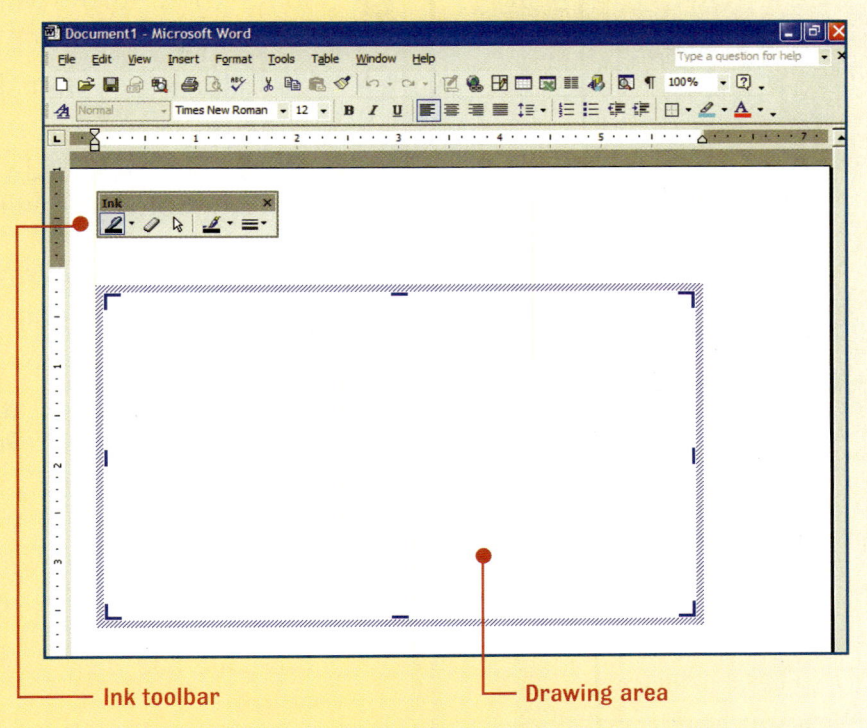

   — Ink toolbar   — Drawing area

3. Tap the **Pen** drop-down arrow in the Ink toolbar to display pen choices.

   Notice that the Ink toolbar also has Eraser, Selection, Pen Color, and Line Thickness buttons.

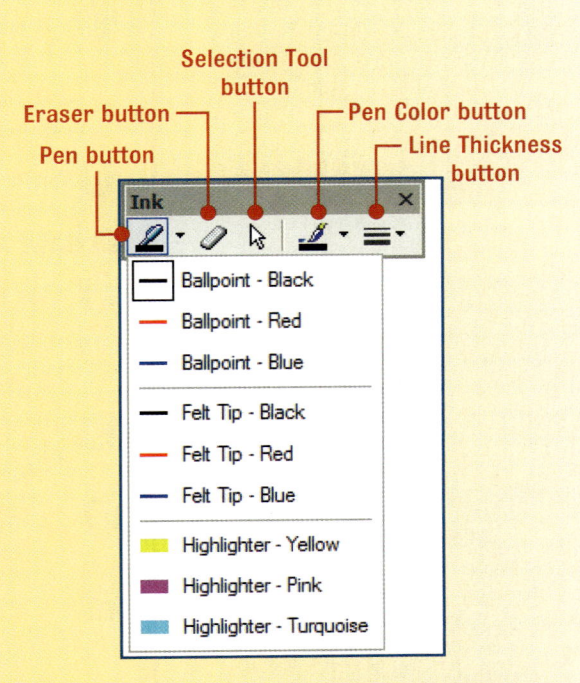

   Eraser button — Selection Tool button — Pen Color button — Line Thickness button — Pen button

   - Ballpoint - Black
   - Ballpoint - Red
   - Ballpoint - Blue
   - Felt Tip - Black
   - Felt Tip - Red
   - Felt Tip - Blue
   - Highlighter - Yellow
   - Highlighter - Pink
   - Highlighter - Turquoise

Suite Smarts

Diagrams and charts created in Journal can be copied and pasted into other Microsoft applications.

Continued

Activity 24

Activity 25

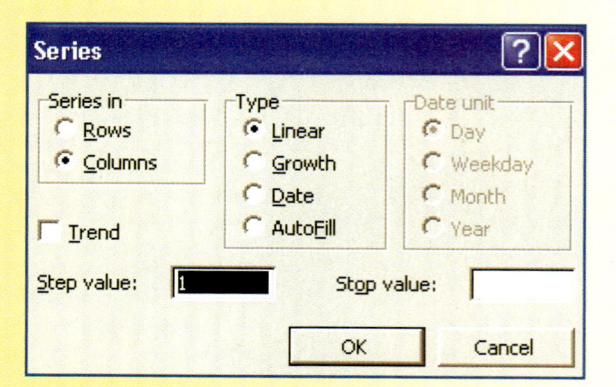

> **DIGI*Byte***
>
> **Fill Series**
>
> Selecting the Edit menu, choosing Fill, and then clicking Series enables you to quickly number a list in a spreadsheet.

Activity 26

Step-by-Step *continued*

Insert Rows and Columns

1. Click Cell A3 and select the **Insert** menu.
2. Choose **Rows**. A row will be inserted above the cell in which you clicked, so what was previously Row 3 will now be Row 4.
3. With Cell A3 still selected, click the **Insert** menu again and choose **Columns**.

   A column will be added to the left of Column A, so what was previously Column A will now be Column B.

Use the Fill Series Feature

You will quickly number a list in a spreadsheet using the Fill Series feature.

1. Click Cell A4 and enter the number 1.
2. Enter 2 in Cell A5.
3. Select Cells A4 through A9.
4. Select the **Edit** menu, choose **Fill**, and then click **Series**.
5. In the Series window, make sure the **Columns** option is selected and click **OK**.
6. Save your file again and close it.

Create a Chart

1. Open the **Air Quality** spreadsheet.
2. Select all the data in your spreadsheet except for the title row (Row 1).
3. Click the **Insert** menu, and click **Chart...** to open the Chart Wizard. Click **Next**.
4. In Step 2 of the Chart Wizard, click the **Rows** option instead of Columns, and click **Next**.
5. In Step 3, enter `Healthy Air Resource Summary` in the Chart Title box, and click **Next**.

Continued

Step-by-Step continued

4. Draw a bar chart similar to the one shown. You might use the chart to compare the temperature in your home town with those of several other cities around the world. Use the Internet to locate temperature data. Add labels to the chart by writing on the screen. Save this graph as *Your Name* **Graph 2**.

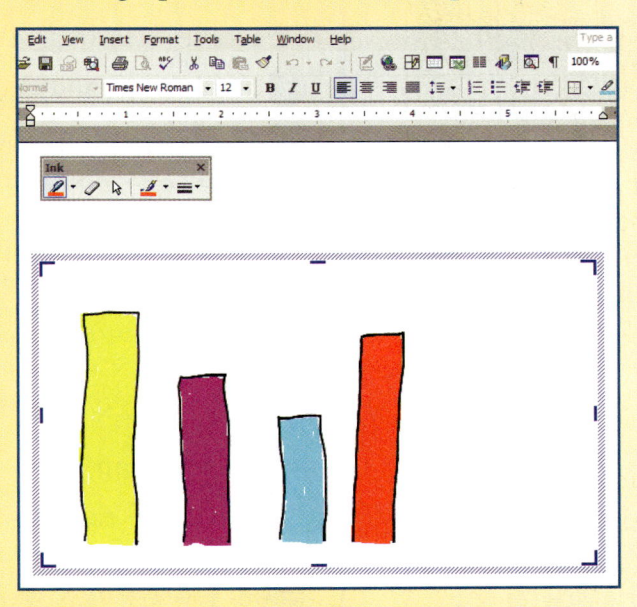

Application 5

Directions At the start of this lesson, in the Digital Dialog, page 212, you selected graphs of current events, such as rainfall, temperatures, unemployment rates, and sports scores from the Internet and newspapers. Choose one of the graphs, and draw the graph in Journal. Add appropriate identification labels to your graph. Paste the graph into Microsoft Word on the Tablet PC. Save the graph as *Your Name* **Graph 3**.

Application 6

Directions Open Word on the Tablet PC. Use the Insert Ink Drawing and Writing button to draw a chart showing the following data.

Student Council Presidential Election Results
Jonah: 251 votes.
Carrie: 203 votes.
Mark: 362 votes.

Choose the type of graph you want to use. Color the chart and add labels as needed. Save the chart as *Your Name* **Graph 4**.

RUBRIC LINK

Go to the Online Learning Center at **digicom.glencoe.com** and click Lesson 7.4, Rubric, to find the application assessment guide(s).

Step-by-Step

In Activities 22–27, you will add the following formatting to your spreadsheet: wrap text in cells, merge cells, insert rows and columns, and use the Fill Series feature. Then you will copy a chart you create in Excel into a Word document.

Activity 22

Wrap Text in Cells

Sometimes text in a cell needs to be wrapped so it will fit in the cell.

1. Start Excel and begin a new spreadsheet.
2. Enter the spreadsheet's title, `President's Budget for Air Quality`, into Cell A1.
3. Enter `Fiscal Year 2004` into Cell B2.
4. Select Cells B2 and C2.
5. The text in these cells will need to be wrapped so it will fit in the cells. Click the **Format** menu and choose **Cells**.
6. In the Format Cells window, click the **Alignment** tab.
7. Select **Wrap Text** and click **OK**.

Fill in the rest of your spreadsheet with the data shown in the spreadsheet below.

	A	B	C
1	President's Budget for Air Quality		
2		Fiscal Year 2004	Fiscal Year 2005
3	Outdoor Air	579,059,000.00	659,876,000.00
4	Indoor Air	48,043,000.00	48,955,000.00
5	Protect Ozone	19,069,000.00	21,814,000.00
6	Radiation	34,859,000.00	34,718,000.00
7	Reduce Greenhouse Gas	106,936,000.00	108,389,000.00
8	Enhance Science & Research	129,017,000.00	130,864,000.00

8. Save the spreadsheet as **Air Quality**.

Activity 23

Merge Cells in a Spreadsheet

1. Select Cells A1, B1, and C1.
2. Select the **Format** menu and click **Cells**.
3. In the Format Cells window, click the **Alignment** tab.
4. Select **Merge Cells** and click **OK**.

> **If you need additional help on this topic, open the Help menu; or refer to the documentation for your hardware.**

> **DIGIByte**
> **AutoSum**
> Use the AutoSum button to add Columns B and C.

Continued

Chapter 10

340

Application 7

Scenario You work for a large manufacturing company as a quality control engineer. One of your responsibilities is to inspect the plant regularly to make sure the equipment is running smoothly. Another responsibility is to ensure that the assembly line is meeting its production quota within the high quality assurance specifications. You are also in charge of developing and enforcing safety guidelines.

Directions Think about how a Tablet PC might be helpful for your inspections, assembly line quota and quality assurance checks, and safety responsibilities.

- How could a Tablet PC save time?
- How could the ability to create drawings help you?
- What types of charts and graphs would you create during your inspections using a Tablet PC?
- Would a Tablet PC make your work more productive? Why or why not?

Answer the questions in a paragraph or two. Be prepared to share your responses with the class.

DIGITAL Decisions

Prom Publicity

Scenario You are a member of the prom decorating committee. The committee members have asked you to design a drawing that will appear on the jewel case cover of a prom DVD that will be sold after the prom. The other committee members would like you to e-mail them a sketch of your design.

Critical Thinking How can you e-mail a rough sketch in a format that you know they will all be able to open? In print or cursive, describe the process you will use to create the drawing. If possible, create a simple design on your Tablet PC and save it as *Your Name* **Prom Design**.

DIGITAL DIALOG

Tablet PC or PDA?

Scenario You work for a large hotel chain as the special events coordinator. The manager of the hotel has asked you to personally greet guests at an upcoming banquet.

In addition to greeting the guests, you have also been asked to survey the guests about what types of activities they would like the hotel to make available for their enjoyment during a visit. To save time, you would like to record the answers electronically; however, a laptop would require you to type and handwriting is quicker. You decide to use either a Tablet PC or a PDA.

Critical Thinking Discuss the advantages and disadvantages of both devices with another classmate. Which one will you use? Why? Explain your decision thoroughly and give examples when you can.

Lesson 10.6 → Integrate Charts with Word Processing

You will learn to:
- **Wrap** text within cells.
- **Merge** cells.
- **Insert** rows and columns.
- **Fill** cells with a series of numbers.
- **Copy** a chart and paste it into a Word document.

Integrating Charts with Word Processing Documents It is often helpful to include a spreadsheet or chart in a memo or letter. Charts can easily be copied into Word documents, where they can provide readers with visual information. Charts summarize information, so they make it easier for readers to understand text. When you copy and paste a chart into a Word document, the original chart is saved in Excel.

In this lesson, you will not only learn to copy a chart created in Excel into a Word document, but you will also learn a shortcut that you can use when entering data into cells. You will also learn some additional ways to format cells, and how to insert rows and columns into your spreadsheets.

✓ Concept *Check*

What are some ways you can use charts when doing your schoolwork? What kinds of data do you use at school that could be inserted as charts into Word documents?

DIGITAL DIALOG

Chart Types

Critical Thinking In a small group, discuss which chart style would be most appropriate to use for each type of data below.

1. Data on average rainfall in five different cities.
2. Data on how much money the top ten box office hits made last year.
3. Data from a science project tracking the growth rate of three different plants.
4. Data comparing the height of a group of students from one year to the next.

For chart style ideas, consider the charts you learned about in Lesson 10.5, pages 335–338.

21ST CENTURY CONNECTION

Learn More, Earn More

No surprise here: college graduates earn more! Just as exercise builds muscle, education builds income. Want proof? According to the U.S. Bureau of Labor Statistics, college graduates over age 25 earn almost twice as much as high school graduates but experience only half the unemployment rate. Compare the earnings and unemployment rates in the table below. Clearly, college graduates earn more, and they enjoy greater job opportunities.

Is a College Education in Your Future?
What are your career goals, and what are the opportunities in the career of your choice?

Failing to Plan Equals Planning to Fail
Planning your education requires some research. Here are some questions to consider:

- What level of education is required in some of the careers in which you are interested?
- What classes do you need to take in high school to qualify for entrance to a school that offers the level of education you want?

What Are the Fastest-Growing Occupations for Different Levels of Education?
The *Occupational Outlook Handbook* is an excellent resource. For example, it lists the fastest-growing job areas by level of education and training.

According to the *Occupational Outlook Handbook* (2004–05 Edition), the top five fastest-growing positions for someone with a bachelor's degree are:

- Network systems and data communications analysts
- Physician assistants
- Computer software engineers, systems software
- Database administrators

The *Occupational Outlook Handbook* has similar lists for candidates with doctoral degrees, master's degrees, and associate degrees, as well as for job candidates who have completed vocational programs.

Activity What are some classes you have taken already that will help you in your education plan?

Earnings and Unemployment for People over 25

Level of Education	Weekly Earnings	Unemployment Rate
College graduate	$896	1.7%
Less than a bachelor's degree	598	2.7%
High school graduate, no college	506	3.5%
Less than a high school diploma	360	6.4%

Source: U.S. Bureau of Labor Statistics

DIGI Byte

Print Preview

Print preview your spreadsheet before you print it to check the alignment of your chart on the page.

RUBRIC LINK

Go to the Online Learning Center at **digicom.glencoe.com** and click Lesson 10.5, Rubric, to find the application assessment guide(s).

Step-by-Step continued

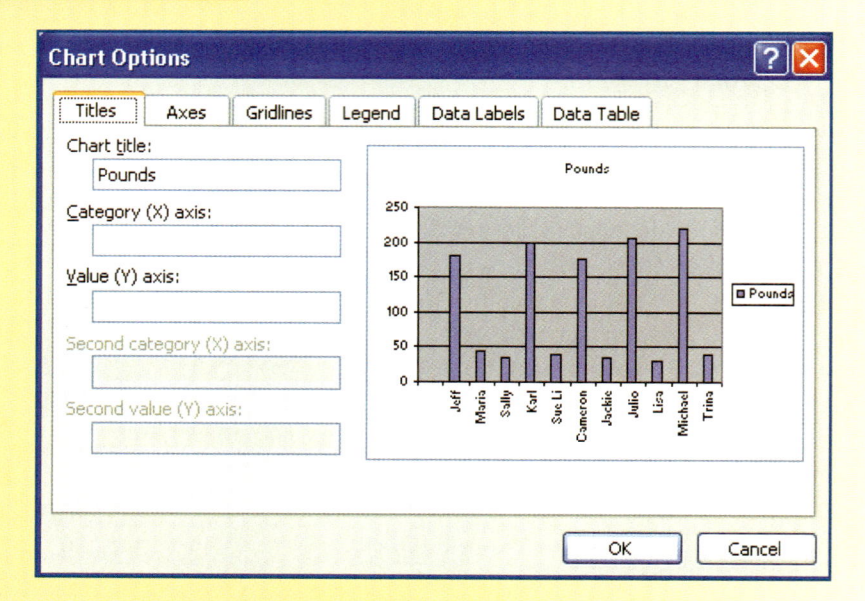

4. Select the elements you want to change and make the changes.

Application 7

Directions Use the **Weight** spreadsheet you created in this lesson to make a new chart. Choose a chart to display the students' names, pounds they can lift, and weight. Create the chart as a new sheet. Analyze the new chart.

- What do you notice about the way this data looks when it is in chart form?
- Compare the two charts you have created from the **Weight** spreadsheet. Which chart do you think is easier to read and why?
- Does the chart show any link between students' weights and how many pounds they can bench press?

Rename the worksheet containing the new chart **Student Information**.

Application 8

Directions Use a search engine to find statistics on the ten costliest hurricanes. Enter the information into an Excel spreadsheet. Create a chart that displays this information. You may need to rearrange the columns in your worksheet so you can accurately chart the data. Do not include the year the hurricanes occurred in your chart.

What information can you quickly determine from the chart? What variations of this data can you use to create other charts?

Lesson 7.5 — Edit Text with Gestures

You will learn to:
- **Delete** text with gestures.
- **Insert** and remove letter spaces with gestures.
- **Insert** a line space with a gesture.
- **Insert** a tab with a gesture.

gesture A simple shape made with the digital pen that sends a command to the Tablet PC to edit or format text.

DIGI Byte

Scratch Out Gesture The Scratch Out gesture is a quick way to erase letters or words from the Writing Pad screen.

Gestures When you edit your own writing on paper, you probably use simple editing marks to quickly indicate changes you wish to make. For example, to indicate to yourself that you want to delete a word, you might just draw a line through that word. You do not need to actually write out an instruction to delete the word.

On the Tablet PC, you can quickly indicate a change by using gestures. A **gesture** is a simple shape made with the digital pen that sends a command to the Tablet PC to edit or format text. The Tablet PC interprets these shapes and makes the changes. Gestures can help you format and edit text more efficiently.

An example of a gesture is the Scratch Out gesture, shown on page 221. This is a large Z mark to erase the letter a.

Working with Gestures To use gestures most effectively, it is best to memorize them so that you do not have to look them up as you work. Touch Writing Pad quickly and with a determined stroke to start a gesture. You may need to practice several times before a gesture works for you.

✓ Concept Check

How do you start a gesture?

DIGITAL DIALOG

Compare Gestures and Graffiti

Teamwork Check to see what gestures are available in Writing Pad. Compare the gestures that can be used in Writing Pad to Graffiti strokes used with a PDA. Work with a partner to create a table displaying Tablet PC gestures on one side and Graffiti strokes on the other. Write answers to the following questions.

- In what ways are gestures and Graffiti strokes similar? How are they different?
- Which gestures and Graffiti strokes are most similar?

Explain briefly why it is necessary to become familiar with both gestures in Writing Pad and Graffiti on a PDA.

DIGIByte

Chart Wizard

As you move through the Chart Wizard, you can click the Back button at any time to return to the previous step and change your selection.

Activity 20

Suite Smarts

Tables from Word can be copied into Excel worksheets.

Activity 21

Step-by-Step *continued*

4. On **Chart Wizard—Step 4 of 4**, select **As new sheet**.

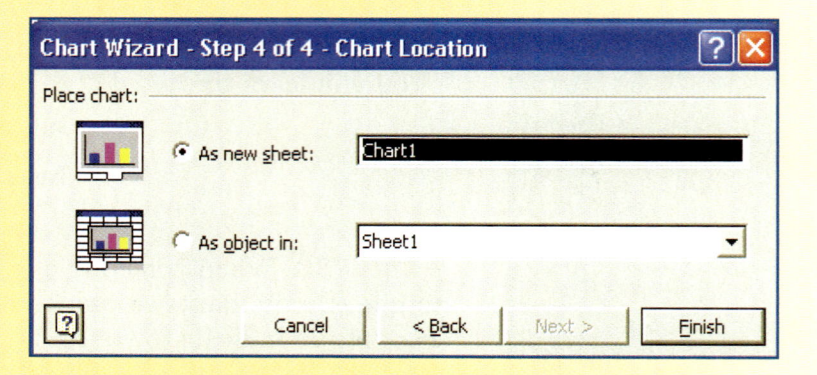

5. Click **Finish**.

   The chart will be displayed in a new worksheet. By default, the worksheet will be named **Chart 1**.

Rename Your Worksheets

1. Double-click the **Chart 1** tab at the bottom of the screen. The text in the tab will be highlighted.

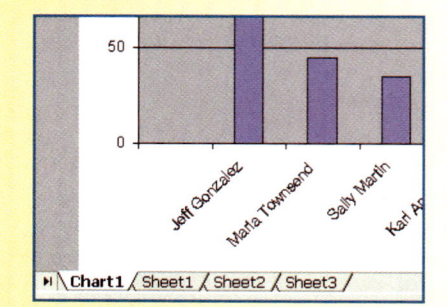

2. Enter `Student Lift Chart`.

3. Double-click a tab to rename it. Rename Sheet 1 **Lift Data**.

   To move back to the spreadsheet where you entered the data, click the tab you just named *Lift Data*.

Change Chart Options

After you create a chart, you can go back and make changes to it.

1. Click the tab you just named *Student Lift Chart*.

2. Click **Chart** in the menu bar. Click **Chart Options**.

3. The Chart Options window will allow you to make changes to your chart and its elements. You can:

   - Add a title to your chart or change the current title.
   - Add gridlines to assist in reading your chart.
   - Move the legend from its default position on the side.
   - Display data labels with percents or values showing.

Continued

Chapter 10

337

Step-by-Step

Activity 28

Use the Scratch Out Gesture

The Scratch Out gesture (shown below right) will erase a letter or number that you write in Writing Pad before it is sent to the document.

1. Open Word and open Input Panel.
2. Tap the **Writing Pad** tab.
3. Write the lowercase letter *a* in Writing Pad.
4. To scratch out the letter before it is sent to the document, draw a large *Z* mark, or swish, over the *a*. Use three or more strokes. Make sure that the tip of the pen stays connected to the screen as you draw the gesture.

Activity 29

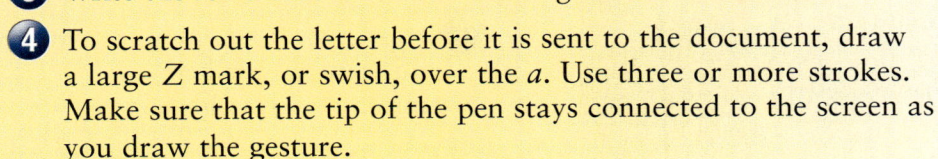

Format a Document with Gestures

1. Write this sentence: `I am testing gestures.`
2. Tap **Send**.
3. Make sure the Writing Pad window is blank before creating a gesture.

Activity 30

Insert a Letter Space with the Space Gesture

1. Place the insertion point in the sentence.
2. In the blank Writing Pad window, make a line to the right with the digital pen to make the Space gesture. Draw the line just above the Writing Pad line. Notice what happens to the sentence on the screen.

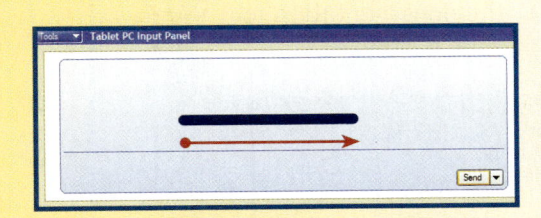

Activity 31

Remove a Character with the Backspace Gesture

1. Place the insertion point in a word.
2. In the blank Writing Pad window, make a line to the left with the digital pen to create the Backspace gesture. Notice what happens to the letters of the sentence on the screen.

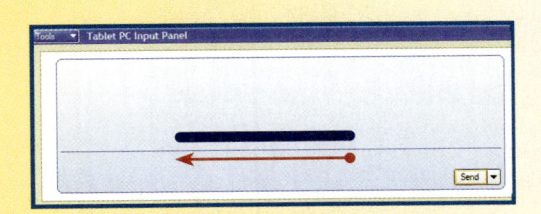

Suite Smarts

Using a gesture on the Tablet PC is similar to using a traditional keyboard shortcut.

Continued

Step-by-Step

In this lesson, you will use a chart to show weightlifting statistics for students in a gym class. Coach Cumberland has recorded his students' weights and the amount of pounds they can bench press. Now he wants to create a graph of their performance and display it on the gymnasium wall. At the end of the semester, Coach Cumberland will chart their weightlifting ability again so the charts can be compared.

Activity 18

Create a Spreadsheet

1. Start Excel and enter the data shown here into a new spreadsheet.
2. Save the file as **Weight**.

	A	B	C
1	Weight Lifting Chart for Bench Press		
2			
3	Student	Pounds	Weight
4			
5	Jeff Gonzalez	180	175
6	Maria Townsend	45	120
7	Sally Martin	35	110
8	Karl Anderson	200	210
9	Sue Li	40	116
10	Cameron Davis	175	166
11	Jackie McDonald	35	105
12	Julio Sanchez	205	197
13	Lisa Caprianni	30	108
14	Michael Schmidt	220	223
15	Trina Jones	40	117

Activity 19

Select Data to Put into a Chart and Create a Chart Using the Chart Wizard

1. Select the data that you want to appear in your chart. Since the students' weights are not needed to graph information about the number of pounds they can lift, do not select Column C. Select the data from Cell A5 through Cell B15.
2. Click the **Chart Wizard** button.
3. Accept the default settings for the first three steps of the wizard by clicking **Next**.

> **?** If you need additional help on this topic, open the Help menu; or refer to the documentation for your hardware.

Chapter 10

336

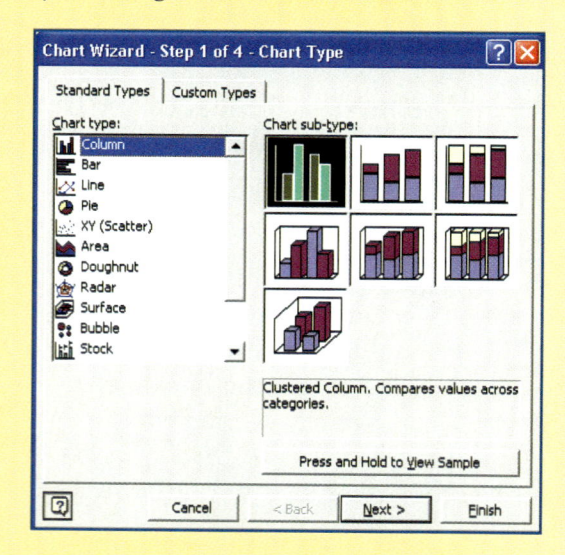

Continued

Activity 32

Step-by-Step *continued*

Insert a Line Space with the Enter Gesture

1. Place the insertion point in the sentence
2. In the blank Writing Pad window, move the pen down and then to the left to make the Enter gesture. Be sure to draw the line just above the Writing Pad line. Notice what happens to the sentence on the screen.

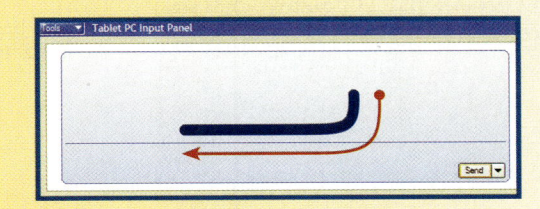

Activity 33

Insert a Tab with the Tab Gesture

1. Place the insertion point on the line below the sentence on the screen.
2. Write this sentence: `I am testing the Tab gesture.`
3. Tap **Send**.
4. Place the insertion point at the beginning of the second sentence.
5. In the blank Writing Pad window, move the pen up and then to the right to make the Tab gesture. Notice what happens to the sentence on the screen.

DIGIByte

Open Input Panel with a System Gesture

If Input Panel does not appear on the screen, you can open it with a gesture. Wave the digital pen close to—but not touching—the screen to open Input Panel.

RUBRIC LINK

Go to the Online Learning Center at **digicom.glencoe.com** and click Lesson 7.5, Rubric, to find the application assessment guide(s).

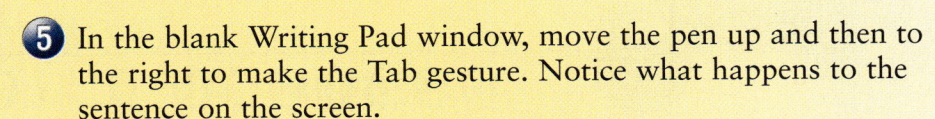

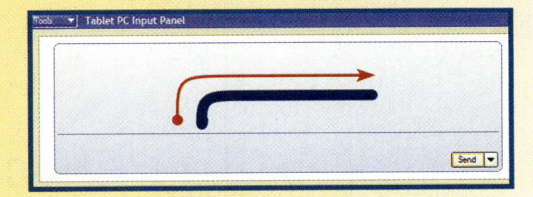

Application 8

Directions Practice writing letters and words in Writing Pad. Write at least two sentences about your favorite Tablet PC feature. Use the Scratch Out gesture to erase a letter or word. Use gestures to insert a letter space, remove a character, insert a line space, and insert a tab in the text.

Tap the **Help** menu and look up gestures. What information do you find? Practice some gestures that may be helpful to you.

Lesson 10.5

Use Spreadsheet Data to Create Charts

You will learn to:
- **Select** spreadsheet data for a chart.
- **Create** a chart.
- **Name** a worksheet and move between multiple worksheets.

Charts Charts are visual representations of numbers and text. For example, data can be displayed in a line, column, bar, pie, XY (scatter), area, doughnut, surface, bubble, or radar chart. Charts help us understand and visualize trends, percentages, and proportions.

Spreadsheet applications such as Excel enable you to easily transform data into chart form. Excel has many chart styles to choose from. The type of chart you use depends on the type of data you want to display. You may need to try several different chart styles before you decide which one best represents the information you want to convey.

Charts can be used to display all types of data. They are commonly used to show trends, such as retail sales over a period of time or the performance of the stock market.

✓ Concept *Check*

What type of data have you seen displayed in a chart? Do you think you could create a timeline with a chart?

DIGITAL DIALOG

Survey

The Athletic Boosters have asked the Future Business Leaders of America (FBLA) club for assistance. At a sports event that the Boosters are hosting this year, they would like to add healthier snacks to the list of items they sell. The Boosters have asked the members of FBLA to survey the students in your school to see which kinds of fruit are the most popular.

Teamwork Work in a small group and create a list of five types of fruit for your survey.

Writing Design a rough draft of how you would enter this information in a spreadsheet. Do you think that creating a chart would help you analyze the results more easily? Why or why not?

DIGITAL Career Perspective

Hospitality and Tourism

Philip Carruthers
Food Service Professional

Ensuring Customer Satisfaction and Thriving Businesses

If you like to interact with many different people on a daily basis and would like to help create a friendly, clean, comfortable, and entertaining atmosphere, then you might find a career in the food service industry to be rewarding. "My favorite activity on the job is discussing current affairs and sports with customers and seeing that they are happy," says restaurant professional Philip Carruthers. "This social time contributes to their positive experience and brings them back to the restaurant," says Carruthers.

The food service industry is highly competitive and includes a range of challenging, fast-paced full- and part-time jobs in service, kitchen, and management. Managers are responsible for overseeing many functions simultaneously.

A manager's functions include communicating with a diverse clientele and staff, often in many languages; scheduling shifts and directing employees; estimating food needs and placing orders; coordinating activities among kitchen dining areas; helping executive chefs develop menus; and adding receipts at closing.

One software program that has made inventory, accounting, and scheduling easier for many businesses is Microsoft Excel. According to Carruthers, "Excel is a user-friendly program that helps me do my job more efficiently. It pulls data from the menu, cash register, and time clock systems automatically and organizes it into preprogrammed spreadsheets, so staff doesn't have to do this manually."

A career in the food service industry is still hard work, even with current digital technology. But the rewards for hard work can be great. Succeeding in the industry takes drive, ambition, and the desire to please people. After all, the food service industry is about customer service. It requires a lot of energy, an outgoing personality, and the ability to be comfortable working with a variety of people from a variety of backgrounds. Food service professionals also need to be risk-takers and have the ability to solve problems quickly.

Training
While fast-food workers and wait staff are sometimes promoted to management positions, most food service companies recruit management trainees from two- and four-year food service management programs. Subjects to study in high school and beyond include nutrition, food planning and preparation, accounting, recordkeeping, and computer science.

Salary Range
Typical earnings are in the range of $27,000 to $47,000 a year.

Skills and Talents
Food service professionals need to be:
- Neat and clean
- Physically fit
- Self-disciplined
- Good leaders
- Able to solve problems
- Attentive to details
- Good communicators
- Able to motivate staff

Career Activity
Brainstorm ways participation in school or community service organizations can help someone prepare for a food service occupation.

Step-by-Step continued

As you enter the amounts into each column, they will be added to the Total column.

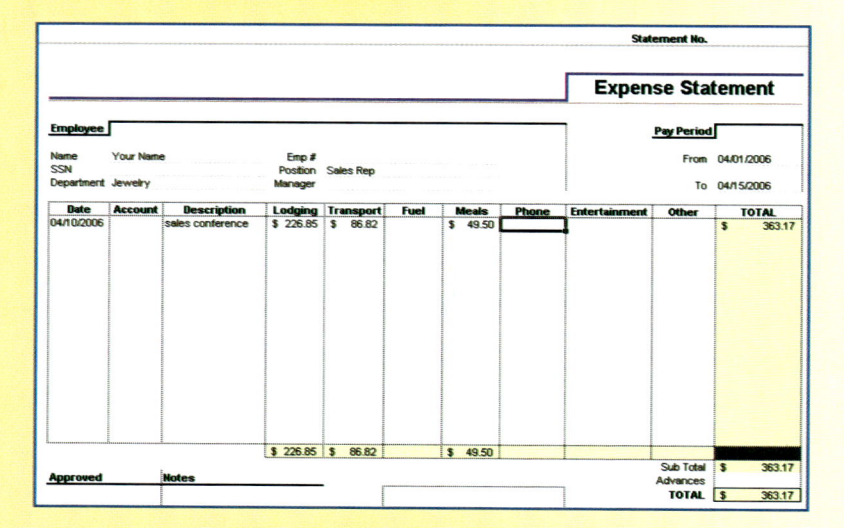

④ Enter your name and other requested information in the header at the top of the Expense form.

This template does not allow you to change or access the formulas in the Total column. The template has those cells locked, or protected.

⑤ Click the **File** menu and choose **Save**. Print preview and then print your April Expenses file.

⑥ Close the file.

Application 6

Open Excel. Open the Sales Invoice template. Enter your name and address as the customer information. Enter the following data in the invoice.

1	128 MB flash drive	$39.99
2	120 GB hard drives	$69.99
3	3.2 megapixel digital cameras	$269.99
2	Color printers	$149.99
	Add 7% sales tax	

What is the total of your invoice? Save your invoice file as Technology.

DIGI Byte

Click and Enter

You can click the mouse in the cell where you want to enter data.

Suite Smarts

Business templates are also available in Word and are often created using tables.

RUBRIC LINK

Go to the Online Learning Center at **digicom.glencoe.com** and click Lesson 10.4, Rubric, to find the application assessment guide(s).

CHAPTER 7

DIGITAL Dimension → Social Studies
INTERDISCIPLINARY PROJECT

Elected Officials

Using the skills you have learned in Chapter 7, you will create a biography on an elected official of your choice. You will include this information in your final presentation in the Unit 2 Lab, pages 258–259.

Investigate Gather personal information on an elected official. You may choose one of the officials you listed in the address book of your PDA. This person can hold a local office or an office at the state or federal level.

Teamwork Work with a partner to search the Internet, magazines, newspapers, and books. Look for information about the elected officials:

- Where did they grow up?
- Where did they attend college?
- In what area is their degree?
- What did they do before they were elected to office?
- Where were they born, and where do they live now?
- What are their hobbies and personal interests?
- Do they have any special accomplishments?

Use the Tablet PC As you read about the elected official you chose, record information using a Tablet PC.

- Take handwritten notes, or use Input Panel with the word processor. Remember, if you write your notes with the word processing feature, your handwriting will convert to typed text automatically.
- Use Journal to graph data you may find or to draw diagrams or pictures that may be informative.
- Include a photo of the elected official. Remember, to get a photo in digital form you may need to scan a picture from a magazine or newspaper.
- Save the notes as **Elected Official Notes**.

Write a Biography Using your notes as a guide, write a biography of your elected official. Save the document as **Elected Official Biography**. Be sure to spell check and proofread your document.

E-Portfolio Activity

Tablet PC Skills

To showcase your skills in writing and in using the Tablet PC, save the document **Elected Official Biography** to your e-portfolio. You could use a USB flash drive to transfer the file from the Tablet PC to your e-portfolio folder.

For help in saving to a USB drive, see Lesson 1.5, Activity 15, page 26.

Self-Evaluation Criteria

In the biography, did you:

- Include information about the elected official's personal information, including hobbies and accomplishments?
- Include a picture?
- Proofread the document?

Step-by-Step

Activity 16

Open and Save a Template

The following steps will introduce you to different templates that are part of the Excel program.

1. Open Excel.
2. Click the **File** menu and choose **New**. In the New Workbook task pane, select **On my computer…**.
3. Click the **Spreadsheet Solutions** tab.
4. Click the **Expense Statement** option and then click **OK**.
5. Select the **File** menu and choose **Save As**. Save this template on your hard drive or on a floppy disk. Name it **April Expenses**.
6. Close this file.
7. Open the **April Expenses** template that you saved in Step 5 above.

> If you need additional help on this topic, open the Help menu; or refer to the documentation for your hardware.

Activity 17

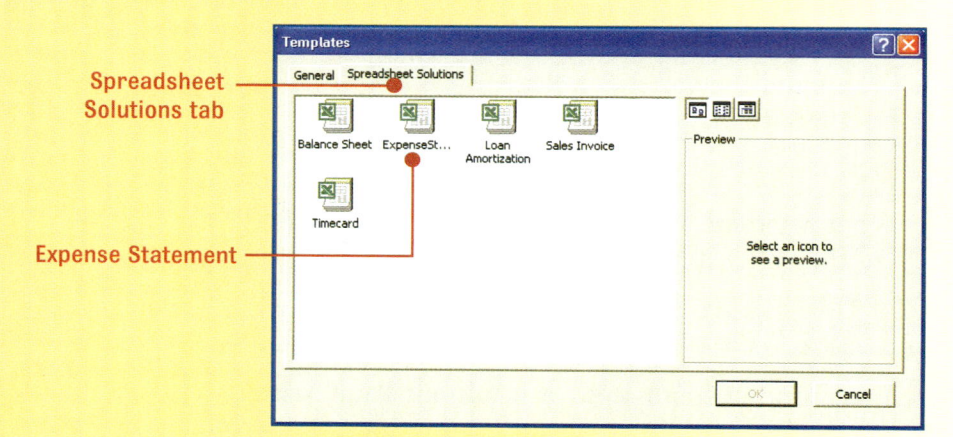

Enter Text into a Template

1. Click the first cell under *Date* and enter today's date.
2. Press TAB twice to select the cell under *Description*. Enter the text `Sales conference`.
3. Click TAB to move to each of the columns after *Description* and enter the following information in the columns.

```
Lodging:      226.85
Transport:     86.82
Meals:         49.50
```

Continued

Chapter 10

333

CHAPTER 7 Review

Read to Succeed PRACTICE

After You Read

Evaluate What You Have Learned Evaluating the topics you learn in a chapter helps you remember them and consider how you can use them in other classes, at home, or at work.

Write down the two topics you have enjoyed most during your study of this chapter's material. Then write down your two least favorite. Use a Tablet PC, if possible. Answer the following questions:

- What characteristics do the two topics you enjoy have in common?
- How are they different from your least favorite topics?
- What was the easiest topic for you to learn?
- What was the most difficult topic for you?

Create a list of the topics and skills you have learned in Chapter 7, and describe how you can use them in areas of your life other than in this class.

Using Key Terms

Tablet PCs are an important tool now and in the future. See how well you understand the terms to communicate with others about Tablet PCs and use them effectively. Match each term with its definition.

- Input Panel (193)
- Writing Pad (193)
- Journal (193)
- Quick Keys (198)
- Selection Tool (206)
- gesture (220)

1. A standard Tablet PC application that saves data as handwriting, drawings, or typed text.
2. A tool that enables you to select text, handwriting, or graphics in Journal.
3. A window where you can enter data using an on-screen keyboard or handwriting and then send the data to a document.
4. A simple shape made with the digital pen that sends a command to the Tablet PC to edit or format text.
5. A group of keys that provide an easy way to move the insertion point through a document.
6. An area of Input Panel that recognizes handwritten input.

225

Lesson 10.4 Use Business Form Templates in Spreadsheets

You will learn to:
- **Choose** an Excel template.
- **Enter** data into an Excel template.

DIGI *Byte*

Template Changes
Changes to templates can easily be made and saved with a new template name.

Templates A template is a file that has a preset format that can be used as a basis for similar types of files. Instead of recreating a particular type of file each time you need it, you can use a template as the starting point. For example, most companies send out invoices to request payment for services they have completed. Instead of creating a new invoice each time they need to send one out, the company can use a template and fill in specific customer information.

Types of Templates Excel comes with a variety of templates, including spreadsheet templates for fundraising and time billing. They are easy to access and can be used many times. Business forms, such as invoices, can be created using templates. Templates can be customized with a company's information. Many other templates are available for free at the Microsoft Web site. Templates are available on topics ranging from gift and grocery shopping lists, workout logs, calorie and fat percentage calculators, and employee training plans. This chapter will concentrate on the business use of templates.

✓ Concept *Check*

What are the advantages of using a template that includes formulas?

DIGITAL DIALOG

Using Templates
Templates can save time in many situations.

Teamwork With other students in your class, discuss uses for templates. Use the following questions to help guide your discussion:

- Where can templates be of use?
- How could templates help with schoolwork?
- How could templates help with hobbies?

Writing Create a list of your ideas to share with the class. Use a PDA or a Tablet PC to record your list, if possible.

Chapter 7 Self-Assessment

Take a moment to review what you have learned in this chapter. Rank your understanding of the topics below.

4 means, "I understand all of this."
3 means, "I understand some of this."
2 means, "I understand very little of this."
1 means, "I don't remember this."

To use a printout of this chart, go to **digicom.glencoe.com** and click on **Chapter 7, Self-Assessment.**
Or:
Ask your teacher for a personal copy.

Rank Your Understanding

Lesson	Topic	4	3	2	1
7.1	• Three ways to use the digital pen				
	• Calibrate the digital pen				
	• Open and close Journal				
	• Enter text with the digital pen				
	• Dock and undock Input Panel				
	• Change screen orientation and brightness				
7.2	• Customize Input Panel options				
	• Alternate between using Keyboard and Writing Pad in Input Panel				
	• Change the number of lines in Writing Pad for entering text				
	• Use Graffiti strokes in Writing Pad				
	• Navigate in a document with Quick Keys				
7.3	• Create and save Journal notes				
	• Use the Selection Tool to select handwriting				
	• Convert handwriting to typed text				
	• Add a flag to a Journal note				
	• Edit Journal notes				
	• Add line space to a Journal note				
7.4	• Create and edit a drawing or graphic in Journal				
	• Copy and paste diagrams from Journal into Word				
	• Create and edit a drawing or graphic in Word				
7.5	• Five ways to edit documents with gestures				

If you ranked all topics 4, congratulations! Consider doing a quick review.
If you ranked yourself 3 or lower in any topic, consider reviewing these topics first.

21ST CENTURY CONNECTION

Be a Peak Performer

Everyone possesses the ability to be a peak performer. This is someone who recognizes and develops his or her positive attributes and skills and focuses on his or her strengths to achieve excellence in whatever he or she does. Review some of the aspects of peak performance below, and ask yourself how you will apply them to your life and career goals.

Believe in Yourself
The first step is to have a positive self-image. Every person has good qualities. Make a list of those qualities in yourself, being as specific as you can. Keep the list with you. If you ever doubt your ability to accomplish something, simply read the list to reassure yourself.

Picture Yourself Doing an Excellent Job
Before you start something, visualize yourself doing exceptional work. See yourself making smart choices, providing clear and useful information, and getting the results you want.

See the Opportunities in Setbacks and Mistakes
Mistakes and obstacles are facts of life. The best way to manage them is to see them as opportunities for learning and growth, instead of barriers to success. When an obstacle arises, simply work around or through it. Treat problems as puzzles that will be fun to solve. When you make a mistake, acknowledge your role in the error, do what you can to correct it, and figure out how you can change your process to avoid making the same mistake again. Instead of giving up on an activity or goal, look for the positive qualities in a situation, and try again.

Be a Creative Risk-Taker
To peak performers, taking risks means attempting to turn a useful, exciting idea into a reality, without being certain whether or not the idea will succeed. It means having enough confidence to try something, without letting fear of failure get in the way.

Make the Most of Digital Tools
The more you know about the technology you use every day, the more benefits you will receive from using it. To use digital tools more efficiently, go beyond learning the basic functions of each.

Traits of a Peak Performer A peak performer exhibits the following qualities and skills:
- A positive, proactive attitude
- Self-respect; respect and compassion for others
- A healthy lifestyle, including a balanced diet, exercise, work, recreation, and rest
- Responsibility toward family and community
- Ability to maximize usefulness of new technology
- Excellent organizational and time-management skills
- Excellent problem-solving skills
- The ability to admit mistakes
- Hobbies and interests outside of main occupation

Activity Write a short paragraph on a Tablet PC about a time when you exhibited one of the peak performance traits above.

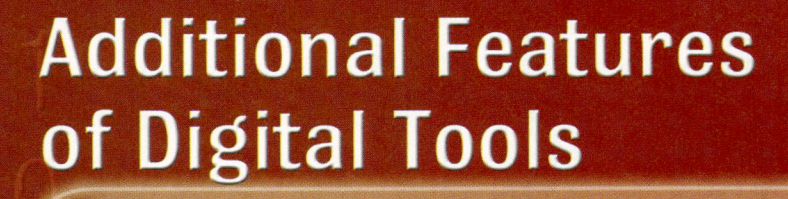

CHAPTER 8
Additional Features of Digital Tools

Lesson 8.1	Add Applications to a PDA
Lesson 8.2	Solve Mathematical Problems Using a PDA
Lesson 8.3	Synchronize Documents with a PDA
Lesson 8.4	Create Document Images with a Tablet PC
Lesson 8.5	Create Templates and Stationery in Journal
Lesson 8.6	Use Write Anywhere on the Tablet PC
Lesson 8.7	Explore OneNote

You Will Learn To:

- **Download** applications for a PDA.
- **Solve** math calculations and create a PDA expense log.
- **Synchronize** Microsoft Office documents with a PDA.
- **Import** Word documents into Journal and add comments and flags.
- **Select** and **create** a new Journal template.
- **Enter** text with the Write Anywhere feature.
- **Take** and **organize** notes with OneNote.

How Can a PDA and Tablet PC Help You Work More Efficiently?

Hundreds of programs can be downloaded to a PDA to help you with anything from geometric equations to guitar chords. Tablet PCs allow you to do many things that previously had to be done on paper. This technology will become part of your daily routine.

Activity 14

Activity 15

DIGI Byte
Another Way to Align
You can also align cells by opening the Format menu, clicking Cells, and choosing the Alignment tab.

RUBRIC LINK
Go to the Online Learning Center at **digicom.glencoe.com** and click Lesson 10.3, Rubric, to find the application assessment guide(s).

Suite Smarts
Word has an AutoFormat feature for tables that works like the AutoFormat feature in Excel.

Step-by-Step continued

Align Cell Contents
1. Select Cells A4 through A8.
2. Click the **Center** button to align the text in the cells to the center.

Apply an AutoFormat
1. Select Cells A2 through D2.
2. Click the **Format** menu and choose **AutoFormat**.
3. In the AutoFormat window, choose **Classic 2** and click **OK**.

   The AutoFormat feature will make changes to the appearance of the selected cells. Save the file.

Application 5

Scenario You are the operations manager for a bank, and Customer Appreciation Day is two weeks away. You have been asked to prepare a list of refreshments that the bank will serve in the lobby for customers on that day. The manager wants you to create a spreadsheet showing what will be served and how much everything will cost.

Directions Enter the data listed below into a spreadsheet. Make sure that the data representing dollar amounts is in currency style. You will decide where in the spreadsheet to begin entering data. Apply formatting to the spreadsheet's labels and values. Label the columns and include a title for the spreadsheet. Choose a font style and size. Align the data in the cells. Add shading and borders. Use formulas to calculate how much the Customer Appreciation Day party will cost. Format the cell borders and shading. Use the AutoFormat feature if you choose. Save your spreadsheet as **Customer Appreciation**.

```
Refreshments for Customer Appreciation Day

3 dozen chocolate chip cookies $2.50 each

4 dozen sugar cookies $2.25 each

3 packages small paper plates $1.25 each

5 packages assorted hard candies $1.50 each

3 cans peanuts $2.99 each

2 packages small paper cups $1.75 each

10 2-liter bottles assorted soft drinks $1.50 each
```

Read to Succeed

Key Terms

- freeware
- shareware
- expense log
- import
- template

Take Notes

What happens when you take notes while you read? Do you understand the material better? One effective learning strategy is to take notes while you read. Try writing electronic notes using your Tablet PC and PDA, or draw a chart similar to the illustration shown here. Use descriptive titles so you will be able to find your notes easily.

Example:

Lesson Number	Page	Notes
Lesson 8.1	229	Freeware is a free software application.
Lesson 8.1	229	Shareware software is distributed free for a trial period, and it can be purchased later.
Lesson 8.2	232	Many math programs are available on the Internet to download on a PDA.

Digital Connection

Help Wanted: Math Skills Required

How many jobs can you think of that use math? In school you can use your textbooks to look up answers to challenging math problems, but what can you do if you are faced with such problems on the job? How can you be sure you have the right answers? A PDA can help you check your math work. There are many math programs you can download to your PDA that will help solve everyday math problems.

Writing Activity

How do you use a calculator at school? What type of calculator do you use? Open the calculator on your PDA. Compare the layout of the onscreen calculator with calculators you use in class. Write a paragraph describing the similarities and differences.

Activity 13

If you need additional help on this topic, open the Help menu; or refer to the documentation for your hardware.

DIGI*Byte*

Borders and Shading

Shading and border formatting can give your spreadsheet a more professional appearance.

Select

Select a cell or group of cells before you format a border or apply shading.

Step-by-Step

Apply Borders and Shading to the Cells in a Spreadsheet

1. Open Excel and the **Benjamin** spreadsheet.

2. Select Cells A2 through D2.

   When you select cells in a spreadsheet, the cell where you begin the selection always remains white. The rest of the cells will turn dark as you drag the mouse over them.

3. Select the **Format** menu and click **Cells**. From the Format Cells window, select the **Border** tab.

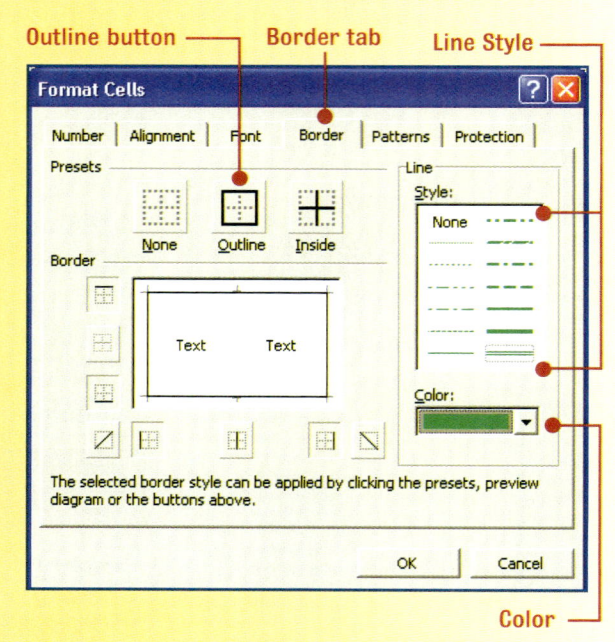

4. Click **Color** and choose **Sea Green**. You will see the color name appear when you allow the insertion point to hover over the colors.

5. Click the double line in the lower-right corner of the Style box.

6. Click the **Outline** button. Click **OK**. It is important that you click the Outline button after you have made all other choices.

7. Select Cells A2 through D2 again.

8. Click the **Format** menu and then click **Cells**. In the Format Cells window, click the **Patterns** tab.

9. Choose a light color, such as light blue, and click **OK**.

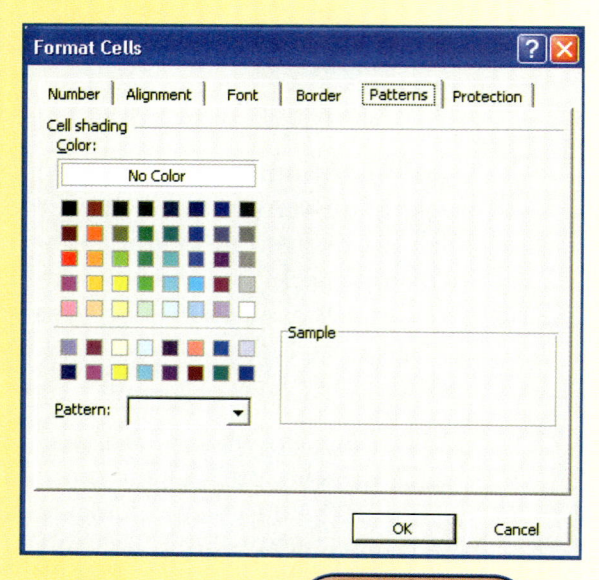

Continued

Chapter 10

329

Lesson 8.1 → Add Applications to a PDA

You will learn to:
- **Search** online for freeware and shareware applications for a PDA.
- **Download** applications to a PDA.
- **Solve** mathematical problems with the calculator feature on a PDA.

Suite Smarts
Check out the books that are available online to download to your PDA. There are books and study guides for all areas of interest.

freeware A software application that you can download and share at no charge.

shareware Software that is distributed free on a trial basis and may later be purchased.

DIGI Byte
Calculators
There are many types of calculators that can be downloaded to your PDA from the Internet.

Applications One of many conveniences you have with a PDA is that software applications are quick and easy to access. There are many applications available on the Internet that can be downloaded to your PDA. A quick search using a search engine will find applications related to just about any subject.

Freeware and Shareware
Many of these applications are **freeware**. A freeware application is a software application that you can download and share at no charge. Another type of application is **shareware**, which is software that is distributed free on a trial basis and may later be purchased. Many shareware applications have a built-in expiration date so you can test the program before you buy it.

In this lesson you will download a loan calculator, which will be used to solve math problems concerning money you would like to borrow.

✓ Concept Check
- Why would you want to use a trial version of a software program?
- What type of program would be helpful to you in your school work?
- What types of programs would be helpful for teachers?

List advantages of downloading applications to a PDA.

Lesson 10.3: Apply Advanced Spreadsheet Formatting

You will learn to:
- **Add** shading and borders to cells.
- **Align** cell contents.
- **Apply** AutoFormat to a spreadsheet.

Dressing Up Spreadsheets Spreadsheets can be formatted so they are easier to read and have a professional appearance. You can add borders and shading to cells, rows, and columns. Borders can be placed around single cells or groups of cells.

AutoFormat You can manually choose which cells to format, or you can use the AutoFormat feature. The **AutoFormat** feature provides predefined shading and border formats that can be applied to cells in a spreadsheet. You can use AutoFormat in any existing spreadsheet to save time when formatting the spreadsheet. In this lesson, you will apply borders and shading manually as well as using AutoFormat.

> **AutoFormat** Predefined shading and border formats that can be applied to cells in a spreadsheet.

✓ Concept Check

Where in a spreadsheet might you use borders and shading, and why?

DIGITAL DIALOG

Discuss Spreadsheet Colors

Teamwork In a small group, use the following questions to guide a discussion on using colors in a spreadsheet. Record answers to the questions as you discuss them in a PDA or Tablet PC, if possible.

- Which colors complement each other?
- Which colors are easy on your eyes?
- If you shade cells, do you need to change the font color?
- How do borders go with the shading of cells?
- What should you consider when coordinating border colors and cell shading colors?
- What colors or widths should you use on borders?

Communication When you are finished, trade your notes with another group. Compare how you answered these questions with how the other group answered them.

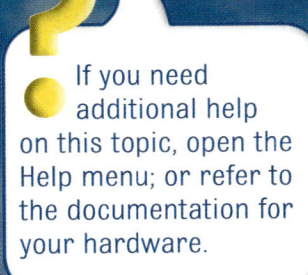

Step-by-Step

Activity 1

Identify the PDA Operating System

Before you begin searching for and downloading software to a PDA, you need to know which operating system the PDA uses.

1. Create a new folder on your desktop computer, and name it *PDA software*. When you download software applications from the Internet, you will save them in this PDA software folder.

2. Determine the operating system and version of software your handheld is using. From the drop-down menu on the PDA, tap **Info**. This option may vary on different models of PDAs.

Activity 2

Download Freeware from the Internet

1. Use a search engine to find free software for your OS.

2. Download the software, and save it to the new PDA software folder on the desktop. You may need a program like WinZip to unzip large files.

Activity 3

Synchronize to a PDA

1. Open the PDA software folder on the desktop. Find the application to be added to the PDA. Double-click the application's file icon. The application is added to the PDA's synchronization program.

2. Place the handheld in its cradle, or connect the PDA to a computer with a USB connection.

3. Press the **Synchronize** button on the PDA or the cradle. The program will be added to the PDA, and you will see the icon for the new program in your applications.

Activity 4

Use a Loan Calculator

1. Download a loan calculator application to your desktop computer. Synchronize it to your PDA.

2. Enter the following data. How much are your payments?

	Loan Amount	Down Payment	Interest Rate	Term (number of years)
A	$14,500	$500	6.5%	6
B	18,500	0	8.9%	5
C	24,700	0	7.25%	6

If you need additional help on this topic, open the Help menu; or refer to the documentation for your hardware.

DIGIByte

AutoSum

AutoSum can be used only when adding a group of cells, not for multiplication.

Editing a Formula

Press F2 to edit a formula. The formula will appear in the cell, where you can edit it. You can also edit the cell in the Formula bar.

Control+Home

Holding down CTRL and then pressing HOME returns the insertion point to Cell A1 from any position in the spreadsheet.

Suite Smarts

In formulas, letters in cell names do not have to be capitalized.

RUBRIC LINK

Go to the Online Learning Center at **digicom.glencoe.com** and click Lesson 10.2, Rubric, to find the application assessment guide(s).

Step-by-Step continued

5. Click the **Currency Style** button $, if necessary. Cell D6 should display the answer, which is $302.00.

Application 3

Directions Open the spreadsheet you created in Lesson 10.1, page 323, and saved as **Benjamin**. In Cell A3, enter the word `Amount`. In Cells A4 through A9, enter the number 2. Enter the word `Cost` in Cell C3. Enter a formula in Cells D4 through D9 that will calculate the cost for buying two of each item. For example, for Cell D4 your formula should answer the question, "How much will it cost to buy two hamburger baskets?" Print the spreadsheet. Write the formulas you entered on the printed spreadsheet next to the cells where you entered them. Save the file.

Application 4

Directions In Excel, click Help, and then click Microsoft Excel Help. Select the Index tab. Key the word `function`, and click Search. Scroll the list of related items and look for information about what functions are and what they do. You should see a list of the functions available in Excel.

In a new worksheet, create a list of ten functions. Include in the list how the functions are written and what they do. Format your list by labeling a column `Function`, labeling the next column `How to write it`, and labeling the third column `What it does`. Once you have completed the list, save and print your spreadsheet. As a class, discuss what the functions on your lists do, which functions you think are most useful, and why they are useful.

Application 1

Scenario You want to buy a new car or truck. At your after-school job you make about $140 per week. You expect the gas for your car to cost approximately $30 per week.

Directions Use the loan calculator on your PDA to figure out which car payment from Activity 4 you will be able to afford. You expect the gas for your car to cost approximately $30 per week. How much will gas and your car payment together cost you each month?

Application 2

Directions Search for another loan calculation program on the Internet. Download this application to your computer, and then synchronize it to your PDA. Reenter the same information you used in Activity 4.

Compare the two loan calculation programs. Is one better to use than the other? Why or why not?

Application 3

Directions There are many types of educational programs that can be downloaded to a PDA. Search the Internet for applications for a PDA. You can search for programs for the Palm OS, Windows CE, or Pocket PC operating system.

How many programs did you find? List and describe five different types of educational programs.

How could these programs help you at school? What programs did you find that interest you for personal use?

RUBRIC LINK
Go to the Online Learning Center at **digicom.glencoe.com** and click Lesson 8.1, Rubric, to find the application assessment guide(s).

Look Who's DIGITAL

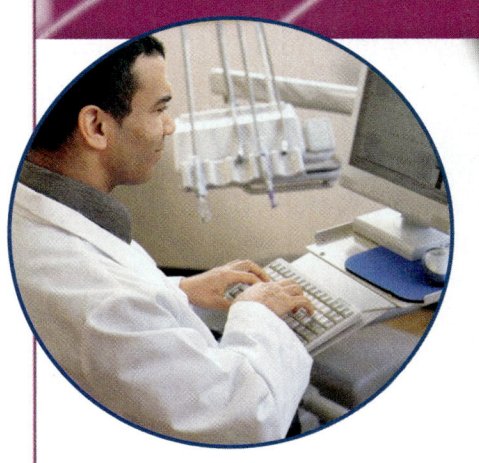

Digital Dentistry

Technology helps today's dentists work and communicate with their patients more effectively. By using an intraoral camera plugged into his computer's USB port, the dentist can show the patient what the outer areas of the teeth and gums look like. This helps the patient understand what treatment is necessary. The dental technician can use a Tablet PC digital x-ray system to check for cavities and immediately display the results. Dr. Henry Ramirez in Santa Cruz, California, can manipulate images he took with an intraoral camera and use 3-D computer-aided design and manufacturing software to create a crown for a patient with a cracked tooth. Then he sends the design over a wireless network to a unit containing tiny robot arms that mill the crown out of a block of porcelain. In 15 minutes it is ready to be seated in the patient's mouth!

Activity Brainstorm ways dentists can use an intraoral camera to help patients.

Suite Smarts

The SUM formula can also be used in tables created in Word.

DIGI Byte

Formula Bar

The formula you enter into a cell is displayed in the Formula bar, not in the cell.

SUM

The formula =sum(B4:B11) could also be written as =B4+B5+B6+B7+B8+B9+B10+B11.

Activity 12

Step-by-Step *continued*

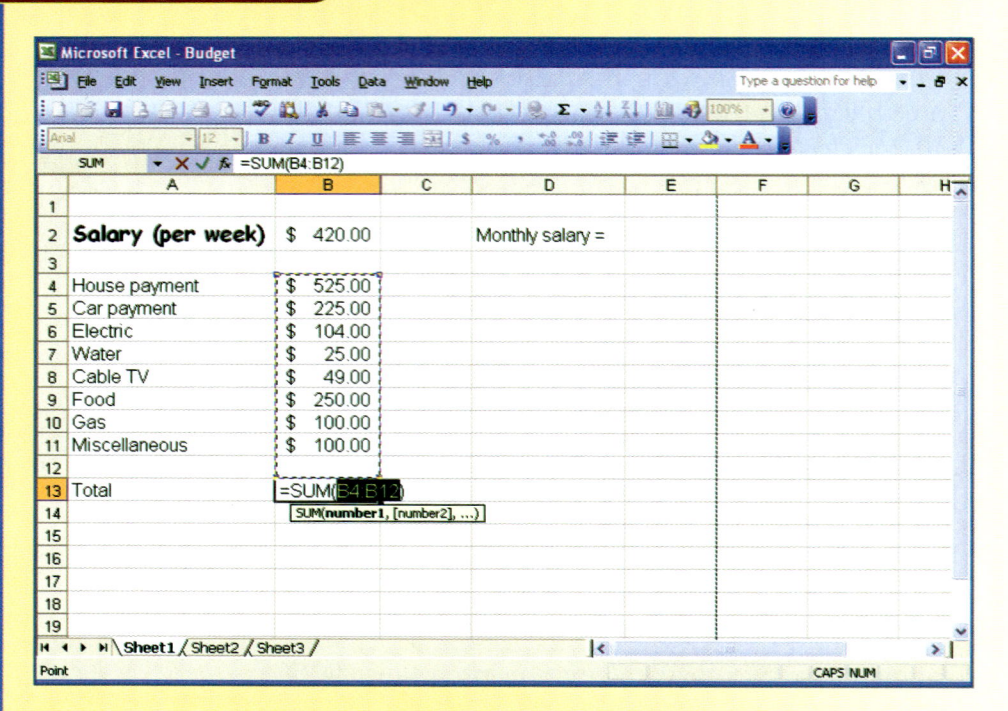

3. Press ENTER to accept this formula.

Enter Additional Formulas into a Spreadsheet

Next you will calculate the total income, or money earned, for the month.

The spreadsheet shows in Cell B2 that the weekly salary is $420. There are roughly four weeks in a month. To find the monthly salary, you will create a formula that will multiply the weekly salary by four.

1. Click Cell D3. Enter the formula =sum(4*B2) and press ENTER.

   The asterisk (*) means *multiply*. As you work through this spreadsheet, ask "What if?" questions about the data. You know the monthly salary is $1,680. Ask yourself, "What if I spend less on food?" and, "How much will I have left over?"

2. Click Cell D5. Enter Balance for Month in the cell and then press ENTER.

3. Cell D6 should be active. Enter the formula =sum(D3-B13). This formula can be translated *as monthly salary minus expenses, or the value of D3 minus the value of B13.*

4. Press ENTER. What is the balance for the month?

Continued

Lesson 8.2 Solve Mathematical Problems Using a PDA

You will learn to:
- **Download** and use a math application for a PDA.
- **Create** an expense log.

Math on a PDA Hundreds of math programs are available on the Internet to download onto your PDA. There are programs for currency conversions, measurement conversions, algebra, loan calculations, spreadsheets, and even graphing.

Adding these programs to your PDA can help solve mathematical problems you encounter both in and out of school. Many of these programs are free. Some of the programs are available in foreign languages.

Expense Log An expense log is a software application in which you can enter and track your daily expenses. Companies usually reimburse employees for travel expenses like gas, hotels, meals, air fare, and taxis. An expense log is an easy way to record these daily expenses.

Deleting Unused Applications If you synchronize a program to your PDA that you do not like or never use, you can delete it so that it does not take up memory on your PDA.

At the top of the Applications screen, tap the Menu button. Choose Delete. In the list of programs, tap the one you want to delete, and then tap the Delete button.

expense log A software application in which you can enter and track your daily expenses.

✓ Concept *Check*

If you worked in a bank, what types of mathematical programs would be helpful to download to your PDA?

DIGITAL DIALOG

Math on a PDA

Teamwork With a small group, search the Internet for math programs to install on a PDA. Find at least eight different programs. Make a list of them on the board, and explain to the rest of your class how the programs can be used.

Activity 10

If you need additional help on this topic, open the Help menu; or refer to the documentation for your hardware.

DIGI Byte

Equal Sign

In a spreadsheet cell, the equal sign (=) identifies the cell as containing a formula that the computer needs to calculate.

Step-by-Step

Enter a Formula into a Spreadsheet

1. Open the **Budget** spreadsheet you created in Lesson 10.1, pages 317–322.

2. Enter the word `Total` in Cell A13.

3. Click in Cell B13.

4. Enter `=sum(B4:B11)` into Cell B13 and press ENTER.

   This formula means *Find the sum of the numbers in Cells B4 through B11.* All formulas must begin with an equal sign.

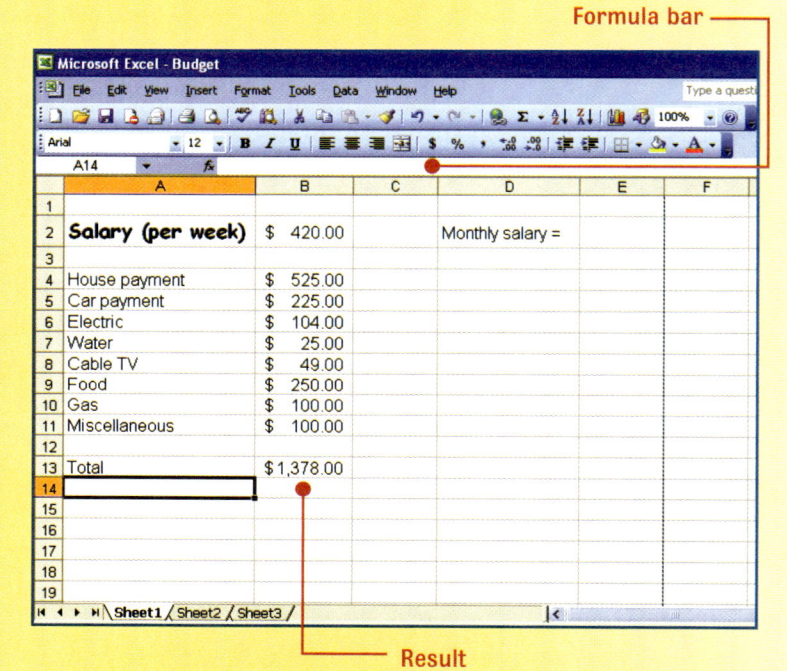

5. The total of the numerical values in Cells B4 through B11 will appear in Cell B13. If the dollar sign does not appear, click the **Currency Style** button $.

Activity 11

Use the AutoSum Feature

1. Delete the formula you entered in Cell B13. To do this, click Cell B13 to make it active, press ←BACKSPACE or DELETE, and then press ENTER. When the formula has been deleted, the cell will be blank.

2. Make sure Cell B13 is still selected. In the toolbar, click the **AutoSum** Σ button. The cells above Cell B13 will be outlined with a dotted line, and Cell B13 will display the formula for adding these cells.

Continued

Activity 5

If you need additional help on this topic, open the Help menu; or refer to the documentation for your hardware.

Step-by-Step

Identify Math Applications for a PDA

1. Search the Internet for a PDA program that will do measurement conversions.
2. Synchronize the application to your PDA.
3. Enter the data below and convert as indicated.

```
15 inches    (convert to feet)
3 miles      (convert to kilometers)
10 inches    (convert to centimeters)
480 yards    (convert to feet)
78 meters    (convert to yards)
4 miles      (convert to yards)
67 feet      (convert to centimeters)
2,500 inches (convert to meters)
2 miles      (convert to yards)
```

Activity 6

Add Expenses

1. Go to the main screen on your PDA.
2. Find the Expense application.
3. Tap to open this application.
4. Tap **New**.
5. Tap **Expense Type**. A drop-down list appears.
6. Tap **Car Rental**. The insertion point appears on a blank line.
7. Enter `149.95`.
8. Repeat Steps 7 and 8 to add another expense.
9. Choose **Mileage** from the menu and enter `207` for the miles.
10. Tap **Details and Vendor** and then enter `FBLA conference`.
11. Tap **City** and enter `Virginia Beach`.
12. Tap the **Who** box next to Attendees.
13. Tap **Lookup**. The Address Book list appears.
14. Tap a name in the address list. Tap **Add**.
15. Close the Expense program.

Suite Smarts

There are different conversion programs available for download.

Lesson 10.2 — Create Formulas in Spreadsheets

You will learn to:
- Enter formulas into a spreadsheet using the Formula bar.
- Enter formulas with the AutoSum button.

Making Calculations One of the main functions of a spreadsheet is to perform mathematical operations. Excel formulas perform these operations on data in cells. Remember that a formula is a mathematical expression. Formulas follow the standard order of operations. This means that the operations will be completed in the order in which they appear in the formula. Parentheses can be used to enclose numbers and operations so they will be calculated first. A great feature of spreadsheets is that the formulas automatically recalculate whenever the values change in cells on which the formulas are based.

Using Functions A **function** is a predefined formula in a spreadsheet. Excel contains a variety of functions. For example, to find the average of data in a spreadsheet, you can simply select Function from the Insert menu, and then find and insert the Average function. You do not need to enter a formula for averaging manually.

In this lesson, you will work with the SUM function. The SUM function adds the values in a group of cells. You will use the SUM function to add the expenses in a budget.

function A predefined formula in a spreadsheet.

✓ Concept Check

When a business provides a service, it sends out an invoice to collect payment for that service. How can a spreadsheet help a business create more accurate invoices?

DIGITAL DIALOG

Formula Ideas

Teamwork In a small group, brainstorm ways you could use spreadsheets and formulas. What types of activities do you do at school or at home where a spreadsheet would help you make calculations?

Writing One person in your group will record the ideas discussed and make a list of the ideas, using a Tablet PC or speech recognition software if possible. Compare your list to the lists of other groups. How many ideas are similar? How many are different?

Application 4

Directions Using the same conversion program, try the following conversions. Use a ruler or tape measure to record the measurements of four objects around you. For example, you could measure a chair, table, or a friend's height.

Enter the measurements in a table similar to the one below:

Item	U.S. Measurement	Metric Conversion
Chair	3 1/2 feet	_____ m
Table	5 feet 4 inches	_____ m
Desk	4 feet 8 inches	_____ m
My Height	5 feet 2 inches	_____ m

With the measurement conversion software, change these measurements to metric units.

Try to do the same with some liquid measurements. Collect labels from different beverages, and try converting the amounts.

Application 5

Scenario You are on a business trip and need to track your expenses.

Directions Open the Expense application. Add the following two expenses:

    Hotel $107.50
    Dinner $14.95

For the hotel, add the vendor *HomeStyle Suites*.

Add two more expenses of your choice.

RUBRIC LINK
Go to the Online Learning Center at **digicom.glencoe.com** and click Lesson 8.2, Rubric, to find the application assessment guide(s).

Digital Decisions

PDAs on the Job

Critical Thinking You work for a large commercial construction company as an estimator. You have to review the drawings and blueprints for upcoming projects that your company will be building.

After the review, you have to create a materials list with approximate costs for the materials. This list will be used to help estimate total construction costs. You must also calculate how much of the material needs to be ordered. Your job involves many mathematical calculations.

Internet Investigation Search Web sites that offer PDA software programs to find out what kinds of applications are available that may help you in your job.

In Memo Pad on a PDA, enter a brief description for each application that you find. The description should include :

- What the application does
- How it would help you on the job
- Cost of the program
- Special features included in the application

Add a brief summary statement with recommendations for the programs that you feel would be most useful in your job.

Application 1

Directions Enter the data below into a spreadsheet. Format the numbers as currency. Make the title bold. Adjust the column width as needed. Begin by entering Benjamin's Restaurant in Cell B2. Leave Column A blank. Enter Hamburger Basket in Cell B4. Enter 4.95 in Cell C4. Continue to enter the rest of the information into the appropriate columns on your spreadsheet.

```
Benjamin's Restaurant
Hamburger Basket          4.95
Chili Dog Special         4.50
Chicken Finger Basket     5.95
Onion Rings               2.95
Milk Shakes               2.50
Soft Drinks               1.25
```

Save your spreadsheet as **Benjamin**. Print the spreadsheet with the gridlines showing.

Application 2

Scenario In this lesson's Digital Dialog, page 316, you planned for the Future Business Leaders of America (FBLA) Club Teacher Appreciation Day breakfast.

Directions Open a new Excel worksheet and create a list in Column A using the five or more items you plan to serve for the breakfast. Include the price of each item in Column B.

Format your spreadsheet so the prices of the items in Column B have dollar signs and so all of the text in Columns A and B is visible. Save your file as **FBLA Breakfast**. Print the spreadsheet with the gridlines showing.

RUBRIC LINK
Go to the Online Learning Center at **digicom.glencoe.com** and click Lesson 10.1, Rubric, to find the application assessment guide(s).

Look Who's DIGITAL

Field Research

What do monkeys, zebras, and lions have in common? All are being studied with the help of technology.

A student recorded observations of woolly monkeys in the Louisville Zoo on a Tablet PC using speech recognition software. She analyzed the monkey calls using PDA sound analysis software.

Princeton scientists and graduate students fitted zebras with GPS and data-transmitting collars. Because the collars swap data every time they find other collars, a scientist only needs to find some of the zebras to download information about the entire herd.

National Geographic scientists use a PDA to control collar-mounted cameras on lions. With "Crittercam" technology, animals can be studied from their point of view.

Activity Write a paragraph about how you would use technology to learn about your favorite animal.

Lesson 8.3 — Synchronize Documents with a PDA

You will learn to:
- **Synchronize** Microsoft Office documents with a PDA.
- **Create** a journal with Memo Pad on your PDA.

Suite Smarts

Microsoft PowerPoint presentations can be shown directly from a PDA that has special presentation software installed. The PDA can be directly connected to an LCD projector to show the presentation.

Synchronizing Microsoft Office Documents Documents created in Microsoft Word, Excel, and PowerPoint on your desktop computer can be synchronized to your PDA. A PDA with Windows CE or Pocket PC operating systems can download Microsoft Office documents when synchronization takes place.

Many new PDAs with the Palm OS operating system can also automatically transfer Microsoft Office documents. Other PDAs may need to have a program called Documents To Go installed before the transfer of files can occur.

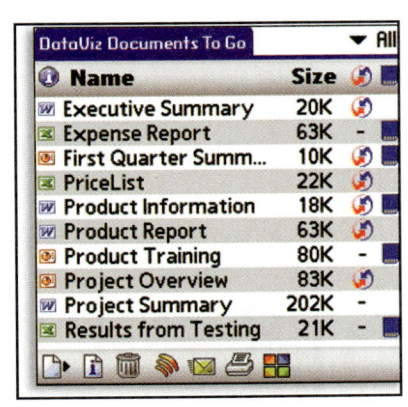

DIGI Byte

File Transfer
Limited menu options are available when you transfer a Microsoft Office document to a PDA.

Editing Files
You can edit Microsoft Office documents on a PDA.

PDA Journal Since you often have your PDA with you, it is a great place to keep a personal journal. You can record your thoughts whenever you like without having to carry a separate notebook.

✓ Concept Check

Why would you want to store Microsoft Office documents on your PDA? Share your ideas with the rest of your classmates. What kinds of documents would you store on your PDA? Are your classmates' ideas similar to or different than yours?

Step-by-Step continued

3. Select the font **Comic Sans MS**, change the font size to **14**, change the font style to **Bold**, and then click **OK**.

Activity 8

Change Column Width

Notice the column does not display the entire contents of Cell A2 since the formatting changes were made. The cell width needs to be changed.

1. Click Cell A2.

2. Click the **Format** menu, then choose **Column**, and **AutoFit Selection**. The column width will automatically change to accommodate the width of the text.

3. Save your spreadsheet again.

Activity 9

Insert and Delete Worksheets

Remember that an Excel file is called a workbook. A workbook can consist of multiple worksheets. Sometimes you may need to insert another worksheet in your workbook, or delete a worksheet.

1. Click **Insert**, **Worksheet** to add a new worksheet to your workbook.

2. While you are viewing this new worksheet, click **Edit**, **Delete Sheet**. The worksheet will be deleted.

Activity 7

If you need additional help on this topic, open the Help menu; or refer to the documentation for your hardware.

Step-by-Step

Transfer Office Documents

Check the model of PDA that you have. If it does not automatically transfer Microsoft Office documents, you will need to download a trial version of a program such as Documents To Go if one is not already installed on your PDA. The following directions describe a Palm operating system running Documents To Go.

1. On the desktop PC, open the **Documents To Go** screen.

2. To add a document to the screen, click the Add Item button. The Add Items window opens.

3. Choose the location of the document to add. Double-click the file name to include it in the list. The next time you synchronize the PDA, the file will be transferred.

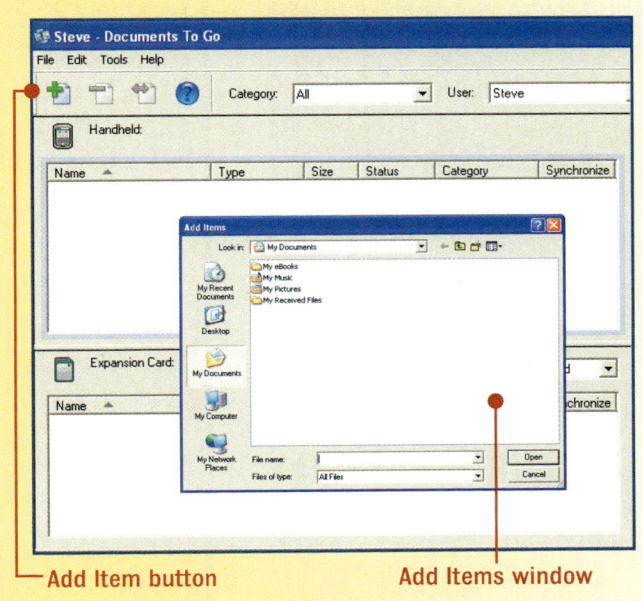

— Add Item button   Add Items window

Activity 8

Personal Journal on a PDA

The PDA can also be used to keep a journal. Memo Pad is an excellent application to use for recording journal entries. You can create a category named Journal to store your Memo Pad entries. When you synchronize your PDA with your computer, the Memo Pad entries will automatically be added to your computer.

1. Open **Memo Pad**. Tap your stylus on the category menu.

2. Tap **Edit Categories** and then tap **New**. Enter `Journal` for the name. Tap **OK**.

3. Create a new memo. Tap the category menu and select the **Journal** category.

4. Enter today's date on the first line of the memo.

5. Skip a line and then enter: `I am going to start recording my journal entries on my PDA.`

6. Tap **Done** to complete the entry.

Suite Smarts

Microsoft Word, Excel, and PowerPoint files can be synchronized to a PDA. However, the PDA versions of these programs have fewer features.

Chapter 8 Additional Features of Digital Tools   *Lesson 8.3*

Step-by-Step continued

Margins From the Print Preview screen you can also change the margins on your spreadsheet.

5 Click the **Margins** button.

6 Place the pointer on the dotted line on the left side of the page. Hold down the left mouse button. On the bottom-left side of the screen you will see the words *Left Margin* and a number telling the width of the margin. Drag the dotted line so the left margin is 1 inch wide. Repeat this step for the top, bottom, and right margins.

You can also change the page margins by clicking the **Setup** button and selecting the **Margins** tab. Use the up or down arrows or key the settings that you want for each margin.

Page Orientation If you have a wide spreadsheet, you can change the orientation of the page from portrait to landscape to show the data on one page width.

7 Click the **Setup** button to open the Page Setup window and then click the **Page** tab. Change the orientation of the page from portrait to landscape. Use Print Preview to look at the change. Change the orientation back to portrait.

8 Click the **Print** button.

Format Text in a Spreadsheet

1 Click Cell A2 (*Salary*) to make it active. Select the **Format** menu and click **Cells**.

2 In the Format Cells window, click the **Font** tab. This is the same window you used in Word to format text.

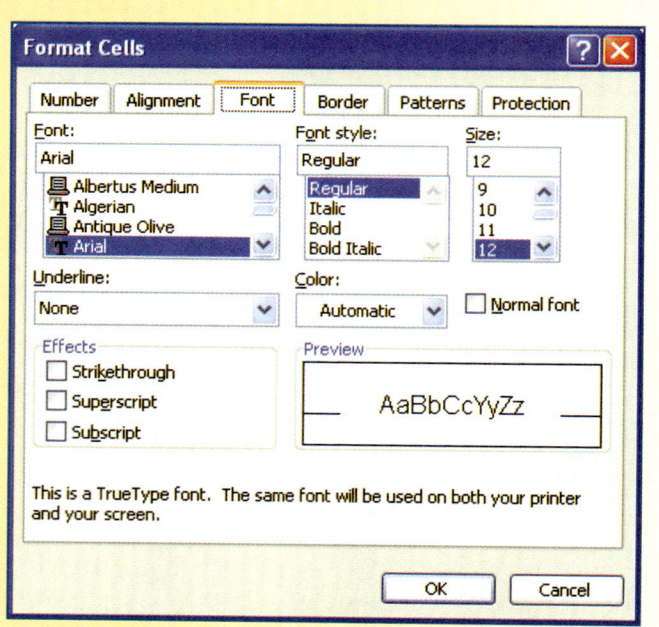

Activity 7

Suite Smarts

You can format the text in cells the same way you format text in a Word document. You can change text color, size, and style.

Continued

Application 6

Directions Learn about some of the many Palm math applications available for PDAs by researching the Palm Web site. Compare and contrast downloadable educational freeware and shareware by creating a Microsoft Word chart similar to the illustration. Your chart should include at least five different math programs. Transfer the file to a PDA.

Program	Description	Cost
Tesselation VI	Draws connections between geometry and art	freeware
Tan Free	Allows students to solve tangrams	freeware
powerOne Personal	Calculates sales tax, markup, and percent change	$9.99, shareware
9.95 Unit Converter	Converts dollars to euros, grams to ounces, etc.	$9.95, shareware
Maths	Solves equations	$20.00, shareware

RUBRIC LINK
Go to the Online Learning Center at **digicom.glencoe.com** and click Lesson 8.3, Rubric, to find the application assessment guide(s).

Digital Decisions

Program Purchase

Critical Thinking The law firm you work for has decided to purchase a PDA for each of the attorneys. You have been asked to review various software programs for the PDAs. The firm partners are not concerned with cost; they feel the benefit of having PDAs with useful programs is very important.

Communication Work with a small group to see what programs you can find. On a Tablet PC, draw a chart that includes the following information for each program that you find:

- Name of the program
- What the program does
- Cost of the program

After you complete your chart, exchange it with another group and compare ideas. How similar or different are your ideas?

Digital Dialog

Storing Documents on a PDA

Teamwork You are going to a technology job fair. There will be hundreds of employers conducting interviews at the job fair. You have prepared a resume and a cover letter. You want to make sure that you have enough resumes and are able to demonstrate your technology skills to potential employers. How can you do this with a PDA? In a small group, discuss how a PDA could be helpful.

Writing Have one person in your group write down the ideas your group discussed on a Tablet PC. Share these ideas with the other groups.

Step-by-Step continued

1. Select Cells **A8** and **B8** and click the **Cut** button, ✂, on the Standard toolbar.

2. Click Cell **A5**. Select the **Insert** menu and choose **Insert Cut Cells**. *Electric* should now be in Row 6 and *Car Payment* should be in Row 5.

Print a Spreadsheet

Print Preview Before you print a spreadsheet, use Print Preview to ensure that the information fits properly on the page. What you see in the Print Preview window is how the printout will appear.

In this activity you will use Print Preview and make adjustments before you print.

1. Click the **File** menu and choose **Print Preview**.

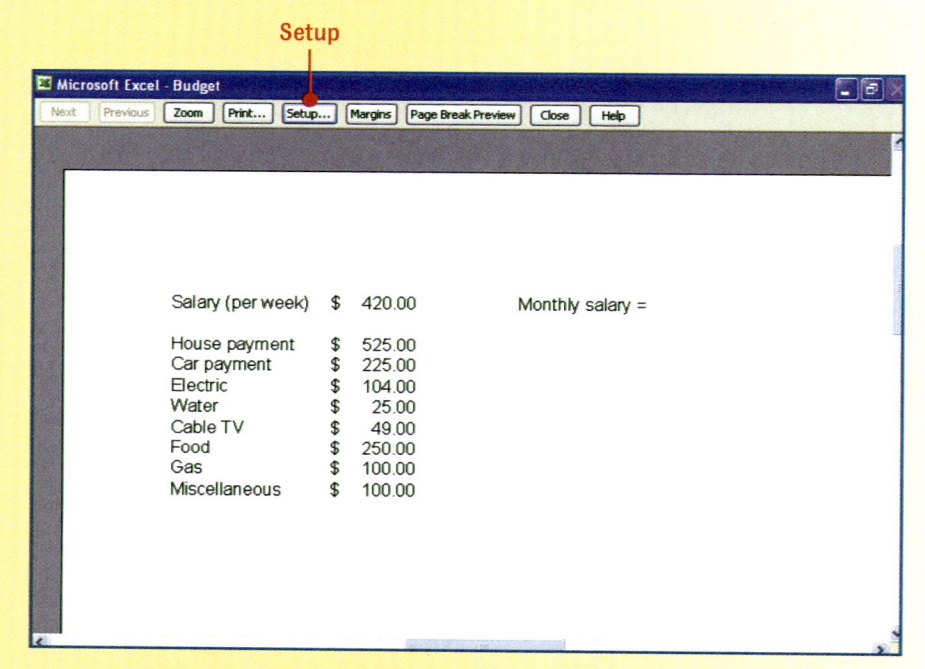

Gridlines Notice that the spreadsheet's gridlines do not show and will not print. To make the spreadsheet easier to read you will set up the gridlines to print.

2. Click the **Setup** button at the top of the Print Preview screen.

3. In the Page Setup window, click the **Sheet** tab and select the check box next to **Gridlines**. A check mark will appear in the box.

4. Click the **OK** button. The spreadsheet's gridlines will appear.

Activity 6

Suite Smarts

Remember that you can use the Undo and Redo buttons that are available in all Microsoft Office applications. Click the Undo button to delete the last action you completed, such as the last word or number that you entered. Click the Redo button if you use the Undo button by mistake.

DIGI Byte

Spell Check
Use the Spell Check feature to look for spelling errors before you print a spreadsheet.

Continued

21ST CENTURY CONNECTION

Work More Efficiently with OneNote

OneNote is an application that allows you to take notes electronically instead of on paper. OneNote can help you organize your notes and save time in many ways. Taking notes is easy because you can enter them anywhere on the page.

You can create different pages and divider tabs to help organize your notes into different subjects. You can customize the notes you create by choosing different stationery styles.

Notes can be password protected when confidentiality is needed. If you don't need to protect the entire note, you can choose to protect only the passage that contains the confidential information.

Input Options By attaching a webcam or microphone to your computer, you can capture and save audio and video in your notes. You can record a phone conversation or part of a lecture or meeting and take notes at the same time. Later, if your notes are not clear, clicking on a note will replay what was recorded while you took that note!

If you are surfing the Internet, you can easily copy and paste information into your OneNote document. Text can be entered in OneNote with a keyboard, or if you have OneNote installed on a Tablet PC, the notes can be entered in your own handwriting.

Drawings can even be entered by using the tools in the drawing toolbar. Even if you do not have a Tablet PC, a pen-input device can be installed on your laptop or desktop computer so you can enter handwritten notes.

Sharing Notes OneNote documents can be e-mailed to others for easy sharing of information. The receiving party does not have to have OneNote in order to open the file you send. This feature makes working in a study group very effective. Members can share their notes and make changes to them, if necessary.

Your notes can even be published to the Internet. If you publish handwritten notes to the Internet, the handwriting is automatically converted to text.

Notes taken on a smartphone or Pocket PC can also be added to OneNote. The notes are entered into OneNote when the device is synchronized with the computer. The same applies to an audio recording which has been saved to a Pocket PC or smartphone.

Activity Write a short paragraph about the way you take notes. Describe how OneNote could improve your method.

Step-by-Step continued

1 Click Cell A5. Press F2 so the insertion point appears in Cell A5. Press ← to move the insertion point to the first e in *electric*. Press ←BACKSPACE to delete the e. Enter E to start the word *electric* with a capital E.

You can also edit data in the formula bar at the top of the spreadsheet. The **formula bar** displays the data within a cell and allows you to edit the contents of the cell.

To replace cell contents from the formula bar, click in the cell you want to edit, press DELETE or ←BACKSPACE, and then press ENTER once you have added new text.

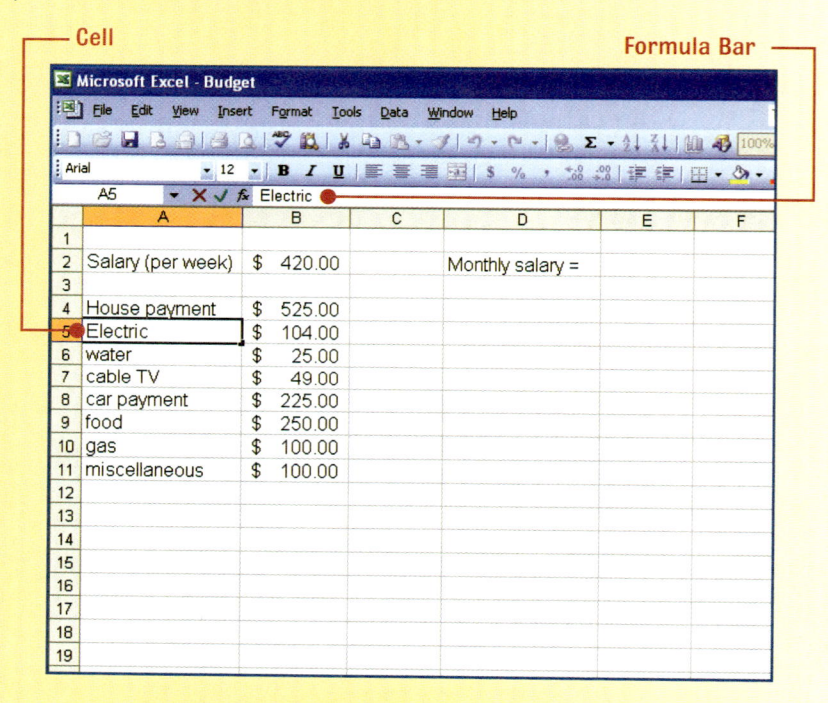

	A	B	C	D	E	F
1						
2	Salary (per week)	$ 420.00		Monthly salary =		
3						
4	House payment	$ 525.00				
5	Electric	$ 104.00				
6	water	$ 25.00				
7	cable TV	$ 49.00				
8	car payment	$ 225.00				
9	food	$ 250.00				
10	gas	$ 100.00				
11	miscellaneous	$ 100.00				

2 Capitalize the first letter of the rest of the items in Column A.

3 Click **File** and **Save** to save the changes to your file.

Since you have already named this file, you can use the Save command rather than Save As.

Move Cell Contents

In Excel, moving data from one cell to another is simple. You can also easily move the contents of one row or column to another.

In this activity, you will move Row 8, containing car payment information, to Row 5, so that it appears after the house payment information in Row 4.

Continued

---

formula bar Displays the data within a cell and allows you to edit the contents of the cell.

Suite Smarts

Remember that the first time you save a file in a Microsoft Office application, you will be asked to name the file and to choose a place in your hard drive to save it. Once this is done, you can simply click the Save button on the toolbar to save future changes to the file.

Activity 5

Lesson 8.4 → Create Document Images with a Tablet PC

You will learn to:
- **Import** Microsoft Word documents into Journal.
- **Mark up** documents in Journal.
- **Flag** Journal documents.
- **Find** Journal flags.

import To open a document in an application other than the one in which it was originally created.

DIGI Byte

Add Comments
Use the Journal to import documents received as e-mail attachments to make comments on them before e-mailing the document back to its originator.

Import and Mark Up Documents The Journal application on the Tablet PC enables documents to be imported as images. When you **import** a document, you open it in an application other than the one in which it was originally created. Using the digital pen, you can mark up and highlight these document images, but you cannot make changes to the text in the documents. The document images can be saved and printed with your markup.

Flags Flags can be added to important documents to make the documents easy to find. Flags can also be added next to paragraphs, charts, or ideas in a document so that you can easily go directly to locations within the document. If you are going to e-mail a document to another person, you may want to place flags next to important information in the document so that the person can be sure to review the important data.

✓ Concept Check

What types of documents do you get from your teachers that you could store on your computer? Do you ever make notes on these documents?

DIGITAL DIALOG

Tablets in the Classroom

Teamwork In a small group, brainstorm how using the Tablet PC in class could be helpful. Give examples of at least three different uses in each of four different subject areas. Think about the tasks you do while in class. Be specific about how you would use the Tablet PC.

Writing Create a diagram on the Tablet PC similar to the one on the right.

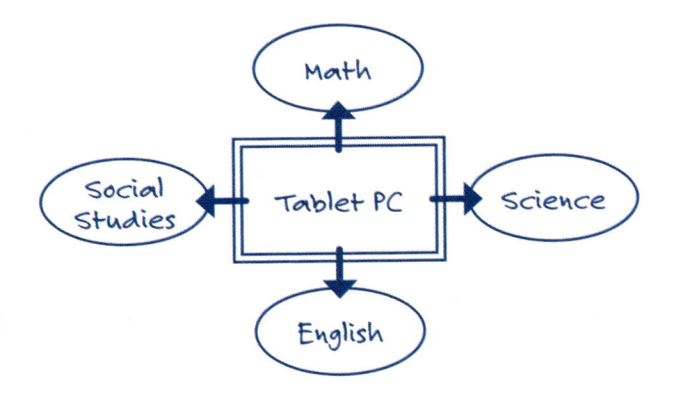

Chapter 8 Additional Features of Digital Tools — Lesson 8.4 — 239

Step-by-Step

1. Select the entire worksheet by clicking on the blank cell above Row 1 and to the left of Column A.

2. Using the drop-down arrow in the Font Size box, change the font from 10 point to 12 point.

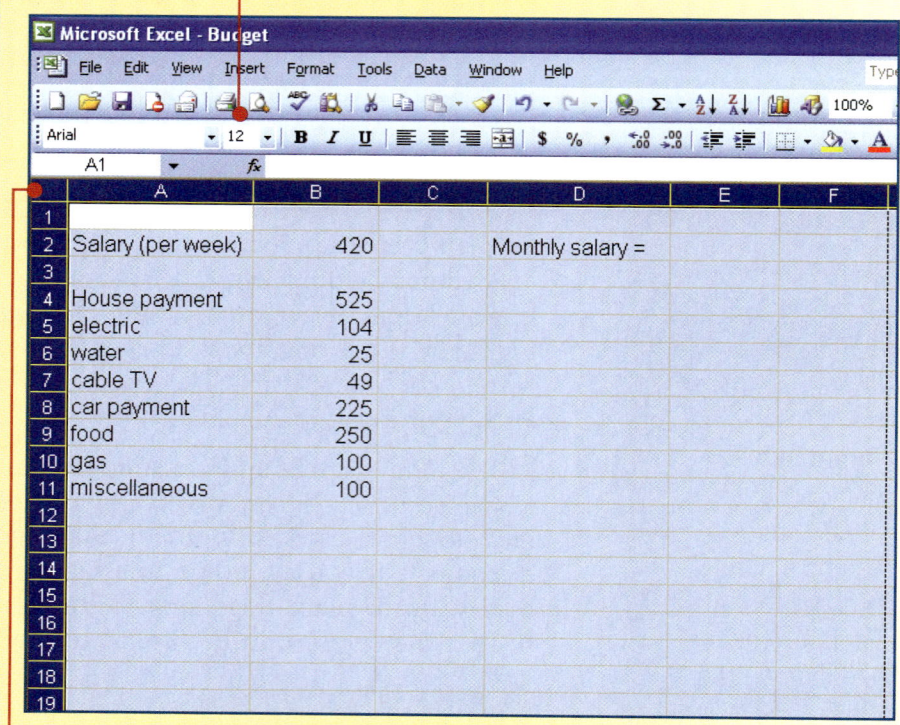

DIGI Byte

Decimal Places

You can adjust the number of decimal places that will display by clicking the **Decrease Decimal** button or the **Increase Decimal** button.

Format Numbers

1. Select the numbers in Column B by dragging from Cell B2 through Cell B11. To do so, click on Cell B2, keep the mouse button pressed, and move the pointer to Cell B11. Then release the mouse button.

2. Click the **Currency Style** button $ and notice what happens to the numbers.

Edit Cell Contents

You do not need to rekey an entire word to change it.

To edit data in a cell, click the cell to select it, press F2, and then move the insertion point where you want to make the change.

You can also select the cell, reenter the data, and then press ENTER.

Continued

Activity 9

If you need additional help on this topic, open the Help menu; or refer to the documentation for your hardware.

DIGI Byte

Ink
Change ink colors and thicknesses as you mark up a document in Journal.

Activity 10

Step-by-Step

Import and Mark Up a Microsoft Word Document

You will open a Microsoft Word document in Journal and add comments to it.

1. Open Journal. Select **File** and then **Import**.
2. Tap **Look In** at the top of the window that opens. Navigate to FBLA Meeting.doc.
3. Tap the **Import** button. The document will appear in Journal.
4. Use the Pen icon on the menu bar to mark up, or take notes on, the document you have imported. You can also use the highlighter button to highlight text in the document.

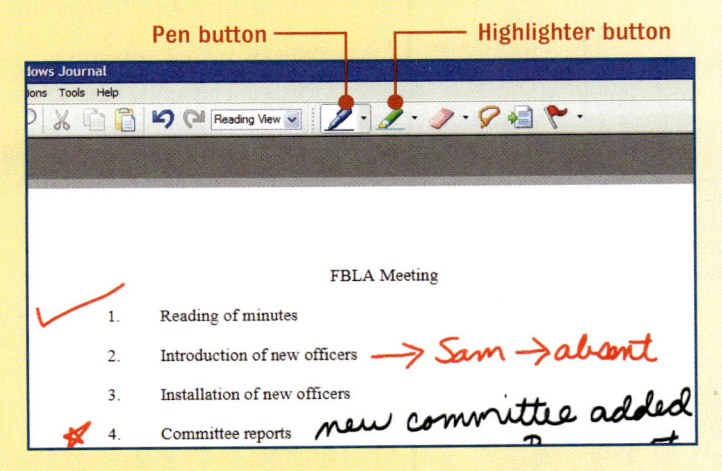

Pen button — Highlighter button

5. Save or print this document according to your teacher's directions.

Add Flags

1. Start a new Journal document.
2. Tap **File** and then **Import**. Navigate to FBLA-PBL.doc.
3. Import FBLA-PBL.doc into the Journal program.
4. Scroll to the paragraph discussing the mission statement.
5. Select the **Flag** button.
6. Move the digital pen to the right margin next to the mission statement paragraph, and tap the digital pen on the screen. A flag will appear. The flag is a graphic that can be resized or moved.
7. Scroll to the paragraph about conferences and place another flag.
8. Tap **File** and then **Save**. Tap **File** and then **Exit**. FBLA-PBL.doc disappears, and a blank Journal note screen appears.

Continued

Step-by-Step

In this step-by-step, you will perform several basic spreadsheet tasks, including entering data into a worksheet, editing data, applying basic formatting to text and the worksheet, moving cell contents, printing a spreadsheet, and inserting and deleting worksheets.

Enter Data in a New Worksheet

1 Start Excel. Select the **File** menu and click **New** to begin a new workbook.

2 Enter the data from the Excel worksheet below into the new worksheet in your workbook. Enter your data in the same cells as you see. For example, enter "House payment" in Cell A4. Widen a collumn by dragging the border of the column heading.

When you enter text in a cell, the text automatically aligns to the left of the cell; numbers automatically align to the right.

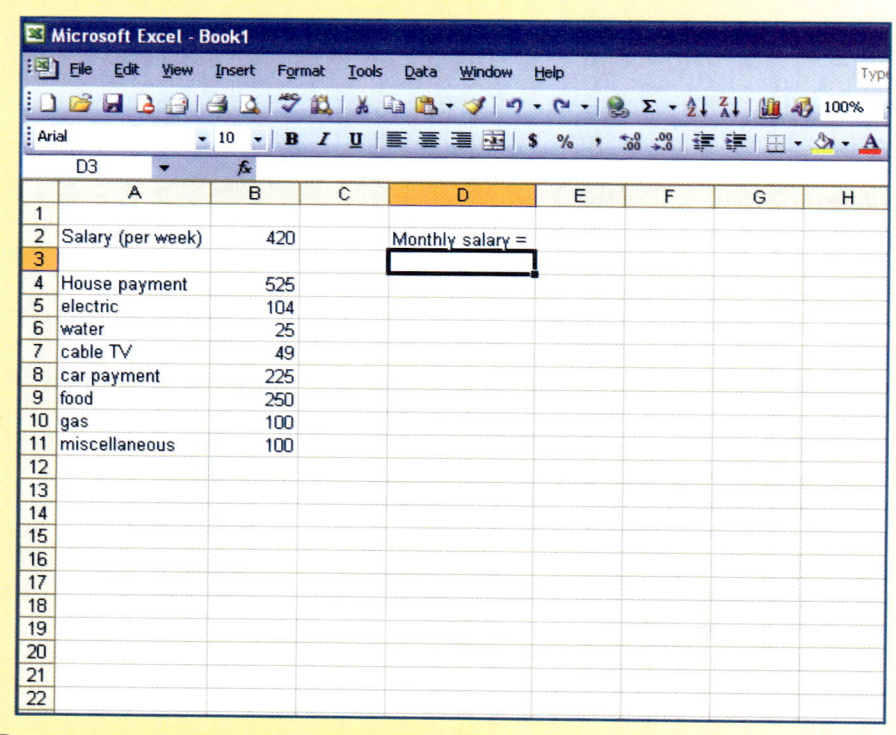

	A	B	C	D
1				
2	Salary (per week)	420		Monthly salary =
3				
4	House payment	525		
5	electric	104		
6	water	25		
7	cable TV	49		
8	car payment	225		
9	food	250		
10	gas	100		
11	miscellaneous	100		

3 Click the **File** menu and select **Save As** so you can name this file. Save the spreadsheet as **Budget**.

Select a Worksheet and Format Text

The default font for worksheets is Arial 10 point. You will change this font for your worksheet.

Activity 1

? If you need additional help on this topic, open the Help menu; or refer to the documentation for your hardware.

Suite Smarts

Most of the buttons on the Standard and Formatting toolbars in Excel are the same as the buttons on the toolbars in Word.

Activity 2

Continued

Activity 11

Suite Smarts
You can also import PowerPoint files into Journal and mark them up.

DIGI Byte

Hovering

Let the digital pen hover over the screen before you perform an action with it. When the pen hovers, it is close enough to the screen to move the insertion point, but it does not actually touch the screen.

Step-by-Step continued

Find Flags

1. Tap **File** and then **Search**. A second toolbar appears below the main toolbar in Journal.

2. Tap the **More** button on the right side of the toolbar. The toolbar expands.

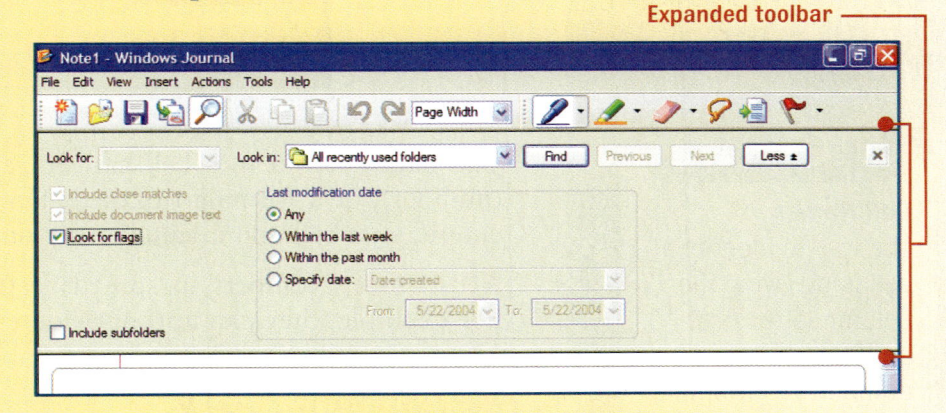

   Expanded toolbar

3. Tap the box next to **Look for Flags**, and then tap the **Less** button to collapse the toolbar.

4. Tap the **Find** button. The toolbar expands again and displays any documents that contain flags. FBLA-PBL.doc appears on the screen.

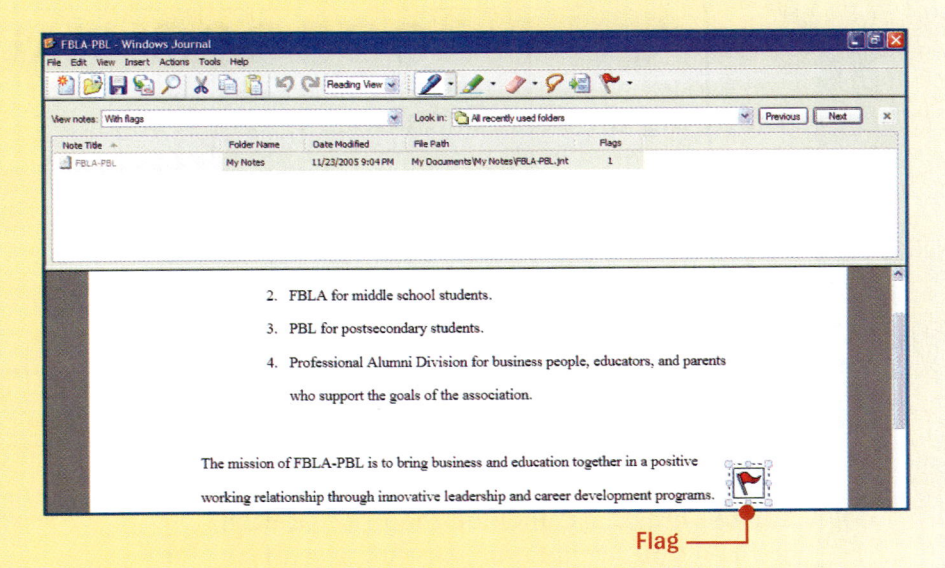

   Flag

5. Tap the flag next to FBLA-PBL.doc, and the document will open to the location where the first flag was placed (the mission statement paragraph).

6. Tap **Next**. The document screen will advance to the flag next to the paragraph on conferences.

label Text that is entered in a cell.

value Any number entered in a cell.

DIGI*Byte*

Cell Names
The way to identify a cell is by giving the column letter first and then the row number. For example, a cell in Column B and Row 29 would be identified as B29.

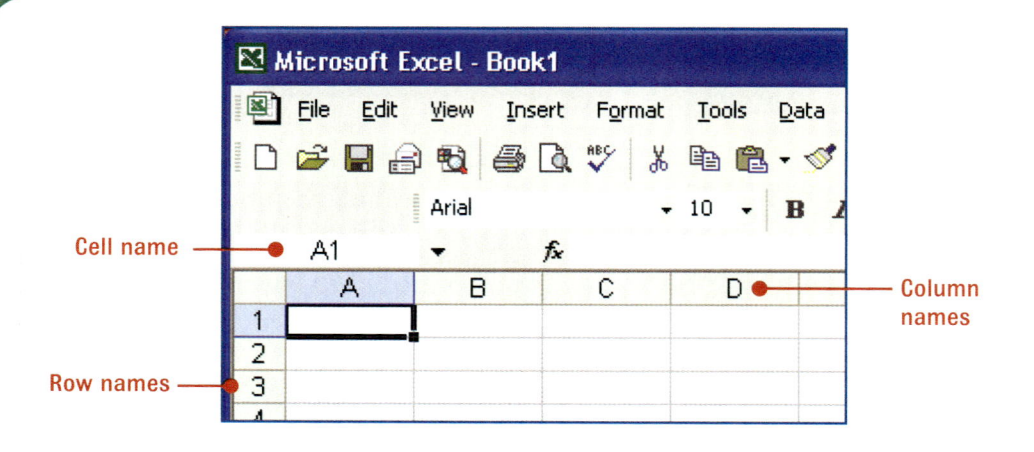

Cell name, *Column names*, *Row names*

The data entered in a cell can be referred to as a label (text), a value (number), or a formula (mathematical expression). A **label** is text that is entered in a cell and a **value** is any number entered in a cell.

Numbers Numbers in spreadsheets can be formatted in many ways. For example, they can appear as dates, currency, or percentages.

✓ Concept *Check*

One activity in everyday life in which people often use addition, subtraction, multiplication, or division is shopping. How could a spreadsheet help you in this activity?

DIGITAL DIALOG

Planning an FBLA Breakfast

Teamwork The Future Business Leaders of America (FBLA) club is planning a breakfast for the teachers of your school on Teacher Appreciation Day. In a small group, discuss what foods and beverages you would like to serve.

Writing Using a sheet of paper, a PDA, or a Tablet PC, make a list of five or more items you would like to serve for the breakfast. Include the price of each item. If you are not sure of the price of an item, make the best guess that you can. Save this information to use in Application 2, page 323, later in this lesson.

FBLA Breakfast Menu

Item	Price
1.	
2.	
3.	
4.	
5.	

Application 7

Directions Create and save a new Word document or retrieve one that you have saved. Import the document into Journal on the Tablet PC. Practice highlighting and making comments in sections of the document.

Application 8

Directions Open Word and import a clip art graphic. Copy and paste the graphic into Journal. Make some notes on the graphic with the pen. Select the graphic and text, and then copy and paste the graphic back into a Word document.

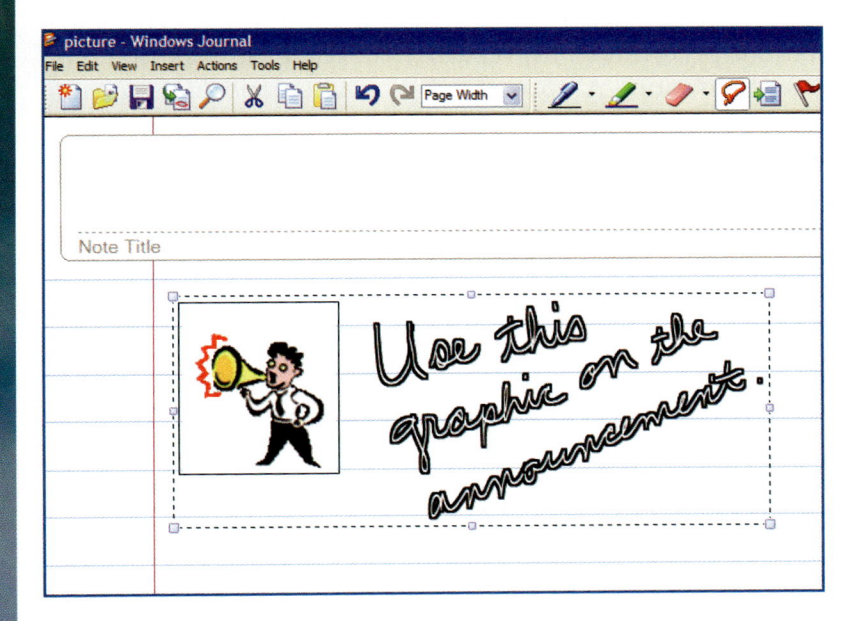

Application 9

Directions Import FBLA-PBL.doc again in Journal. Add another flag next to one of the paragraphs. Save this note.

RUBRIC LINK
Go to the Online Learning Center at **digicom.glencoe.com** and click Lesson 8.4, Rubric, to find the application assessment guide(s).

Digital Decisions

Tablet PC Art

You are the technology director for a large advertising firm. Your firm specializes in digital artwork that is sold to companies that advertise heavily on the Internet.

The directors have decided to purchase Tablet PCs for each department, and you need to decide which types of Tablet PC to purchase for the departments.

Internet Investigation Find a partner and search the Internet for manufacturers of Tablet PCs. Find out what options are available on the Tablet PCs.

Find the answers to the following questions:

- How much memory comes with each type of Tablet PC?
- What sizes are the hard drives that are available from the manufacturer?
- What programs are installed on the Tablet PC?

Include answers to the following questions:

- What is the screen size on each model?
- What type of processors are available?
- What is the maximum battery life on each model?

Critical Thinking Why would a Tablet PC be a good tool for an artist? Write a paragraph that includes at least three reasons.

Lesson 10.1 → Format Basic Spreadsheets

You will learn to:
- **Format**, edit, and delete numbers and text in cells.
- **Move** cells.
- **Change** margins and page orientation.
- **Print** a spreadsheet.
- **Change** column width.

spreadsheet A software application used to list, analyze, and perform calculations on data.

formula A mathematical expression, such as adding or averaging, that performs calculations on data in a spreadsheet.

Why Use Spreadsheets? A spreadsheet is a software application used to list, analyze, and perform calculations on data. The most common calculations performed in spreadsheets are addition, subtraction, multiplication, and division. A formula is a mathematical expression, such as adding or averaging, that performs calculations on data in a spreadsheet. You can use formulas to perform operations such as averaging numbers, calculating loan payments, or calculating interest on investments. In addition to performing calculations on data, spreadsheet applications can help represent the data in a chart.

A spreadsheet simplifies tasks. Instead of calculating sums by hand or with a calculator and then entering the total on paper, you can create a spreadsheet that automatically performs calculations. Instead of having to start your work over when you want to see the results using different values, you simply enter a different number. Spreadsheets are ideal for comparing your options and projecting the outcome of scenarios. As you research costs for each option, you can enter each figure into your spreadsheet and instantly see how it affects your budget. Used in this way, a spreadsheet becomes a decision-making tool.

Worksheets and Workbooks When you start the Excel spreadsheet application, you will see a grid of columns and rows. This is a worksheet. The Excel file is referred to as a workbook and can contain multiple worksheets. You can add more worksheets to your workbook.

Planning a Spreadsheet Before you enter text and numbers into a spreadsheet application, it is helpful to create a rough draft of the spreadsheet layout on a piece of paper. A spreadsheet consists of columns and rows. Columns are named with letters of the alphabet, and rows are identified by numbers. Columns and rows intersect to form cells. You enter data in the cells. Cells are identified by their column letter and row number, so the first cell in a spreadsheet would be in Column A, Row 1, and would be identified as Cell A1.

Lesson 8.5 → Create Templates and Stationery in Journal

You will learn to:
- **Use** one of the templates installed with Journal.
- **Create** and **save** a new template.
- **Create** stationery notes with a custom background.

template A file that has a preset format that can be used as a basis for similar types of files.

What Are Templates Used For? A **template** is a file that has a preset format that can be used as a basis for similar types of files. Templates usually contain basic information the user needs to include in all documents based on the template. For example, in a letter template information is included to place name, address, phone numbers, or e-mail address. Some templates also contain graphics or special background effects.

Templates Save Time You can save time at school, at home, and on the job when you use templates.
- Templates make it easy to create documents that all use the same basic format.
- Journal comes with several template designs you can use.

Why Is It a Good Idea to Use Stationery? All businesses have customized stationery for sending letters and memos. Stationery usually contains the name, mailing address, e-mail address, and phone and fax numbers of the business. Many companies use a particular font or text size to identify their company.

✓ Concept *Check*

Think of two types of documents you could use frequently for school.
- What tasks would you use this template for?
- Why do you think companies use logos on their stationery?

DIGITAL DIALOG

Custom Stationery
Why might it be important to use different stationery backgrounds for various applications?

Teamwork Work with another student and list at least five reasons why it is useful to change stationery backgrounds.

Read to Succeed

Key Terms
- spreadsheet
- formula
- label
- value
- formula bar
- function
- AutoFormat

Make Connections

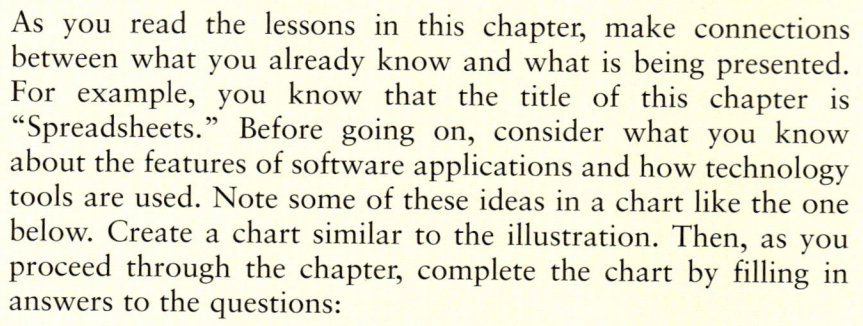

As you read the lessons in this chapter, make connections between what you already know and what is being presented. For example, you know that the title of this chapter is "Spreadsheets." Before going on, consider what you know about the features of software applications and how technology tools are used. Note some of these ideas in a chart like the one below. Create a chart similar to the illustration. Then, as you proceed through the chapter, complete the chart by filling in answers to the questions:

- What? —Identify the lesson.
- When? —Describe when this feature can be used.
- Why? —Explain why you would use this feature.
- How? —Note key steps in the process to help you remember them.

What do you already know about spreadsheets?			
What?	**When?**	**Why?**	**How?**
I can format spreadsheets.	I can format cells with bold or italic.	I can emphasize headings.	I click the bold or italic button.

DIGITAL CONNECTION

Making Charts

Your science teacher has asked you to survey the students and faculty at your school about which products they recycle. After you tally their responses, you will create a chart. As you discuss this project with a classmate, you realize that collecting the data will not be too difficult, but you wonder how you can make an eye-catching chart that will accurately display the results. What is the best type of chart to represent the data on product recycling: a bar chart, a pie chart, or a line chart? You will find that a spreadsheet is a great tool for creating charts.

Writing Activity

Describe how you would create the chart if you did not have a computer. What type of chart would you use?

Activity 12

Step-by-Step

Use Templates

1. Open Journal.
2. Tap **File** and then **New Note From Template**. The template files are stored in My Notes.
3. Select the template you want to use.

Activity 13

Personalize Stationery

In the following activity, you will create a custom stationery design for a Journal note.

1. Open Journal.
2. Tap **File** and then **Page Setup**. The Paper tab will appear.
3. Tap the **Style** tab. Choose the line style that fits your handwriting size. Select line and paper colors.
4. Tap the **Background and Title Area** tab.
5. In the Picture drop-down menu, choose **Psychedelic.jpg**. In the Position drop-down menu, choose **Stretch**. The design will cover the entire page.
6. Tap **Apply** and then **OK**.

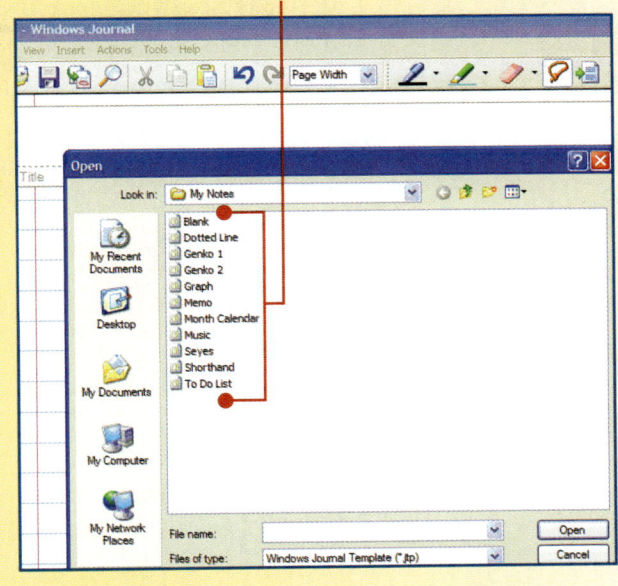

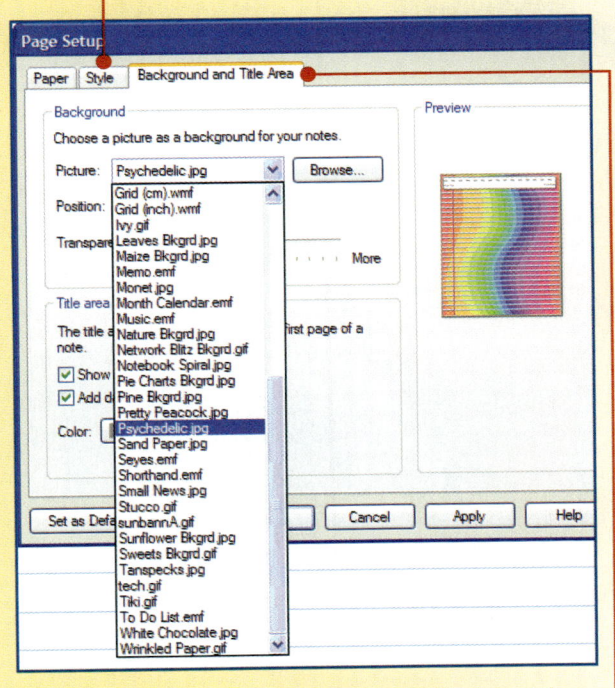

If you need additional help on this topic, open the Help menu; or refer to the documentation for your hardware.

Suite Smarts

Templates can be modified and saved to meet specific needs. For example, you could create a different template for taking notes in each of your classes.

Chapter 8 Additional Features of Digital Tools

Lesson 8.5

244

CHAPTER 10

Spreadsheets

Lesson 10.1   Format Basic Spreadsheets

Lesson 10.2   Create Formulas in Spreadsheets

Lesson 10.3   Apply Advanced Spreadsheet Formatting

Lesson 10.4   Use Business Form Templates in Spreadsheets

Lesson 10.5   Use Spreadsheet Data to Create Charts

Lesson 10.6   Integrate Charts with Word Processing

Lesson 10.7   Integrate Spreadsheets with Word Processing

Lesson 10.8   Use Online Spreadsheet Templates

You Will Learn To:
- **Format** and **edit** text and numbers in cells.
- **Enter** and **edit** formulas.
- **Format** spreadsheet cells.
- **Use** spreadsheet templates and **create** charts.

When Will You Ever Use a Spreadsheet?

When you have a list of information, such as items and prices, and need to quickly perform mathematical calculations, a spreadsheet is the right tool for you. You can enter data and any mathematical formulas in a spreadsheet, and the answer will be calculated for you.

Application 10

Directions Open Journal, and then open the Month Calendar template. Customize the template for the current month by adding the name of the month and the date numbers. Enter the following information on the correct dates:

```
4th:   Team debates
9th:   Swim practice
12th:  Swim meet
18th:  Half day
19th:  No school
27th:  Report due
30th:  School pictures
```

If any of these events fall on a weekend in the month you have chosen, go to the next school day for the date of the event. Add two other events on any dates. Save the document with the name of the current month.

Application 11

Scenario You work as a marketing specialist for a home health care company. All of the health care specialists use Tablet PCs to record patient information when they do home visits.

The marketing director has asked you to create a new stationery design for the specialists to use when they record notes on their visits.

Directions Open Journal and create a new stationery design. Customize the line width of the stationery on your Tablet PC so that it fits your handwriting size and style.

RUBRIC LINK
Go to the Online Learning Center at **digicom.glencoe.com** and click Lesson 8.5, Rubric, to find the application assessment guide(s).

Digital Decisions

Templates for Medicine

Scenario You work in Dr. Thompson's office. All new patients have to fill out information forms on their first visit. Dr. Thompson asks you to apply your skills with Tablet PCs to help.

Teamwork Choose a partner to work with. Review the templates in the Journal program.

Writing Design a template for a simple information form for patients. Use the Convert Handwriting to Text feature for some of the data. You can begin with one of the templates already created and alter it to fit your needs.

Patients need to fill in information such as:

- Today's date.
- Name, address, and home and work phone numbers.
- Date of birth and age.
- Medical history.

After you have created the template, answer the following questions:

- How could a template on a Tablet PC save time in the doctor's office?
- Do you think this would be a good use of technology?
- Why or why not?

How did the other groups in your class respond?

CHAPTER 9

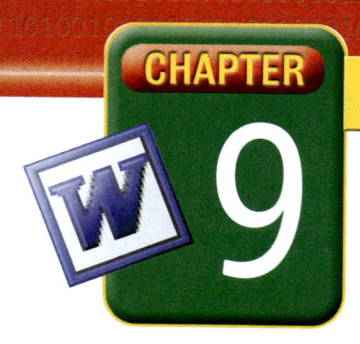

Self-Assessment

Take a moment to review what you have learned in this chapter. Rank your understanding of the topics below.

4 means, "I understand all of this."
3 means, "I understand some of this."
2 means, "I understand very little of this."
1 means, "I don't remember this."

To use a printout of this chart, go to **digicom.glencoe.com** and click on **Chapter 9, Self-Assessment.**
Or:
Ask your teacher for a personal copy.

Rank Your Understanding

Lesson	Topic	4	3	2	1
9.1	• How to apply bold, italics, or underline				
	• How to change fonts				
	• How to cut, copy, and paste text				
9.2	• Format paragraphs and lines				
	• Use Spell Check effectively				
	• Replace words using the Thesaurus feature				
9.3	• Change margins				
	• Add and edit headers and footers				
	• Add bullets and numbering				
9.4	• Create and edit tables				
	• Format cells and lines				
9.5	• Identify business forms				
	• Create and format business forms with tables				
9.6	• Create a letter with speech recognition				
	• Use block style				
9.7	• Insert, edit, and format text boxes				
9.8	• Insert and format graphics				
	• Change the positioning of a graphic				
9.9	• Synchronize word processing documents with a PDA				
	• Edit a word processing document on a PDA				
9.10	• Search for and use online business templates				

If you ranked all topics 4, congratulations! Consider doing a quick review.
If you ranked yourself 3 or lower in any topic, consider reviewing these topics first.

21ST CENTURY CONNECTION

PDAs and Probes

In aquatic environments, pH and temperature are key components in the stability of aquatic systems. The pH is a measure of the acidity or alkalinity of a chemical solution.

Each water system, whether it is a lake or a stream, has unique properties in relation to pH and temperature. Temperature and pH readings can affect other aquatic measurements, like available oxygen and the level of nutrients in an aquatic system. It is important to understand the relationships of these elements and the role they play within aquatic ecosystems. The chemical nature of these aquatic systems is one of the many variables used to determine the health of ecosystems as a whole.

Probes Collect Data Quickly Temperature and pH measurements show some of the basic principles involved with aquatic ecosystems. They also serve as a simple assessment of the surrounding watershed. Data of this kind can be tedious and time consuming to collect. Often field ecologists need a fast but accurate means of collecting this type of data. To do this, ecologists use probes that connect to PDAs. These probes collect fast, accurate data for the aquatic ecosystem.

PDA Database The PDA collects digital readings from the probe and places them into an organized database. This automatic process of database organization has great value for ecologists. The database reduces the use of paper and makes it easy to input field data for further analysis.

PDA Probes and You Just as field ecologists study environments, you can conduct experiments, test hypotheses, and quickly record results in your PDA. Varieties of probes are available as attachments to PDAs.

Software associated with these probes allows you and your classmates to explore, research, record, and analyze data. The traditional lab is extended to your home, in the field, on the bus, and in museums. For example, you can graph temperature changes over time in ponds. You could monitor carbon dioxide levels. These records can be taken at different times of day or on different days or at different ponds for long-term data collection spanning, hours, days, and months. The data is beamed to your teacher's PDA or stationary computer.

> **Activity** Think of a scientific experiment in which you could use a probe with a PDA.

Chapter 9 Review

Read to Succeed PRACTICE

After You Read

Select Technology Brainstorm ways to use the new technology skills you are learning to make studying more fun and interactive. Remember, you can use the applications on your desktop, laptop, Tablet PC, or PDA. Choose from the strategies below, and add your own.

- **Bulleted or Numbered List** Use a word processing program to outline what you have learned in each chapter. In this chapter, you learned to make bulleted and numbered lists. You can use your outline as a map to help you remember information, quickly study before a test, and even to prepare for writing a report or giving a presentation.
- **Thesaurus** Enter the key terms from the chapter in a word processor. Also enter any words in the chapter that are unfamiliar to you. Use the thesaurus language tool to look up alternative meanings for each word. Where the key term has two words, look up each separately. Choose the replacement word you prefer and enter it next to the key term.
- **Tables and Text Boxes** Create graphic organizers in your word processor using tables and text boxes. Make tables to chart related information in the chapter. Select text boxes to organize the topics you are learning.

Using Key Terms

Word processing is the most used of any application. Make sure you are familiar with this important application by matching each key term with its definition.

- margin (265)
- alignment (270)
- default (270)
- header (277)
- footer (277)
- column (283)
- row (283)
- cell (283)
- merge (290)

1. The intersection of a column and a row in a table.
2. A setting that was determined at the time the program was created.
3. A part of a table that extends across the table.
4. An area for text or other objects that will appear at the bottom of every page in a document.
5. To combine two or more cells to create one cell.
6. The blank edge that borders the area in which text and objects can be placed on a page.
7. The way that text or objects are lined up across a page.
8. An area for text or other objects that will appear at the top of every page in a document.
9. A part of a table that extends up and down the table.

Lesson 8.6 — Use Write Anywhere on the Tablet PC

You will learn to:
- **Use** the Write Anywhere feature.
- **Explain** Sticky Notes

Write Anywhere The Write Anywhere feature creates a transparent writing surface over most of the Tablet PC screen. Write Anywhere lets you write at any location on your screen. The text is inserted wherever the insertion point is located. You can print or write in cursive, and your handwriting will be converted to typed text in the currently active program.

With Write Anywhere, you can write in any order or any direction on your screen. You can even write over typed text that is on the screen. You can use the Quick Keys pad to tap Tab or Enter. When you pause, the text is inserted into the document. The delay time between writing on the screen and sending the handwriting to the document can be adjusted through the Options menu on the Input Panel.

If you have trouble getting the pen to write on the screen, try holding the pen in a more upright position. Holding the pen at an angle may not send a strong signal to the screen.

Also, clean the screen of the Tablet PC frequently. A dirty screen can interfere with the operation of the digital pen. Use only recommended cleaning cloths.

Write Anywhere can be used in other Office programs as well as in Word. However, the on-screen writing features work best in a word processor, such as Note Pad, Memo Pad, or Word.

Sticky Notes The Sticky Notes accessory functions like paper sticky notes, allowing you to stick notes on the desktop. To use Sticky Notes, tap the Start menu, choose Programs, and tap Sticky Notes. You can minimize Sticky Notes while working in other applications to keep them easily accessible on the Taskbar.

✓ Concept Check

When you take notes in class, do you ever find it necessary to go back and insert more notes? How could Write Anywhere help you in that situation?

CHAPTER 9

DIGITAL Dimension → Science
INTERDISCIPLINARY PROJECT

E-Cycling

Using the word processing skills you have learned in this chapter, create a flyer about products that are being collected for e-cycling and the companies that collect them.

Products Collected for E-Cycling Create a table of information in your word processor that you can share with your community about products that can be e-cycled.

List names and addresses of places to contact that will recycle electronic products. Search the Internet, newspapers, phone books, and recycling centers to gather the information. Include a list of common e-cycling products and the prices paid for the materials they contain.

Consider the following data for your table:

- Company name.
- Company address.
- Company phone number.
- Types of products collected for e-cycling.
- Amount paid for products.
- Any other facts you think are relevant.

Emphasize Important Data After you create your table, emphasize important information in your table by using the formatting skills you have learned. For example, consider formatting the borders and shading selected cells. These features also make your table easier to read. Save the document as E-Cycling Table.

Create a Flyer Using your word processing program, create an exciting and informative flyer to get the word out about e-cycling in your community. Include your e-cycling table as well as text and graphics to make your flyer attractive and easy to read and the data easy to understand.

Save the document as **E-Cycling Flyer.** Print preview and print the flyer.

E-Portfolio Activity

Word Processing Skills

To showcase the new word processing skills you have learned in this chapter, open a new document and create a bulleted list of those skills.

Give a title to your document.

Include a graphic or photo.

Save the document as an HTML file to your e-portfolio.

Self-Evaluation Criteria

In the document E-Cycling Table, have you included:

- Company names, addresses, and phone numbers?
- The names of e-cycling products?
- The amount paid?
- Other relevant facts?
- Table formatting?

In your flyer, have you included:

- Your e-cycling table?
- Informative text?
- Attractive graphics?

Activity 14

? If you need additional help on this topic, open the Help menu; or refer to the documentation for your hardware.

Step-by-Step

Use Write Anywhere

1. Open Word and start a new document.

2. Open the Input Panel. Tap the **Write Anywhere** button to activate the Write Anywhere feature. The Write Anywhere line appears across the screen wherever the digital pen is placed.

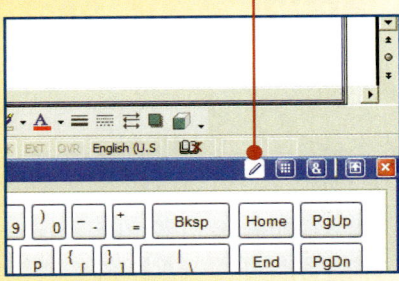

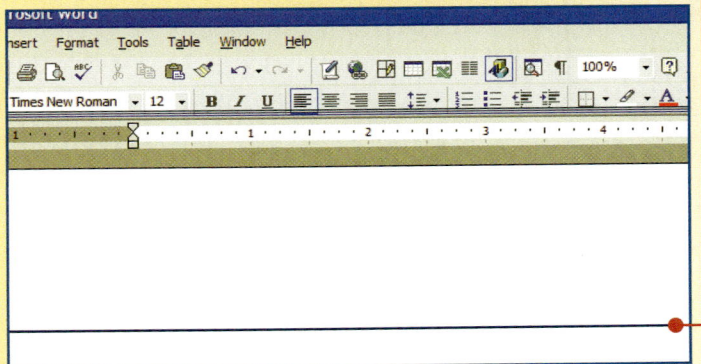

3. The blinking insertion point shows in your document. Tap the screen to move the insertion point where you want the new text to appear. Use the digital pen to write on the line.

   When you pause after you have written or printed something, your handwriting will be converted to typed text, and inserted into your document.

   You can move the line anywhere on the screen to write. The text will always be inserted wherever the insertion point is located.

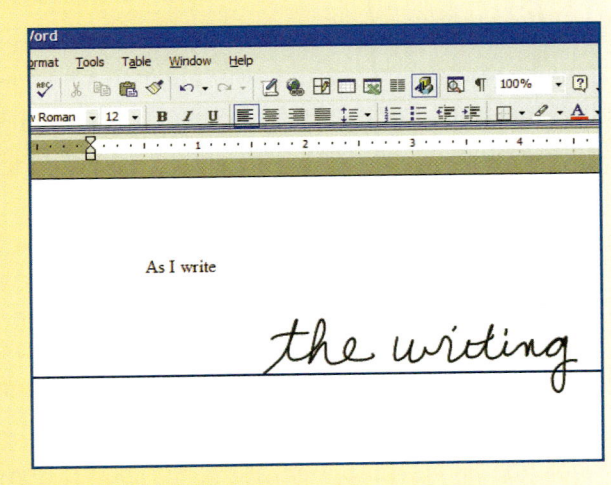

4. Tap the **Write Anywhere** button on the Input Panel to turn off the Write Anywhere feature.

Chapter 8 *Additional Features of Digital Tools*

Lesson 8.6

248

DIGITAL Career Perspective

Financial Services

Mark Jolley
Insurance Representative

Bringing Financial Security to Clients

Notebook computer, cell phone, fax machine, e-mail, desktop computer—all of these technology devices are important to Mark Jolley, an insurance representative. "I carry the notebook computer with me on all my appointments," says Jolley. "It helps me give clients a more accurate picture of their financial situation." Jolley travels to visit clients. He meets with them in their home or office. Some of his clients are businesses, and others are families. He helps all of his clients to make the best investment choices they can with the money they earn.

Jolley collects financial information from clients and enters it into a special software program on his computer. "The software gives clients the opportunity to view a more accurate picture of their financial situation," adds Jolley. Then the client can make a reasonable decision about investing. For example, Jolley is able to advise families on the insurance investment they should make to provide for the family if the main wage earner were to lose the ability to work.

E-mail keeps Jolley in touch with clients as he travels. He can quickly answer any questions from clients. "I use the notebook computer to gather data from clients so I can give them the best possible plan for their future," notes Jolley.

"It is important for me to stay on top of current trends and data. I attend courses occasionally so I can be sure to advise clients of the best plan for their situation," says Jolley. He had to learn new software programs during the training he received for his license. The software has to be updated from time to time to assure that it is providing current information for Jolley and his clients.

What does Jolley like best about his job? "I like the freedom to work with many different people; some clients are young, some are nearing retirement. Even the businesses I work with employ people of all ages and backgrounds," says Jolley.

Training

A high school education is required, but a college degree in business or economics is preferred to work as an insurance representative. Subjects to study in high school and beyond include math, economics, psychology, sociology, public speaking, and computer applications, as well as business courses. A state license is obtained by taking a related course and passing an examination.

Salary Range

Typical earnings are in the range of $28,800 to $64,400 a year.

Skills and Talents

Insurance representatives need to be:

- Flexible
- Enthusiastic
- Confident
- Disciplined
- Hard working
- Willing to solve problems
- Effective communicators
- Good time managers

Career Activity

What studies and activities in high school can help you develop the talents needed by an insurance representative?

Application 12

Directions Open a new Word document. Activate the Input Panel and turn on Write Anywhere. Make sure the insertion point is showing at the top of the document.

Use the digital pen to enter the following text in the Word document using the Write Anywhere feature.

```
The black horizontal line that
appears when you turn on Write
Anywhere is there to guide you and
help you write on a straight line.
However, it is not necessary to
place your handwriting directly on
this line.

Tap the Enter key on the Input
Panel keyboard when you want your
text to start on a new line.
Continue writing with the Write
Anywhere feature.
```

RUBRIC LINK
Go to the Online Learning Center at **digicom.glencoe.com** and click Lesson 8.6, Rubric, to find the application assessment guide(s).

Digital Decisions

Tablet PC Tutorial

Internet Investigation Work with a partner. Examine the interactive tutorial program installed on the Tablet PC. You can view animations and practice guided activities as you learn. Tap **Start**, **Programs**, **Tablet PC**, and then **Tablet PC Tutorial**. Take turns progressing through the tutorial.

Critical Thinking If you come upon a feature that is new to you, write it down and explain it briefly.

When you have finished, compare what you have learned with what your classmates learned. Share any new ideas from your tutorial notes for using the Tablet PC.

DIGITAL DIALOG

Tablet PCs at Work and School

Teamwork Prepare a brief presentation on the features of a Tablet PC including the features you have learned about in this class. You may want to review Chapter 7, pages 190–226. You can also search the Internet for Tablet PCs and see what additional features you can find. How would you use some of these features on the job? Would they have a use in school? In a small group, discuss ways to use Tablet PCs at work and school.

Writing List the advantages of using Tablet PCs compared to using a laptop or desktop computer. Are there also any disadvantages? Be ready to explain your answers.

Step-by-Step continued

2 In the description area, enter:

```
Set up and install ten computers in offices.
Connect computers to office network.
Install Microsoft Office on computers.

Amount:  $3595.00

Email:   Enter your e-mail address.
```

Create a New Folder From the Save As Window

Folders help to keep your files organized. As you prepare to save a file, you may decide it would be helpful to save it in a new folder that you have not created yet.

1 Choose **File** and then **Save As**. Click the **Create New Folder** icon.

2 In the New Folder window, enter **Invoices** and click **OK**.

3 Save your document as **Research Triangle Invoice** in the folder you just created.

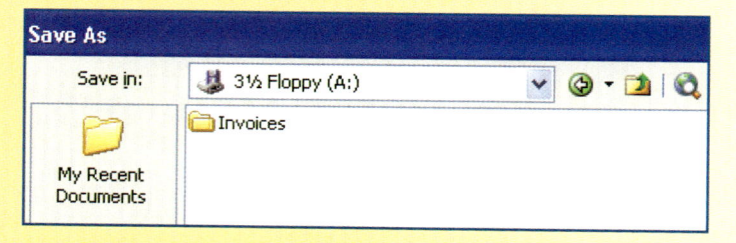

4 Print your invoice when you are finished.

Activity 45

DIGIByte
Adding New Folders
You can add a new folder to a floppy disk or hard drive from the Save As window.

RUBRIC LINK
Go to the Online Learning Center at **digicom.glencoe.com** and click Lesson 9.10, Rubric, to find the application assessment guide(s).

Application 12

Directions Go to the Microsoft templates Web page that you bookmarked in Activity 43. Look at the templates in the Your Career category. Which of these templates would you use if you were looking for a job? Choose one to download and fill in.

Lesson 8.7 Explore OneNote

You will learn to:
- **Create** a new note.
- **Rename** divider tabs.
- **Flag** notes.

Your Digital Notebook OneNote is a Microsoft Office application that works like a digital notebook. The note-taking area resembles a piece of virtual notebook paper. You can type or write or draw anywhere on the page, just like you do on a real sheet of paper. You instantly capture electronically all the ideas you want to remember—in one place, not in many different notebooks or on separate sheets of paper. Never again will your notes be lost or misplaced.

You can create sections to organize your notes into different subjects. OneNote pages are easily located by divider tabs that appear on the right side of the screen—just like tabs on pages in a notebook. These features make taking notes easy without needing to create folders and saving separate files as you would in other programs, such as Word.

Your notes are stored in the My Notebook folder. OneNote makes it easy to find your notes, too. After you have entered notes, you can add a flag to the file so it will be easy to search for later. You can also search for text in notes.

Works with Other Applications OneNote works with Outlook calendar to insert details from the calendar into a note with just a few clicks. OneNote also works with other Office products, such as PowerPoint, Word, and a Pocket PC handheld computer. You can easily drag and drop or synchronize notes into these programs. For example, if you drag pictures and text from the Internet into OneNote, the URL is automatically added to your notes for easy reference. You can also share your notes through e-mail or publishing them to a Web site.

OneNote is compatible with Microsoft Office 2003, not with prior versions of Office. When OneNote is used on a Tablet PC, you can enter notes in your own handwriting. In this lesson, you will use OneNote on a desktop PC.

✓ Concept Check

How do you locate your notes in a notebook? Do you use dividers or self-adhesive flags to organize? How do you track Internet sources as you are researching on the Internet? Describe how you could use these methods in OneNote.

DIGI Byte

Audio and Video Files
Audio and video files can be created in OneNote and attached to notes you create.

Activity 43

If you need additional help on this topic, open the Help menu; or refer to the documentation for your hardware.

Step-by-Step

Download Business Templates

1. Click the **Help** menu, and then click **Microsoft Office Word Help**.
2. In the **Search for** box, enter "download templates." Click **Search**.
3. Click the first result, which should begin with **Microsoft Office Templates**. You will be taken to a page on the Microsoft Web site. Bookmark this page in your browser.
4. In the Template categories in the lower center of the page, click **Orders and Inventory**. On the new page, click **Invoices and Purchase Orders**.
5. Scroll down the list to review the type of documents you can download.
6. Click **Purchase order with unit price**.
7. Look at the preview of the template. What information relating to the order must be filled in on this document? List six items other than your company name and address and the Ship To name and address.
8. Click the **Back** button in your Internet browser.
9. Click **Services invoice with total only**, and then click the **Download Now** button.
10. Save the form to a floppy disk or a folder on your computer.

Activity 44

Fill in a Business Template

1. Fill in the form you just downloaded with the following information.

```
Your Company:  Data Connections
    Address:   2510 Hwy 18 South
               Suite 201A
               El Paso, TX 79901
    Phone:     888-555-1016  Fax:  888-555-2020
    Bill to:   Research Triangle
               4815 Mountain Grove Road
               Richmond, VA 23219
    Phone:     888-555-9827
    For:       Network Setup Services
    Invoice #: A202
```

Suite Smarts

Templates for business forms can be created in Word, Excel, and Publisher.

Continued

Step-by-Step *continued*

Activity 15

If you need additional help on this topic, open the Help menu; or refer to the documentation for your hardware.

Enter and Move Text

When using OneNote, you can enter text anywhere you want. It is easy to move text from one location to another.

1. Open the OneNote application. Click anywhere on the screen and key your first and last name.
2. Click in a different spot on the screen. Key your address.
3. After you enter text, you can move it to a new location on the page. Click the top edge of the box with your address in it, and drag the address box so that it is directly below your name.

Activity 16

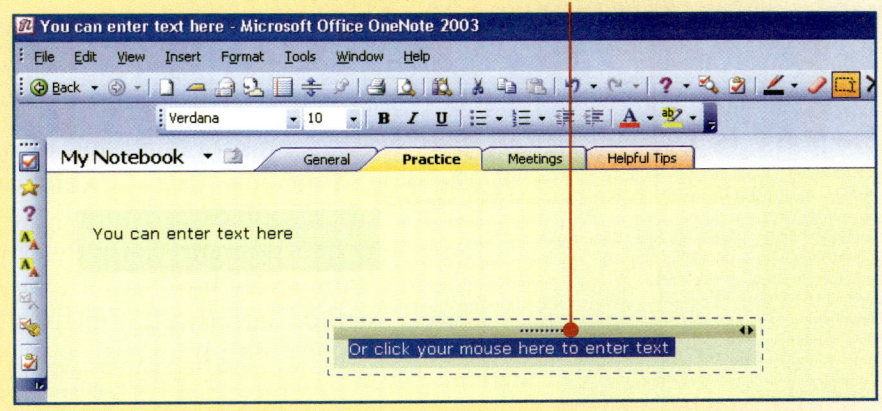

Rename Divider Tabs

The divider tabs at the top of the screen can be renamed manually.

The divider tabs on the right side will automatically match the title you give to the note.

1. Right click on a divider tab at the top of your notebook.
2. Click **Rename**. Enter `Class Notes`.
3. Click the General divider tab. This screen has a title area at the top. Enter the text `FBLA Meeting` in the title area. Notice that the tab on the right now says FBLA Meeting as well.

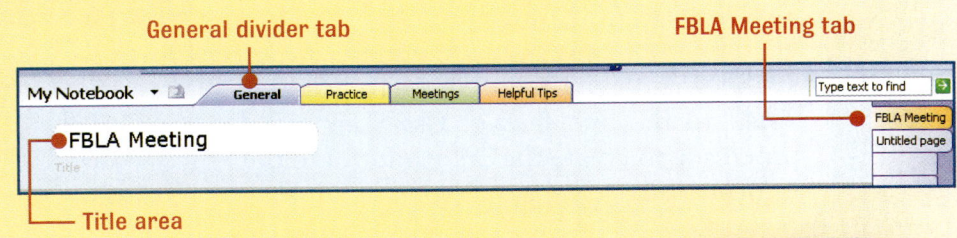

Continued

Lesson 9.10: Use Online Forms and Templates

You will learn to:
- **Locate** Microsoft templates on the Web.
- **Use** a Word template to create a document.

Uses of Online Templates Employers today use the Internet for almost every aspect of doing business. Templates make it easy for employers to create similar forms.

Online Templates Are Easy to Access Although templates can be e-mailed to employees at different locations, a Web site can give employees easy access to these documents whenever they need them. Document sharing via the Internet is very convenient and reliable. Business templates and forms can be organized in an easy-to-use manner on a Web site.

Business Forms The types of forms that a company uses depends on the type of business that the company is in. However, some forms related to human resources are found in nearly every business. For example, interview questionnaires, job descriptions, job applications, and reference check forms are helpful in all types of business. Other common types of forms include letterheads, memos, purchasing forms, fax cover sheets, business plans, and budgets.

✓ Concept Check

How could human resource forms be used on the Internet?

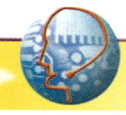

DIGITAL DIALOG

Forms

The computer consulting firm you work for has asked you to join the committee to review online business forms they can use.

Internet Investigation Enter "business forms" in a search engine, and review some of the results.

Writing Create a chart similar to the one shown. Use your word processor to create the table and fill in the information.

Name of Form	Purpose of Form

Chapter 9 Word Processing — Lesson 9.10 — 306

Activity 17

Step-by-Step continued

Flag Notes

Flags are used to quickly locate information within your notebook. They serve several different functions. For example, you can create a summary of items that have been flagged.

1. Go to the General divider tab. Key the text FBLA Meeting in the title area.

2. Click in the note area and enter the text:

   ```
   Discuss fundraiser activity.
   Need to raise money to cover trip to state competition.
   Get fundraiser activity approved by principal.
   ```

3. Move the insertion point to the end of the last line of text.

4. Click **Format, Note Flags**.

5. Click **Question**. A purple question mark will appear at the beginning of this line.

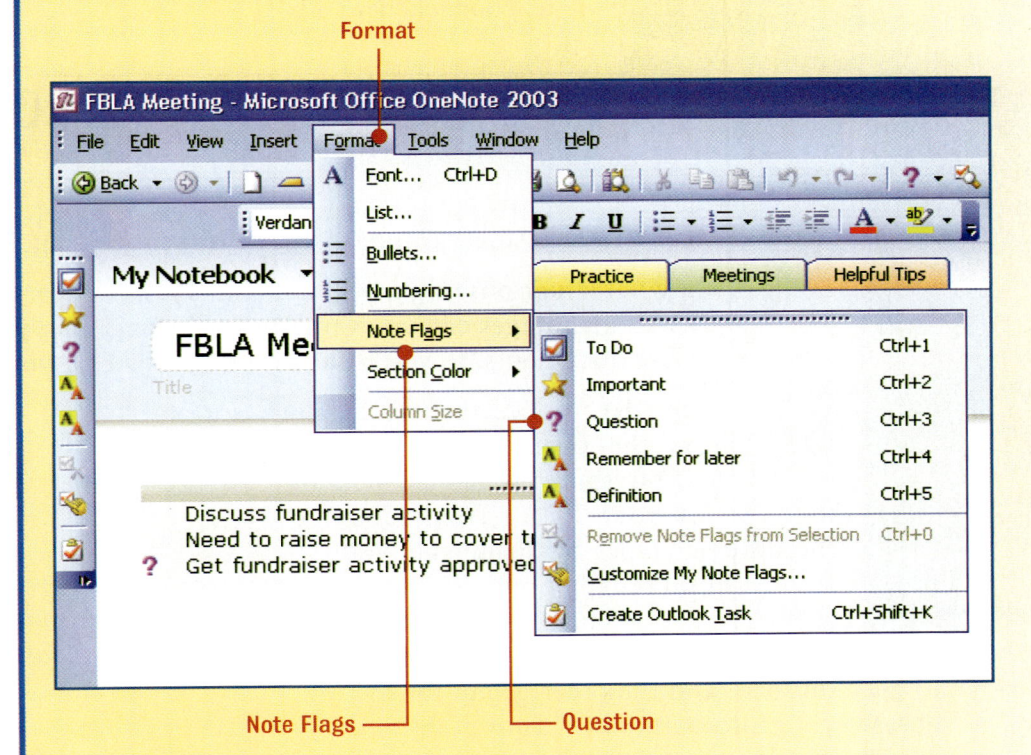

Continued

Activity 41

Step-by-Step

Synchronize a Document

1. Open the *Your Name* **E-Waste** document that you saved in Lesson 9.3.
2. Add this document to the program that synchronizes documents to your PDA.
3. Connect the PDA to the computer and synchronize the document.

Activity 42

Edit a Word Document on a PDA

1. Open the Word document on the PDA.
2. Add a sentence to the document.
3. Save the document.

RUBRIC LINK
Go to the Online Learning Center at **digicom.glencoe.com** and click Lesson 9.9, Rubric, to find the application assessment guide(s).

Application 11

Directions Open one of the other documents you have created in this chapter and synchronize it to your PDA. Add a paragraph to the document. Save the document. Then beam the document from your PDA to your teacher's PDA.

DIGITAL DIALOG

PDAs

Teamwork In a small group, discuss word processing documents on PDAs.

- In what situations might you use your PDA for word processing?
- What advantages can you think of for having such portability?
- What documents do you think you might edit on your PDA?

Writing Reproduce the chart shown and fill in the ovals with the types of documents you think would be helpful to have on your PDA. Think about the different types of documents you studied in this chapter and add ovals as needed. Exchange your chart with another group when you are finished. Did the other group list the same information? Add any additional information that your group discussed to the chart of the other group.

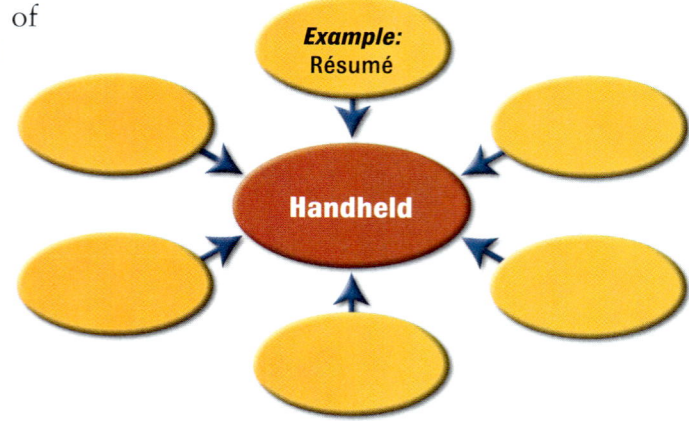

Example: Résumé → **Handheld**

Step-by-Step continued

6 Click the **Note Flags Summary** button on the toolbar.

A summary screen appears to the right and lists the items that have been flagged.

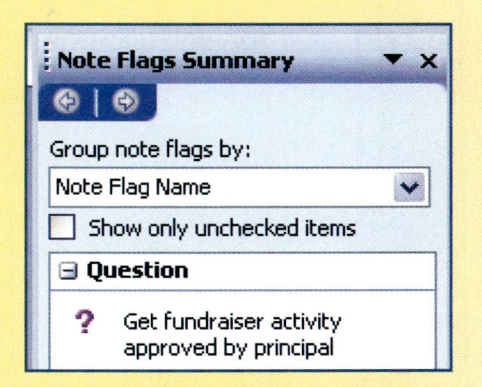

RUBRIC LINK

Go to the Online Learning Center at **digicom.glencoe.com** and click Lesson 8.7, Rubric, to find the application assessment guide(s).

Application 13

Directions Open OneNote. Select one of the divider tabs at the top of the page. Change the name of the divider tab to Jobs. On this new page, key the title Summer Employment Possibilities.

Click below the title and enter the names of four local businesses where you would like to submit an application for summer employment. Below each company or organization name, enter two reasons you would like to work for that company. Add a flag next to your top job choice.

DIGITAL DIALOG

Technology in the Future

Teamwork Create a small group of students to explore some of the features of OneNote. Go to the Microsoft Web site at **www.microsoft.com** and enter "OneNote tour" in the search box. As you take the tour, take notes.

Communication Prepare a short presentation for the class to discuss the features that you learned about and how you could use them. Answer the following questions in your presentation:

- How could OneNote help you with your school work?
- How can it help with Microsoft Outlook e-mail?
- Do you think OneNote can be used in a business? Explain why or why not.

Divide the presentation responsibilities among your group.

Lesson 9.9

Use a PDA for Word Processing

You will learn to:
- **Synchronize** a Word document with a PDA.
- **Edit** a Word document on a PDA.

Portable Word Processing A PDA can store word processing documents to edit, view, and share. The flexibility and quick access to word processing documents that a PDA can provide can be very timesaving for busy people, such as business travelers. For example, a traveling salesperson may want to be able to access client information, databases, and correspondence quickly and easily. A PDA can help to organize all of this information.

Additional Software Some PDAs need an additional program installed to be able to synchronize documents created on a desktop computer. Check the operating system of your PDA to find out whether you need additional software to transfer word processing files.

Beam or Synchronize Word documents on your PDA can be easily beamed to another PDA. Best of all, if you need to make changes to the word processing documents stored on your PDA, you can perform simple editing on the PDA, save the changes, and then synchronize the edited document back to your computer.

Formatting with PDAs Simple formatting features, such as bold and italic, can be used on most PDAs with a word processing document.

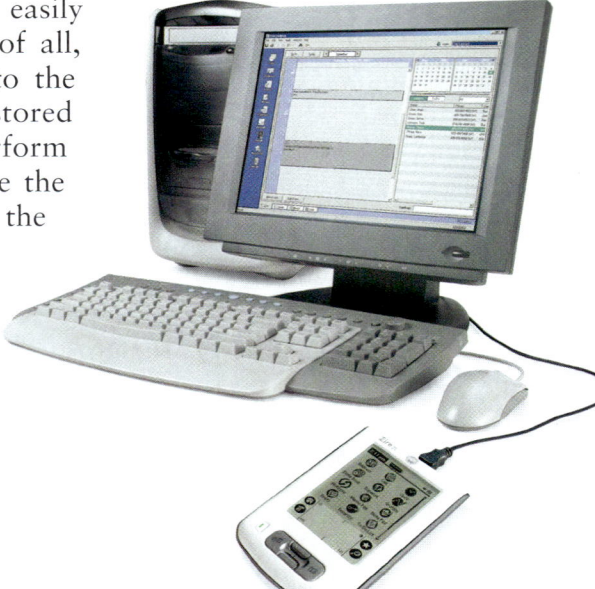

✓ Concept Check

What kind of word processing documents might be helpful to carry on your PDA? Why? In which of your current life situations might portable word processing be of great benefit to you?

Suite Smarts
Excel and PowerPoint documents can also be synchronized from a computer to a PDA.

Chapter 9 Word Processing Lesson 9.9 **304**

DIGITAL Career Perspective

Government and Public Administration

David Curry, Jr.
Public Safety Officer

Keeping Students Safe

"I am not confined to my office with a desktop computer. In emergency situations, like if a student is injured, I do not have to lose valuable time asking other office personnel to look up vital information," says David Curry, Jr., a public safety officer.

There are over 2,300 students at the high school where Officer Curry is assigned. "The PDA makes my job easier. When I'm dealing with a student, I have instant access to his record and parent contacts. I can look up student schedules on the move. It is my lifeline to all the information I need."

E-mail is also an important communication tool for Curry and his partner. Curry states, "It's the easiest and quickest way for us to pass on information to staff members, and it gives another alternative for parents and students to reach us for advice or help."

Curry attends classes periodically to stay on top of changing laws and new technology. "Some of the training I receive involves recertification every few years. I also attend any school trainings related to safety and discipline," states Curry. The school does training on DTRAC, a program used to track discipline procedures. The school is required to report certain infractions to the state every year.

What does Officer Curry like best about his job? "I get to make a positive impression on adolescents and help them learn how to appropriately communicate problems to law enforcement. They learn to understand legal consequences in a way that would prevent future infractions."

When asked to describe some of his job responsibilities, Curry mentioned that he attends most after-school home functions, such as athletic events, dances, pep rallies, and graduation exercises. He also mentioned daily responsibilities such as patrolling the parking lots to ensure students' safety, supervising traffic before and after school, and assisting the school administration with student disruptions and truancy issues. He feels that being a role model to students in term of attitudes and respect is his most important responsibility.

Training

A high school education is required, and federal and state agencies require a college degree. Post-high school study in law enforcement-related subjects is helpful. Officers usually undergo a 12- to 14-week academy training course. Courses to take in high school and beyond might include foreign languages, computer science, finance, accounting, and physical education.

Salary Range

Typical earnings are in the range of $32,000 to $53,000 a year.

Skills and Talents

Public safety officers need to be:

- Honest
- Responsible
- Physically fit
- Able to handle stressful situations
- Personable
- Good communicators
- Able to adapt to new technologies

Career Activity

How do you think participating in sports and learning a foreign language might be helpful in a career in public safety?

Step-by-Step *continued*

2 Now you will change the background color of the image. Choose the **Colors and Lines** tab. In the Fill section, click a new color in the **Color** drop-down menu. Click **OK**.

You can also resize the image by choosing the Size tab in the Format Picture window.

Move an Image in Front of Text

The image that you inserted is now formatted in line with the text of the document. It can be moved like any text in the document: by pressing the Enter key to move down the page, or by aligning the picture to the left, right, or center of the page. To move the image more freely, you will change its formatting so that it will float above the text.

1 Right-click the image and select **Format Picture** from the drop-down menu.

2 Click the **Layout** tab and select **In front of text**.

3 Click **OK**. Drag the image to the lower right-hand corner of the page.

Activity 40

DIGI*Byte*
Save Often
Save your document frequently as you add graphics.

RUBRIC LINK
Go to the Online Learning Center at **digicom.glencoe.com** and click Lesson 9.8, Rubric, to find the application assessment guide(s).

Application 10

Directions Add five images to a new Word document. Resize and move the images so they are arranged in a circle. Change the background shading on one of the images. Set the image to appear in front of text. Save the document as **Image Circle** and then print your document.

DIGITAL DIALOG

Search for Clip Art

Teamwork You are working for an advertising agency that has been hired to create a promotional flyer about recycling for a new recycling company in town. Work with a partner to find some free clip art on the Internet that will be appropriate for the flyer.

Writing Create the flyer, adding text about the new recycling company as well as the pictures you chose.

Critical Thinking Compare your design with others in your class. How did you use clip art to get your message across? How did the text format help?

CHAPTER 8

DIGITAL Dimension → Social Studies
INTERDISCIPLINARY PROJECT

Elected Officials

Using the skills you have learned in Chapters 7 and 8, you will enter political information about your state in a Tablet PC.

Draw Your State Map Your start-up activity in Unit 2 Digital Dimension, page 123, was to print a map of political districts in your state. Now, applying the skills you have learned by using a Tablet PC, redraw this map in the Journal application. Choose appropriate stationery for the background.

Mark Your District After you have drawn your state, locate the area on the map that represents where you live. Using the pen and highlighter features of the Journal program, mark an area to represent your location. Include a title for your document and a legend that explains what each mark means. See the example below.

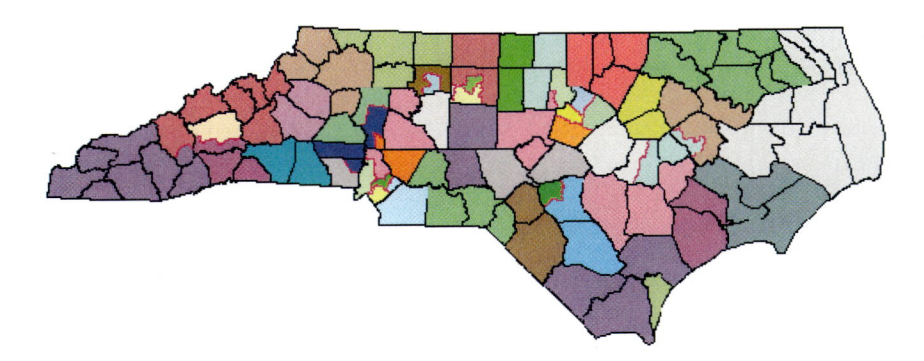

Investigate Include the following information about your district. You should be able to find this information on the Internet.

- Registered voters for each party
- Population of your district
- Voter turnout during last election

Save your document as **State Districts**.

E-Portfolio Activity

Writing and Drawing on the Tablet PC

To showcase your skills in handwriting and drawing on the Tablet PC, save your State Districts document to your e-portfolio folder. You could use a USB flash drive to transfer the file from the Tablet PC to your e-portfolio folder. Remember that the free Journal Viewer application can be used to view and print Journal notes from a computer that does not have the full Journal application installed.

Self-Evaluation Criteria

Have you included:

- A hand-drawn map of the political districts in your state?
- A way to identify the district where you live?
- The number of registered voters in your area?
- The population of your district?
- Results of voter turnout in the last election?

Step-by-Step

Activity 37

Add Clip Art to a Document

1. Start Word.
2. Select the **Insert** menu, and choose **Picture** and then **Clip Art**.
3. In the Insert Clip Art task pane, enter "recycle" in the Search For box and click **Search**.
4. Scroll to view the images that appear in the Results section of the task pane.

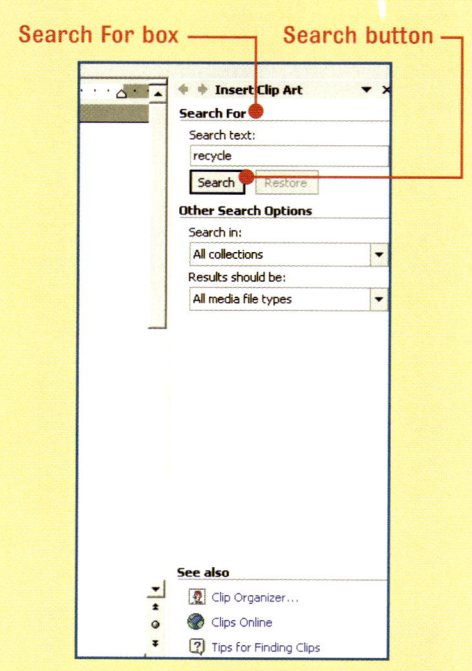

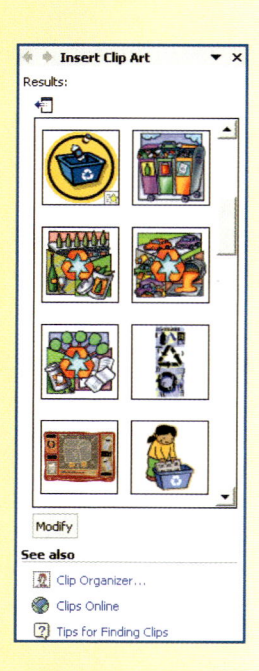

5. Choose an image, and click it to insert the clip art into your document.

Activity 38

Resize an Image

1. Click to select the image in your document.
2. Drag one of the corner sizing handles to make the image slightly larger.

   The pointer takes the shape of a double-headed arrow when a sizing handle is selected.

Activity 39

Edit an Image

1. Right-click on the clip art image and choose **Format Picture** from the drop-down menu.

DIGI Byte

Resizing

Always use corner sizing handles to resize an image so that the image will keep the correct proportions.

Continued

Chapter 9

CHAPTER 8 Review

Read to Succeed PRACTICE

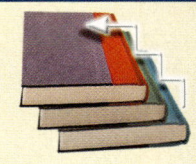

After You Read

Design a Study Template Use what you have learned in this chapter to create a template to help you study. As you know, a template is a file that has a preset format that can be used as a basis for similar types of files. Be creative!

You can use your study templates over and over again. Perhaps you want to design a template for inputting functions and their commands, or a template that organizes a chapter outline. Use stationery, clip art, and other ideas in this chapter to help you.

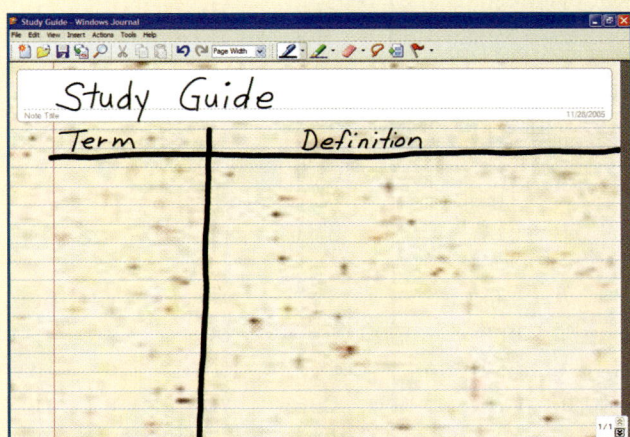

Using Key Terms

The Tablet PC and PDA can increase your productivity in many ways. To check your familiarity with some of these ways, write each sentence on a sheet of paper, Tablet PC, or PDA and fill in the blanks.

- freeware (229)
- shareware (229)
- expense log (232)
- import (239)
- template (243)

1. A software application in which you can enter and track your daily expenses is a(n) _____.

2. To open a document in an application other than the one in which it was originally created is to _____.

3. A(n) _____ is a file that has a preset format that can be used as a basis for similar types of files.

4. _____ is a software application that you can download and share at no charge.

5. Software that is distributed free on a trial basis and may later be purchased, is called _____.

256

Lesson 9.8 Insert and Format Graphics

You will learn to:
- **Insert** a graphic.
- **Resize** a graphic.
- **Position** a graphic on the page.

Graphics Adding graphics to a word processing document helps a reader to get a visual understanding of what the text is describing and makes the page more eye-catching. Clip art images are electronic illustrations that can be inserted into a document. You learned about clip art in Lesson 4.3, pages 100–103.

Where to Find Clip Art Clip art is usually included with a software package. Clip art can also be found on hundreds of Web sites, and much of it is free for non-profit use. This type of clip art can be downloaded and used in documents as long as the documents are not for sale. You can also buy clip art from Web sites. Microsoft Word comes with a large library of clip art for all types of occasions. When you use clip art, make sure that you are not violating any copyright laws. Review Chapter 3, pages 73–74, for information about copyright laws.

DIGIByte
File Formats
The most popular clip art file formats are JPG, which is often used for photos, and GIF, which is often used for drawings.

✓ Concept Check
Why might you want to add graphics to a word processing document?

DIGITAL DIALOG

Clip Art

Teamwork Choose a partner and together browse the Clip Art Gallery of Microsoft Word. Identify which of the categories contain images you might find useful at school, for a Web site, or for personal correspondence. Make a list of three categories that can help you in your schoolwork right now. You can come back to these categories later and view all of their clip art images.

Discuss With a small group, consider two topics:
What are the pros and cons of adding graphics to school reports? Do you think they add to the meaning of a report or are a distraction?

Discuss the advantages and disadvantages of using the clip art that is built into a word processing program versus downloading an image from the Internet.

CHAPTER 8 Self-Assessment

Take a moment to review what you have learned in this chapter. Rank your understanding of the topics below.

4 means, "I understand all of this."
3 means, "I understand some of this."
2 means, "I understand very little of this."
1 means, "I don't remember this."

To use a printout of this chart, go to **digicom.glencoe.com** and click on **Chapter 8, Self-Assessment.**
Or:
Ask your teacher for a personal copy.

Rank Your Understanding

Lesson	Topic	4	3	2	1
8.1	• Search for freeware applications				
	• Download freeware to a desktop PC				
	• Synchronize a PDA to a desktop PC to install freeware on the PDA				
8.2	• Use a PDA to convert measurements				
	• Create an expense log				
8.3	• Synchronize Word documents from a desktop PC to a PDA				
	• Create a journal entry using Memo Pad on a PDA				
8.4	• Import Word documents into Journal				
	• Mark up imported documents in Journal				
	• Add flags to Journal documents				
	• Search for flags in Journal documents				
8.5	• Open a template in Journal				
	• Create a template in Journal				
	• Use stationery in Journal				
8.6	• Use the Write Anywhere feature of the Tablet PC				
8.7	• Enter and move text in OneNote				
	• Rename divider tabs in OneNote				
	• Add flags in OneNote				

If you ranked all topics 4, congratulations! Consider doing a quick review.
If you ranked yourself 3 or lower on any topic, consider reviewing these topics first.

digicom.glencoe.com

21ST CENTURY CONNECTION

Successful Verbal and Written Communications

Your career success will depend greatly on your ability to communicate well. Before you deliver your message, visualize yourself interacting with people in ways that show both self-respect and consideration for others.

Choose the Right Delivery Method for the Message Decide if something that should be said in writing, in person, or by phone. Is the comment or question short and not urgent or private, or is it more complex and confidential?

Verbal Skills Think about the way you speak and listen both in and out of school. You could also do listening role-plays with a friend.

To be a good listener:

- Look at the person speaking and focus on his or her words. Notice facial expressions and changes in vocal tone.
- Do not interrupt the speaker or be distracted by sounds outside a conversation.

To think before you speak:

- Envision, or even practice, what you are going to say before you say it.
- Before speaking, reflect on what you know about the audience.
- Think about a work area before speaking and adjust your tone of voice to that space. Many people share open workspaces, so it is best to speak quietly if others work nearby.
- In routine conversations, keep eye contact with listeners, form words clearly, and speak at a medium, understandable pace.

Writing Skills So much information is available today that your audience may read only what is most important to them. The best way to get your message across is to keep it simple and write it in a way that will grab and hold their attention.

- Write clearly, and stay on point. Be sure all document content relates to your reason for writing the document.
- Proofread your document for spelling, punctuation, and grammar before sending the document.
- Read from a hard copy to catch errors you miss on screen.
- Do not rely on the spelling and grammar check functions in your word processing application alone.
- Get answers to questions in more complete dictionaries or grammar guides.
- Write in a tone that fits your audience.
- When writing to your employer, your customers, or your coworkers, show the same respect you would like to receive from them.
- How would written communication skills help people succeed in those jobs?

Activity
Skim the job ads in a newspaper or at an online job search site.
- How many ads list a requirement for good communication skills?
- How would verbal communication skills help people succeed in those jobs?

UNIT 2 LAB

Digital Dimension → Social Studies
INTERDISCIPLINARY PROJECT

Elected Officials

Background In Unit 2 you have been gathering information about elected officials:

- In Chapter 5 Digital Dimension, page 150, you gathered information about four elected officials.

- In Chapter 6 Digital Dimension, page 187, you chose one of these elected officials to interview and created interview questions.

- In Chapter 7 Digital Dimension, page 224, you took notes on a Tablet PC about the elected official of your choice.

- In Chapter 8 Digital Dimension, page 225, you sketched a map of the political districts in your state on a Tablet PC.

Presentation on Your Chosen Elected Official

You will be preparing a class presentation using the information you have gathered. The goal of the presentation is to inform your audience about your chosen elected official.

In creating this presentation, consider the digital input skills you have developed in this unit. You may want to combine the use of speech recognition with on-screen writing to develop your presentation. You could also develop your presentation in a PDA.

E-Portfolio Activity

Showcase Presentation

Include your Unit 2 showcase presentation in your e-portfolio. If you created an electronic presentation, save it to your e-portfolio. If you made a digital video or sound recording, save it to your e-portfolio.

Continued

Step-by-Step continued

10. Select the text box again. Notice the black squares that appear in each corner of the text box when it is selected. These squares are called sizing handles. Sizing handles are used to change the size of an object. Drag one of the corner selection handles to resize the text box.

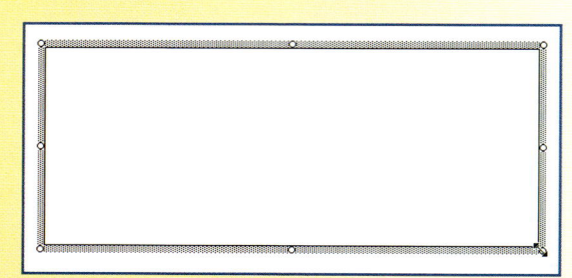

Change a Border and Fill with Color

1. Right-click the border of the text box.
2. In the drop-down menu, choose **Format Text Box**.
3. In the **Color** box in the **Colors and Lines** tab, choose **Orange**.
4. In the **Style** box, choose **3 pt**.
5. In the Fill Box, choose light yellow.
6. Move the Transparency slider to 50%.
7. Click OK.

Enter and Edit Text in a Text Box

1. Click inside the text box and enter `This is a text box.`
2. Select the text you just typed by dragging across it. Change the formatting of the text. Refer to Lessons 9.1 and 9.2, pages 264–275, to review formatting text.
3. Enter your name in the text box.
4. Save and print your document.

Application 9

Directions In a new document, create a text box and enter your name and address in it. Resize the box as necessary to fit the information. Make your name bold.

Give the text box a one point (1 pt.) purple border and fill it with a light color of your choice.

Activity 35

Activity 36

RUBRIC LINK

Go to the Online Learning Center at **digicom.glencoe.com** and click Lesson 9.7, Rubric, to find the application assessment guide(s).

UNIT 2 LAB *continued*

Follow these steps to complete your presentation on your chosen elected official:

Step 1 Prepare an attention-getting opening to your presentation. This could be a quote from the interview you conducted or a surprising fact about your chosen elected official.

Step 2 Prepare the body of your presentation on the following three topics:

1. Biographical information on your chosen elected official, such as job title, political affiliation, home town, hobbies, and when his or her term of office expires
2. Highlights of the interview you conducted, including quotes from the elected official.
3. How the elected official voted on two separate issues. Include how you would have voted on these issues and explain why.

Step 3 Prepare a closing for your presentation. To help make your presentation memorable, encourage members of your audience to take action that relates to the information you have presented.

For example, if you feel strongly about the responsibility of citizens to vote, you could encourage your audience to register to vote when they are 18 years old.

Step 4 Give your presentation a catchy title. Follow the self-evaluation criteria for guidelines to format your presentation.

Step 5 Save a record of your presentation to your e-portfolio. If you have not prepared an electronic presentation, consider arranging a digital video recording of your presentation.

Self-Evaluation Criteria

In your presentation, have you:

- Included an attention-getting opening?
- Included the following three topics in your presentation: biographical information, highlights of the interview, how the elected official voted on two issues?
- Included a closing to encourage audience members to take action?

Activity 34

Step-by-Step

Create a Text Box

1. Start Word. Open a document you have worked on recently. Place the insertion point at the beginning of the document.

2. Select the **Insert** menu and choose **Text Box**. If you have the Drawing toolbar open, you could also choose the **Text Box** icon in that toolbar.

3. Move your pointer inside the drawing area, which is labeled *Create your drawing here*.

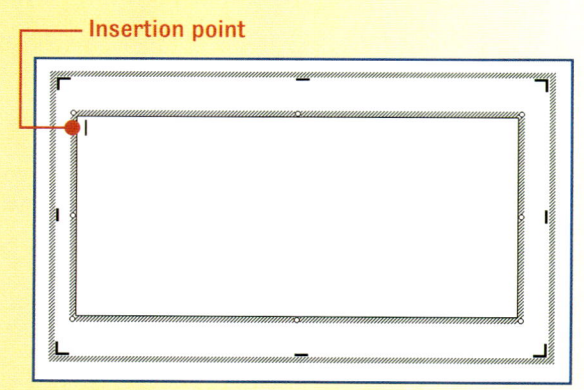

   Your pointer will take the shape of a crosshair.

4. Drag from the upper left-hand corner to the lower right-hand corner of the box.

   A text box will be created within the drawing area, and the insertion point will appear inside this text box.

5. Click on the border of the text box to select the text box.

6. Choose the **Edit** menu and click **Cut**.

7. Place the insertion point in another location in the document where you want the text box to appear. Select **Edit** and then **Paste**.

8. Click on the border of the text box and drag it to a new location. Select Edit and Paste again. Now you have two identical text boxes. Select one of them and press DELETE.

9. Now that you have created and moved the text box, you will delete the original drawing area. Click the outside border of the drawing area. Press DELETE.

Continued

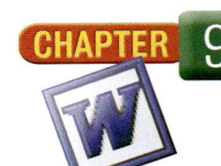

UNIT 3
Digital Communication Tools in the World of Work

CHAPTER 9 Word Processing

CHAPTER 10 Spreadsheets

CHAPTER 11 Databases

CHAPTER 12 Presentations

Curriculum Connections
- Language Arts—Reading
- Language Arts—Writing
- Science
- Technology

Lesson 9.7 — Create and Edit Text Boxes

You will learn to:
- **Create**, resize, and delete a text box.
- **Edit** and format text in a text box.

Text in a Box In newsletters, flyers, reports, and even letters, it is sometimes helpful to separate short paragraphs or statements from the main text of the document. To do this in word processing, you can insert text in a text box. A text box is an area that can hold text and that can be moved independently of the main text in a document. Text boxes are often used to hold text that should be highlighted for a reader.

Graphic Objects Text boxes are just one of the graphic objects that can be inserted from the Drawing toolbar in Word. Other objects include lines, arrows, ovals, WordArt, clip art, and AutoShapes.

Like text, several of the graphic objects can be formatted to change their appearance on the document. You can change the thickness, style, and color of lines and shapes. You can make shapes transparent or filled with a color, pattern, or texture. You can arrange shapes to be in front of or behind other shapes or the main text. It is even possible to have text show through a shape or to have text wrap around a shape you have drawn on the page. The Drawing toolbar gives you control over the artistic aspect of document layout.

DIGIByte
Formatting in a Text Box
Text within a text box can be formatted just like any other text in a document.

✓ Concept Check

What types of information do you think might be placed in a text box? Look through this book to see what kind of text boxes you can find. How are the boxes formatted? How are colors used? What information is contained within the text boxes?

DIGITAL DIALOG

Text Boxes

Teamwork Find a partner to work with and discuss various ways you could use text boxes. Think about the business forms you prepared earlier in this chapter. Could you use text boxes on those forms? If so, where and how would you use them? Are there any other business documents or forms that you could use text boxes with?

Writing Dictate your answers with speech recognition.

DIGITAL Dimension → Science
INTERDISCIPLINARY PROJECT

Link what you will learn about business computer skills in Unit 3 with science by investigating e-cycling. In your project you will investigate and gather data on e-cycling. You will finish with a presentation to the class.

E-Cycling

What Is E-Cycling? Everyone has heard of recycling, but what is e-cycling, or electronic recycling? E-cycling is finding an environmentally friendly and cost-effective solution to the problem of disposing of discarded electronic hardware. This includes computers, TVs, and phones.

How Can You Learn About E-Cycling? An agency known as the U.S. National Recycling Coalition has been formed to help with e-cycling. This agency predicts that by 2007 the average North American personal computer will be replaced every two years. That means that about 500 million of them will be obsolete. Computers and other electronic hardware contain hazardous materials that many of us are unaware of, such as lead and mercury. The raw materials in each computer are only worth about $4.00. Old computers are hazardous and should not be dumped into our landfills. So what can be done? You will investigate and report on what is being done now.

Digital Dimension Activities

Start-Up
What is your state doing about e-cycling? Search the Internet for e-cycling programs within your local area or state. Make a list of the e-cycling locations you find. Include their address, phone number, e-mail, and Web site URL in your list. You may want to add these to your PDA address book. If you do not find any information for your state, choose a neighboring state.

Look Ahead...
Continue building your project in the following activities:

Chapter 9 .... p. 310
Chapter 10 .... p. 356
Chapter 11 .... p. 390
Chapter 12 .... p. 423
Unit 3 Digital Dimension Lab .... pp. 426–427

RUBRIC LINK

Go to the Online Learning Center at **digicom.glencoe.com** and click Lesson 9.6, Rubric, to find the application assessment guide(s).

Application 8

Directions Use speech recognition to dictate a memorandum, or memo, to let the science teachers at your school know about the invitation you sent to Dr. Harvey from the environmental science class to speak about recycling.

Ask the other science teachers if they would like to include their classes in the recycling project.

You may want to begin the memo with the following text.

```
Memorandum
To:     Science Department
From:   Your Name
Date:   October 23, 20—
Re:     Guest Speaker on Recycling
```

Create the body of the memo in your own words. Be sure to include all the necessary information mentioned in the directions for this memo.

DIGITAL DIALOG

Investigation and Awareness

As a follow-up to the invitation to Dr. Harvey that you prepared in the last activity, your class is going to learn more about recycling. In order to be prepared for his visit, take the following steps:

Teamwork In a small group, prepare a list of environmental companies that recycle products. Note what kinds of waste each company accepts.

Internet Investigation Search the Internet for information about recycling companies. Find out what the companies recycle and what they pay for recycling. Look for companies located in your area.

Writing Dictate a memo to the teachers at your school about your recycling initiative. Create announcements to post in science classrooms about the recycling program. Be sure to list the products that can be collected for recycling.

Communication Brainstorm ways your class could work with school clubs and community groups to encourage recycling locally. Consider how you could support fund raisers or contests, and how you could present what you have learned about the importance of recycling.

Prepare a list of three questions to ask Dr. Harvey at the end of his presentation on recycling.

CHAPTER 9
Word Processing

Lesson 9.1	Use Basic Text Editing and Formatting
Lesson 9.2	Use More Advanced Editing and Formatting
Lesson 9.3	Add Headers, Footers, Bullets, and Numbering
Lesson 9.4	Create and Edit Tables
Lesson 9.5	Create Business Forms
Lesson 9.6	Dictate Letters with Speech Recognition
Lesson 9.7	Create and Edit Text Boxes
Lesson 9.8	Insert and Format Graphics
Lesson 9.9	Use a PDA for Word Processing
Lesson 9.10	Use Online Forms and Templates

You Will Learn To:
- **Format** and **edit** text.
- **Insert** lists, tables, text boxes, and graphics.
- **Write** a business letter using speech recognition.

When Will You Ever Use Word Processing?
With a word processing application, you are limited only by your imagination. Web pages, reports, books, flyers, invitations, and newsletters are just a few of the things you can create.

262

Activity 33

If you need additional help on this topic, open the Help menu; or refer to the documentation for your hardware.

Step-by-Step

Dictate a Letter with Speech Recognition

You may need to refer to Chapter 6 for a quick review of speech recognition commands before you begin this practice. You may also refer to Appendix R, page R2, on correct formatting procedures for a letter.

1. Start Word, then start speech recognition in Dictation mode.
2. Dictate the following letter. Use the New Paragraph and New Line commands as necessary. Since this is a block style business letter, every element will be left-aligned.

```
October 23, 20--
Dr. John Harvey
Environmental Science Dept.
Central College
1900 W. Innes Blvd.
Salisbury, NC 28144

Dear Dr. Harvey:

I would like to invite you to speak to our environmental science class. We are interested in beginning a campus-wide recycling program.

I understand that Central College recycled many of the materials left over from building your new environmental science building. The environmental science class would like to hear about the process that was used to recycle these materials.

The class would also like to take a field trip and visit your new environmental science building at Central College. Could you arrange for a tour and some time to discuss recycling with our class?

Thank you for your assistance. You may call my teacher, Ms. King, at school at 888-555-9023 if you have any questions.

Sincerely,

Your Name

Environmental Science Class Secretary
```

DIGIByte

Left Alignment

In a block style letter, all paragraphs align on the left margin, even the date and closing.

Read to Succeed

Key Terms

- margin
- alignment
- default
- header
- footer
- column
- row
- cell
- merge

Predict

Taking a moment to predict what you will learn makes learning easier! Predicting what you will learn in a lesson gives you a boost. That boost will translate into more success and eventually higher earnings.

Every day, take ten minutes to preview the lesson you will work on the following day. Skim read through the lesson title and paragraph headings. Predict what you will be learning. Also, predict how learning these topics will help you in your life at school and in the future. Can you use this new information in other parts of your life: other subjects, in school-related clubs, at home, and at work? Create a graphic organizer similar to the illustration as you work through this chapter. Use Journal in Tablet PC to create your chart, if possible.

Example:

Use Basic Text Editing and Formatting	
Paragraph Headings	Predict
Explore Print Layout and Normal View	I will learn about the different ways to use word processing for business and personal purposes.

DIGITAL CONNECTION

Creative Word Processing

The Environmental Science Club is sponsoring an awareness campaign for Earth Day. The creator of the winning entry will receive a movie pass and pizza for two at the local movie theater. Acceptable entries include posters, flyers, newsletters, or brochures. You think, "If only I were good at drawing!" Your friend reminds you of the graphics you are learning about in computer class and asks, "Why can't you create a flyer using word processing?"

Writing Activity

Write a short paragraph describing how you can create a poster or flyer on paper without using a computer.

Lesson 9.6 Dictate Letters with Speech Recognition

You will learn to:
- **Dictate** a business letter with speech recognition.
- **Create** and dictate a memorandum with speech recognition.

Communicating in Business Business letters are a special form of communication. They differ from friendly letters because they are often directed to people you do not know. A good business letter clearly tells the reader the purpose of the letter in the first paragraph. Consultant Patrick Burne explains that business letters should be as helpful as the topic allows. He says it is best to use a conversational tone, as though you were talking aloud to your reader. In a business letter, you are usually trying to convince someone to take action on your behalf. You will get a response only if your message is clear. Mr. Burne identifies seven qualities of effective letters that he calls the seven Cs of letter writing. The best letters are:

- Clear
- Concise
- Correct
- Courteous
- Conversational
- Convincing
- Complete

Companies usually send letters on stationery that has their logo and address information already printed at the top. This type of business stationery is called a letterhead. You can create a letterhead using the header and footer skills you learned in Lesson 9.3.

Block Style Business letters are usually formatted in block or modified-block style. In block style, all paragraphs are aligned left and there is a line space between paragraphs. In modified-block style, the paragraphs and some other elements, such as return address, are indented. However, many companies also adopt their own style for letters.

DIGI*Byte*
Spacing
The text in a business letter is single-spaced with double-spacing between paragraphs.

✓ Concept *Check*

What does block style mean? Refer to Appendix R, pages R1–R15, to view business letter formats.

Chapter 9 Word Processing

Lesson 9.1 — Use Basic Text Editing and Formatting

You will learn to:
- **Create** a new document and enter text.
- **Cut,** copy, and paste text.
- **Format** text font by choosing style, typeface, size, and color.
- **Use** Format Painter to make formatting easier.
- **Save** and print a document.

DIGIByte

Get Noticed
To help a reader notice text in a document, you can bold, italicize, or underline the text.

Home Row
Whenever you use the keyboard to enter data, remember to begin with your hands on the home row keys. You can view the home row keys and finger in the Keyboarding Skills Appendix, pages A-1–A-47.

Why Use Word Processing? Word processing enables you to create many types of documents and easily make changes to those documents on the computer. Incorrect text or punctuation can easily be deleted, moved, or replaced. Word processing applications can even check for spelling, grammar, and common typing mistakes. With these and other features, word processing applications can make proofreading, which is the process of reading what has been typed and making any needed corrections, quick and easy. Word processing applications, such as Microsoft Word, also make it easy to format text, or change its appearance.

In this lesson, you will use Word to perform basic text editing, such as deleting or moving text, as well as basic text formatting, such as changing the text typeface, size, color, and style.

✓ Concept Check

Where have you seen bold, italicized, or underlined text used? Why was the text formatted in this way?

DIGITAL DIALOG

Help Menu

Teamwork Work in a small group. From the Word Help menu, select Microsoft Word Help. If the Office Assistant appears, right-click the Assistant, make sure that Use the Office Assistant is not selected, and click OK. Click the Index tab. Key the word *format* into the Type keywords box and click the Search button.

Writing Review the list that appears to find information on applying decorative underlining and embossed, engraved, outlined, or shadow formatting to text. Write a brief description of each of these types of formatting.

Application 7

Directions Create a purchase order form using a Word table. Merge, split, and shade cells as necessary to create a design that makes your data easy to read and that is attractive to look at.

You may use a format that is similar to the invoice you created in the Step-by-Step Activities.

Fill in the purchase order with the following information:

```
TO:    Center Office Supply
       901 S. Broadway Blvd.
       Your City, Your State 22222
       Phone: 888-555-3333
       Fax: 888-555-7444

FROM:  Enviro-Health Corp.
       P.O. Box 44
       Your City, Your State 34443
       Phone: 888-555-9999
       Fax: 888-555-6666

Items ordered:

3  PDAs                    $299.99   $899.97
10 Headsets with
   microphone              $10.99    $109.90
4  USB Flash
   Drives 256 MB           $59.99    $239.96
Total                                $1,249.83
Shipping & handling                  $100.00
Total                                $1,349.83
```

RUBRIC LINK
Go to the Online Learning Center at **digicom.glencoe.com** and click Lesson 9.5, Rubric, to find the application assessment guide(s).

DIGITAL Decisions

Plan a Recycling Program

Investigation Your class has decided to investigate what it would take to begin an e-waste recycling program. Interview teachers and other classmates to see if they would participate in a recycling program. Ask them for suggestions for collection sites on campus.

Take Action As a class, assign the following roles to students.

- Find out what your school currently does with old computers, print cartridges, and other electronic devices.
- Survey possible areas as collection locations.
- Contact local authorities and/or businesses to see if they have a recycling location in place where you could send the collected devices.
- Organize an awareness campaign.
- Prepare flyers and brochures to display at your school.
- Investigate the possibility of including a nearby community in your collection efforts.

Set a date to meet again and discuss the information everyone has collected. Can your class conduct a successful e-cycling program? Why or why not?

Activity 1

? If you need additional help on this topic, open the Help menu; or refer to the documentation for your hardware.

margin The blank edge that borders the area in which text and objects can be placed on a page.

Step-by-Step

In the following activities, you will apply basic formatting and editing in a word processing program.

Explore Print Layout and Normal View

1. Start your word processing application. Enter the following text.

> Scientists have been studying our environment for many years. They are concerned about how people interact with the earth and how our future resources may be affected. A growing concern for environmentalists is how we dispose of products we no longer use.

You can use the Undo and Redo buttons that are available in all Microsoft Office applications. Click the Undo button to delete the last action you completed, such as the last word or number that you entered. Click the Redo button if you use the Undo button by mistake.

2. You can view a Word document in Print Layout View. The entire page is displayed as it will look on a printer, including the margins and edges of the page. The **margin** is the blank edge that borders the area in which text and objects can be placed on a page. To switch to Print Layout view, click the **Print Layout View** button in the bottom-left corner of the application window.

3. You can also view a Word document in Normal View, in which only the area within the margins shows and page breaks are indicated by dotted lines. To switch to Normal View, click the **Normal View** button in the bottom-left corner of the application window.

4. Make sure your document is in Print Layout View.

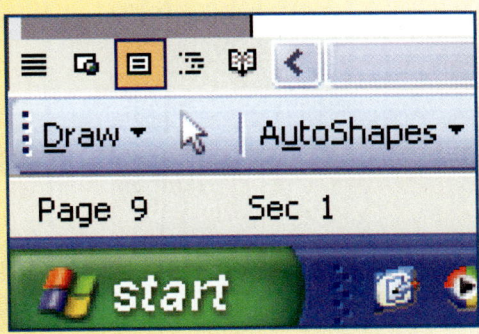

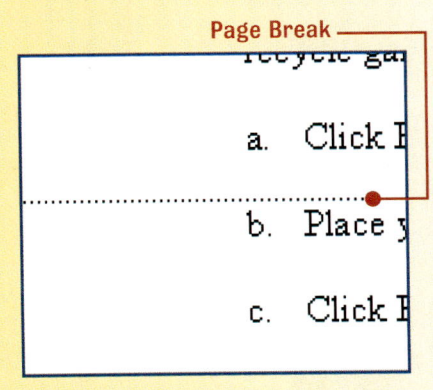

Continued

Chapter 9

265

Step-by-Step *continued*

Activity 29

Change Row Height

1. Click inside the last row, select the **Table** menu, and choose **Table Properties**. Choose the **Row** tab.
2. Check the **Specify height** box and click the up arrow until the window reads **0.3"**. Row height is measured in inches. Clike **OK**.

Activity 30

Change Column Width

1. Select the cells in the last column from the heading TOTAL down to $837.98. Select the **Table** menu, and choose **Table Properties**. Choose the **Column** tab.
2. Check the **Preferred width** box and click the down arrow until the window reads **1.2"**. Click **OK**.

Activity 31

Automatically Resize to Fit Contents

Columns and rows can be set to automatically resize based on cell contents.

1. Choose the **Table** tab in the **Table Properties** window and click **Options** in the bottom-right corner.
2. Select **Automatically Resize to Fit Contents**.

Activity 32

Split a Table

To break a table into two tables, you can split the table.

1. Place the insertion point in the **Total** row and click the **Table** menu.
2. Choose **Split Table**.

DIGITAL DIALOG

Starting a New Business

Internet Investigation You have just been hired as the office manager for a new environmental recycling company. One of your first tasks is to assemble and organize the different business forms the company will need. You will use the Internet to do your research.

Work in a small group. To get started, use a search engine to find business forms. You may also visit the Web sites of some office supply companies. These companies may have an online search engine to help you locate preprinted business forms that are available for purchase.

Create a table that includes five forms you think the company will need. Be sure to include the price of the forms and whether the supply companies will print your company's name and logo on the forms.

Step-by-Step *continued*

Activity 2

Indent Text Using the Tab Key

Often a tab, which is typically five spaces, is inserted in front of the first line of a paragraph. To indent a paragraph, always use the Tab key; never use the spacebar.

1. Place the insertion point at the beginning of the first sentence of the paragraph you keyed, to the left of the word *Scientists*.
2. Press TAB. The first line of the paragraph will automatically be indented by five spaces.

Activity 3

Select Text and Make Text Bold

You must select text before you can change its appearance. When you select text, you indicate to the computer that the text has been chosen so that operations, such as text formatting, can be applied to it. Selected text displays with a highlighted effect on the screen.

1. Select text by holding down the left mouse button and dragging across the word *future* in the second sentence. The word will appear to be highlighted in black, indicating that it is selected.
2. Click the **Bold** button on the Formatting toolbar.

— Bold

Activity 4

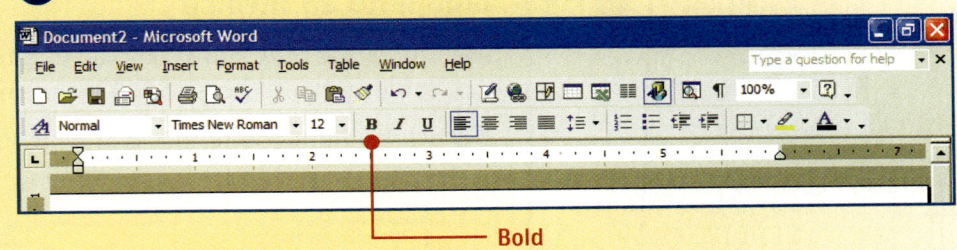

Make Text Italic and Underlined

1. Select the word *our* in the first sentence.
2. Click the **Italic** and the **Underline** buttons on the Formatting toolbar.

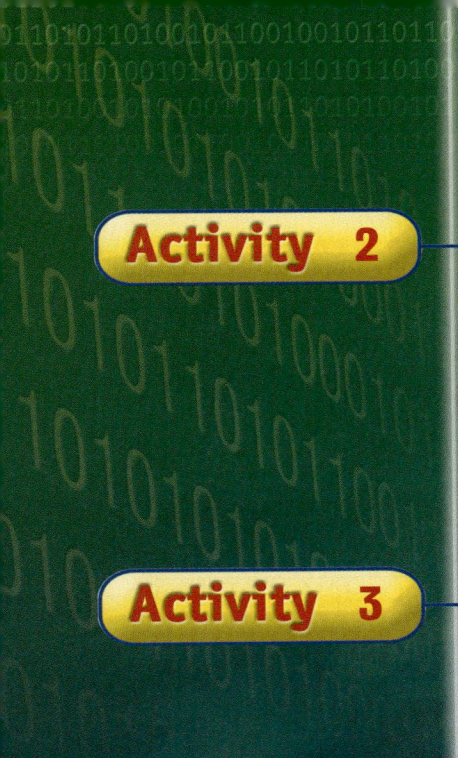

Italics — Underline

Scientists have been studying *our* environment for many years. They are concerned about how people interact with the earth and how our **future** resources may be affected.

DIGI*Byte*

Go To Help

Use the Help menu if you do not remember how to apply a type of formatting.

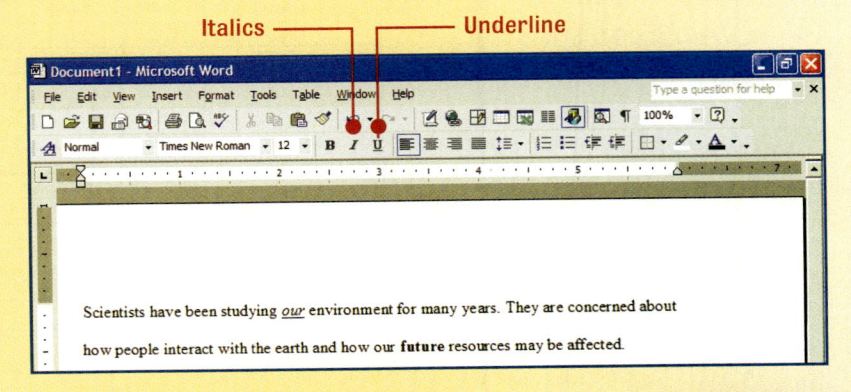

Continued

Step-by-Step continued

4. Select all of the cells in the second, third, and fourth rows. Select the **Table** menu, and choose **Merge Cells**. Shade the new cell a light gray. See Activity 24, page 287, for directions on shading.

5. Fill in the other two cells in the top row as shown below.

Office Products, Inc. P.O. Box 555 Your City, Your State 55555 Phone 888-555-8080 Fax 888-555-9090	INVOICE NO.	802346
	DATE:	Today's Date
	TERMS:	5/10, N/30

Split Cells

1. Click **Table, Select, Cell** to select the new gray cell. Select the **Table** menu, and choose **Split Cells**. In the **Number of columns** box, enter 1. In the **Number of rows** box, enter 3. Click **OK**.

2. Select the third row and change the shading to white.

Complete an Invoice

1. Enter the text shown in the example below to complete the invoice table. Review Lesson 9.4 if you need help formatting the table.

2. Right align the text in the Unit Price and Total columns.

Office Products, Inc. P.O. Box 555 Your City, Your State 55555 Phone 888-555-8080 Fax 888-555-9090	INVOICE NO.	802346
	DATE:	Today's Date
	TERMS:	5/10, N/30

TO: Environmental Recycling
       1234 Main Street
       Your City, Your State 44444		

QUANTITY	DESCRIPTION	UNIT PRICE	TOTAL
2 boxes	Recycled computer paper	$22.99	$45.98
1	Laser printer	$499.00	$499.00
1	Black printer cartridge	$109.00	$109.00
1	Color printer cartridge	$109.00	$109.00
	TOTAL		$762.98
	SHIPPING & HANDLING		$75.00
	TOTAL DUE		$837.98

Activity 27

Activity 28

Continued

Step-by-Step continued

Activity 5

Delete Text

1. Select the word *many* in the first sentence.
2. Press DELETE.

Save and Save As

Activity 6

1. Select the **File** menu and choose **Save As**. Save this document on the computer desktop by clicking the Save In drop-down arrow. Name the file **Recycle**.

2. Place a floppy disk in the computer's disk drive and select **File** and then **Save As** again. Choose **3½" Floppy (A:)** from the Save In list and save the file to the floppy disk.

The first time you save a file, the Save As window will automatically appear. Now that this file has been saved and named, you can click **File** and then **Save** to save any changes that you make to it in the future without having to make selections in the Save As window. If you need to change the file name or save the file to a new location, you must choose **File** and then choose **Save As** again.

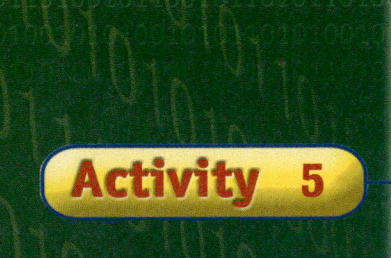

DIGI*Byte*

Floppy Disk

When you save to a floppy disk, you can place the floppy disk in a different computer and still be able to access your files.

Activity 7

Copy, Cut, and Paste Text

Cut and copy are features that allow you to move or copy a word, sentence, paragraph, or even a page of text to a different location. To cut text is to delete the text from its original location and store it in the computer's memory so that it can be pasted in another location. To copy text is to leave the original text intact but place a copy of the text in the computer's Clipboard, which is a temporary memory holder. When the computer is turned off, the contents of the Clipboard is cleared and is therefore lost if it has not been pasted. To paste text is to insert cut or copied text in a new location.

1. Open the file **Recycle** if you do not already have it open. Enter the following text as a new paragraph.

Suite Smarts

My Documents The default location for saving files in MS Office is My Documents. Be sure to check the Save In location before saving a file so that you will know where to look for it.

```
Many people recycle garbage. Recycling is good
for the environment. Refuse collection companies
give residents containers to sort recyclable
garbage for disposal. Recyclable materials
include plastic, glass, aluminum, and paper.
Plastic and glass containers must be rinsed and
sorted before they can be recycled.
```

Continued

Activity 25

Step-by-Step

Create a Table for an Invoice

When a business performs a service or delivers a product, it sends an invoice to the customer stating the amount owed. An easy way to create an invoice form is to create a word processing document containing a table.

1. Start Microsoft Word.

2. Insert a table that has 4 columns and 15 rows. See Lesson 9.4, page 284, to review the steps for creating a table.

Activity 26

merge To combine two or more cells to create one cell.

Merge Cells

You can **merge**, or combine two or more cells to create one cell. This can make the data easier to read.

1. Drag to select the first two cells in the first row.

2. Select the **Table** menu and choose **Merge Cells**.

3. Click in the new cell you just created and enter the following information.

```
Office Products, Inc.
P.O. Box 555
Your City, Your State 55555
Phone 888-555-8080
Fax 888-555-9090
```

Continued

Suite Smarts

Like text, graphics can also be cut, copied, and pasted.

Activity 8

Activity 9

Activity 10

Step-by-Step *continued*

2 Select the words *Many people recycle garbage*. Select the **Edit** menu and click **Copy**. Place the insertion point at the end of the paragraph. Choose **Edit** and then **Paste**.

3 Select the words *Recycling is good for the environment*. Choose **Edit** and then **Cut**. Place the insertion point at the end of the paragraph. Choose **Edit** and then **Paste**.

Use the Format Painter

1 Select the word *recycle* in the first sentence. Click the **Bold**, **Italic**, and **Underline** buttons on the Formatting toolbar.

2 Click the **Format Painter** button on the Standard toolbar.

3 Use the Format Painter on all occurrences of *recycle*, *recycling*, or *recyclable* in the paragraph.

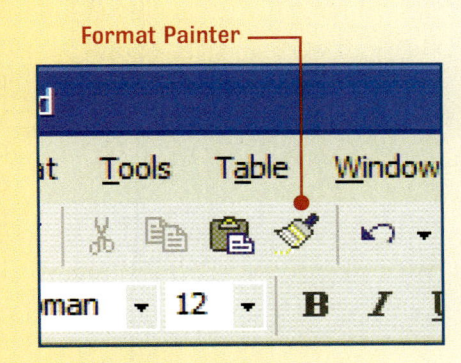

Print Preview and Print

1 Select **File** and **Print Preview** to see how the document will print.

Using Print Preview in a multipage document is especially helpful. This feature allows you to see where the page breaks are and make sure graphics and other objects appear as you want them.

2 Choose **File** and then **Print** to print the document.

A default printer is set to be used with Word. If on a network, you may be able to choose another printer. The Print window also allows you to choose how many copies you want to print. You can choose to print individual or multiple pages.

Format Text Font

The set of characteristics, including size and typeface, that determines how text appears is called a font. The font includes the typeface, such as Arial or Times New Roman, which determines how characters are actually shaped. The font also includes size and style. Font size is measured in points; the larger the number of points, the larger the font size. Font style includes such attributes as italic or bold. For example, *Times New Roman 12 pt bold* is a complete font name.

1 Select each occurrence of *recycle*, *recycling*, or *recyclable* by holding down CTRL and double-clicking each word.

Continued

Lesson 9.5 Create Business Forms

You will learn to:
- **Identify** forms used in business.
- **Create** business forms with tables.
- **Merge** and format table cells.

Forms Used in Business Businesses use a variety of forms to make work more efficient for employees. Forms help to keep information organized. Some companies require employees to fill out forms even for very small tasks, such as using the copy or fax machines. Employees often need to fill out expense report forms to get reimbursed for business expenses such as travel. With word processing software, you can easily customize the formatting and design of business forms.

In this lesson, you will examine a few basic forms that most businesses use. These forms are invoices and purchase orders.

Concept Check

What types of activities may businesses require forms for? Have you ever ordered anything through the mail or Internet? Did any forms come with the item you ordered? If so, what information was on the forms?

DIGITAL DIALOG

Templates for Business

Teamwork With a partner, examine the business form templates that are provided with Word. To do this, select the File menu and choose New. In the New Document task pane that appears on the right, select General Templates. Look at the templates that are available online as well. To do this, click Templates on Microsoft.com in the New Document task pane.

Discussion As you review the templates, discuss the following questions:
- What are some of the similarities between the documents?
- What is unique about the templates?
- What other types of business forms not discussed in this lesson are available?
- Why do you think these templates are important to businesses?
- Is there anything you would change about one or more of the forms? If so, describe the changes.

Share your findings with the class.

DIGI Byte

Typical Text Size

Most text in published materials, such as newspapers or magazines, is 12 point.

Tool Tips

To locate the name of a button on a toolbar, hover your pointer over the button until the ToolTips box appears displaying the button name.

Points per Inch

72 points is approximately equal to 1 inch.

RUBRIC LINK

Go to the Online Learning Center at **digicom.glencoe.com** and click Lesson 9.1, Rubric, to find the application assessment guide(s).

Step-by-Step continued

2. Click the **Font** drop-down arrow on the Formatting toolbar to see the available typefaces. Choose Comic Sans.

3. Click the **Font Size** drop-down arrow on the Formatting toolbar and choose 36.

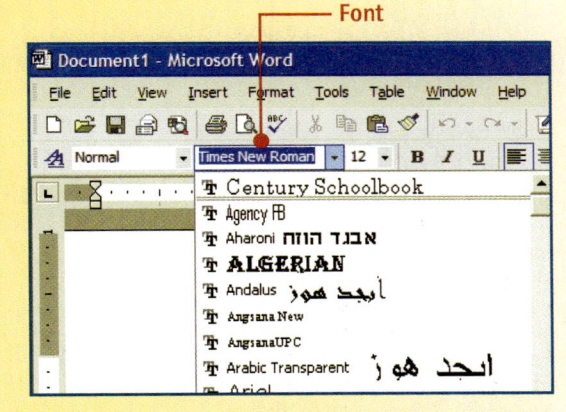

4. Click the **Font Color** button on the Formatting toolbar and change the text color.

5. Select the **Format** menu and choose **Font**.

6. Choose other text attributes, such as SMALL CAPS, Strikethrough, Superscript, Subscript, Shadow, Outline, Emboss, Engrave, or ALL CAPS.

Application 1

Directions Create a new word processing document. Enter the following text on separate lines, as shown:

```
School Recycle Plan
Collect aluminum and plastic containers.
Bag plastic and aluminum to be collected.
Post signs around school about recycling.
Place collection containers around campus.
Contact local collection agency to collect bags
of recycled materials.
```

Select the first line and apply bold style and change the typeface, size, and color. Indent the first line. Delete the first line, then Undo that action.

Think about the order in which the events you just keyed might occur. Select the first word in each line and apply italic or underline style. Use Format Painter to make this formatting easier. Use cut and paste or copy and paste to put the events in an order that makes sense. View the document in Print Layout and Normal views. Save the document as **List**. Print preview and print your document.

DIGITAL Career Perspective

Arts/Audio/Video Technology and Communications

Julie Tam
Web Site Designer

Creating Web Sites

"I love the variety of my job, the combination of logical, creative, and business tasks," says Julie Tam, a freelance Web site designer. "There are times when you're meeting people and times when you're working alone; times when you're proofreading code and times when you're creating the design of the site." As a Web designer, Tam is simultaneously an artist, a computer programmer, and a businessperson.

Tam has been designing Web pages since 1995 and maintains several corporate Web sites, as well as a variety of small business and not-for-profit sites. During her time in the Peace Corps, Tam read an article on the Internet, which was sweeping the nation. She knew then that she would be part of what is now a revolution. She began her career as a technical writer and translator, and progressed to writing and content development and finally webmaster before she launched her own business.

Tam uses a wireless network, digital camera, and scanner in her office and a laptop computer when she meets with clients. During her initial client consultation, when she identifies the client's objectives, she shows a PowerPoint file containing examples of other Web sites she has created, to help the client make choices about styles, colors, and layout.

Back in the office, she makes a flowchart that shows the structure of the Web site and how all the pages will connect. Then she designs sample pages with special drawing and painting software. Next she programs the site using whatever programming language is best suited for the function of the site. Tam takes courses "all the time" to stay current in her field. In addition to HTML, she is familiar with many other programming languages that add even more to Web sites.

What does Tam like best about her job? "Overall," she says, "my primary objective is to create a site that meets the client's needs and is easy for everyone to use, including people with disabilities. It also has to be pleasing to the eye and quick to load. This is a challenge I really enjoy."

Training
A bachelor's degree in graphic design or computer science is required to work as a Web designer. Courses to take in high school and beyond include both math and art. Classes in business would be useful for anyone who might want to work independently.

Salary Range
Typical earnings for this field are in the range of $40,000 to $85,000 a year.

Skills and Talents
Web designers need to be:

Creative

Good problem-solvers

Effective listeners

Able to develop trusting business relationships

Attentive to detail

Flexible

Lifelong learners

Career Activity
How do you think studying a wide variety of subjects, including math and art, would be helpful in a career in Web design?

Lesson 9.2 — Use More Advanced Editing and Formatting

You will learn to:
- **Change** line spacing.
- **Change** text alignment.
- Use spell check.
- **Change** margins and page orientation.

DIGI Byte

Double Spacing
All rough drafts should be double spaced so that proofreading and editing can be done easily.

alignment The way that text or objects are lined up across a page.

default A setting that was determined at the time the program was created.

Formatting and Readability The text formatting that you are learning about in this chapter can improve the appearance of a document. Formatting can also make text easier to read and can focus a reader's attention on important information. For example, creating space between paragraphs can make a document much easier to read.

Alignment To draw attention to items of equal importance, you can format the items to the same alignment. **Alignment** is the way that text or objects are lined up across a page. The objects can be aligned at the center of the page or at the right or left edges of the page. Justified alignment lines up text at both the right and left sides of the page.

Business Documents Business documents such as letters, memos, and contracts are usually single-spaced. Business reports and reports created for school are usually double-spaced. The **default**, or the setting that was determined at the time the program was created, is single-spaced formatting.

✓ Concept Check

What instructions for formatting a report or essay does your teacher often give you? Perhaps you are told to double-space and use one-inch margins all around. How can you be sure to do these tasks correctly?

DIGITAL DIALOG

Formatting Effects

Teamwork Work with a partner. Look through magazines or books to examine formatting effects, such as font color, size, and style, as well as the space between lines and around the page. Take note of the alignment of text and graphics on the page.

Writing Make a list or chart of what you find. Discuss the effects of the formatting choices. Do the effects help you to read the text or to get your attention as you read? Write a short summary of the points you discuss.

Activity 24

Suite Smarts

You can choose to apply shading to the entire table or to just one cell. To apply shading to one cell only, from the Shading tab of the Borders and Shading window select the Apply to list and choose Cell.

RUBRIC LINK

Go to the Online Learning Center at **digicom.glencoe.com** and click Lesson 9.4, Rubric, to find the application assessment guide(s).

Step-by-Step continued

Add Shading to Cells

1. Make sure that the insertion point is inside one of the table cells. Select the **Table** menu and click **Table Properties**. In the Table tab, click the **Borders and Shading** button. Select the **Shading** tab.

2. In the color palette, choose light yellow. Choose a light shade so the text will be visible in the table. Click OK twice

3. Check your table in Print Preview and then print.

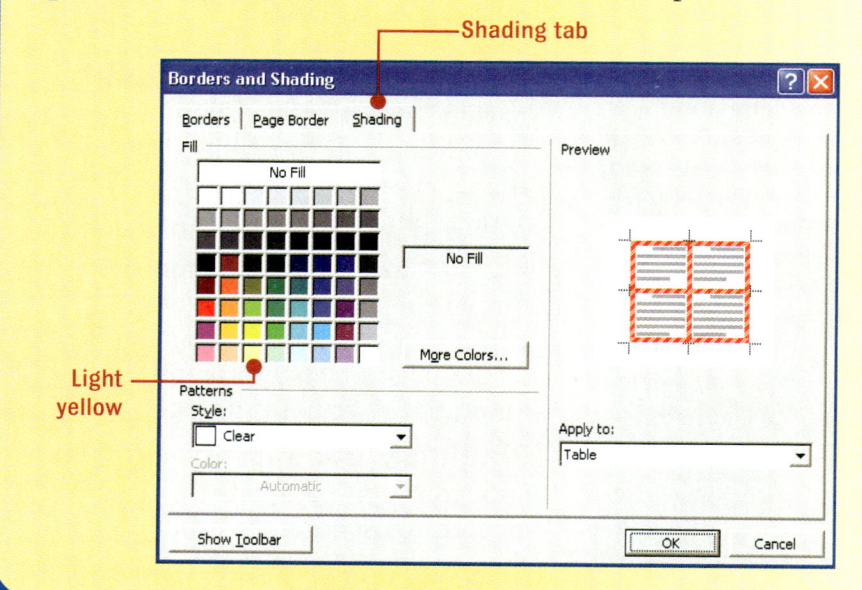

Application 6

Directions Use the insert or draw method to create a table with five rows and two columns. Make the border a thick solid line. Shade the right-hand column in light green. Enter the text below in your table. Make the heading bold. Delete the table.

```
Bills Passed in Select States for E-Waste or
E-Cycling
Arkansas       SB 807, Enacted 4/9/01
California     SP 1523, Introduced 2/20/02
               SB1619, Introduced 2/21/02
Florida        SB1922, Introduced 2/6/02
Georgia        HB2, Passed the House, in the
               Senate 2/5/2002
```

Step-by-Step

In this exercise, you will use more formatting features, such as line spacing and text alignment. You will learn how your word processing application can help you to edit your work by using the Spelling and Grammar spell-check feature.

Change Line Spacing

1. Start Word.

2. Enter the following paragraph. Do not use the Enter key between lines and do not change the line spacing. The paragraph should appear single-spaced.

> More than 315 million computers are obsolete. These computers contain an estimated 1.2 billion pounds of lead, 2 million pounds of cadmium, 400,000 pounds of mercury, and 1.2 million pounds of hexavalent chromium.

3. Select the paragraph by dragging the mouse from the beginning to the end of the paragraph.

4. Select the **Format** menu and choose **Paragraph**.

5. Choose **Double** from the Line spacing drop-down list. Click **OK**.

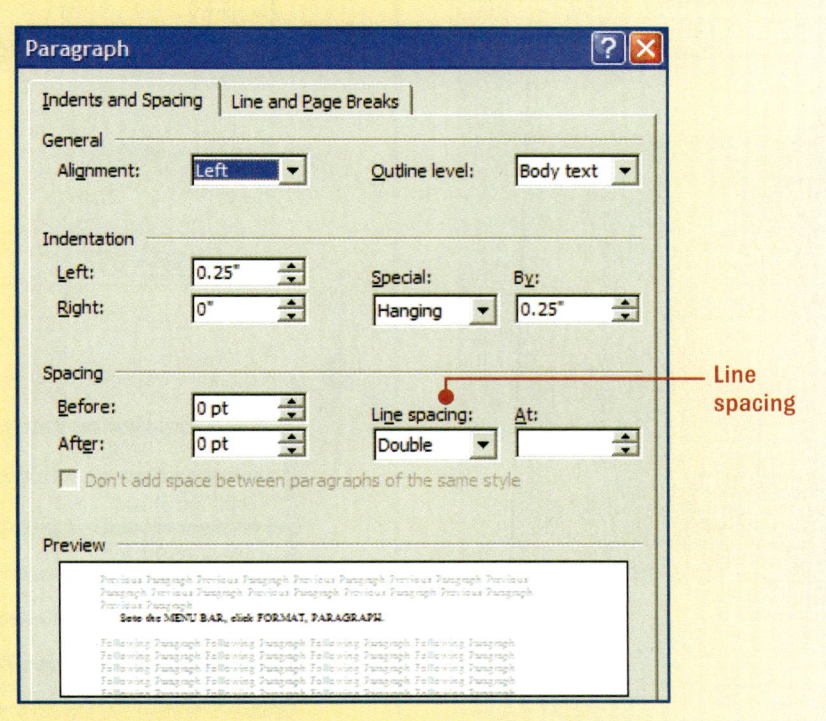

Activity 11

? If you need additional help on this topic, open the Help menu; or refer to the documentation for your hardware.

Continued

Chapter 9

Activity 23

Step-by-Step *continued*

Change the Border

1. Select the **Table** menu and click **Table Properties**.
2. Click the **Borders and Shading** button.
3. In the **Style** list, choose the line and shadow style near the bottom of the list.

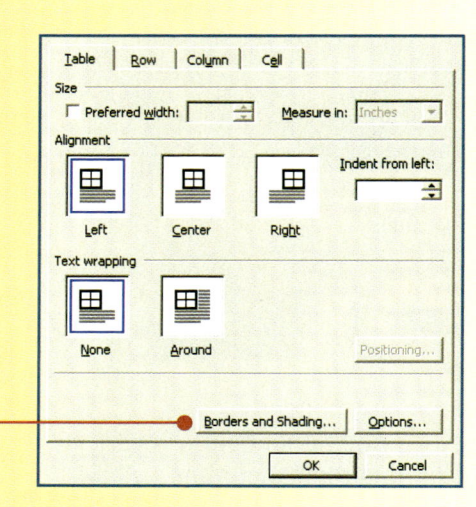

4. In the **Color** list, choose sea green and click **OK** twice.

State	Legislature Web Site
North Carolina	http://www.ncga.state.nc.us
South Carolina	http://www.lpitr.state.sc.us
Your state	*Web site*
Bordering state	*Web site*
Bordering state	*Web site*

DIGI*Byte*

Insertion Point Placement

Make sure the insertion point is inside your table before you select the Table menu if you want to edit your table.

Continued

Chapter 9

Step-by-Step *continued*

Activity 12

DIGI *Byte*

Show/Hide

Showing formatting marks is helpful when editing a document. These marks enable you to see the position of tabs, paragraph returns, and spaces.

Show and Hide Formatting Marks

In Word, you can show formatting marks, such as tabs, spaces, and paragraph marks, by clicking the Show/Hide button.

1. Make sure that formatting marks are showing by clicking the **Show/Hide** button ¶.

2. Click the **Show/Hide** button again when you want to hide the marks.

Space

More than 315 million computers are obsolete. These computers contain an estimated 1.2 billion pounds of lead, 2 million pounds of cadmium, 400,000 pounds of mercury, and 1.2 million pounds of hexavalent chromium. ¶

Paragraph mark

Activity 13

Align Text

1. Place the insertion point anywhere in the paragraph you keyed in Activity 11, page 271.

2. Select **Format** and then **Paragraph**. Notice the Alignment list at the top of the window.

3. Choose **Right** in the Alignment list and click **OK**.

4. Repeat Steps 2 and 3 above, but choose **Centered** and then **Justified** and finally **Left**. Which alignment makes the text easiest to read?

Activity 14

Use the Spelling and Grammar Feature

1. To demonstrate how the Spelling and Grammar spell-check feature works, go back to the first sentence and rekey the word *More* as *Mor*.

2. Click to place the insertion point before the paragraph.

3. Click **Tools** and choose **Spelling and Grammar.**

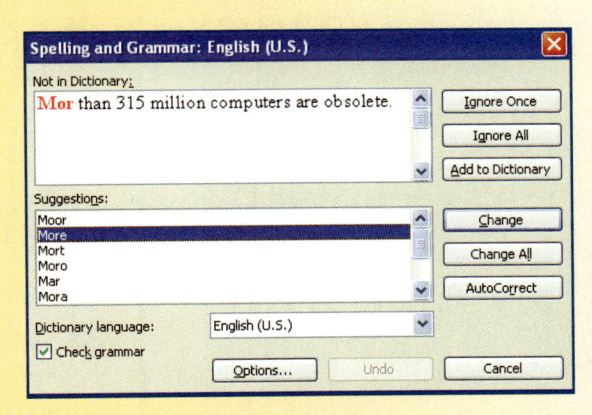

Continued

Step-by-Step *continued*

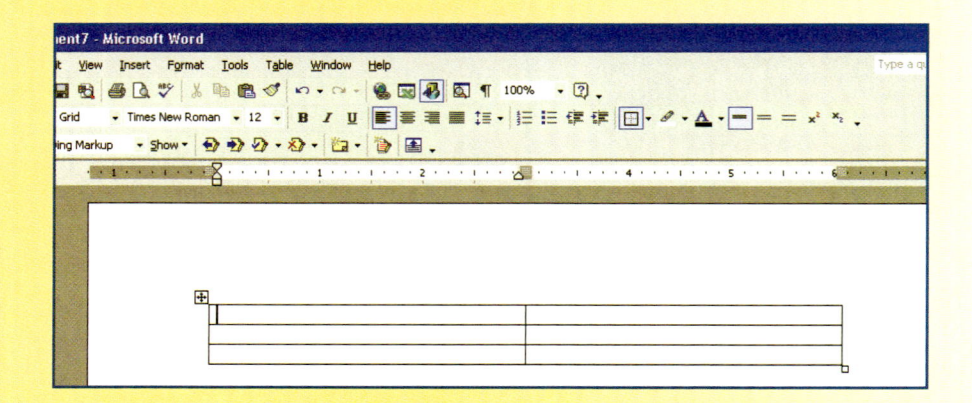

⑤ Notice that the insertion point automatically appears inside the first cell. As you key, the text will move to the next line in the cell. When you need to move to the next cell, press `TAB`. Enter the following text in the cells. Apply the same formatting that you see in the table below.

State	Legislature Web Site
North Carolina	http://www.ncga.state.nc.us
South Carolina	http://www.lpitr.state.sc.us

⑥ Save the document as **State Legislatures**.

Add and Delete Rows

① Click to place the insertion point in the last cell in the left-hand column.

② Select the **Table** menu and choose **Insert**. Select **Rows Below**. Notice that another blank row appears at the end of the table.

③ If it is not already selected, click the new row to select it. Click **Table** and then **Delete** and **Rows.**

If you wanted to delete a column, you would select the column and choose **Table** and then **Delete** and **Columns**.

④ Insert three more rows below the last row.

⑤ Add legislature Web site information for the state you live in as well as two states that border your state. To find the Web sites for the legislature in your state and the bordering states, enter the name of the state and the word "legislature" in a search engine.

DIGI*Byte*

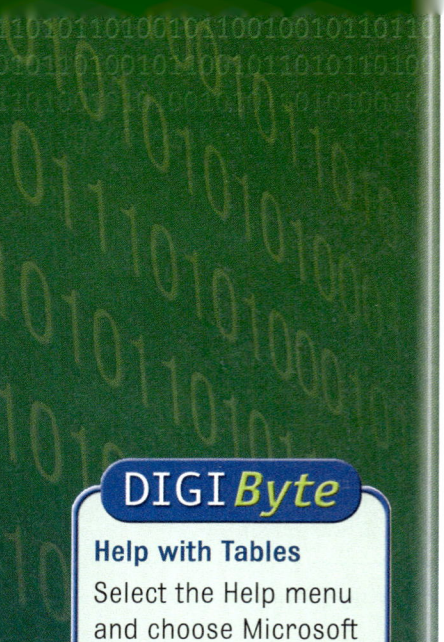

Help with Tables

Select the Help menu and choose Microsoft Word Help. Click the Index tab. Key the word *table* and click the Search button. Click through the list of information to learn more about creating and editing tables.

Activity 22

Continued

Chapter 9

285

Suite Smarts

Use the Add to Dictionary button to add a word or name you use often to the dictionary so that spell check will recognize it.

The Thesaurus feature allows you to search the Internet for additional words if you are unable to find one in the list first provided by your software.

Activity 15

Step-by-Step *continued*

In the Spelling and Grammar window, words that are possibly misspelled are displayed in red and shown within the context of the sentence. Suggestions of replacement words are displayed at the bottom of the window.

4. Select the correct spelling for *More*, and click the **Change** button.

    The **Ignore All** button is used to skip proper names or uncommon words that are not in the computer's dictionary.

    In the paragraph you just keyed, note how the number 400,000 appears. The green squiggly underline indicates a possible grammatical error.

> More than 315 million computers are obsolete. These computers contain an estimated 1.2 billion pounds of lead, 2 million pounds of cadmium, 400,000 pounds of mercury, and 1.2 million pounds of hexavalent chromium.

5. View any grammar suggestions for the paragraph to accept or ignore them.

Use the Thesaurus Language Feature

Word also has a Thesaurus feature to help you find alternative words. In the paragraph you just keyed, the word *estimated* is used in the second sentence.

1. Click in *estimated* and choose the **Tools** menu. Click **Language** and then **Thesaurus**.

2. Choose **projected** from the list of alternative words. Click **Back**.

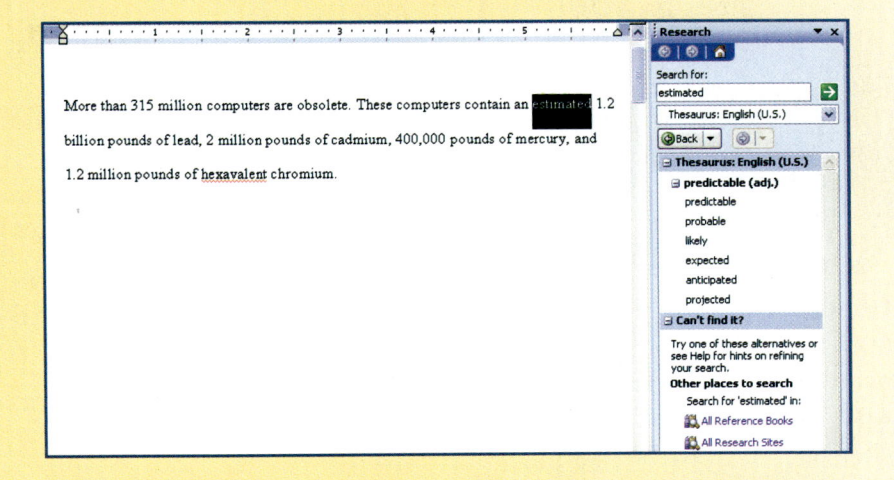

Continued

Step-by-Step

Activity 20

If you need additional help on this topic, open the Help menu; or refer to the documentation for your hardware.

Insert and Delete a Table

1. Start Microsoft Word.
2. Select the **Table** menu.
3. Click **Insert** and choose **Table**. In the **Number of columns** box, choose 2. In the **Number of rows** box, choose 3. Click **OK**. Title column 1 "Fruits" and column 2 "Vegetables." Add appropriate examples in each column. 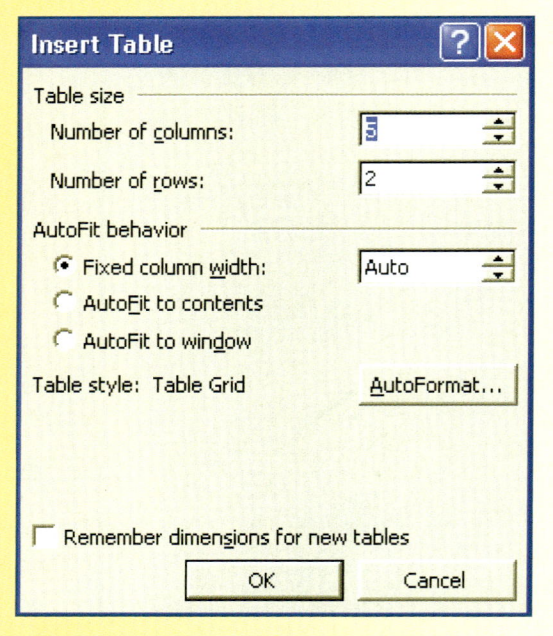
4. To select the entire table, click inside the table, choose the **Table** menu, and click **Select** and then **Table**. Choose the **Edit** menu and click **Clear** and then **Contents** to delete the words in the table.
5. Select the table. Click the **Table** menu, **Delete**, and then **Table** to delete the entire table.

Activity 21

Draw a Table and Enter Text

1. Choose the **Table** menu and select **Draw Table**.

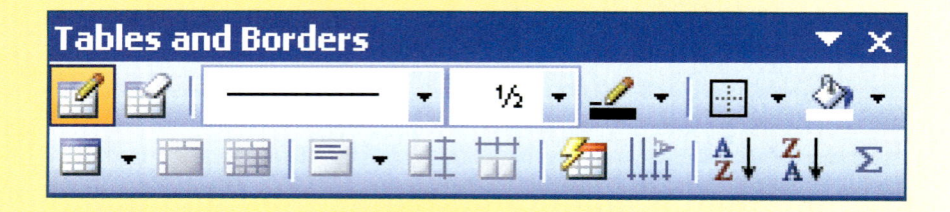

2. Click the **Draw Table** button in the **Tables and Borders** toolbar. Notice that the pointer takes the shape of a pencil.

3. Click and drag in your document to create the table outline. Click to draw one vertical line and two horizontal lines so that the table has two columns and three rows.

4. Click the **Draw Table** button again to turn off the drawing tool.

DIGI Byte

Editing Cells

Click the insertion point inside a cell you wish to edit to make changes.

Continued

Step-by-Step *continued*

Set Margins

One way to make a document easier to read is to set the width of the margin, or the blank area that borders the portion of the page in which text and objects can be placed. The default left and right margin settings in Word are 1.25 inches. The default top and bottom margins are 1 inch. Depending on the document, you may want to provide more, or less, working space. You can do this by setting the margin width.

Activity 16

1. Click the **File** menu, and choose **Page Setup**.
2. Change the left and right margins to 1 inch and click **OK**. How many lines of text do you have in the document now?
3. Change the left and right margins to 2 inches. How does this change affect your paragraph?

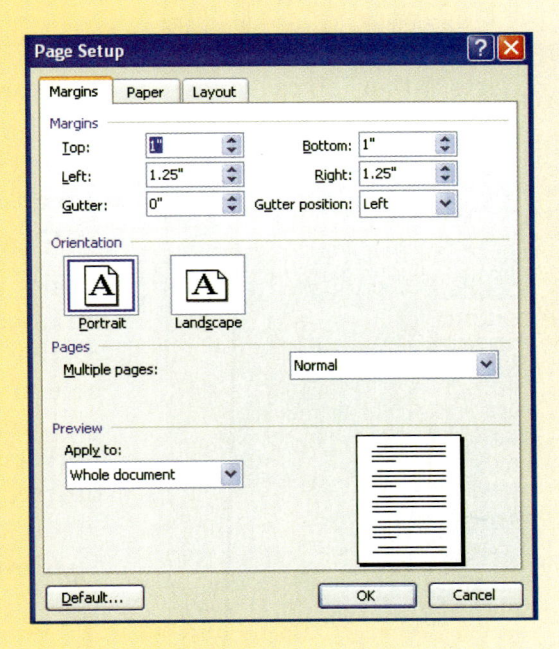

Set Page Orientation

Depending on the type of information your document contains, you may want to change the page orientation, or the determination of whether the page appears tall (portrait), or wide (landscape). For example, if the document contains a wide table you may want to switch the orientation from portrait to landscape.

Activity 17

1. Select the **File** menu and choose **Page Setup**.
2. Under the Orientation heading, choose the **Landscape** icon and then click **OK**.
3. Switch to Print Preview to see how the orientation change affects your document.
4. If you need to print on larger paper, you can change the paper size. Click the **Paper** tab of the Page Setup window; note you can change the paper size from letter, which is the default size, to legal. You can also choose **Custom size** and specify the width and height of your paper.

DIGI*Byte*

Landscape

Landscape orientation is often used for special documents, such as certificates, or for documents that contain wide tables, so that all data can be displayed.

Continued

Lesson 9.4 — Create and Edit Tables

You will learn to:
- **Create** a table.
- **Add** text to a table.
- **Insert** rows in a table.
- **Apply** borders and shading to table cells.

DIGIByte
Lists Within Tables
Bulleted and numbered lists can be used inside a table.

column A part of a table that extends up and down the table.

row A part of a table that extends across the table.

cell The intersection of a column and a row in a table.

Tables in Word Processing Tables help to organize information in word processing, making it easier for readers to grasp relationships and make comparisons between items. For example, notice how quickly you are able to understand the information in the table below about holiday clothing purchase trends.

Tables help to organize information in word processing. A table is made up of columns and rows of information. A **column** extends up and down a table and a **row** extends across a table. A **cell** is the intersection of a column and a row in a table. Text can be formatted differently within each cell of a table, and cells can also be shaded different colors. Table borders, which are the lines around the cells in the table, can be set to different thicknesses and colors as well.

Buying Clothing as Holiday Gifts

	This year	Last year
Women	66%	60%
Men	57%	51%

✓ Concept Check
Can you think of a way that tables could be used in a report?

DIGITAL DIALOG

Tables

Teamwork In a small group, discuss how tables can be used to organize data. Look through a textbook from another class. Do you find any tables?

Discuss Make a list of the tables you find and the type of information contained in the tables. Do you see any tables that have only one column? Be prepared to discuss what you find with your classmates.

Application 2

Scenario You have been asked by your teacher to prepare a short public service announcement regarding recycling electronics to be read over the PA system at your school.

Directions Create a new word processing document, and enter the following two paragraphs:

> In 1998, recycling firms handled 275 million pounds of computer gear that had been thrown away. This accounts for just 11 percent of the estimated 20 million central processing units that became obsolete that year.
>
> The Environmental Protection Agency conducts an Electronic Product Recovery and Recycling Roundtable, where businesses, government, and other officials brainstorm ways to deal with the disposal of obsolete computers.

Use the spell-check feature to check for typing errors and correct any in your paragraph. Select a word and use the thesaurus to change to a new word that has a similar meaning and that is appropriate in the context of the sentence.

Show formatting marks.

Change the line spacing and margin settings of your document, and change the page orientation to landscape. Change the alignment of the first paragraph.

Save your document. Print preview and print your document.

RUBRIC LINK

Go to the Online Learning Center at **digicom.glencoe.com** and click Lesson 9.2, Rubric, to find the application assessment guide(s).

TECH ETHICS

Exchanging Data

Scenario The marketing firm David works for purchased PDAs for its sales representatives. Employees were told to use the PDAs for personal needs to become familiar with them.

As David walked into his office one morning, he overheard a conversation between two coworkers about a new game one of them had purchased for the PDA. It was a game David would very much like to have, but could not afford to purchase because of recent car repairs.

As David walked out of the office for lunch, he noticed the coworker's PDA on a desk. Since no one was around, David beamed the game to his own company PDA, thinking, who would know?

Teamwork Create a group with two other students to discuss how this situation should be handled. Discuss the questions below and write a recommended plan of action for this incident. Defend your response with good reasons.

- Is what David did ethical? Explain your answer.
- Was the coworker out of line to purchase a game for a PDA that was bought for him by his company? Why or why not?

Application 4

Directions Create a new document. Enter the information below, creating a bulleted list. Select the list and change the bullet style. Save the document as **Bullets** and print it.

> Our school is going to launch a recycling campaign. We have decided to collect the following materials:
> - Newspapers
> - White paper
> - Aluminum cans
> - Cardboard
> - Plastic
> - Glass containers

Application 5

Scenario You are in charge of arranging your school's recycling campaign. You begin the planning of the project by creating a list of important things that will need to be done and putting the items in a logical order.

Directions Create a new document and enter the following "to do" list, using numbers instead of bullets. Save the document as **Numbers** and print it.

> School Recycle Plan - To Do List
> 1. Place collection containers around campus.
> 2. Post signs around school about recycling.
> 3. Collect recyclables in specified containers.
> 4. Bag recyclables that have been collected.
> 5. Contact local collection agency to collect bags of recycled materials.

Digital Decisions

Formatting

Teamwork In a small group, look through recent issues of the *USA Today* newspaper and identify 5-10 uses of text formatting, alignment, and bullets. Note how formatting is used to make sections easier to read, to add emphasis to important points, to catch the reader's attention, or to group similar items together.

Internet Investigation Access the online version of *USA Today* by searching for "USA Today" online. Visit several pages of the Web site and identify 5-10 uses of text formatting, alignment, and bullets. Just as you did with the print version, notice how the Web site's use of formatting affects your experience as a reader.

Write a paragraph comparing the uses of formatting elements in the print and online versions of the newspaper.

- Are the print Life and Sports sections organized the same way as the online sections?
- Are the front page of the paper and the home page of the Web site alike in the layout and format?
- Which medium do you think has the more effective format?

Be prepared to explain the reasons behind your opinions.

21ST CENTURY CONNECTION

Flexible Thinking

What would you say if someone told you that by thinking flexibly you could do better at math, be more successful in your career, and feel more optimistic? The way you think can have an amazing influence on your life. Flexible thinking is being able to see multiple solutions to a problem and being able to see an issue from different perspectives.

Try a Different Approach

Professor Lillie Crowley has studied how college students approach mathematics problems. She has found that while students still must learn mathematical rules and procedures, the most successful ones think about math as a puzzle that needs to be solved. They see more than one way to approach the problem and pick the easiest!

Many Different Applications At a job, with your classmates, and at home with your family, flexible thinking can improve communication, reduce conflict, and improve productivity. How often are parents told, "You just don't understand"? Sometimes one generation can have a hard time seeing the point of view of the other. Stay calm and try to come up with another way to explain your point of view. If you are working with a group on a project that is not progressing, brainstorm other ways to get the job done. In the future, you will probably collaborate with colleagues to find better ways to market a product, cut costs, or improve customer service. Being able to think flexibly will make you the kind of employee needed in today's economy.

Think Positively

Research shows that your attitude determines how well you cope with adversity. Inflexible thinkers believe that bad events cause them to act in certain ways. People trapped in this kind of thinking tend to blame themselves and give up easily.

Flexible thinkers recognize that their beliefs about adversity affect how they feel, and consequently what they do. Being able to see that bad situations are not permanent and that other parts of your life are okay is flexible and healthy thinking. Thinking like this will help you be more resilient and able to bear life's challenges.

The next time you are confronted with a thorny math problem, a conflict at home, or a disappointment, remember to exercise flexible thinking. With patience the light bulb will go on and you will discover the solution you need.

> **Activity** Bring in a kitchen utensil from home. To practice flexible thinking, brainstorm ways the item could be used in other rooms in the house.

Suite Smarts

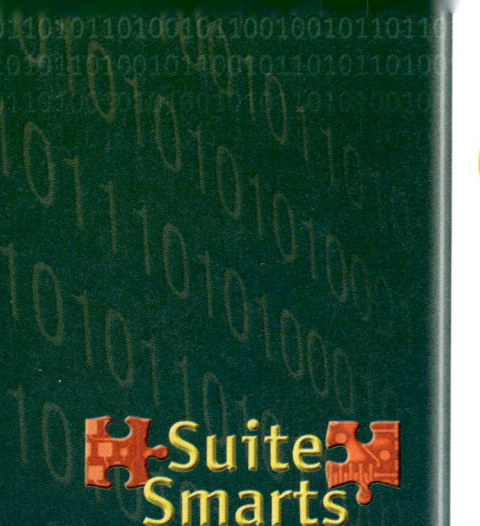

Using the Bullets and Numbering feature, you can choose to number paragraphs with numerals, letters, or Roman numerals.

Step-by-Step continued

4. Select the **Format** menu, and choose **Bullets and Numbering** again. Choose **None** and click **OK**.
5. Save the document as **Bullets and Numbering**. Do not close the document.
6. Enter the text `Task Priority`, and press the ENTER Key. 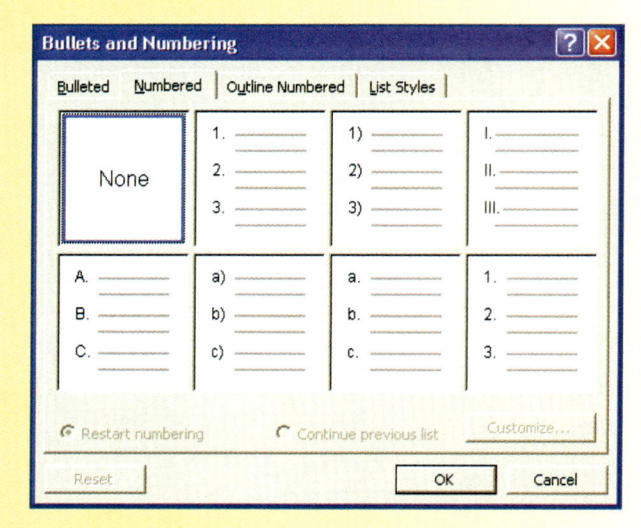
7. Open the **Bullets and Numbering** window again.
8. Click the **Numbered** tab.
9. Select a number style that displays numerals followed by a period and click **OK**.
10. Enter the following information, allowing Word to automatically create the numbering.

```
Reply to important voice mail.
Answer important e-mail.
Prepare packages to send.
Check inventory.
Compare pending orders to inventory level.
Notify customer fulfillment department of product shortfall.
```

11. Save and print the document.

Application 3

Directions Open the *Your Name* **E-Waste** document. View the header area, and change the formatting of your name. Add the date and time to the header. Change the footer by choosing a different page format. Save and close the document.

RUBRIC LINK

Go to the Online Learning Center at **digicom.glencoe.com** and click Lesson 9.3, Rubric, to find the application assessment guide(s).

Lesson 9.3 — Add Headers, Footers, Bullets, and Numbering

You will learn to:
- **Add** and edit headers and footers.
- **Insert** page numbers and dates.
- **Format** bulleted or numbered lists.

Other Ways to Make Documents Easier to Read In Lessons 9.1 and 9.2, pages 264–275, you learned that text formatting as well as margins and page orientation can help make a document easier to read. In this lesson, you will learn some additional ways to improve readability.

Headers and Footers One way is to repeat text, such as the title of the document, on each page if a document has two or more pages. Other objects, such as total page count, can make the document easier to read. A **header** is an area for text or other objects that will appear at the top of every page in a document. In multipage documents, headers sometimes contain the author's name. A **footer** is an area for text or other objects that will appear at the bottom of every page in a document. Page numbers are often included in footers. In books, chapter numbers and chapter names are often displayed in a header or footer.

Bulleted and Numbered Lists Another way to make a document easier to read is to create bulleted or numbered lists. Lists help to organize groups of similar information. Placing a bullet, which is a small icon, at the beginning of each item in an unordered list, enables a reader to quickly understand the items. Numbers can be used to create lists in which the order of the items is important, such as in step-by-step instructions.

header An area for text or other objects that will appear at the top of every page in a document.

footer An area for text or other objects that will appear at the bottom of every page in a document.

✓ Concept Check

In a multipage report, what type of information do you think it would be useful to display in a header or footer? Why?

DIGITAL DIALOG

Bulleted and Numbered Lists

Teamwork Discuss with another student the differences and similarities between bulleted and numbered lists. Why would you use these features? What purpose do they serve? To record your points, draw a Venn diagram, or two overlapping circles. Label one circle "Bullets" and the other "Numbers." List the points in the appropriate segment. Write points common to both in the overlapping segment.

Activity 19

Step-by-Step continued

Create Bulleted or Numbered Lists

Items in a list can be either complete sentences or short phrases. In this activity you will create a bulleted list and a numbered list.

1 To create a bulleted list, click the **Format** menu and choose **Bullets and Numbering**. There are several bullet styles to choose from. An important rule is to avoid mixing styles in a single document. Whatever style is used at the beginning should be used throughout.

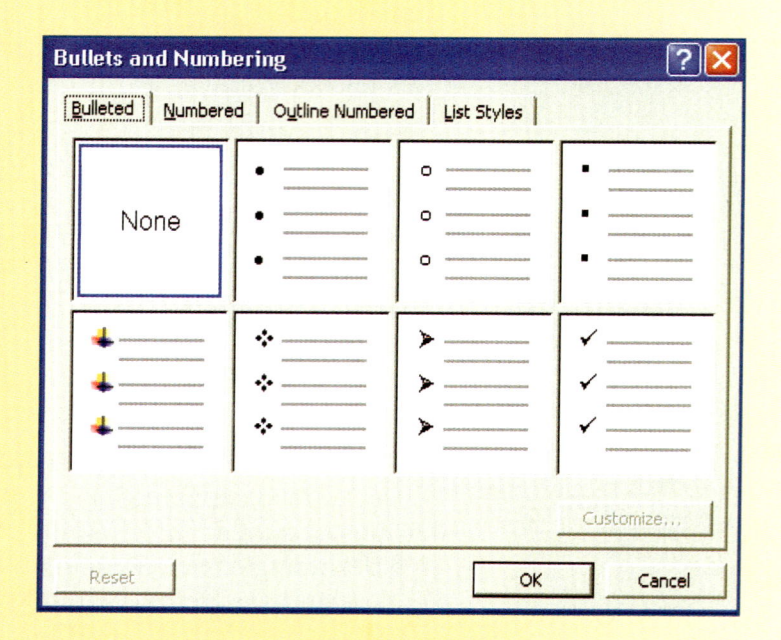

2 Choose the arrow bullets and click **OK**.

3 Enter the following information. Do not press the Enter key until you reach the end of each item.

Each time you press the Enter key, the text insertion point will move to the next line and automatically insert a bullet.

➢ Approximately 2,054,800 tons of electronic waste are deposited in landfills each year

➢ Almost 250 million computers will become obsolete over the next five years

➢ An average of 4 pounds of lead can be found in TVs and computer monitors

Continued

Suite Smarts

The first time you save a file in a Microsoft Office application, you will be asked to name the file and to choose a place in your hard drive to save it. For future changes to the file you can simply click the Save button on the toolbar.

Activity 18

If you need additional help on this topic, open the Help menu; or refer to the documentation for your hardware.

DIGI*Byte*

Footnotes
Do not place footnotes in the footer. Footnotes are created in a different way.

Step-by-Step

Create Headers and Footers

In this activity, you will open a multipage document named E-Waste, and add a header and a footer.

1. Open the **E-Waste** file and save as *Your Name* **E-Waste**.
2. Select the **View** menu, and choose **Header and Footer**.

   The text on the page will appear gray and a box bordered by a dotted line will appear at the top of your screen showing the header area.

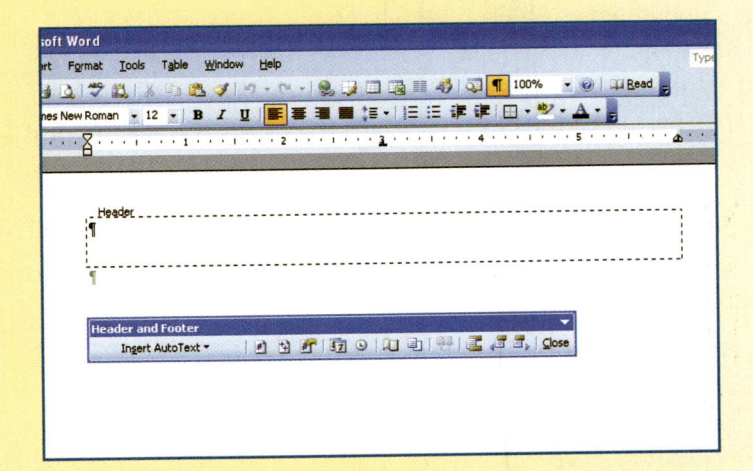

3. Key your name in the header.
4. Click **Format, Paragraph,** and then right-align the text.

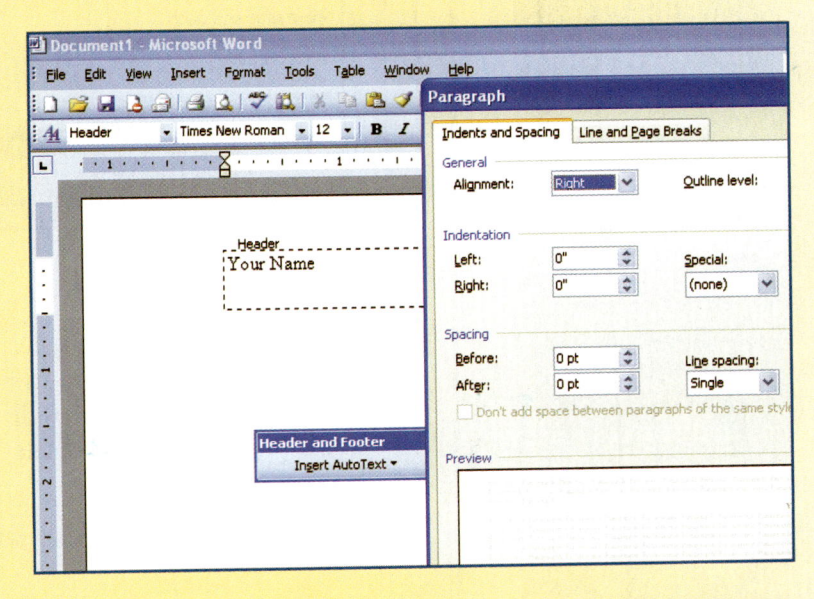

Continued

Chapter 9

Step-by-Step continued

5. Click the **Switch Between Header and Footer** button on the Header and Footer toolbar to switch to the footer.

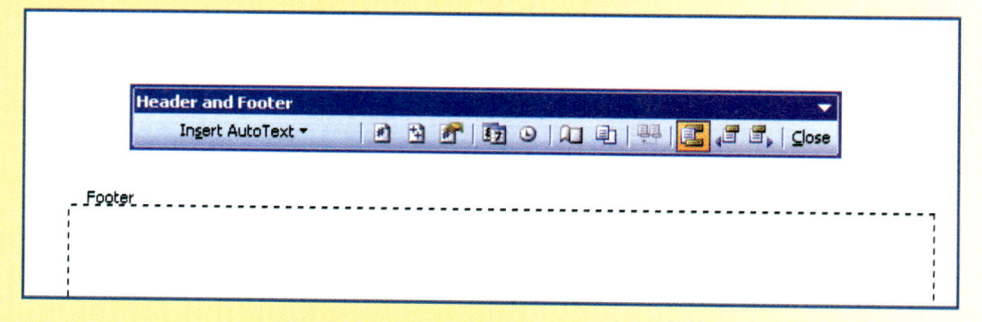

6. Select the **Insert AutoText** list on the Header and Footer toolbar.
7. Choose **Page X of Y** from the list.
8. Align the footer text to the right.

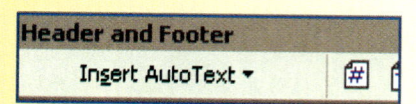

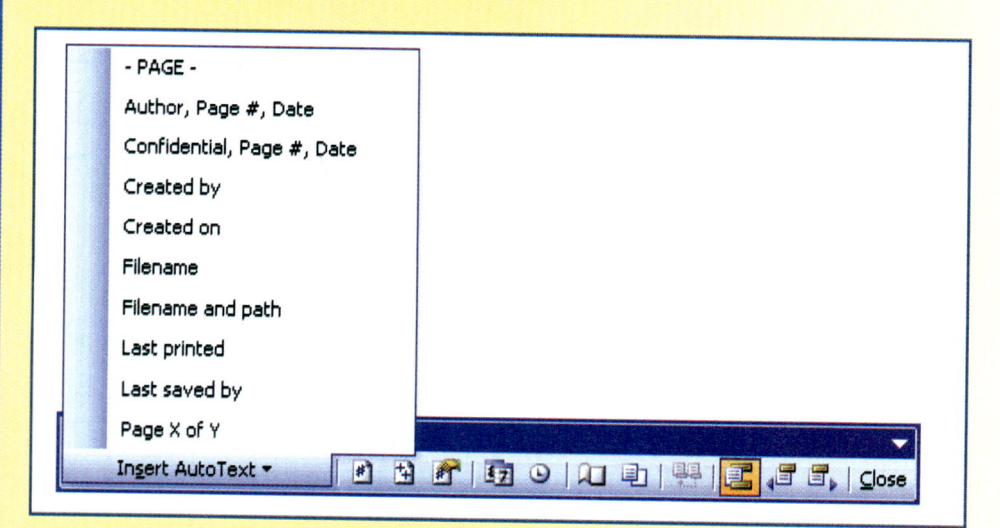

9. Click the **Close** button on the Header and Footer toolbar to return to the main area of the document.

   While you are working in the main area of your document, the header and footer text will appear gray. To switch to the header and footer areas, select the **View** menu and choose **Header and Footer.**

10. Save and print the document.

Continued

DIGI*Byte*

Formatting in Headers and Footers

Text that appears in the header and footer area of a document can be centered or aligned right or left. It can also be formatted with any text effects, in a similar way to text that appears in other parts of the document.

Suite Smarts

- The date and time can be inserted into a header or footer. To do this, select the Insert menu and choose Date and Time.
- You can press the Enter key after you add text to the header or footer to insert more lines of text.
- Headers and footers can be added to a document at any time.

Chapter 9

279